The Evolutionary Mechanism
of Human Dysfunctional Behavior

The
EVOLUTIONARY
MECHANISM

of
Human
Dysfunctional
Behavior

Relaxation of Natural Selection Pressures throughout
Human Evolution, Excessive Diversification of the
Inherited Predispositions Underlying Behavior, and
Their Relevance to Mental Disorders

A GENE-CULTURE COEVOLUTION THEORY

IVAN FUCHS

Radius Book Group
New York

Distributed by Radius Book Group
A Division of Diversion Publishing Corp.
443 Park Avenue South, Suite 1004
New York, NY 10016
www.RadiusBookGroup.com

For more information, email info@radiusbookgroup.com.

Library of Congress Control Number: 2018958126

Hardcover ISBN: 978-1-63576-601-1
eBook ISBN: 978-1-63576-602-8
Manufactured in the United States of America

10 9 8 7 6 5 4 3 2 1

Cover design by Erin New, Neuwirth & Associates
Interior design by Elyse Strongin, Neuwirth & Associates

Radius Book Group and the Radius Book Group colophon are registered trademarks
of Radius Book Group, a Division of Diversion Publishing Corp.

"Developing a cohesive story is far more instructive than merely compiling a list of unconnected facts."

—ERNST MAYR, *What Evolution Is*

"Besides facts, there are explanations and stories that pull the facts together into accounts that make sense."

—PETER DEAR, *The Intelligibility of Nature*

Contents

3

4

5

6

The Effects of Psychotropic Drugs in the Context of the Present Theory   321

7

Discrete Clinical Disorders in Evolutionary Perspective   373

8

Summary and Implications   531

Introduction

0.1. Opening Remarks

"Read not to contradict and confute, nor to believe
and take for granted, nor to find talk and discourse,
but to weigh and consider."

—FRANCIS BACON (*Esquire,* 1857, p 439)

As an unknown author writing on an unknown subject in the domain of psychiatry, or at least a subject that is completely unexploited by psychiatric theories on mental disorders, I find it useful to provide a somewhat more detailed exposition of the main ideas of this text than is usually found in an introduction. This book is an attempt to explain some of the phylogenetic mechanisms that brought about in humans the *innate liability* of developing varied forms of dysfunctional or maladaptive behavior, the graver of which are categorized as mental disorders. These behavioral patterns, while widespread in humans, so far as I know are nonexistent among wild animals living in their natural habitats. In the animal kingdom, inherited programs or predispositions to behave in a certain way (i.e., instincts) induce adaptive forms of behavior to ensure survival and reproduction. According to evolutionary logic, it is reasonable to assume that natural/sexual selection would weed out any inherited variation that predisposes to dysfunctional or maladaptive behavior of considerable severity.

Humans, however, live in an artificial environment, built by culture and traditions that have a far-reaching impact on the ways they interact with that environment's physical, biological, and social elements, ways that are very different from those seen in social animals living in a natural habitat. The degree of success of human populations in a biological sense (e.g., survival and reproduction) as well as in cultural terms (achievements in science, technology, arts, social organization, military power, etc.) is unimaginable in other animals, and the ensuing ecological dominance has led to the spread of human populations to almost every part of the globe.

On the other hand, these same human populations exhibit a high prevalence of dysfunctional forms of behavior with a clear genetic foundation. If we include those with attenuated (subclinical) severity, surely we can argue that individuals who, during their lifetimes, do not suffer from episodes of dysfunctional behavior or enduring dysfunctional personality traits are a rarity. Even mental disorders that are known to lower survival and reproductive rates (like schizophrenia) do not show a progressive decrease of prevalence over time in epidemiological studies (Adriaens & De Block, 2011, p. 6). Moreover, mental disorders in their more fundamental characteristics seem to be quite similar across populations and cultures. Since the ways genes are transmitted from one generation to the next determine the genetically underpinned aspects of behavior (both normative and dysfunctional) and, as mentioned earlier, this transmission is regulated by natural selection, we have to ask: What happened to natural selection during human phylogeny (i.e., over the course of human evolutionary history)?

The thesis of this book is that a longstanding, comprehensive, and progressive relaxation of natural selection pressures with regard to behavior accounts for the basic phylogenetic mechanism underlying human beings' great behavioral diversity. This excessive diversity is the basis of the biological and cultural prosperity of human populations, but it is also responsible for the high prevalence of dysfunctional forms of human behavior. This idea is so central to the present work that, at the risk of stating the obvious, I will attempt to state it even more explicitly.

The basic mechanisms of Darwinian selection comprise "processes that generate variation, the selection of *limited* numbers of variants, and the propagation of the selected variants" (Plotkin, 2007, p. 63, emphasis mine). The key word here, from the point of view of the present theory, is "limited." It is reasonable to assume that when a selection pressure (say, predation) increases for a considerable time, only the optimal variants of a prey animal (more precisely, those traits related to successful avoidance of predation) will be transmitted to the next generation. And conversely, when selection pressures relax, individuals with less-than-optimal variants will succeed in transmitting those genes to the next generation, a process that, in the case of traits with polygenic inheritance (discussed later), will lead to the widening of the respective spectrum of quantitative variation around the optimum. For the sake of clarity (and at the risk of oversimplifying), it is useful to imagine changes in the intensity of various selection pressures on behavior throughout human evolution in three overlapping stages:

1. *The relaxation of selection pressures coming from the natural environment* (i.e., external to the group of conspecifics) during human evolution began about two million years ago. Primitive stone implements improved defense against dangerous predators and enhanced hunting techniques; the use of animal hides as clothing protected against the vicissitudes of the climate; the control of fire protected against dangerous animals and cold and enabled the cooking of food that was difficult to digest in a raw state, etc. The increasing ecological dominance that resulted inevitably led to progressive relaxation of natural selection pressures. This trend seems to have continued and even accelerated up to the technological advances of the present day.

2. It must be hypothesized, however, that parallel with the relaxation of natural selection pressures, at the early stages of human evolution, *an intensification of selection pressures took place for innate behavioral predispositions that improved the abilities of human beings in the areas of social coexistence and cooperation,* which in turn ensured better survival and reproductive chances for the whole group. I have in mind such traits as the inherited foundations for improved communication (linguistic abilities); improved "theory of mind" (the ability to guess others' concealed emotions, intentions, beliefs, etc.); improved abilities to cooperate, even among those with widely different physical and mental endowments; the ability in certain circumstances to accept the priority of the group's interests over the individual's; the capacity to compete and settle clashes of interests in ways that do not seriously undermine social coexistence; and so on. It seems clear that prehistoric nomadic hunter-gatherer groups could not afford the burden of individuals, including those genetically predisposed to mental frailties, who could not function well together on a permanent basis.

This text will refer to the selection pressures discussed in point 1 as *Natural Selection Pressures* (NSPs) and those in point 2 as *Intragroup Natural Selection Pressures* (IGNSPs). In a rudimentary form, IGNSPs can also be seen in highly evolved social animals, like chimpanzees. (This subject will be discussed in some detail in section 4.6.)

3. At a later stage of human evolution, especially following the agricultural revolution beginning some 10,000 years ago, and after the genetic foundations for successful social coexistence had already become well ingrained (fixated) in the human genome, the IGNSPs began to loosen. In my opinion, the relaxation of IGNSPs with regard to behavioral traits has played a decisive role in cultural evolution. The increasing complexity of social life, the division of labor, and the progressive multiplication of what this text will refer to as "social niches" aroused the need for individuals with specialized mental, behavioral, and physical abilities, and with talents or aspirations (which have, as a rule, an inherited foundation) in one or another restricted field of the social or occupational domain. These improved abilities in certain domains frequently came with poorly developed mental or physical abilities in other, unrelated, fields. Individuals with a developed imagination, the capacity for abstract thinking, and other specialized cognitive or artistic talents (like the painters of prehistoric cave-paintings, for example, or the mathematicians, astronomers, inventors, and scientist-philosophers of the ancient world), but possibly with more modest socializing abilities, practical sense, and physical prowess, began to fulfill important social functions and thus were protected by the behavioral code of the community. Similarly, older, intelligent members of the community with important accumulated knowledge and experience but in frail physical condition received the same social protection. This relaxation of selection pressures, similar to those stemming from harmful natural influences, became progressively more accentuated, a trend that continues. Consider that more and more specialized tasks and occupations require special mental or physical abilities (some of which may be partly inherited).

We posit that further relaxation of the IGNSPs and the consequent increase in the diversification of the emotional, cognitive, and physical constitution of humans, the increased ability for cooperation among differently endowed (and genetically unrelated) persons, and increased social mobility (allowing for a better match between innate predispositions and the requirements of specialized social niches) permitted human communities to become more and more efficient at exploiting the diversified abilities and talents of their members. At the same time, they were able to minimize the fitness-reducing consequences of the drawbacks of their mental and physical constitutions (a development reflected in cultural traditions, mores, and laws concerning interpersonal transactions). The progressively improving exploitation of natural resources, especially after the introduction of agriculture and the domestication of animals, allowed for increased economic prosperity, which could now support the burden of growing numbers of individuals with dysfunctional behavior, particularly when the dysfunction was mild or transient (mild depressive episodes, dissociative reactions, etc.).

It may be argued that the relaxation of IGNSPs approached its climax with further technological and cultural progress, especially with the advent of modern, technologically advanced democracies. Their increased concern for individual needs has led to humanitarian ideologies that sanctify individual human life *irrespective* of the extent to which it contributes to the group's successful functioning, or alternatively represents a burden on its resources.

To conclude this general discussion of selective pressures and relaxations in human evolution, I cite the well-known American biologist and pioneer of sociobiology, Edward Wilson: "To my knowledge no evidence exists that the human genome is changing in any overall direction. . . . The big story in recent human evolution is not directional change, not natural selection at all, but homogenization through immigration and interbreeding. . . . Its main consequence is the gradual erasure of previous racial differences. . . . It also *increases the range of individual variation within the population and across the entire species.* Many more combinations of skin color, facial features, talents, and other traits influenced by genes are now arising than ever existed before. . . . *Variance increases, the extremes are extended, new forms of hereditary genius and pathology are more likely to arise*" (Wilson, 1998, pp. 271–273, emphasis added).

This subject will be taken up in greater detail in chapter 4. However, an important distinction concerning mental disorders should be mentioned in advance. While ontogenetic factors may influence such variables as the time of onset, course, gravity, possible secondary complications, and manner of expression (especially verbal) of the characteristic symptom-atology, the *essential, distinct* attributes of discrete mental disorders are defined by underlying *phylogenetic predispositions*. No known specific environmental conditions or influences (psychological or somatic) can reliably lead, by themselves, to discrete mental disorders (including even grave traumatic situations). This subject will be discussed in some detail in the section on schizophrenia (see 7.2.8.9.2.).

0.2. The Four Comprehensive Instinctive Mechanisms Relevant to Mental Disorders

As a clinical psychiatrist in full-time practice up to my retirement, I do not possess sufficient knowledge to elaborate on these complex matters. My intention here is only to place the presentation of my own, more re-stricted, topic into a wider, gene-culture coevolutionary setting.

During my residency in psychiatry, I was impressed by the prevalence of instinctive motives in mental disorders on one hand, and the scarcity of primary cognitive disabilities on the other. Moreover, it struck me that most mental disorders, and certainly the major ones in general psy-chiatric practice, do not affect a single instinctive category, like feeding, sexuality, parenting, dominance/submission, territoriality, or the sleep-wakefulness cycle, but rather affect most of them simultaneously. For example, in the depressive or manic episodes of bipolar disorder, all these instinctive motives may be quantitatively affected. During the transition from the preschizophrenic "model child" behavior into the characteristic schizophrenic psychosis, most behavioral attributes fun-damentally change. Therefore, the genetic predisposition underlying these (as well as other) common psychiatric disorders cannot consist of alterations in one circumscribed instinctive motive as a result of di-rectional selection inherent in the human social environment (or of a mismatch between this environment and a discrete instinctive motive). Nevertheless, among mental health professionals with an interest in evo-lution, this is a favored way of reasoning, which they use to attempt to explain the phylogenetic mechanisms underlying mental disorders. This approach is evident in Freud's overemphasis on sexuality, or in the field of evolutionary psychology and psychiatry. Martin Brüne, a recognized expert in evolutionary psychiatry, for example, interpreted manic and depressive episodes as exaggerated dominant and submissive forms of behavior, respectively (Brüne, 2008, pp. 211–216). This topic will be discussed in detail in chapter 3, in the sections on evolutionary psychology and psychiatry.

Being intensely preoccupied with this puzzle for several years, namely that mental illness categories cannot be explained by invoking a single circumscribed instinctive motive, I eventually arrived at *four comprehensive phylogenetically evolved mechanisms* that influence or modulate at the same time a considerable number of specific instinctive predispositions (or, in lower animals, innate behavior patterns).

In my opinion, these mechanisms, or more exactly, their extreme poles on a diversified scale (as a result of relaxation of natural selective pressures), constitute the inherited foundations of most common mental disorders. The present work focuses on these four general mechanisms, their excessive diversification under relaxed selective conditions, and the relevance of their diversification to discrete forms of psychopathology. It will be useful to give a concise description of each of these mechanisms.

0.2.1. Seasonal Variations in Instinctive Intensities

The first mechanism subjected to relaxation from natural selection pressures (RfNSPs) in humans, relevant to the present theory, consists of seasonal variations in the intensity of life-supporting and reproductive instinctive activities at higher latitudes of the planet. The direct cause of these oscillations is, of course, variations in the abundance of life-supporting environmental resources (light, temperature, abundance of vegetation and prey animals, etc.) dependent on the alternation of seasons—or, more precisely, winter inactivity and intense life-supporting and reproductive activity in warmer seasons. We have, I think, very good reasons to believe that this evolutionary mechanism underlies the episodic nature of several common and frequently serious mental disorders (bipolar and some unipolar affective disorders, schizoaffective disorder, some personality disorders like cyclothymic disorder, and others). The mechanism of seasonal variations in instinctive intensities has, of course, lost its fitness-enhancing function during human evolution as a result of technological progress and has become diversified as a result of RfNSPs. Nevertheless, clinical evidence (inclusively an efficient form of therapy, light therapy) supports this claim.

Please note that this claim about the relationship between seasonal variations in instinctive intensity and clinical disorders refers only to the *episodic nature,* or ebb and flow, of specific aspects of these disorders— those connected directly to the overall instinctive energies fueling behavior and intrapsychic events, and not to the entire nosological category. Clinical disorders are first of all descriptive entities with more or less pronounced empirical regularities with regard to their age of onset; constellation, or combination of symptoms or syndromes; characteristic response to different treatment modalities; course; and degree of inheritance. They cannot be regarded as homogeneous units of dysfunctional behavior induced by one specific inherited predisposition. Therefore, it should be understandable that frequently a specific, comprehensive instinctive mechanism may account only partly for the peculiarities of a whole nosological category. Sometimes it may explain almost the whole clinical picture (as in winter depression; or seasonal, good prognosis, and uncomplicated manic or hypomanic episodes), while in other instances (like schizoaffective disorder or recurrent acute schizophrenic episodes alternating with remissions) several innate phylogenetic mechanisms and their longstanding interaction with ontogenetic influences are needed in order to account for the whole clinical picture. This subject will be detailed in chapter 7.

0.2.2. Three Forms of Transformation of Active Instinctive Behavior in Frustrating Conditions

The second evolutionary mechanism affecting a wide variety of instinctive predispositions whose RfNSPs are related to psychopathology in humans, like the one just discussed, is well known in animals possessing more complex behavior. When one of the life-supporting or reproductive instinctive motives (which will be called *active instincts* in this text in order to distinguish them from innate behavioral tactics employed against dangerous or harmful environmental conditions, which will be called *reactive instincts*) is specifically frustrated by an environmental agent (lack of a resource, competition, loss of an immature offspring), the original behavior induced by the respective instinctive motive may be transformed into one of its three *forms of frustration.* For want of a better alternative, I refer to these three forms as: a) *displacement and vacuum behavior,* b) *aggression,* and c) *dysphoria.* These denominations are used in an unusually broad sense, referring to normative as well as dysfunctional human and animal behavior, and to their subjective, emotional aspect in humans.

0.2.2.1. Displacement and Vacuum Behavior

One of the basic distinctions ethologists use for categorizing instinctive behavioral elements in animals is that between the *fixed motor pattern* (FMP) and the proper environmental component or agent toward which this motor pattern has to be directed in order to fulfill its aim of survival or reproduction. (The technical term for such behaviors is *fitness-enhancing.*) Since these environmental agents actually release the FMP, ethologists call them *releasers.* Their term for the innate mechanism that ensures the recognition of the appropriate releaser (without acquired knowledge), and leads to the release of the FMP toward it, is the *innate releasing mechanism* (IRM). Animals actively search for the needed releasers, and the environmental niche in which they live has to provide them on a regular enough basis for the respective animal population to survive. However, a widespread phenomenon occurs in animals higher on the evolutionary scale in circumstances when the proper releaser is absent or unavailable. In this situation, the respective active instinctive behavior may be released toward another agent in the environment that may or may not resemble the original fitness-enhancing releaser. Ethologists call this phenomenon *displacement.* In more extreme circumstances, the instinctive behavior may be released without the presence of any agent at all that can be called the releaser—a phenomenon the great Austrian ethologist Konrad Lorenz named behavior *in vacuo*, or vacuum behavior. It seems evident that, in both cases (displacement and vacuum behavior), the fitness value of the instinctive behavior is greatly reduced or lost altogether. Therefore, it is reasonable to presume that natural selection kept its frequency and intensity in animals (living in natural conditions) at a low level to minimize the damage to fitness.

In humans, the natural selective pressure is so relaxed that it is difficult to imagine human behavior without massive displacement of instinctive motives toward biologically maladaptive releasers or even ones that are nonexistent in natural conditions. Think, for example, of the great variety of environmental agents that are capable of arousing and releasing behavior fueled by sexual drives in humans, from sensual aesthetic pleasure in the human body expressed in visual arts to extreme examples of sexual drive disorders like object sexuality—the arousal and discharge of sexual behavior toward an object such as a skyscraper or the former Berlin Wall. Sexual drives certainly can be discharged in humans in vacuo, as evidenced by the frequent practice of masturbation in both sexes. Even more characteristic of human nature, persons can "emotionalize" (that is, direct certain elements of instinctive activity toward) almost anything the cultural environment offers (such as sporting activities, whimsical fads, or enthusiasm for a political, religious, social, or ecological program) regardless of whether it does or does not have some fitness-enhancing value in the biological sense.

0.2.2.2. Aggression

In contrast to displacement (and dysphoria), instinctive aggression in mature animal and human behavior has an obvious fitness-enhancing function: the neutralization, removal, annihilation, etc., of an environmental impediment lying between the life-supporting or reproductive instinctive aim and its appropriate releaser. Aggression is so important for fitness in the natural world if discharged properly, and so destructive if discharged in undesirable amounts or circumstances, that surely natural selection controls it tightly in its form, intensity, tenacity, and other parameters, ensuring a positive balance between the gain from its discharge and its cost (for example, energy expenditure, possibility of eliciting counteraggression, and so on). In humans, on the other hand, aggressive drives are greatly diversified in their intensity and form as a result of the relaxation of selective pressures; aggression—actual or imaginary, verbal or physical, overt or covert, individual or group—pervades most areas of normative life as well as psychopathology.

0.2.2.3. Dysphoria

The dysphoric transformation is the most difficult to discern in mature wild animals in nature, but is by no means nonexistent. (I will provide examples of it in section 5.2.3.) In the framework of the theory being proposed here, dysphoria is an intense, adaptively useless (in mature animals) "protest" against an active instinctive frustration when the animal (or human individual), for some reason, cannot use aggression or displacement effectively in order to get rid of the frustrating agent and the accumulating instinctive tension.

The subjective experience of this state is an intense mental agony and despair, agitation, and helplessness. In very young animals and human infants, this is the most widespread response to instinctive frustration (caused, for example, by hunger, physical discomfort, or unavailability of the mothering figure) and has an obvious fitness-enhancing role (mobilization of parental help). Because this state strongly suppresses other, more useful instinctive behavior patterns (life-supporting or reproductive ones, as well as the alertness to possible environmental danger), it is reasonable to presume once again that natural selection in mature animals living in the wild strongly selects against dysphoria when it is unduly strong or longstanding.

However, in mature humans, owing to RfNSPs, dysphoria of undesirable intensity or tenacity is widespread in normative behavior as well as in mental disorders. But beware of an important (and overlooked) distinction concerning psychopathology. Both comprehensive mechanisms hitherto discussed—seasonal alternation of overall instinctive intensity and the forms of response to active instinctive frustration—can produce mental conditions considered in clinical psychiatry as depression. The first produces depression with psychomotor retardation (the most characteristic example being winter depression), while frustration of strong active instinctive intentions may induce a very different form of depression, with agitated and possibly aggressive (aggression against oneself or others) features. Possibly the most extreme and consequently the most easily apprehensible situation in this respect in normative behavior is the mental state of a mother who has suddenly lost a child, or that of an unexpectedly deserted lover. While in clinical psychiatry the pathological intensities of both states are called depression, in our theoretical framework they are thought to be induced by entirely different evolutionary mechanisms (see sections 5.1. and 5.2.).

0.2.3. The Differentiated/Diffuse Scale of Instinctive Drives

The third comprehensive mental mechanism that influences a wide variety of specific instinctive activities is, I think, the most concealed, in the sense that it is the most difficult to sense intuitively in another person's instinctively motivated behavior or to deduce by evolutionary reasoning. Nevertheless, the diversification of this inherited mechanism is central to understanding human nature in general, and its elaboration constitutes the first step on the difficult road toward understanding the phenomenon of schizophrenia and its genetically related spectrum disorders.

A well-recognized evolutionary trajectory exists in the animal kingdom according to which creatures low on the evolutionary ladder behave mostly in accordance with rigidly preprogrammed, genetically fixed behavior patterns. As we ascend this ladder, however, we find that the behavior of animals becomes more and more flexible, and most of the time the instincts manifest themselves as a more or less clearly defined *intention, propensity,* or *inclination* to behave in a certain manner rather than as a strictly preprogrammed action or reaction. (I find no better words to describe this kind of instinctive activity than those that designate private emotional experience.)

This development "makes room" for flexible modification of behavior according to experimentation (that is, learning from the outcomes of prior organism-environment interactions) and personal experience, or learning directly from more mature conspecifics. Rigid, preprogrammed instincts induce a very selective perception of, and interaction with, discrete agents of an optimally predictable, rather restricted environmental niche, one that is able to provide the releasers the animal needs. More diffuse and flexible instinctive intentions induce a less selective perception of the environment, more openness to new impressions, trial-and-error learning, and new strategies for problem solving. In short, more diffuse, "open program" instincts, to use evolutionary theorist Ernst Mayr's computer analogy, enable the animal to cope flexibly with a more complex natural and—more importantly, in humans—social environment.

In other words, while more diffuse instinctive intentions render the motor behavior they induce more tentative, less strictly defined genetically, and more open to modification, at the same time the choice of environmental releasers toward which this behavior is directed will become less clear-cut, more variable, and more susceptible to and even dependent on guidance coming from the natural—and especially the social—environment.

It is now well recognized that this development in human beings has advanced so far, compared to even the most intelligent social animals, that until the advent of evolutionary theory it was unanimously believed that humans and other animals are fundamentally different kinds of creatures (animals being driven by instincts while humans are guided by reason). Even today, the controversy about the extent to which humans behave according to instinctive motives as opposed to social influences is far from settled.[1]

Taking for granted this evolutionary trajectory in the nature of instincts from very rigid and differentiated toward more diffuse and tentative, I will propose the hypothesis that, while in wild animals the place of each species' instinctive behavior on the differentiated/diffuse scale is quite fixed and tightly controlled by natural selection, in humans a pronounced relaxation of selection pressure also took place in this domain. The result is an extended spectrum of diversity. On one end of this spectrum are individuals who possess instinctive motivations that are *more than normatively differentiated,* leading to a more selective, less flexible perception of the social environment. Such an instinctive predisposition induces a direct, more rigidly goal-oriented route toward narrower, more biologically relevant goals, while frequently disregarding social norms or even the law.

On the other end of the spectrum are individuals with *instinctive motivations that are too diffuse.* It can be presumed that such persons will have difficulty selecting from among the multitude of socially provided complex (and often contradictory) influences and sources of information those ones that will afford them a good chance to attain their biological goals. Such people may be overwhelmed by a multitude of unselectively absorbed stimuli (a phenomenon well recognized in schizophrenics). Their overt behavior, influenced by these more than normatively diffuse instincts—even when their learning abilities are good—will be more insecure, tentative, confused, or bizarre in a social setting, making their interactions with others, especially on an emotional and intentional level, highly problematic. (This difficult subject will be further clarified and elaborated on in chapters 5 and 7.)

Normatively differentiated (or normatively diffuse) drives give at least an overall direction to the intentionality or motivation in a biological sense (that is, securing material means for physical sustenance, personal territory, or shelter; negotiating or fighting for a social status; finding a mate; bringing up children, etc., while taking into account the opportunities and constraints that the social environment offers for the realization of these goals).

0.2.4. Active-Reactive Behavior Dichotomy and Interaction

The fourth and last general mechanism affecting a wide variety of instinctive drives that underwent the effect of RfNSPs during human evolution, as hypothesized in this theory, refers to the interaction and interrelatedness of two already highly complex behavioral categories. This subject can be best introduced by beginning with some concise statements about the phenomenon of life in general.

What characterizes living organisms in the widest sense of the word, according to a fairly broad consensus among experts in this field, are: (1) they possess a relatively stable, organic, macromolecular structure continuously repaired and reconstructed; (2) these activities of repair and reconstruction depend on the chemical processes of metabolism, which transform the raw material resources provided by a suitable environmental niche into components of the organism or use them as sources of energy; and (3) these organisms have the ability to reproduce themselves (Morange, 2008, pp. 35–45).

All three capabilities (we refer here to their behavioral aspects when appropriate) are partially inherited and *autonomous* (in the sense that they are not induced by external stimuli; instead, to realize them the organism actively and selectively initiates interactions with environmental factors). The behavioral element serving these aims of the organism, as well as its accompanying physiological processes and emotional experiences (at least in humans), will be called in this text *active behavior*. However, another behavioral category exists, with quite different aims. Individual life, besides being autonomous, is also fragile, and the ultimate fate of many organisms is to sooner or later become part of a food chain, raw material for other organisms' survival needs. Competition for resources, predation, deleterious weather conditions (heat, cold, too much or too little water), difficult terrains to move about on, etc., pose inconveniences and dangers to the life processes of organisms. Every living organism has to develop defensive strategies to fend off these and other environmental threats. Examples in the living world range from plants secreting toxins and growing thorns, to unicellular organisms with cilia that enable them to move away from liquids with undesirable chemical composition, to the alarm calls of certain highly evolved social animals at the approach of a predator.

It is important to clarify that the present theory divides undesirable environmental influences and the behavioral response to them into three separate categories (contained by two more inclusive units) of which only the third is referred to as "reactive behavior." We need this categorization because the resultant behavioral responses are widely different: The first refers to regularly recurring, *predictable environmental influences* harmful to the life processes, such as seasons that are scarce in life-supporting resources (e.g., few daylight hours, temperatures that are too high or too low, drought, etc.) against which organisms evolved suitable defense mechanisms, both behavioral and somatic, including hibernation, migration, shedding of the coat, and synchronization of the reproductive cycle with the cycle of the seasons. A special case of this kind of evolutionary adaptation was discussed under section 0.2.1.

The second kind of possible environmental harm refers to *unpredictable, frequently sudden, potentially harmful or dangerous environmental influences*, which in the present context have to be divided into two different subcategories:

1. The first subcategory, mentioned in section 0.2.2, consists of harmful environmental influences that *differentially hamper or frustrate a particular circumscribed active instinctive striving in progress*. These harmful influences lead to a change in the behavioral expression of the instinctive motive (displacement, aggression, or dysphoria) or in the nature of the targeted environmental agent.

2. The second subcategory consists of possible harmful influences, occurring at random, that do not target a specific active striving in progress. Instead, they may endanger the animal's bodily integrity, or life itself—for example, a falling rock in a mountainous region or a charging predator. The behavioral patterns designed to deal with, fend off, or neutralize these kinds of deleterious environmental influences are referred to, in our theoretical framework, as "reactive behavior." In the animal kingdom, reactive behavior comprises three universally occurring tactics: freezing, avoidance or flight, and counteraggression. The respective accompanying subjective experiences (symbolically shareable only in humans) are *prompt, goal-directed arousal; fear or anxiety,* and *anger.*

Obviously, both active and reactive behavior imply interactions with the environment. In the professional literature, both, when successful, are unfortunately called adaptations to the environment, despite the great differences between them regarding the preservation and perpetuation of individual life.

Students in the field of behavioral sciences have long recognized this dichotomy of active and reactive behavior. Ethologist Wallace Craig (1918) defined this dichotomy as appetites versus aversions. Lorenz noted the same phenomenon: "While appetence continues until a certain stimulus is reached, aversion continues until a certain stimulus has been got rid of" (Lorenz, 1982, p. 130; see also Craig, 1918). In the context of human social adaptation, J. Piaget's contrasting of "assimilation" ("the active attempt on the part of the child to mould and adapt the external environment to his own needs") with "accommodation" (the "process of adapting oneself to the external environment") expresses a similar idea (Piaget, 1953).

Serving basic biological goals, these two behavioral categories are tightly regulated by natural selection, along with the forms of interaction between them. For example: An environmental danger may strike unexpectedly and suddenly, while organismic processes are slow by comparison and frequently recur rhythmically; as a rule, the gratification of organismic processes may also be postponed for a while. Therefore, natural selection built into the nervous system of animals an almost universal priority for the discharge of reactive over active behavior, suppressing the active one when the simultaneous discharge of both forms is impracticable. Think, for example, of an herbivorous animal detecting an approaching predator. It will stop grazing instantaneously and, depending on the circumstances, will freeze, flee, or counterattack.

It will be argued that during human evolution both behavior forms became very complex, and their interrelationship even more so—possessing, besides instinctive incentives, a vast learned "superstructure"; sometimes they become fused or synthesized (that is, the same behavior, like working, serves active as well as reactive motives). In addition, these two behavioral categories and, even more importantly for our subject, their interrelationship and power balance were subjected to RfNSPs during human evolution, resulting in a wide spectrum of diversification with far-reaching consequences for both normative behavior as well as psychopathology.

This active/reactive instinctive behavioral dichotomy is valid for both simple and evolved animals, solitary or social. However, in evolved social animals, and especially in humans, an additional kind of inter-relationship evolved with the social environment, one that complicates the rather simple picture thus far described. The need to adapt to group life led to a new form of reactive behavior with inherited foundations, that of *submission to or conforming with* the rules of social coexistence, leading to far-reaching modifications of the simpler active and reactive behavioral predispositions and their interaction. As an extreme example of how individual and group interests interact in human reactive behavior, think of an average soldier's behavior in combat. (This subject will be further elaborated in section 5.4.3.)

I am fully aware that the causal chain between *comprehensive* instinct-modulating mechanisms and overt behavior is less apparent than the link between *circumscribed* instinctive motivations and the overt behaviors they induce. (Consider that even this latter relationship, like the impact of sexual motives on behavior, may be quite obscure in humans. It required psychoanalytic theory to bring this relationship fully to the surface.) Perhaps the existence of that complex, concealed causal chain is one of the reasons why the etiology of many psychiatric disorders remains so elusive.

I will try my best to be as clear as I can in the exposition of my subject. However, the reader should take into account that the discussion of a field that straddles biology, evolution, normative and dysfunctional behavior, and the effect on them of subjective experience has to move on a more abstract and more generalized level than is customary in scientific discourse.

The next few pages relate to some personal experiences during the maturation of the present work. I am well aware of the undesirability of such personal material, especially in a work concerning scientific matters. Such a work has to speak to the reasoning abilities of readers instead of trying to impress them emotionally. It may be that the short passage that follows originates mostly from an internal need of mine, and the uninterested reader may skip it without missing anything important pertaining to the theory itself.

0.3. Some Personal Remarks

The mostly intuitive realization of the behavioral implications of evolutionary reasoning, which occurred to me while I was a young resident in psychiatry, gave me numerous moments of elation. However, when I tried to examine my crude intuitions critically, put them in words appropriate for scientific parlance, write them down, and seek (quite unsuccessfully) helpful comments on them from my teachers, I began to realize the true magnitude of the task before me. The problem of justifying a theory built on the *hypothetico-deductive method* led to feelings of uncertainty that accompanied me for many years. This method, a comprehensive construct built intuitively on the implications of former theoretical formulations and supported only by indirect, circumstantial evidence, contrasted dramatically with most explanations in psychiatry, which were primarily conscious, logical, one-step generalizations built on experimental data (for example, the catecholaminergic theory of schizophrenia and affective disorders) or generalizations built on statistical regularities of clinically observable material (that is, an *inductive* method of theory building). The controversial nature of that obscure concept "the inherited incentives of human behavior," which does not lend itself to direct experimentation using scientific methodology (besides being fraught with ideological connotations), as well as the lack of integrated work on the effects of the RfNSPs in animal and human behavior, made me realize how lonely the fate of those who are lured to follow excessively their own independent, intuitive thinking can be. However, I found in one book a passage that greatly encouraged me to continue my investigations in my own way: "One of the most frequent observations [of the research on personality traits related to creativity] is that creative people, and perhaps especially creative scientists, tend to display (and very likely require) *a stubborn intellectual autonomy and independence of judgment, which makes them less willing than most to be influenced by group opinions and pressures*" (Butcher, 1968, p. 113, emphasis added).

In any event, possessing now the perspective of almost forty years of pondering my original ideas, checking them against insights and understanding obtained in working with my patients as a clinical psychiatrist, against my own mental events, and above all by reading scientific literature relevant to the subject, I can say the following:

While my clinical and life experience, as well as my reading, enriched in detail and clarified considerably the present theory, and even modified it at some points, neither factual data nor rational reasoning contradicted any of its original ideas. On the contrary, I found that a considerable amount of material from the literature indirectly supported it, as will be shown throughout this text. I hope my judgment was not selective to an extent that made me completely blind to contradictory material, but that is for the reader to judge.

Second, while I found some hints in the scientific literature on the possible role of the RfNSPs and consequent diversification in shaping human behavior, I found no material that attempted to investigate the subject methodically and in detail or relate it to psychopathology. These considerations brought me to the decision to write down my ideas on this subject as clearly, systematically, and accessibly as I could and to try to make them available to that segment of the public interested in the evolutionary mechanisms underlying dysfunctional human behavior.

0.4. Concise Summary of the First Four Chapters

The first four chapters of this text on the subject of scientific theory, the nature of the instincts, the scientific branches or domains relevant to the present theory, and relaxation of selection pressures in general are meant to provide preliminary material. They will make clear to the reader, I hope, how I understand these topics and make use of them throughout this work, as well as provide some general knowledge about these domains for the nonprofessional. In particular, they will facilitate the understanding of the following three chapters, which contain more detailed arguments and supporting data for the four comprehensive instinctive mechanisms previously mentioned and their relevance to psychiatric disorders.

The first chapter deals with the subject of scientific theory in general and biology and psychiatry in particular. My intention to build a theory using the hypothetico-deductive method led me to try to understand more about various kinds of scientific explanations and especially about the two opposite methods of theory building, the inductive and hypothetico-deductive. My goal was to determine whether or not the latter method could justifiably be employed in attempting to explain inherited aspects of mental disorders along the lines of evolutionary reasoning, a field in which direct experimentation is largely impracticable.

Thus, chapter 1 will begin with a short historical discussion of these two methods of theory building and their interrelationship in Western thought from the ancient Greeks onward. The work of two great philosophers of science of the twentieth century, Thomas Kuhn (1922–1996) and Karl Popper (1902–1994), argued strongly in favor of the centrality of comprehensive, hypothetical theories of science that in turn guide the selection of relevant scientific observations and experiments and the interpretation of their results. Their work convinced me that explanations of mental disorders cannot be approached by any method other than the hypothetico-deductive one.

Since my theory is built on the foundation of evolutionary theory (in its modern "neo-Darwinian" form), it was important for me to be sure that this foundation is sufficiently firm. Therefore, the next topic in this chapter is a discussion (to the extent my knowledge allows) of the strong as well as the contentious aspects of evolutionary theory (including later developments in cellular and molecular biology). Following that is a short overview of the three main theoretical and methodological approaches in modern psychiatry—the clinical, the biological, and the psychodynamic, in order to put my own theoretical approach into wider perspective.

Chapter 2 contains a more detailed discussion of the concept of instinct. How did the instinctive nature or instinctive aspects of behavior evolve from the simplest animals to the most complex? How did the understanding of the concept, as well as the role of instinct in behavior, evolve in the thinking of Western scientists and philosophers beginning with the ancient Greeks, through the Middle Ages, up to the Darwinian revolution? This chapter discusses in some detail the important contribution of the science of ethology (the study of animal behavior) to the understanding of instinct's adaptive role in behavior. In addition, it will take up the contentious nature of the concept of instinct in human behavior and attempt to make clear how this concept is understood in the framework of this theory. Three different aspects of instinctive behavior will be described—the behavioral, the physiological, and the subjective experiential. Since each of these aspects is approached by a very different methodology, and each has its own relationship to the purely scientific one, some confusion in this field seems almost inevitable.

The next topic in this chapter is the instinct-learning interaction. Here I try to convince the reader that instinctive behavior and learned behavior, far from being mutually incompatible (think of the nature-nurture controversy), are inseparably interdependent. In the animal world, no learning is possible outside the domain of instinctive activity (with very few exceptions), while instinctive motives, especially in humans, can be modified in far-reaching ways, suppressed, or channeled by social learning into domains remote from the purely biological ones.

The last part of chapter 2 deals with the subjective experiential aspects of instinctive activity and the possibilities of communicating verbally about these experiences. In spite of the difficulty of approaching this field with purely scientific methods, in my opinion the subjective experience of instinctive activity contains data indispensable for a satisfactory understanding of instincts' role in human behavior (including dysfunctional behavior). Therefore, a method—the three-step empathizing mechanism, which seems to me to ensure the most precise possible grasp of another's subjective experience—will be proposed.

In chapter 3, I briefly discuss some theories and data pertaining to disciplines relevant to the concerns of this work that approach instinctive behavior from an evolutionary perspective. First, in the field of primatology, I examine the various social organizations and behavior patterns (especially the sexual ones) of two closely related species of apes, the chimpanzee and the bonobo, in light of selective constraints and relaxations in their natural habitats. Second, I give an account of the field of evolutionary psychology, which deals with the effect of inherited predispositions on normative human behavior, as well as of the field of evolutionary psychiatry, which tries to interpret mental disorders in accordance with principles of evolutionary thinking. In order to complete the picture of the scientific domains relevant to the present work, I concisely summarize theories on cultural evolution, gene-culture coevolution, and the phenomenon called niche construction.

Chapter 4 deals with the topic of relaxation of natural selection in general in the animal world. The biological mechanism of inheritance and its various forms (monogenic/polygenic, dominant/regressive), as well as the various kinds of natural selection (directional, purifying, stabilizing, diversifying), will be discussed. In this context, I stress the special relevance of polygenic inheritance and of the normalizing (or stabilizing) form of natural selection with regard to behavioral traits, since these genetic mechanisms enable the evolution of graded continuums.

This chapter also details a specific path of evolution in humans, as compared to other animals—namely, the reorganization of instinctive behavior as a result of changing environmental circumstances. In other animals, changing environmental conditions led to relaxation of the selection pressures of the former natural environment and the intensification of the selective pressures of the new one. A consequence was the transition from one form of more or less rigid instinctive behavior to another. In humans, however (with regard to behavioral predispositions), the relaxation of natural selective pressures (external to the group of conspecifics) was followed by intensification of intragroup natural selection pressures leading to improved socializing abilities and cooperation and further relaxation of the selective pressures coming from the natural habitat as a result of increasing ecological dominance. This evolutionary path led to a more intense diversification and malleability of the inherited foundations of behavior and to increased modifiability by knowledge acquired during ontogeny. The complex and quickly and unpredictably changing nature of the social milieu precluded natural selection from building narrowly preprogrammed adaptive behavior patterns.

The next theme in this chapter concerns the various causes, besides the effect of the RfNSPs, leading *to genetic diversity* in a population: sexual reproduction, chromosomal recombination during gamete formation, mutation in the gametes, random drift of neutral mutations, and outbreeding. The various consequences of RfNSPs—diversification, atrophy of organs and behavior patterns, allocation shift—are discussed. Examples include widely different organisms living in relaxed selective conditions (the RNA virus, domesticated animals, and evolved social animals). This section stresses how intense intragroup selective pressures in humans has led to new kinds of behavior with an inherited foundation (cooperation, highly improved communicative abilities, as well as "theory of mind" abilities). In addition, I argue that diversified behavioral abilities in humans has led to diversified social environments ("social niches") and to progressively increasing social mobility, with individuals of various mental and physical makeups tending to gravitate toward the social niche best suited to their physical and mental constitution.

The concluding section of this chapter enumerates the various mechanisms that influence *gene transmission* in humans from one generation to the next, both those pertaining to different forms of natural selection (including intragroup natural selection), as well as those dependent on various social and cultural practices that do not follow the rationale of natural selection.

The following three chapters, as previously noted, contain the specific theoretical constructs this work was written to expound in the first place: the detailed description of the four comprehensive instinctive mechanisms proposed to underlie normative and dysfunctional behavior (summarized in section 0.2.); how they are affected by the main psychotropic drug categories; and the relevance of these comprehensive instinctive mechanisms (at the extremes of their diversified continuums) to common mental disorders.

To summarize, it may be said that the present theory differs from former attempts to account for human dysfunctional behavior using evolutionary logic in the following two ways:

1. Instead of concentrating on *discrete, circumscribed instinctive motives* underlying dysfunctional human behavior (like Freudian psychoanalytic theory, for example, or most hypotheses in evolutionary psychiatry), the present work argues for the centrality of *comprehensive instinctive mechanisms* that act on a wide variety of specific instinctive motives at the same time. It seems reasonable to presume that the excessive diversification of these comprehensive mechanisms (more precisely, the extreme ends of the resultant diversified continuums) affects the adaptive and fitness-enhancing qualities of behavior more profoundly than the excessive diversification of one discrete instinctive predisposition (feeding, sexuality, parenting abilities, etc.).

2. Instead of stressing the role of directional selection in bringing about instinctively underpinned forms of normative and dysfunctional behavior (the usual line of reasoning in evolutionary biology, evolutionary psychology, and psychiatry), or arguing on behalf of a mismatch between discrete instinctive motives that evolved in our distant evolutionary past and the adaptive requirements of the modern social environment, the present theory emphasizes the centrality of

the *relaxation* of natural selection pressures. This relaxation led to wide, quantitatively diversified, gradated scales of the comprehensive instinctive predispositions. The extreme ends of these scales, it is hypothesized, represent the inherited predispositions underlying most common mental disorders.

1

Scientific Theory and Methodology

1.1. Inductive versus Hypothetico-Deductive Theories

It seems necessary to include this section because of the problematic status of the hypothetico-deductive theory in psychiatry as well other scientific fields. Endocrinologist Hans Selye, famous for his work on stress, describes this situation:

> Right now, especially in North America, we see an entirely unwarranted overemphasis upon fact-finding, accompanied by what often amounts to an actual disdain for theories and interpretations ... the prejudice against "mere theorizing" has become so serious in the biologic sciences that many an investigator who describes facts, makes a special point of emphasizing ... that he does not attempt to interpret their meaning. What is the value of facts without meaning? (Selye, 1964, p. 278)

In psychiatry, the situation is more complex. As I detail at the end of this chapter, psychiatry lies at the interface of somatic and mental phenomena. While those trends in psychiatry that emphasize the somatic aspect are busy in fact-finding, facilitated by modern medical technology, those trends focusing on the mental phenomena (unobservable subjective mental experience)—the psychodynamic or existentialistic schools, for example—produce theories that cannot be tested with the scientific method.

Since the present theory's subject matter is the evolutionary interpretation of the inherited predisposition to dysfunctional behavior, no possibility exists, in my view, of achieving very much progress through the piecemeal integration of observational or experimental data. Instead, I propose here a hypothetical theoretical explanation based on evolutionary reasoning, as well as indirect circumstantial evidence. I have taken care, however, to formulate my proposals in a way that permits the derivation of implications that can be tested by the scientific method. These I will mention throughout, but especially in chapter 8.

The scientific enterprise, as is well known, is essentially built on two kinds of activities, the collection or creation of reliable data on one hand, and on the other, the conceiving of theories that explain the order, lawfulness, and causal or interactive mechanisms hidden behind the data. Put in an oversimplified form, the transition from one activity to the other may happen in one of two basic ways. The first is the *inductive generalization* of reliable results of particular scientific observations or experimentation. The generalization here is a one-step, conscious, logical mental process. An example from psychiatry would be the following: Drugs that suppress the manifestation of acute psychotic symptoms in schizophrenia depress dopamine (as well as additional neurotransmitters') activity at the neuronal synapses, while drugs that increase catecholaminergic activity (dopamine, noradrenaline) at certain brain synapses may exacerbate acute psychotic symptoms in schizophrenia and in other psychotic conditions and may induce them in normative persons. From these data, one can arrive at the following conclusions:

1. *A cautious generalization*: Dopamine activity has a certain as-yet-unclarified role in the accentuation/attenuation of acute psychotic symptoms in schizophrenia and other psychotic disorders.
2. *An exaggerated generalization* (for example): Dopamine overactivity, presumably of genetic origin, is the direct cause of the whole clinical picture in schizophrenia.

Most expert opinions lie somewhere in between these extreme possibilities, although we can discern movement from more daring conclusions shortly after the discovery of these phenomena toward more cautious interpretations with the passing of time and accumulation of new related evidence.

The second route between data and theory is a more complex and tortuous one. The inductive method is built on data derived from experimentation, which tries to keep to a minimum the number of variables of the phenomena studied; ideally, scientific experimentation tries to isolate one single phenomenon and see how it behaves under various controlled conditions. The *hypothetico-deductive method of theory building*, at its initial stages, requires the scientist (who has suitable cognitive inclinations) to unselectively absorb a wide range of phenomena and existing theories and hypotheses in a field of inquiry and, often, beyond it. In psychiatry, as will be elaborated later, it also requires the absorption of "data" coming from a kind of emotional resonance induced by an individual's behavior and verbalizations. I came upon a quotation that expresses this stage of theory building well: "Saturate yourself through and through with your subject . . . and wait" (Lloyd Morgan, quoted in van den Hoff, 1995, p. 20). (For a more detailed description of this phase, see Beveridge, 1961, pp. 76–78.)

The next stage (which hopefully comes during *disconnection* from the conscious effort to solve the problem) differs considerably from the inductive method. While in the latter the mental process that formulates the theoretical generalization consists of a conscious, one-step mental operation and conforms to the laws of logic, in the hypothetico-deductive method, the mental process leading to the theoretical formulation is of an obscure, poorly understood, unconscious ordering work of the mind (or intuition, in the sense used by creative scientists). Clearly, no other way exists, when a great amount of information of various kinds has to be put in order to arrive at a small number of regularities or laws. Albert Einstein remarked: "There is no logical way to the discovery of these elemental laws. There is only the way of intuition, which is helped by a feeling for the order lying behind the appearance" (Einstein, quoted in van den Hoff, 1995, p. 23; and in Selye, 1964, p. 47). But unlike the inductive method, the hypothetico-deductive approach is not completed with the intuitive sensing of a scientific law. In fact, it may be argued that the more strenuous work is yet to be done.

While it is difficult to give a precise quantitative evaluation, it appears that the majority of the intuitions of those scientists who are able to experience them ultimately turn out to be incorrect (Beveridge, 1961, p. 72). Therefore, following the surfacing of an intuitive idea a process of testing has to follow. The stages of the testing of a new theory will be detailed later when discussing Karl Popper's views of the hypothetico-deductive method. Here, I want to stress the indispensable requirement that scientists in their respective fields must try to test the theory by determining first which of its implications can be experimentally tested.

In fact, it seems that the main function of a hypothetico-deductive theory is exactly that—to generate experimental scientific work that may substantiate, and even more importantly, may refute it. This process of refutation of theories may be simple and short or may continue indefinitely. Darwin's evolutionary theory, for example, considered one of the most successful theories in the history of science, has generated an enormous amount of scientific work in its more than 150 years of existence, yet it still is subject to much controversy among leading scientists. At one extreme of this debate, the great geneticist Theodosius Dobzhansky declared, "in biology nothing makes sense except in the light of evolution" (Dobzhansky, 1970, pp. 5–6), while at the other, Australian molecular biologist Michael Denton, whose views will be detailed in section 1.3., considers Darwinian evolutionary theory, particularly what he deems its more pretentious claims, as "the great cosmogenic myth of the twentieth century" (Denton, 1986, p. 358).

By now, the reader may have begun to wonder why I felt the need to enter in such detail into a topic that pertains to the field of philosophy of science, of which I have only modest knowledge. Besides trying to define and exemplify the kind of theory that one will encounter in this book, as well as to put this theory in a larger perspective, I have a more concrete aim. Put succinctly, my intention here is to try to convince the reader, particularly my open-minded colleagues in the behavioral sciences, that a comprehensive theory is as important in psychiatry as in any other scientific discipline, even if its claims are speculative to a considerable degree, even if it is based on circumstantial rather than direct evidence, and even if it may ultimately prove to be partially or entirely incorrect. These arguments will be elaborated further in this section when discussing the views of Popper and Kuhn.

In the following, I will sketch a brief historical survey of the changing relationship between the inductive and hypothetico-deductive methods of theory building in Western thought, hoping to convince the reader of the increasing centrality of the hypothetico-deductive method in twentieth-century scientific enterprise.

1.1.1. A Brief Historical Survey of the Relationship between the Inductive and Hypothetico-Deductive Methods

In ancient Greek culture we can discern a clear dissociation between theorizing based on contemplative observation of natural phenomena (unchecked by scientific experimentation) and accumulating pragmatic useful knowledge on how the physical and biological world behaves in such applied domains as agriculture, breeding domesticated animals, sailing, building, perfecting weaponry, etc. These latter preoccupations had a low social prestige and were considered unfit for freeborn citizens of the city-states. In fact, it was believed that continuous preoccupation with them led to the degeneration of body and soul (another example of unchecked theorizing). Therefore, these activities were left mainly to the slaves. A special word was coined for this doctrine: *banausia*— meaning the habits of an artisan, which in the ancient Greek context carried the negative connotation of vulgarity, or bad taste; a freeborn citizen would not have gotten his hands dirty doing it (Fowler, 1962, pp. 11, 21, 31–32). The dissociation between methodical observation of certain phenomena (which was sometimes very accurate; idem, p. 17) and the naive theoretical explanations given for them can be exemplified by the four personality types identified by the ancient Greeks (sanguine, choleric, melancholic, and phlegmatic). These categories are still in use in some circles, and the present theory will also make use of them in chapters 5 and 7. The theoretical explanations provided for these personality types—that they are caused by irregularities in various body fluids (blood, bile, etc.)—are of course entirely unfounded. However, in certain other cases such speculative ideas are considered by philosophers of science as anticipations of great scientific discoveries, some more than two thousand years later. For example, Thales's idea that the earth rides on water like a ship and in consequence somehow gives birth to earthquakes "strangely anticipates the modern theory of continental drift" (Popper, 2002, p. 185); Anaximander's explanation of the earth's ability to preserve its distance from other celestial bodies anticipates the theory of gravitation; Heraclitus's conviction that "everything is in flux, and nothing is at rest" anticipates the contemporary understanding of physical, chemical, and biological processes; the "atomists" Leucippus and Democritus anticipated nuclear physics (idem, pp. 183–206).

What is of even more interest for us is that Aristotle anticipated evolution, including the descent of humans: "In 'Historia Animalum' Man is placed at the top of Scala Natura (directly above the Indian elephant), and is accorded superior intellectual powers, but none qualitatively distinct from those of other species" (Birney & Teevan, 1961, p. 166). However, Aristarchus, an Ionian scientist from the third century BC, was a notable exception. He based his theoretical proposals on observation and making "careful observations . . . of the size of the earth's shadow on the moon during a lunar eclipse," anticipated the heliocentric system, and postulated that the stars are distant suns (Hawking & Mlodinov, 2010, p. 21).

It seems that the great problem of ancient Greek science was not with bold theorizing (sometimes in the right direction) but rather with testing and thus sorting out promising from untenable theories with the aid of experimentation. According to historian of science Peter Dear, this situation—that is, the dissociation between practical knowledge and speculative, unsubstantiated theoretical assumptions (whether materialistic in nature like the ancient Greeks or supernatural and religious like the contemplations of the scholastics in the Middle Ages)—"began to change during the seventeenth and eighteenth centuries" (2006, p. 10).

The philosopher and statesman Francis Bacon (1561–1626) was influential in catalyzing the move toward an experimentally oriented science, accepting only strictly inductive generalizations (idem, p. 10). Bacon argued on behalf of a scientific methodology that consists of rigorous observation of phenomena, and any theoretical inferences drawn from the data obtained have to be very cautious generalizations that do not surpass what is justified on the basis of the facts—and even these conclusions have to be constantly checked by reference to other facts. He maintained that human understanding cannot surpass what is observable (Fowler, 1962, pp. 42–43).

Consequently, Bacon strongly condemned those mental inclinations and cultural influences that may lead the scientist's judgment astray, including the presupposition that a greater order exists in nature than is actually demonstrable; reverence for the opinion of authorities in a certain field, or the opposite, the tendency to discard traditional beliefs and customs indiscriminately and to try to impress with novelties; using words with an imprecise or ambiguous meaning or language that speaks to the emotions rather than to reason; and the tendency to disregard facts that contradict one's cherished beliefs (Quinton, 1980, pp. 35–36; Fowler, 1962, p. 42). Bacon labeled these noxious influences on scientific thinking "idols" and, of course, rejected the ancient Greek belief that "the dignity of the human mind is lowered by long and frequent intercourse with experiments" (quoted in Quinton, 1980, p. 56).

Bacon's views, while considered revolutionary in his time, are viewed with ambivalence today, even criticized or ignored by some theorists (Quinton, 1980, pp. 81–84). His method is especially embraced by what Thomas Kuhn calls "normal science," that area of scientific activity that tests circumscribed aspects or implications of a theory.

Bacon is criticized for ignoring or rejecting the unique abilities needed for scientific theory building. He argues on behalf of a mechanical, inductive kind of generalization that can be mastered by any methodical scientist who follows its logical rules, ignoring the importance of the scientist's critical judgment, imagination, and creativity. Even more important, it is now clear that observation and experimentation cannot be done randomly but must have a target chosen by some kind of selective reasoning from the multitude of observable natural phenomena. Observation and experimentation are used to test a hunch, hypothesis, or theory—or more precisely, some implication of it, that can be formulated in such a way that observation or experimentation can provide a certain measure of approval or disapproval: "all observation involves interpretation in the light of our theoretical knowledge" (Popper, 2002, p. 30).

By the nineteenth century, induction had become the standard scientific method. A good example would be the teachings of French chemist Antoine Lavoisier (1743–1794), the discoverer of oxygen and nitrogen in the atmosphere: "we ought to form no idea but what is a necessary consequence, and immediate effect, of an experiment or observation" (Lavoisier, 1790, p. xvi). Darwin himself, trying to conform to this view, which was dominant in Victorian England, declared in his autobiography that he "worked on true Baconian principles, and without any theory collected facts on a whole-sale scale" (Barlow, 1969, p. 119). This approach was augmented by his superior ability in noticing "things which easily escape attention, and in observing them carefully" (idem, p. 141).

However, Darwin's conformity with pure inductive principles was far from being consistent. Later in his autobiography we encounter the following confession: "From my early youth I have had the strongest desire to understand or explain whatever I observed—that is, to group all facts under some general laws." And: "I have steadily endeavored to keep my mind free, so as to give up any hypothesis, however much beloved (and I cannot resist forming one on every subject), as soon as facts are shown to be opposed to it" (idem, p. 141).

Might it be that this exceptional, creative mind was able to move freely between meticulous observation, careful inductive generalization, bold hypothetico-deductive theorizing, and impartial verification, while matching the method of inquiry to the specific task at hand? (For more about Darwin's scientific method, see also: Dear, 2006, pp. 100–103; Medawar, 1969, pp. 10–11; Howard, 1982, pp. 91–96; Mayr, 1982, chapters 9–11, Mayr, 1991, pp. 9–11, 104–106).

By the beginning of the twentieth century, the central role of hypothetical theory in scientific activity was being increasingly acknowledged. The *logical positivist school*, a school of the philosophy of science in the first half of the twentieth century, established in Vienna, influenced by the radical empiricism of British philosopher David Hume, still "required that *every theoretical term in a scientific theory must be provided with an explicit definition composed entirely of observational terms*" (Klee, 1997, p. 33, emphasis in the original).

The American philosopher of science Ernest Nagel, who has a logical positivist background, describes a more complex but still recognizable connection between abstract notions of a theory and experimental data, which he named "correspondence rules" (Nagel, 1961, pp. 93–105). However, he admitted that: "The ways in which theoretical notions are related to observational procedures are often quite complex, and there appears to be no single schema which adequately represents all of them" (idem, p. 94).

Rudolf Carnap, a leading positivist, argued that assertions in science are tested "with the help of protocol sentences." Protocol sentences refer to "the contents of immediate experience, or the phenomena; and thus [to] the simplest knowable facts" (quoted in Popper, 1968, p. 96).

Nevertheless, by the second half of the twentieth century, the verifiability of scientific hypotheses and theories encountered an increasing difficulty. It became evident that verifiability and falsifiability were not equally possible: There exists "an *asymmetry* between verifiability and falsifiability," and: "theories are . . . *never* empirically verifiable" (idem, 40–41, emphasis in the original).

Verifiability became viewed as increasingly unreliable due partly to the human mind's propensity to interpret phenomena in a way that is favorable to its presuppositions. This is nowhere more evident than in the fields of psychology and psychodynamic psychiatry and in the social sciences. Indeed, Popper criticized Marx, Freud, and Adler primarily according to this consideration: "I found that those of my friends who were admirers of Marx, Freud, and Adler, were impressed by a number of points common to these theories, and especially by their apparent *explanatory power*. These theories appeared to be able to explain practically everything that happened within the fields to which they referred . . . you saw confirming instances everywhere: the world was full of *verifications* of the theory . . . unbelievers were clearly people who did not want to see the manifest truth; who refused to see it, either because it was against their class interest, or because of their repressions which were still 'un-analyzed' and crying out for treatment" (2002, p. 45; emphasis in original). To me, these lines contain echoes of Francis Bacon's concept of "idols," mentioned previously.

Another formulation of the idea of the unequal weight of verifiability and falsifiability was provided some thirty years earlier by the British physicist, mathematician, and astronomer Sir James Jeans: "Nature may answer our question by shewing us a phenomenon which is inconsistent with our hypothesis or by shewing us a phenomenon which is not inconsistent with our hypothesis. She can never shew us a phenomenon which proves it; one phenomenon is enough to disprove a hypothesis, but a million do not suffice to prove it. For this reason, the scientist can never claim to know anything for certain, except direct facts of observation" (quoted in Fowler, 1962, p. 98). Even the reliability of these "direct facts of observation" was questioned by Karl Popper.

If verification or definitive validation of a theory by direct experimentation is impossible, only one way remains to distinguish a useful scientific theory from an erroneous, unscientific, or metaphysical one. This is the route of testing those implications of the theory that do lend themselves to experimentation with scientific methodology.

It can be argued that Popper's hypothetico-deductive method of testing theories is in a sense the exact opposite of Bacon's inductive method of arriving at scientific laws and theories. It encourages far-reaching speculative thought and imagination (condemned by Bacon as an "idol"), and rejects the usefulness of simple, one-step inductive generalizations. Popper transfers the empirical testing from the beginning to the end of the process leading to theoretical laws. In contrast with Bacon, he negates the possibility of ultimate knowledge of nature and consequently embraces the method of successive refutation of theories (instead of their verification) when more sophisticated technology or better contrived experiments show that they no longer account for newly generated data or empirical regularities. In fact, the usefulness of a new theory lies to a great extent in its ability to generate new, goal-oriented experimentation that may yield results that contradict its predictions and therefore necessitate its modification or replacement with a new theory that better accounts for the discordant facts. At the end of his book *The Logic of Scientific Discovery*, Popper summarizes his view of scientific practice as follows: "Out of uninterpreted sense-experiences, science cannot be distilled, no matter how industriously we gather and sort them. Bold ideas, unjustified anticipations, and speculative thought, are our only means for interpreting nature: our only organon, our only instrument, for grasping her. . . . Those among us who are unwilling to expose their ideas to the hazard of refutation do not take part in the scientific game" (1968, p. 280).

At the same time, Popper questions the value of one-step logical generalizations from observational data. In his opinion, these "ad hoc" theories (like the catecholaminergic explanations of schizophrenia) do not transcend by very much the data that led to their formulation and in consequence cannot be tested independently of it. Although the probability of their being valid is higher than that of bold, comprehensive, speculative theories, their explanatory power is restricted and their "theoretical interest" small (Popper, 2002, pp. 15–16).

A more detailed account of Popper's theory seems to me superfluous here (the interested reader can consult the books that served as sources for this discussion). Instead, I want to briefly address two related themes: the impossibility of pure observation and the stages through which theories are tested.

As mentioned before, Popper turns upside down the traditional stages of observation, generalization, and theoretical interpretation. In his view, no such thing as pure observation exists in science: "Observation is always selective. It needs a chosen object, a definite task, an interest, a point of view, a problem . . . it presupposes similarity and classification" (Popper, 2002, p. 61). And in conclusion: "through our theories . . . we learn to observe, that is to say, *to ask questions* which lead to new observations and to new interpretations. This is the way our observational knowledge grows" (idem, p. 335, emphasis in original). Medavar (1969, pp. 27–28) quotes views in the same spirit from Nietzsche and Kant. In summary, the reason for scientific experimentation and observation is always to try to solve a problem, either a practical problem or a moot point in a theory, or to find out whether the facts support or contradict a hunch or hypothesis (Popper, 1972, p. 258). Or as Darwin put it: "All observation must be for or against some view" (idem, p. 259).

The testing of a theory follows several steps. First, it is ascertained that the theory has *logical coherence*; that is, its conclusions possess "internal consistency of the system." Second, it is ascertained that the theory has the logical form of an *empirical theory*; that is, it does not belong to a formal or theoretical science (mathematics and logic) and is not a metaphysical theory. Third, it has to be ascertained that the new theory, because it accounts better for accepted factual knowledge, is preferable to already existing theories in the field (and can explain, if possible, the lesser efficiency of the former theory's explanations); that is, it constitutes a *scientific advance*. Fourth, we have to test the *empirical implications* of the theory (Popper, 1968, pp. 32–33).

This last step needs some more detailed discussion. We have seen previously that even direct observation does not yield pure data unbiased by reasoning, and that theories (being abstract constructs) are never directly verifiable (idem, p. 40). Theories can be tested only if they indicate some testable phenomena that may contradict them—to use Popper's word, phenomena that are "forbidden" by the theory. The more such types of phenomena are indicated by the theory the better, because the theory's refutation is therefore easier, and failure to refute it by experimentation is also more impressive (idem, pp. 112–113).

Another problem raised by Popper in this respect is the *objectivity* of the "forbidding" statements of the theory. Because objectivity and subjectivity are problematic terms, we require a definition of how these concepts are used here. Popper defines objectivity, following Kant, as knowledge that can be tested and understood by any reasonable person in the possession of the necessary professional knowledge of the field, "independently of anybody's whim." In consequence, "I shall therefore say that the *objectivity* of scientific statements lies in the fact that they can be *inter-subjectively tested*" (idem, p. 44, emphasis in original).

This last requirement in the exact sciences seems resolvable, since their conclusions rest ultimately on phenomena that are directly observable (with or without the aid of specific instruments), and which can be corroborated by different persons—for example by joint observation of the same phenomenon—relatively easily.

The present theory, however, must also deal with *subjective* experiences, like emotions or intentions, which indicate instinctive motivations or predispositions and which, unlike sensory perceptions, cannot be easily and precisely corroborated or communicated with. Consequently, the requirement for "inter-subjective testability" poses quite a serious problem. I will attempt to deal with this subject in the section of the next chapter on the empathizing process.

In his landmark work, *The Structure of Scientific Revolutions*, Thomas Kuhn analyzes scientific activity from a different viewpoint than does Popper. While Popper's approach is purely philosophical (epistemological), arguing clearly for the primacy of theory, with the role of experimentation relegated to trials that could refute it, Kuhn sees scientific enterprise in a more pragmatic way, as a social institution, the joint effort of communities of highly trained specialists in a given field. In this enterprise, he seems to attribute equal weight to theory; the rules governing its application in experimentation; instrumentation; and even significant accidental discoveries. Kuhn attributes more importance to the special ways these elements of scientific work are intertwined than to the priority of any one of them (Kuhn, 1970, pp. 46, 109), labeling a sharp distinction between fact and theory as "exceedingly artificial" (idem, p. 52).

Practitioners in a mature scientific field agree about the fundamental issues in their field; that is, in Kuhn's term, they share a common "paradigm." A paradigm comprises "law, theory, application, and instrumentation together," which "provides models from which spring particular coherent traditions of scientific research" (idem, p. 10).

During "normal science"—that is, when the paradigm is unchallenged—the scientists' work consists of "puzzle solving," the aim of which is "to articulate the paradigm theory, resolving some of its residual ambiguities and permitting the solution of problems to which it had previously only drawn attention" (idem, p. 27). This phase of scientific activity is piecemeal, cumulative, circumscribed problem-solving work. The history of a scientific field consists, according to Kuhn, of periods of "normal science" punctuated by what he calls "scientific revolutions." Research during normal science may produce data incompatible with the prevailing paradigm (that is, "anomalies"). If these inconsistencies cannot be solved by modifying the paradigm, they lead to a crisis in the respective scientific field. The crisis in turn leads to the loosening of the dominant theory's hold on interpreting the available data, to proliferation of alternative theories, and to less coherent, less goal-oriented research. The crisis is resolved by discarding the old paradigm and the acceptance of a new one. This stage involves a fundamental change in interpreting existing data—a turbulent period in the history of the respective scientific field leading to psychological stress, confusion, and tension between scientists who still stick to the old paradigm and those who accept the superiority of the new one. The former tend, as a rule, to consist of the older generation whose lifelong work is embedded in the old paradigm, while the latter are, as a rule, the young or those who are new to the respective field.

Kuhn, in opposition to Popper, accepts verification as well as falsification of theories (idem, p. 80). Nevertheless, the agreement between these two great thinkers is more important in my view than their differences. Both refuse to admit an exclusively cumulative nature of scientific progress (that is, progress through the accumulation of "objective" data and careful inductive generalization, idem, p. 96), and both argue in favor of successive profound reformulations of scientific understanding when experimental data and theoretical formulations clash, recognizing those later events as the main moving force of scientific progress.

Kuhn has important things to say about sciences that are still in the "pre-paradigmatic stage" of their development—that is, scientific fields in which no agreement about fundamental issues exists among the practitioners. Since psychiatry is such a science, Kuhn's opinions in this respect are important for us. (A more detailed discussion of this point will be postponed until the third section of this chapter.)

Before leaving our historical detour into the vicissitudes of observation, experimentation/theory interaction, I would like to mention one further point. Popper's and Kuhn's illustrations for their philosophical arguments are taken largely from the field of exact empirical sciences—primarily physics, chemistry, and astronomy. *However, the subject matter of the present theory, the inherited aspects of (dysfunctional) human be-havior, differs essentially from the exact sciences in that it deals with a topic that is absolutely inaccessible to direct scientific experimentation.* The events of the vast spans of time during which the animal kingdom evolved can be reconstructed only very partially (and that is all the more true with respect to behavior) using meager, indirect evidence.

The subjective experience of humans with regard to their inherited incentives to behave (which provides our intuitive knowledge of these incentives) cannot be directly corroborated by observational data on overt behavior. In consequence, the study of the innate incentives of human behavior will necessarily require more complex methods and reasoning than the study of directly observable phenomena. This topic will be further elaborated in the next and following chapters.

1.2. The Risks and Gains of Building on Knowledge Pertaining to Scientific Fields Outside One's Expertise

The reader has probably already noticed that I build my arguments mainly on facts and expert opinions that do not pertain to my field of expertise. The present chapter deals with topics belonging to the phi-losophy of science. The following three chapters will also deal mostly with scientific fields outside psychiatry: biology, ethology, genetics, evolutionary theory, evolutionary psychology, primatology, and so on. Working full-time as a clinical psychiatrist throughout my professional career, I naturally have restricted knowledge in these and other fields, as well as in highly specialized domains within the field of psychiatry. As evolutionary psychologist and psycholinguist Steven Pinker remarks: "Nowadays we specialists cannot be more than laypeople in most of our disciplines, let alone neighboring ones" (Pinker, 1997, p. x).

Nuclear physicist Richard Feynman strongly warns against the prac-tice of borrowing ideas from unrelated scientific fields: "In talking about the impact of ideas in one field on ideas in another field," Feynman cau-tions, "one is always apt to make a fool of oneself. In these days of spe-cialization there are too few people who have such a deep understanding of two departments of our knowledge that they do not make fools of themselves in one or the other" (Feynman, 1998, p. 3).

I certainly do not count myself belonging to these "few people." However, it seems that anyone who desires to approach a phenomenon in a wider, theoretical perspective has no alternative than to try to intertwine knowledge from one's own profession (a kind of knowledge acquired through direct contact with the respective phenomena, direct contact with one's teachers' and colleagues' ways of thinking and professional practices, in addition to familiarity with the professional literature) and relevant knowledge of neighboring or even unrelated fields. This latter kind of knowledge is necessarily acquired from written sources (as a rule, from books) in which the respective field is presented in a concise, accessible form according to the personal point of view of the author.

While I am aware of the risk of making a fool of myself from time to time as Feynman cautioned, I have no choice but to incur this risk.

The central topic of the present theory being comprehensive categories of instinctive behavior or predispositions that evolved during the phylogeny of diverse animal species (including humans), I cannot discuss them without sketching first the wider context of their evolution according to the relevant literature. Moreover, a more general intention of this work is the aspiration to establish tight interconnections between psychiatry and the evolutionary theory of behavior, since I believe that the inherited aspects of mental disorders cannot be interpreted in any other way.

The aspiration to unite different scientific fields, or even all the sciences, into one convergent body of knowledge is a recurrent theme (dream?) of scientists from different specialties. Anthropologist Ernst Becker urged the creation of a "science of man" in his book *The Structure of Evil, An Essay on the Unification of the Science of Man*, calling for an integration of the critical ideas of psychology, sociology, psychiatry, history, and philosophy (1976). Biologist Edward Wilson dedicated a book to the same theme: *Consilience: The Unity of Knowledge*. Here he puts his dream in the following passionate words: "Thanks to science and technology, access to factual knowledge of all kinds is rising exponentially while dropping in unit cost. . . . What then? The answer is clear: synthesis. *We are drowning in information, while starving for wisdom.* The world henceforth will be run by synthesizers, people able to put together the right information at the right time, think critically about it, and make important choices wisely" (1998, p. 269, emphasis added). Even Richard Feynman, who in the quotation cited earlier cautioned so pungently against blending ideas from different scientific fields, just several pages later in the same book—when discussing Faraday's discovery that electricity has both physical and chemical properties—welcomes this meeting between two different scientific fields with the following enthusiastic words, that this moment signaled: "one of the most dramatic moments in the history of science, one of those rare moments when two great fields come together and are unified" (Feynman, 1998, p. 14).

Finally, I would like to quote American anthropologist Marvin Harris, this time on the subject of human behavior: "Without a strategy aimed at *bridging the gap between specialties and at organizing existing knowledge along theoretically coherent lines*, additional research will not lead to a better understanding of the causes of lifestyles. If we genuinely seek causal explanations, we must have at least some rough idea about where to look among the potentially inexhaustible facts of nature and culture" (Harris, 1974, pp. vii–viii, emphasis added).

It seems that we have here a no-win situation that compels us to compromise. We have to sacrifice detailed, firsthand knowledge in the fields outside our specialty in order to obtain a wider perspective on our subject of inquiry.

1.3. The Suitability of Neo-Darwinian Evolutionary Theory as "Background Knowledge" for the Present Theory

The present theory is built on the assumption that at least the basic relevant ideas of evolutionary theory are valid. It would be useless to argue about the role of the relaxation of selective pressures without being convinced that natural selection pressures played a central role in evolution.

In Popper's view, building new ideas on existing ones is inevitable. He names this existing knowledge "background knowledge." Though Popper uses the word in a much wider sense than I will use it here, and includes commonsense knowledge and even inborn predispositions for acquiring knowledge, I think the concept is also adequate for the present work's purposes: "*the growth of all knowledge consists in the modification of previous knowledge* [emphasis in original] . . . knowledge never begins from nothing but always from some *background knowledge—knowledge which at the moment is taken for granted*—together with some difficulties, some problems" [emphasis added] (Popper, 1973, p. 71).

From this reasoning, the following question emerges: Do the ideas of evolutionary theory—ideas that "at the moment [are] taken for granted" and on which I intend to build my own ideas—have a scientific status strong enough to give to such an additional "theoretical superstructure" the needed stability, thus precluding its refutation at its preliminary stages of testing?

After the "new synthesis" in the thirties and forties, Darwinian evolutionary theory became an immensely successful scientific theory underlying most of the experimental work and theoretical interpretations in biology and related sciences. The new synthesis refers to the synthesis of the original Darwinian thesis with new Mendelian and, later, molecular knowledge on genetic inheritance, as well as with other new developments (such as Hamilton's theory of kin selection).

The great geneticist Theodosius Dobzhansky's often quoted dictum that "in biology nothing makes sense except in the light of evolution" (Dobzhansky, 1970, pp. 5–6; Denton, 1986, p. 154; Lieberman, 2006, p. 1) reflects well the almost unconditional acceptance of evolutionary theory. Darwinian ideas were employed to explain phenomena far beyond what Darwin himself intended, such as the evolution of life from organic and inorganic matter according to evolutionary principles (Denton, 1986, pp. 249–271); the cultural transmission of discrete packages of information analogous to the transmission of inherited information by genes (Dawkins's "meme" hypothesis, discussed in chapter 3); and the mechanism of social processes (social Darwinism) (Denton, 1986, p. 70).

However, seen more soberly, it turns out that the claims of evolutionary theory are heterogeneous with regard to their scientific validity. Its more modest claims (restricted variations of traits inside a species' evolutionary trajectory, or the evolution of one or more new species from a mother population [speciation, adaptive radiation] are strongly supported by experimental and observational data (Denton, 1986, p. 70). The more pretentious claims, however, that the whole diverse living world came into being by an uninterrupted evolutionary process, in which new and more complex forms evolved in small steps from more primitive ones solely according to the mechanism of natural selection, was problematic from the beginning and came under increasing criticism, especially by modern embryology and cellular and molecular biology. I will highlight only some general considerations here, to the extent they are relevant to the question of whether evolutionary theory is reliable enough to serve as "background knowledge" to my own theory. I found the Australian molecular biologist Michael Denton's book *Evolution, a Theory in Crisis* (1986) very useful for this purpose, and the interested reader is advised to consult the original text.

After the "evolutionary synthesis" between 1936 and 1947 (Mayr, 1982, p. 119), and especially after the arrival of molecular genetics and modern embryology, it became a common practice to divide Darwinian evolutionary theory into two separate domains—*microevolution*, or "the special theory of evolution," and *macroevolution*, or "the general theory of evolution" (Denton, 1986, p. 44). I will begin with macroevolution, the far-reaching and contentious claims of evolutionary theory lacking direct support from, or (partially) contradicted by, existing scientific data.

The transitional forms of organisms between the larger typological groups (class and above—that is, between fish, amphibians, reptiles, birds, and mammals) were very rare in the paleontological record of Darwin's time and remained so afterward. These bigger taxonomical groups (extant and extinct) show clear demarcations *without* transitional forms (Denton, 1986, pp. 157–195). Transitional forms, however, are indispensable, and were expected to turn up in great numbers according to a theory that presumes gradual, small-step transitions from lower to higher organisms. Even the small number of such paleontological findings, like the Archaeopteryx (thought to be transitional between reptiles and birds) or the rhipidistian fishes (thought to be transitional between fish and terrestrial animals), when closely examined, turned out to possess well-developed organs or features characteristic of a well-defined class of organisms rather than being truly transitional (Denton, 1986, pp. 172–180; Mayr, 1982, p. 613).

The alternative to actual paleontological findings for solving the riddle of continuous evolution in small steps from one class of animals to another would be an imaginary hypothetical reconstruction of a series of transitional forms of animals between two classes with the requirement that *all the transitional forms have to be fully functional,* that is, to be able to solve all the survival and reproductive tasks in the organism's special environmental niche. However, such a hypothetical exercise proved to be impossible (Denton, 1986, pp. 199–230). For example, it is difficult to imagine how the forelimb of a (terrestrial) reptile became the feathered wing of a bird through small evolutionary steps, with each step fully functional and conferring an adaptive advantage on the animal (idem, pp. 202–209). (For an opposing view on the evolution of wings—namely, that such a hypothetical reconstruction is possible, see Dawkins, 1987, pp. 89–90.)

Even more problematic is finding these transitions, or imagining them hypothetically, on the microscopic level, that is, on the level of the living cell. The advance of molecular biology produced a complex picture of the structure and functioning of a cell, again with a complete lack of transitional forms between different unicellular organisms (for example, between prokaryotic and eukaryotic ones, that is, cells without or with a nucleus, respectively). Biochemist Michael Behe (1998) gives a vivid description of the extraordinary complexity of the cell mechanisms (even of a not very central part of it—the cilium and bacterial flagellum), and the unimaginability of their step-by-step evolution (pp. 39–45, 59–73). On the other hand, molecular biologists J. Klein and N. Takahata (2002) have no reservations about interpreting the molecular record as proof of phylogeny on evolutionary lines from the most primitive to the most evolved life forms (pp. 156–157, 164–167).

It seems to me, after all, that traditional evolutionary theory can be applied directly only when thinking about whole organisms or populations and not about parts of them (organs, tissues, cells, molecules). It seems clear that natural selection pressures act ultimately by modifying these constituents of the *organism*. However, *no competition (struggle for life) exists at the "sub-organismic" level but rather one finds cooperation and complementarity for the fitness of the organism as a whole.* This subject is detailed in the following section under the heading "internal selection."

Another traditional argument in favor of macroevolution is the strikingly similar anatomical construction of corresponding organs in animals from different vertebrate classes: for example, the vertebrate alimentary canal, forelimbs, and kidneys. This phenomenon is called *homology* and constitutes one of the main arguments in favor of macroevolution. The similarity of genetic specification, as well as of the embryologic development is, however, an indispensable requirement for establishing homology (Denton, 1986, pp. 142–155). However, embryological as well as genetic studies have found that, contrary to expectations, these homologous organs frequently develop from different embryological tissues and are under the control of different genes.

Finally, we have to note here mechanisms for bringing about new life forms that are different from natural selection: the horizontal transfer of genetic material, and the creation of new cell forms through symbiosis. Horizontal or lateral gene transfer means the incorporation of genetic material from a dying or decomposed cell into the genome of a cell of a different kind. That is, in certain circumstances the incorporated genetic material, instead of being decomposed, recombines with genetic material of the host cell, leading to an altered genome. In the case of symbiosis between two different cells, it is hypothesized that over time the simpler cell, through a process that can be called retrograde evolution, becomes a cell organelle of the host cell. Strong indications exist that the mitochondria (the energy-producing structure) of eukaryotic cells developed in this way. The mitochondria have an entirely different kind of DNA than the host-cell nucleus, with a different, three-dimensional format (not linear but circular) that is transmitted exclusively through the maternal line (Behe, 1998, p. 188–189; Klein & Takahata, 2002, pp. 49, 60–61).

Another mechanism that is not natural selection in the traditional sense is called *internal selection*. As expected in the case of any highly complex living system, selection exists against those mutations (or chromosomal recombinations during meiosis or fertilization) that lead to a considerable disturbance in the smooth cooperation of the interacting parts of the respective organismic unit. This kind of selection was recognized by Darwin at an organismic level (Denton, 1986, p. 308). An interesting discussion of internal selection can be found in Lancelot Law Whyte's *Internal Factors in Evolution* (1965). This topic will also be examined in chapter 4 in the section on the various kinds of selection.

In general, it seems evident that at many levels of complexity of living matter—the molecular and cellular levels, and at the level of the organs and tissues—the principle of cooperation or complementarity between the constituent parts of the organism is the only one operative, while the principle of competition and survival of the fittest is nonexistent. At the level of interaction between organisms of different species, reciprocal gain may also be important, as in the case of symbiosis. In the case of conspecifics living in interacting social groups, like chimpanzees or humans, the principle of cooperation and complementarity is of equal importance to the principle of intragroup competition. In other words, when cooperation is vital (that is, ensures improved transmission of genes to the next generation), natural selection chooses those variants best suited for cooperation rather than competition.

Two further considerations complicate the macroevolutionary controversy, both of which speak in favor of the hypothesis that macroevolution indeed happened. The first is the great similarity of the basic design of all living forms. "Molecular biology has . . . shown that the basic design of the cell system is essentially the same in all living systems on the earth from the bacteria to mammals. In all organisms the role of DNA, mRNA and protein are identical. The meaning of the genetic code is also virtually identical in all cells" (Denton, 1986, p. 250; see also Mayr, 1982, pp. 827–828). Another example of this, with psychiatric implications, is the existence in protozoa of the neurotransmitters norepinephrine, dopamine, serotonin, GABA, glutamate, etc., the targets of most psychoactive drugs in the human brain (Brüne, 2008, p. 60; Healy, 2009, p. 150).

The second consideration is the incontestable fact that in paleontology the chronological succession of life forms goes from simple toward more complex organisms: "The fossil record revealed that the history of life on earth was overall one of progress from simple to more complex types of life. The first organisms to appear in the fossil record are simple invertebrates and simple plants such as seaweeds and algae; later the more complex vertebrates appear—fish first, then amphibians, followed by reptiles and mammals. Moreover, even within particular groups, the more specialized types tended to occur later" (Denton, 1986, p. 52).

The precise way that this macroevolution occurred is not of immediate concern to our present discussion. Although it seems clear that macroevolution from simple toward complex organisms did indeed take place, it seems equally clear that the mechanism of natural selection alone as proposed by the neo-Darwinian evolutionary theory is insufficient as an exclusive explanation of this process. The similarity of the basic design of all living organisms, as well as their successive appearance in the paleontological record, from simple toward progressively more complex, will suffice for evolutionary theory to serve as adequate "background knowledge" for the present work. Even more important, if we take for granted that evolution did indeed occur and that natural selection was a central factor in shaping the optimization of behavioral strategies (above all in their innate aspects), it is not important for our purpose here to know the specific mechanisms through which evolution led to the inherited behavior patterns or predispositions of the ancestors of modern humans. Our central topic is what happened to these inherited behavior patterns during subsequent human evolution, which is characterized by increasing ecological dominance, progressive relaxation of natural selective pressures, and consequent diversification.

The *microevolutionary* claims of Darwinism (or the special theory of evolution), in contrast to the macroevolutionary ones, seem to have an entirely different degree of scientific validity. The improvement of domestic animal stocks by artificial, human-made selection regarding desirable properties was already well known to the ancient Greeks—in fact, it was probably known much earlier, from the beginnings of agriculture and the domestication of wild plants and animals about 10,000 years ago (idem, pp. 44–45). "The variation of domestic animals provided Darwin not only with evidence of the power of selection but also with irrefutable evidence that organisms could indeed undergo a considerable degree of evolutionary change" (Denton, 1986, p. 45).

Evolution at the species and speciation level, in contrast to the big leaps of macroevolution, is also well substantiated by the paleontological record (Denton, 1986, pp. 57, 182–184). The actual occurrence of microevolutionary changes in nature has been scientifically documented. The British peppered moth that changed its color to a darker one as a result of industrial pollution (as a means of camouflage against predators) is a well-known example (Denton, 1986, pp. 79–81; Ehrlich, 2009, pp. 27–29). The resistance of bacterial cultures to antibiotics, acquired through natural selection, is commonly observed in hospital laboratories. And concerning human evolution there is a wealth of transitional forms in the palaeontological record between apes and anatomically modern humans (Mithen, 1996, pp. 19–25; Ehrlich, 2000, pp. 78–100).

The question whether the theory of evolution can serve as adequate "background knowledge" in Karl Popper's sense, that is, whether it can provisionally be taken for granted and built upon for our present purpose, must be answered in the affirmative. This is due to the fact that most of the ideas of the present theory can be accommodated within the framework of microevolution, while those that bridge behavioral issues across bigger taxonomic categories can claim validity from the "unity of the basic design" of organisms, as well as from the sameness of the ultimate biological incentives of behavior (namely, survival and reproduction) across the whole animal kingdom. The far-reaching release of instinctive behavior from natural selective pressures and consequent diversification in humans as a result of technological and social-cultural progress, which is the central idea of this theory, obviously refers to events within the evolution of one species only, that of Homo sapiens. Evolutionary psychologists argue that the evolution of unique human instinctive predispositions (such as language acquisition, reciprocal altruism, theory of mind abilities, and characteristic mating behavior, which will be detailed in chapter 3) evolved during the Pleistocene era (from approximately 1.6 million to 10,000 years ago). However, these instinctive predispositions are built upon those of the apes (most relevant here are the chimpanzee and bonobo), which are built in turn upon the foundations of mammalian instincts in general, and so on.

All four comprehensive instinctive mechanisms discussed in this theory—1) seasonal variation in instinctive intensity as the result of changing amounts of environmental resources for the sustenance of life processes; 2) basic changes in active instinctive behavior in frustrating conditions; 3) the progress of instinctive activity from well-differentiated behavior patterns toward more diffuse predispositions; as well as 4) the active/reactive behavior dichotomy and interrelationship—are relevant to at least the whole vertebrate subphylum. The presumed continuity of these basic behavioral incentives from animals with simple and more complex behavior up to humans is supported by the aforementioned sameness of the basic design in all living systems (including the genetic apparatus), as well as by the successive appearance of living forms in paleontological and molecular records from simple ones in the distant past to more complex ones. However, the diversification of these instinctive characteristics, well beyond that observed in nature as a result of gene-culture interaction, as well as the relevance of that diversification to mental disorders, naturally refers to humans only, and in consequence must be seen as a microevolutionary event.

1.4. Theoretical and Methodological Approaches in Psychiatry

As mentioned previously, philosopher of science Thomas Kuhn discussed some of the characteristics of the immature sciences that, in his terminology, were still in a "pre-paradigmatic" stage of their development. In the preface of *The Structure of Scientific Revolutions*, he recalls the following experience: "Spending the year in a community composed predominantly of social scientists confronted me with unanticipated problems about the differences between such communities and those of the natural scientists among whom I had been trained. Particularly, I was struck by the *number and extent of the overt disagreements between social scientists about the nature of legitimate scientific problems and methods* . . . somehow, the practice of astronomy, physics, chemistry, or biology normally fails to evoke the *controversies over fundamentals* that today often seem endemic among, say, psychologists or sociologists" (Kuhn, 1970, vii–viii, emphasis added).

Kuhn's characterization of a science that is in its pre-paradigmatic stage of development applies to a large extent to contemporary psychiatry. Lacking a "generally accepted view," there exist a "number of competing schools and sub-schools. . . . At various times all these schools made significant contributions to the body of concepts, phenomena, and techniques . . . though the field's practitioners were scientists, the net result of their activity was something less than science. *Being able to take no common body of belief for granted, each writer . . . felt forced to build his field anew from its foundations*" (idem, p. 13, emphasis added).

I should note that in formulating the present theory, I adopted this same practice. For example, in the first four chapters of this text I discuss in detail the merits and drawbacks of the inductive versus hypothetico-deductive methods of theorizing and the instinctive foundations of human behavior, as well as instinct-learning interaction; I also propose a contentious route by which humans communicate their subjective experiences.

Kuhn goes on to remark that "in the absence of a paradigm . . . all of the facts that could possibly pertain to the development of a given science are likely to seem equally relevant. As a result . . . fact-gathering is a far more nearly random activity than the one that subsequent scientific development [guided by a paradigm] makes familiar" (idem, pp. 12–15).

This observation seems especially relevant to our concerns. None of psychiatry's different approaches differentiates between the direct, primary behavioral consequences of the innate predisposition to a certain clinical disorder and its secondary, tertiary, etc., repercussions, whose relevance to the original genetic inclination is progressively less and less certain. (This topic will be discussed in detail in chapters 6 and 7.)

It seems to me useless to describe here the various schools of psychiatry.[1] Instead, I wish to briefly discuss here the three basic theoretical approaches used, each demanding a different methodology and which, to my mind, are the main expressions of the existence of "competing schools and sub-schools." The three main approaches are:

1. The practice *of atheoretical observation, description, and classification*—that is, no attempt is made to develop a theory of causality. Instead, this approach seeks only to formulate empirical regularities or laws. This method characterizes most aspects of what is generally called "clinical psychiatry."
2. The *reductionist scientific method* of the medical sciences using one-step inductive generalizations in theory building—that is, biological psychiatry.
3. The "*psychodynamic*" *method* of sensing a person's unobservable intrapsychic events by an intuitive route (detailed at the end of the second chapter). This approach is prolific in producing theories, but they are of a kind that cannot be tested with the scientific method.

1.4.1. The Descriptive, Atheoretical Approach in Psychiatry

The atheoretical, descriptive approach in psychiatry studies dysfunctional behavior impartially, "objectively," and from a distance (in terms of emotional involvement). In its approach to fact-gathering, it aspires to be like the emotionally detached observation of animal behavior of many ethologists, or even to the observation of the "behavior" of inanimate objects in exact sciences. This approach seems unavoidable at the beginnings of a new science, when the phenomena under study must first be outlined in detail, precisely described, categorized, and systematized into more comprehensive units, which contain characteristic constellations of aspects of the phenomena (Quinton, 1973, p. 284). In psychiatry, this approach led to the nosology of mental disorders, with each mental-disorder category containing frequently co-occurring signs, symptoms, and syndromes. The observation of, experimentation with, and follow-up to these categories led ultimately to the recognition of regularities, generalizations, or empirical laws.

This approach in principle seems to be not unlike the Baconian inductive method of cautious generalizations on the basis of observationally obtained material.

This method obviously yielded important results. In mental disorder categories, regularly reappearing symptom complexes with a characteristic time of onset, characteristic ways of making social functioning inefficient, a characteristic course of the disorder, and some regularities in the way the disorder responds (or refuses to respond) to different treatment modalities were found. In addition, some characteristic clustering patterns of these symptom complexes were found in genetically related individuals.

Another very desirable consequence of the descriptive observational method is that it neutralizes subjective bias and enables precise communication between mental health professionals, as well as producing good inter-rater correlations in research.

The achievements of the empirical approach in psychiatry, however, as generally acknowledged, are partial and have considerable shortcomings. First of all, the extent to which clinical psychiatry has succeeded in adhering to the detached observational method—indeed, whether it is even possible to adhere to it most of the time—is highly questionable. For example, is it appropriate in an account of a psychiatric examination to describe precisely the state of the muscles participating in a facial expression (that is, the purely observational aspect of the respective behavior), or do we automatically interpret the facial expression as a reflection of an internal emotional or intentional state? This second alternative clearly transcends impartial observation. Such symptoms, syndromes, or disease categories as euphoria, depression, anxiety disorders, affective disorders, and incongruent or blunted affect are by no means observational phenomena but, at least partly are designations referring to unobservable subjective mental states. (The proposed mental mechanism leading to these interpretations will be detailed in the section on "the empathizing mechanism" in chapter 2.)

Furthermore, the observable behavior does not represent an immutable phenomenon analogous to the behavior of inanimate objects or most animals. Human behaviors that are similar looking (such as pretending as opposed to earnestly meaning something) may be the "final common pathway" of widely different underlying mental mechanisms or incentives.

The inefficiency of the purely observational method is reflected in serious inconsistencies in clinical psychiatry. The border between the nosological categories is by no means clear-cut. Symptoms and syndromes thought to be specific to a clinical category can frequently be found in other, unrelated disorders (like obsessions or affective symptoms in schizophrenia), and the clinical disease may change over time from one nosological category to another (for example paranoid personality disorder into delusional disorder, persecutory type). It is well known that psychoactive drugs are not specific to circumscribed mental disorders and a considerable proportion of individuals do not respond to drug therapy in spite of having the characteristic, clinical disorder the respective drug is meant to ameliorate.

1.4.2. Method and Theory in Biological Psychiatry

Biological psychiatry studies the somatic substrate of behavior regulation and mental events. Its methodology consists of the exploration of brain structures as well as neurophysiologic- and neurochemical-level functions. This strategy is based on the scientific reductionism used in the study of biological systems in general, a strategy that has proven to be a very powerful one in natural sciences. Neurology, neuropsychology, and biological psychiatry have successfully correlated many mental events with brain anatomy, histology, and neuronal functioning and clarified to a considerable extent the effects of psychotropic drugs at the synaptic level. Without this achievement, the development of new psychiatric drugs (beyond the first representatives, which, as a rule, were discovered accidentally) would have been impossible. However, the greatest problem with biological psychiatry, in my opinion, is that instead of correlating brain structures and functions with discrete phylogenetically evolved (adaptive or maladaptive) behavioral and mental mechanisms, it tries to correlate brain structures and physiology with whole clinical disease entities.

It is argued throughout this work that whole mental disease categories do not represent circumscribed, phylogenetically evolved behavioral mechanisms (with a well-definable brain structure and function underlying them) but are rather the final outcome of a longstanding interaction between discrete innate behavioral predispositions and ontogenetic influences, both physical and social. Contrary to this reasoning, it seems that contemporary clinical psychiatry had expected the reductionist strategy to provide the decisive breakthrough in discovering the causes of mental disorders (and in consequence more effective treatment methods). While it is always risky to make generalizations of this kind, it still seems valid to say that, unlike the 1970s, when the breakthrough was still expected from a comprehensive paradigm, at the beginning of the twenty-first century the same hopes are directed toward advances in biological psychiatric research based on the newest medical technology. Here are two relevant quotations: "The behavioral sciences, in particular psychiatry, are still at a preparadigmatic level because of the uncertainties regarding the body-mind relationship—uncertainties that reflect the difficulty of defining their focus, their methodology, their boundaries, and their reciprocal relations" (Mora, 1980, p. 4). And twenty-five years later: "Psychiatry today, as did psychiatry in the 1980s, lacks valid diagnostic tests, innovative treatments, or an understanding of the basic pathophysiology of any of its major disorders. Despite a century of neuropathological studies in schizophrenia, the location and the nature of the lesion causing this illness are not yet known. Genomics and neuroimaging have resulted in hundreds of findings, but none have yet changed how clinicians diagnose and treat patients with mental disorders" (Insel & Fenton, 2005, p. 4060).

At a more general, epistemological level, it seems quite clear that the elucidation of brain functioning at a lower level (molecular, cellular, neurophysiologic, etc.) of the *reduction/emergence hierarchy* of the organization of living matter can never compensate for a lack of understanding of emergent properties at a higher level of organization (on the level of organ and tissue interaction, organism/(social) environment interaction, and especially the mind-body relationship).[2] As an illustration, consider the following science fiction scenario: Imagine that alien creatures, with a more developed technology than ours but ignorant of the higher-level organization of life processes on Earth, were to study the human reproductive organs. With the aid of their developed technology, they discover the reproductive organs' cellular, molecular, and even atomic composition. However, without gaining knowledge of sexual reproduction from sources other than technology, they will have no chance of understanding the role of the reproductive organs at the organismic, interpersonal, and population levels. Therefore, for optimal scientific understanding, it is necessary to grasp the reduction/emergence hierarchy at all relevant levels.

While this requirement in psychiatry seems closer to a fairy tale than to a serious scientific program, consider that in many medical specialties—in cardiology, for example—many steps of this reduction/emergence scale are understood quite well. The molecular composition of the contractile material of heart muscles, the material substrate influencing the strength and rhythm of the heartbeat, the mechanism determining the unidirectional flow of the blood, the role of the cardiovascular system in the context of the whole organism's life processes (that is, its role in the metabolism, oxygenation of tissues and organs, excretion, etc.) are well clarified.

This standpoint is not a new one. Philosopher of science David Hull puts it in the following way: "Simpson believes that the reductionist and compositionist [the equivalent of emergence] models of explanation are not competing methods of explanation but complementary. Pure compositionist explanations are as incomplete as pure reductionist explanations" (Hull, 1974, p. 132; see also: Simpson, 1964, pp. 109–110, and Mayr, 1982, pp. 62–64).

The same reasoning regarding the relationship between organismic functions at a physico-chemical level and their evolutionary origins seems valid. In the words of evolutionary biologist Graham Bell, "It will often be very useful to understand physical principles in order to understand the context of evolutionary change; and it will often be very useful to understand evolutionary principles in order to understand any aspect of the physics or chemistry of organisms. . . . [But] no knowledge of physical principles, no matter how profound or detailed, can lead to an understanding of evolution and vice versa" (Bell, 2008, p. xiii).

1.4.3. The Psychodynamic Method and Theorizing

As we saw when discussing the descriptive method of clinical psychiatry, certain aspects of an individual's mental condition cannot be approached by detached observation alone. Depression, anxiety, euphoria, etc., are not directly observable phenomena but are inferred from behavior by a mental operation that will be called in this text "the empathizing mechanism." To approach this mental mechanism with a better chance of furthering our understanding in this field requires first a more elaborate discussion of the topic of instinct; therefore, the issue of sensing unobservable aspects of the patient's mental functioning will be postponed till the end of the next chapter.

The theories based on this intuitive method (for example the psychoanalytic, psychodynamic, or existential ones), while they may appeal to the "intuitive grasping" of professionals or laypersons, can be neither directly validated nor disproved by the scientific method (as a result of the impossibility of observing the respective subjective phenomena and sharing the results of that observation with others).

In summary, in this section it was argued that psychiatry has three dissociated approaches to mental disorders, each possessing a different methodology and approach toward theoretical understanding.

1. *Clinical psychiatry,* using the objective data-gathering method whenever possible, tries to arrive at empirical (statistical) regularities in such domains as the clustering of symptoms and other clinical characteristics of a disorder, its distribution among blood relatives, etc. It does not attempt to formulate causal explanations.

2. *Biological psychiatry* studies mental disorders using the scientific methodology of reduction of the mental and behavioral phenomena to co-occurring, and most probably causally related, somatic (brain) events. Its theoretical explanations, which are of the inductive type, maintain that the somatic events are the cause of the clinical mental disorders.

3. *The psychodynamic school* of psychiatry tries to understand mental disorders and their causation by an enigmatic method of sensing the individual's subjective experience. Its theoretical explanations cannot be tested by scientific methodology because the phenomena they describe are not directly observable.

Considering this situation, the following question seems justified: How can the discipline of psychiatry progress? Is it possible to appease, integrate, or intertwine the three approaches to dysfunctional behavior, or do we have to choose one of them—the most promising one?

Thomas Kuhn argues (with regard to scientific fields in general at their pre-paradigmatic stage) in favor of the latter alternative. In his opinion, the transition from the pre-paradigmatic stage of a scientific field to the paradigmatic one involves: "the triumph of one of the pre-paradigm schools, which because of its own characteristic beliefs and preconceptions, emphasized only some special part of the too sizable and inchoate pool of information" (Kuhn, 1970, p. 17). During this process, however, while the ensuing "puzzle solving" of the "normal science" leads to "a more efficient mode of scientific practice," valuable insights of the refuted paradigm schools will be lost (idem, pp. 178–180).

I have to confess that I am reluctant to imagine a similar route concerning future progress in the field of psychiatry. To understand this reluctance, it is important to take into account a unique difference between psychiatry and other empirical sciences (exact, biological, and psychosocial)—namely, that *psychiatry lies at the interface between somatic and mental phenomena.*

At the initial stages of psychiatry's development—that is, when "mapping" the boundaries of the studied phenomena by methodical observation, classifying the relevant dysfunctional forms of behavior, and finding empirical regularities—the "epistemological gap" between the mental and somatic did not interfere with the professional routine, or was at least much less apparent. However, when attempts to arrive at a true theoretical understanding began, that is, in attempting to understand the mental and brain mechanisms underlying dysfunctional behavior, this unresolved (and maybe unresolvable) schism between the mental and somatic became an insurmountable obstacle. Therefore, in accordance with this "epistemological gap," the attempts at theoretical explanations in psychiatry became split into two irreconcilable trajectories. While both used clinical disease categories as reference points, one approach searched for explanations using the methodology and theoretical framework of the exact sciences, while the other employed a kind of "mental resonance" (which will be called in this text the "empathizing mechanism") in order to sense the concealed mental processes underlying an individual's dysfunctional behavior.

I, myself, have no idea whether the schism between mental and somatic phenomena can be effectively bridged. On one hand, it seems clear that a *bidirectional causal interrelationship* exists between them. Emotions and other mental events are accompanied by specific neurochemical, electrical, and other activity in circumscribed brain areas in a quite predictable manner, as well as by physiological changes in the rest of the body. And, conversely, influencing localized brain activity by physical, chemical, or mechanical means (for example, drugs, brain simulation via intra-cerebral electrodes, prefrontal lobotomy, etc.) is accompanied quite predictably by some alterations in subjective experience. (This causal interrelatedness between mind and body was acknowledged even by Descartes, in spite of his views concerning substance dualism [Kim, 2011, pp. 34–35]). (For a detailed description of the close interrelationship between neurophysiology, neuroanatomy and neuropathology on one hand, and discrete mental functions—normative and disordered—on the other, see Ian Glynn's book *An Anatomy of Thought: The Origin and Machinery of the Mind*, 2000.)

However, the nature of the leap from the somatic to the mental, and vice versa, is incomprehensible to the human mind. This is true even in the case of simpler mental phenomena than the subjective experience of emotions, intentions, or cognitive processes. For example, the way in which sensory processing involving neurochemical and electrical events at the sensory organs, in brain neurons, and in nerve fibers and synapses gives birth to the subjective sensation of seeing colors and forms, hearing sounds or music, or feeling pain, remains a mystery.

I am aware of three lines of reasoning that seek to explain why mental phenomena cannot be approached by the scientific method:

1. The first line of reasoning concerns the philosophical notion of *solipsism*. According to this concept, subjective experiences can be directly sensed only by the person to whom they belong and cannot be shared with others with the degree of precision that directly observable phenomena can. However, this requirement is indispensable in the scientific method. (Solipsism is further discussed in section 2.3.)

2. If we postulate that mental events are the *ultimate function* of the organ we call mind/brain, we have to admit that this function differs fundamentally from the functions of other body tissues and organs. It has no objectively identifiable and measurable properties, unlike the functions of other organs (muscles, endocrine glands, skin, respiratory apparatus, etc.). Therefore, mental events cannot be translated into the language of mechanics, biochemistry, biophysics, or indeed into the language of any scientific field. Mental events (unlike the function or products of other organs) have no known material composition or physical properties, cannot be localized in space, and cannot affect or interact directly with anything in the material world. We presuppose that they contribute greatly to getting acquainted with our environment and our own mental constitution and thus have an

all-important guiding effect on behavior, but without knowing how this cause-and-effect sequence actually works.

3. It seems very problematic when the same organ (the mind) is both the target of the inquiry and the instrument that contrives and carries out the inquiry and interprets the obtained data. Austrian-born economist and philosopher F. A. Hayek argues in this respect that *"'any apparatus . . . must possess a structure of a higher degree of complexity then is possessed by the objects' which it is trying to explain"* (emphasis added; quoted in, Popper & Eccles, 1986, p. 30; see also von Hayek, 1952, p. 185). If we translate this reasoning into the domain of our concerns, we can argue that the mind cannot comprehend those emergent evolutionary events that brought about its very existence.

These considerations are only indirectly relevant to our discussion, however. Since the main theme of this text is the interpretation of the phenomenon of dysfunctional behavior from the perspective of evolutionary theory, we already possess a widely accepted theory— something akin to Kuhn's concept of a paradigm. In the framework of this theory, we will attempt to interpret the diverse phenomena pertaining to the domain of psychiatry. For this quest we will use any data or expert opinion from any one of the three dissociated schools of psychiatry to the extent it may prove profitable.

Afterthought

Finally, I would like to mention a categorization of scientific theories different from that treated in the first section of this chapter. In this categorization, theories are divided into theories of unit formation; theories of classification; and theories of causality (Selye, 1964, pp. 285–292; see also: Quinton, 1973, pp. 284–285, 307–313). According to this categorization, *clinical psychiatry* deals with the first two kinds of theories— unit (or concept) formation and classification in order to define signs, symptoms, syndromes, and clinical disorder categories, as well as more comprehensive entities, like personality and anxiety disorders, affective disorders, or the schizophrenia spectrum. Biological and psychodynamic psychiatry, on the other hand, attempt to formulate theories of causality (biological psychiatry by employing the abovementioned reductionist method, and psychodynamic psychiatry by a scientifically unapproachable conceptualization of intrapsychic and interpersonal events such as malfunctioning of the sexual drive or the behavior of the "schizophrenogenic" mother's deleterious effect on the child).

The present theory's basic conceptual units and their classification are represented by the four comprehensive instinctive mechanisms and their subtypes or components, while the causal factors concern natural selective pressures throughout the evolution of the animal kingdom and their excessive relaxation during human evolution. The final clinical entities are defined as the outcome of the interaction between extreme variants of these comprehensive instinctive mechanisms (or their subtypes) and ontogenetic influences.

2

The Concept of Instinct

2.1. Instinct, General Remarks

Two good reasons exist to include a chapter on the concept of instinct in this text. First, since the main theme of this theory is the relaxation of natural selection pressures on the inherited aspects of human behavior, we must examine more closely what is meant by "the inherited aspects of human behavior." Evolutionary theory being our main "background knowledge," as argued in the previous chapter, we have to presuppose that this inherited aspect of human behavior evolved from inherited aspects of animal behavior. Therefore, we must put this subject in evolutionary perspective.

The second reason for including a general discussion of instincts follows naturally from the first: If we wish to examine what happened to the inherited aspects of behavior during human (natural and cultural) evolution, we must try to roughly outline what kind of inherited, instinctive predispositions may have existed at the beginning of human evolution in our apelike ancestor, predispositions that the progressive relaxation of selection pressures, induced by cultural, technological, and social progress, acted upon. I have tried to organize this subject in a way that facilitates clearer exposition of the ideas that relate to the present theory. I hope this discussion succeeds in clarifying to a certain extent why the concept of instinct became so problematic and contentious.

Behavior enables the organism to employ adaptive strategies to changing environmental conditions at a much quicker pace and in a much more complex and flexible manner than somatic adaptations. We can appreciate how much quicker if we compare these two evolutionary routes. For example, the transition of the basic body plan and physiology from aquatic adaptation to a terrestrial and aerial one (reflected by the timescale of the successive appearance of classes of animals in the paleontological record: fish, amphibians, reptiles, birds) or within the class of mammals, from terrestrial to aquatic (as in cetaceans: whales, dolphins) is estimated in millions of years by the molecular clock hypothesis.[1] The British peppered moth's ability to change color in industrialized environments, mentioned in section 1.3 as a microevolutionary event, probably took about fifty years (Ehrlich, 2000, p. 27). The catching of a prey by a mammalian carnivore, on the other hand, may take several minutes, while the flight of the prey animal at the approach of a predator may be initiated in fractions of a second.

In terms of evolutionary reasoning, it seems clear that very *slow somatic* adaptations (basic body plan, for example) are designed to deal with the "permanent" (extremely long-lasting) features of the environment. The *predictable* resources and dangers, as well as periodic changes in the environment (circadian or seasonal changes, for example), are dealt with mainly by *instinctive behavior*, that is by preprogrammed sensory and motor patterns, as well as by periodic adaptive changes in body features (like shedding of the coat or skin, adapting the reproductive cycle to the seasonal changes; hormonal and behavioral changes adapted to the circadian rhythm, etc.).

On the other hand, *unpredictable* (to a degree that makes it impossible for a rigid hereditary behavioral mechanism to evolve), quickly changing, or complex features of the environment important for the life processes of the organism are dealt with by behavior that is made even more flexible through *learning* at different levels of complexity, from simple habituation to the complex, symbolic features of human cognition (Odling-Smee, Laland, & Feldman, 2003, p. 255). "While some adaptations enable animals to cope with relatively invariant aspects of the physical world, others have evolved to enable animals to interact successfully in the rapidly shifting social world" (Barkow, Cosmides, & Tooby, 1995, p. 396).

These three modes of adjusting to the environment (somatic, instinctive, and learned) are strongly interrelated, intertwined or welded together in anatomy, physiology, and behavior. The first two mechanisms are genetically transmitted to successive generations, while in the third one, inheritance builds only the machinery that ensures *the capability* to learn, while the *content* learned during ontogeny cannot be genetically transmitted and, in animals, perishes with the death of the individual organism. In humans, however, cultural transmission makes possible the transmission of accumulated knowledge, changes to the physical environment (buildings, roads, etc.), inheritance of personal property, etc., effects that have a considerable influence on the fitness of successive generations.

As we have seen, evolutionary theory claims that human behavior evolved from complex animal behavior, and complex animal behavior evolved from simple animal behavior. The ubiquitously present dysfunctional forms of human behavior (whose more severe forms are categorized as mental disorders), and especially their inherited aspects, also have to be understood according to the logic of evolutionary theory—that is, as an outcome of changing selection pressures and relaxations in the special conditions of the human cultural environment.

It is well known that the concept of instinct in human beings is problematic, so much so that many professionals in the field of behavior prefer to avoid it. This problematic, contentious status of instinctive behavior or instinctive predispositions illustrates well Thomas Kuhn's claim that the behavioral and social sciences are still in a "pre-paradigmatic" state of development. They are characterized by "controversies about the fundamentals," the existence of "a number of competing schools," the gathering of facts is a "nearly random activity," and the need of "each writer . . . to build his field anew from its foundations," a practice I have adopted, too, in the exposition of the present theory.

Kuhn's observation seems to be valid not only for psychiatry and the psychological and social sciences but also for the study of instinctive behavior in animals as well: "Whenever any two biologists attempt to discuss the problem of instincts, it soon becomes apparent that they do not understand each other, because they do not speak the same language. Each connects a different meaning to the word instinct" (Lorenz, 1975, p. 129).

With respect to human instinctive motivations, the dissension among professionals grows even greater. The behaviorist school of psychology questions the usefulness of the concept of instinct altogether: "Actual observation . . . makes it impossible for us any longer to entertain the concept of instinct" (Watson, 1957, p. 136; see also Skinner, 1976, pp. 51–79; Kuo, in Birney & Teevan, 1961, pp. 22–25).

American philosopher and psychologist William James (1842–1910), on the other hand, argues that humans possess the greatest number of instincts in the animal kingdom (Watson, 1957, p. 110). The list of behavioral manifestations he claims represent human instincts clearly consists of a mix of heterogeneous phenomena. Several of James's instinctive categories are motor patterns with an innate foundation, like climbing, without an accompanying emotional or intentional element, while other categories are simply the emotional components of instinctive phenomena, like anger, fear, or love. Still other instincts identified by James, like modesty or shame, seem to be emotions characteristic of humans only (as far as we can tell), because they need the maturation of self-consciousness in a social setting. A succinct and clear critique, to my level of understanding, of the various approaches to instincts and to instinct-learning interaction is contained in the first five chapters of Lorenz's *Evolution and Modification of Behavior* (1965).

The word *instinct*, having fallen into disfavor in more recent literature, has been replaced by other terms. For example, in what became known as *affective science*, the words *emotion* or *affect* frequently took on the general meaning of instinct, and it has been argued that emotion or affect in turn induce other aspects of instinctive activity, such as behavior and physiological changes, as well as related cognitive contents (Ekman & Davidson, 1994, pp. 15–16, 20–22, 52, 79, 89, 113, 123–24, 132). The words *emotion* and *affect*, however, according to my understanding, carry the undeniable connotation of subjective mental experience that is not directly observable. While no doubt a causal relationship between emotions, somatic events, and behavior can be found, it is of an incomprehensible nature, making the use of the word *emotion* instead of *instinct* unprofitable, to my mind.

In the framework of evolutionary psychology, the inherited behavioral and cognitive mechanisms, frequently with a localizable brain area as their structural base, have received various designations: "Many psychologists have been forced by their data to conclude that both human and nonhuman minds contain . . . a large array of mechanisms that are . . . *functionally specialized, content-dependent, content-sensitive, domain-specific, context-sensitive, special- purpose, adaptively specialized* and so on" (Barkow, Cosmides, & Tooby, 1995, p. 93, emphasis added). These expressions probably refer to complexes of human behavior in which an underlying instinctive motive or striving holds together relevant affective, intentional, and cognitive contents.

Evolutionary psychologist and psycholinguist Steven Pinker reverted to the use of the word *instinct* in his book *Language Instinct*, in order to convey more forcefully the idea that the human mind possesses a discrete innate predisposition or developmental program for language acquisition. This program enables small children to learn correct grammar without being explicitly taught it by the social environment.[2] Pinker's use of the term *instinct* is probably a useful way to bring home to the reader his intended message. In any case, it exemplifies the various ways the word is used and the necessity to go back "to the fundamentals," in Thomas Kuhn's parlance, and define the way I intend to use it for the exposition of my own, very different, topic.

In this work, instinct will be used as an umbrella term covering various aspects of the concept of instinct. I prefer to see instinctive activity as a biologically unitary phenomenon separated into different aspects as a result of the constraints of human understanding, which has to approach it with various methodologies, some scientific in nature, some not. This unitary nature of instinctive activity is most evident when we consider the close interrelationship between the behavioral and physiologic aspects of instinctive activity. Many physiological correlates of instinctive activity (from piloerection to cardiovascular changes) are understood as preparing or supporting the body in the successful execution of the instinct-induced behavior. Another example supporting this interpretation involves certain hormonal activities. Oxytocin, for example, induces at the same time the physiological, behavioral, and subjective experiential aspects of the same instinctive domain—the domain related to giving birth and maternal activities, and their reflection in the subjective experience (Panksepp, 2005, pp. 250–252). Instinctive behavior can in fact be reduced by scientific exploration to changes in the functional state of various organs and tissues: muscles, skeleton, cardiovascular and respiratory systems, neural coordination, endocrine changes, etc.

However, the scientific exploration of behavior *in its entirety* uses a very different methodology than the study of the various organs and systems of the body, as we have seen in the reductionism/emergence discussion of methodological approaches in psychiatry.

Besides the physiological and overt behavioral aspects of instinctive activity, three additional ones exist and will be briefly discussed—the *intentional, energetic,* and *emotional.*

2.1.1. The Intentional Aspect of Instinctive Activity

Intentionality describes a "feature of the mental—[which is] that mental states are *about,* or are *directed upon,* objects that may or may not exist or have contents that may or may not be true" (Kim, 2011, p. 24, emphasis in the original). However, human intentionality, in my opinion, is deeply rooted in the evolution of the animal kingdom. This idea may become more apparent by looking at ethology's division of instinctive activity into the "*innate motor pattern*" and "*innate releasing mechanism.*" The latter concept means that *the instinctive behavior is directed by an innate brain mechanism toward the appropriate environmental agent in order to fulfill its fitness-enhancing function, frequently with the animal having no, or a very restricted, degree of acquired knowledge about the nature of the respective agent.* In other words, animals and, in a more malleable form, humans have some kind of innate, inherited "knowledge" or guiding or "teaching" mechanism (Lorenz, 1965) in relation to aspects of the environment that play a critical role in survival or reproduction, and which are targeted intentionally and differentially by instinctive strivings. It is argued in this text that, in the complex human cultural environment, the same (originally instinctive) intentionality may be directed toward targets that are very distant from the natural, fitness-enhancing ones.

2.1.2 The Energetic Aspect of Instinctive Activity

The *energetic aspect of instinctive activity* refers to the instinct-induced behavior's strength and tenacity, its periodical waxing and waning, and its power relationship versus other concurrent instinctive activities.

In spite of the great importance we humans attribute to learning through memorizing, memorizing per se plays, in my opinion, no direct causative role in modulating the intensity and tenacity of behavior when it is unrelated to some intrapsychic, originally biological motive. Think for example of the great amount of material we have to memorize during our formal education. The learned material that bears no special importance for us, or if not used from time to time during our daily activities, is simply forgotten or stored for long periods without having any role in influencing the behavior.

Therefore, it seems clear that the energetic aspect of behavior cannot be learned but is conferred to behavior by an underlying instinctive motive. This is the opinion of most theorists on this subject of whom I am aware, like the early theorist of instincts McDougall (1966, pp. 17–19), as well as Freud (1966, pp. 20–22), and the leading ethologists Konrad Lorenz (1966, pp. 23–27) and Nico Tinbergen (1966, pp. 28–33). Others, however, like R. A. Hinde (1966, pp. 34–45) and psychologist H. W. Nissen (1966, pp. 46–51) criticized this idea.

Some pathological mental states also speak to the accuracy of the above reasoning. Think, for example, of a bipolar patient whose behavior differs immensely between the manic and depressive phases in the amount of its energetic charge, while all the ontogenetically acquired (learned) material stored up in the patient's brain remains essentially the same. The energetic aspect of instinctive activity will be further discussed in section 5.1.

2.1.3. The Emotional Aspect of Instinctive Activity

While intentionality and the energetic aspect of instinctive activity cross, at least partially, the border between the directly observable physical realm and the unobservable mental one, our real challenge in this respect comes from the phenomenon of emotions. While the energetic aspect of behavior can be observed and quantitatively appreciated to a certain extent, we also have a subjective feeling about it. The object toward which an intention is directed can be observed in animals and, frequently, in humans, too. However, in humans, the object may also be abstract or symbolic, in which case it is not directly observable. We also certainly have a subjective experience, a particular feeling for the objects toward which instinctive intentions are directed. Some objects excite us and give us a pleasurable feeling (a sexual object, a beloved child), while others irritate us or leave us unaffected.

Emotion, affect, and *mood* are more obscure and problematic aspects of instinctive activity. They are defined in the context of this work as *the reflection of the earlier aspects of instinctive activity (behavior, physiology, intentionality, and the energetic aspect) in our subjective, private, mental experience.* The emotional experience evidently cannot be learned from an agent outside the individual; therefore, it has to be an organic part of our innate mental equipment, although its behavioral expression can be voluntarily modulated—accentuated, concealed, or faked, at least in humans and, probably, to a certain extent in chimpanzees and other apes as well (see section 3.1). "Few individuals need any formal training to know what it means to feel happy, sad, frustrated and angry, scared, full of lust . . . or to identify the environmental events that trigger these feelings" (Panksepp, 1994, p. 20; see also Gangestad & Simpson, 2007, pp. 49–50).

If the intuitive conviction that emotions are innate and cannot and need not be learned is not sufficient, we find further support for it in the work of ethologist Irenäus Eibl-Eibesfeldt, who proved that children born deaf and blind possess the same repertoire of emotional facial expressions as their healthy counterparts. Eibl-Eibesfeldt also studied the expression of emotions cross-culturally and concluded that in all the cultures he studied the emotional repertoire is identical (Lorenz, 1982, p. 11).

Emotional experience can be expressed in words and communicated to others, albeit in a much less precise and direct manner than communication of material obtained through observation. This subject will be detailed in section 2.3.

2.2. The Evolving Concept of Instinct, and Instinct–Learning Interaction

The involvement of instinct in how humans view themselves or try to define what has been called "human nature" has been and still is powerfully influenced by ideological views and religious beliefs in addition to scientific understanding. It will help to get a clearer picture of this contentious topic if I sketch here (to the extent my knowledge permits) a brief historical account of how the concept of instinct evolved in European thought. I will also give a hypothetical account of how emerging scientific data and ideas regarding instincts can help reconstruct the evolution of instinctive behavior, beginning with simple organisms up to complex social animals and finally humans. This mental exercise may also help us build a hypothetical picture of instinctive behavior in the immediate ancestors of anatomically modern humans. This line of reasoning will be continued in chapter 4 with an accent on the effect of the relaxation of natural and intragroup natural selection pressures (RfNSPs and RfIGNSPs) on instincts amid emerging, specifically human, social/cultural conditions.

As we saw in the previous chapter, the concept of instinct, like several other important scientific ideas, was anticipated by the ancient Greek philosopher-scientists in order to explain more evolved animals' adaptive and intelligent behavior, in spite of not being rational creatures and lacking a "soul," unlike men and gods. This man/beast dichotomy was developed further by scholastic theology. In the thirteenth century, "Albertus Magnus . . . removed Man from the natural scale [of Aristotle], holding that he is unique in possessing the gift of reason and an immortal soul. Animals lacking reason 'are directed by their natural instinct and therefore cannot act freely'" (Birney & Teevan, 1961, p. 166). Four centuries later, "René Descartes and his followers aggressively restated the existence of a man-brute dichotomy" (idem, p. 167). However, a hundred years after Descartes, the Swedish taxonomist Linnaeus "included man in his mammalian order Primates and made it very clear . . . how close he thought man was to the anthropoid apes" (Mayr, 1982, p. 438).

The evolutionary approach to the topic of instinct begun with *The Origin of Species,* in which Darwin claims that the principle of common descent and the mechanism of natural selection in animals account for the evolution of structural aspects of the body to the same extent as innate incentives of behavior: "I can see no difficulty in natural selection preserving and continually accumulating variations of instincts to any extent that was profitable. It is thus, as I believe, that all the most complex and wonderful instincts have originated" (Darwin, 1859/1976, p. 245).

Ethological theory formulates the same idea even more unequivocally: "Ethology, or the comparative study of [animal] behavior, is based on the fact that *there are mechanisms of behavior that evolve in phylogeny exactly as organs do, so that the concept of homology can be applied to them as well as to morphological structures*" (K. Lorenz, 1982, p. 101, emphasis in the original). Darwin in his *The Expression of Emotions in Man and Animals* painstakingly demonstrated that not a single newly evolved facial muscle was needed in order to express humans' richly nuanced emotional states.

The phenomenon of instinct was defined traditionally as a hereditary behavioral program that induces rigid, stereotypic behavioral patterns characteristic of animals of the same species (at least of the same sex) and has a role in survival and reproduction. Consequently, the respective behavior pattern cannot be induced by any form of learning (Alcock, 1979, 51–55). It seems obvious that instincts, according to this definition, can account for animal behavior only partially (more in simple animals, less in evolved ones), and almost not at all regarding human behavior.

A very important breakthrough in clarifying innate, genetic influences on animal behavior, which in my understanding opened the door for their extrapolation to human behavior, was achieved by ethology in the form of new additions and refinements regarding the theory of instincts. The three most important from this point of view are the following:

1) Instinctive behavior was divided into a *preprogrammed motor pattern* (resembling the traditional definition of instinct), as well as into another, genetically defined, neural mechanism that enables the animal to recognize, without acquired knowledge, the proper environmental agent (or situation) toward which the instinctive motor pattern has to be directed in order to fulfill its survival or reproductive potential. It was found that this environmental agent is recognized instinctively by some of its very simple or partial characteristics, that is, they "are of much simpler nature than our knowledge of the potential capacities of the sense organs would make us to expect" (Tinbergen, 1966, p. 104). For example: "To see a newly hatched turkey . . . hide in cover and crouch at its first sight of a hawk flying over . . . is indeed impressive. To see turkey chicks react in the same manner to a fat fly slowly crawling along the ceiling . . . is actually disappointing" (Lorenz, 1982, p. 158).

The ethologists, as we have seen, call the motor aspect of instinctive activity the "fixed motor pattern" (FMP), while the neural mechanism that enables the recognition of the proper environmental stimulus, which releases the FMP, was called the "innate releasing mechanism" (IRM). In my opinion, this dichotomy defines the *intentional character* of instinctive activity (that is, innate behavior directed toward an object, a characteristic of the object, or a situation with adaptive significance). The other very important conclusion drawn from the existence of IRMs, in my view, is that this mechanism seems to be the strongest proof that animals (including humans) do have innate "knowledge" of circumscribed elements of the environment with fitness-enhancing value.

2) It was recognized, however, that the FMP directed toward the appropriate environmental agent ("releaser") by the IRM, in more evolved, mature animals could account for only a small part of their behavior. A chain of FMPs directed toward successive releasers until the drive achieves its final goal is encountered frequently in very simple or very immature animals living in a highly predictable environment. Prechtl and Schleidt, for example, describe such a chain of FMPs in (still blind) newborn kittens. A chain of FMPs and IRMs assure that the kittens will first find the mother's body, then find the teat, then begin to suckle, and at the same time stimulate the mammary gland with their forelegs in order to increase its milk output (Lorenz, 1982, p. 192).

In more evolved mature animals, however, in most instances the releaser desired by an instinctive need has to be searched for, sometimes over long distances and in a complex terrain. Thus, it was recognized that the behavioral sequences from the arousal of an instinctive need until the desired releaser is contacted may be much more complex than the FMP/IRM dichotomy may suggest. The search for the desired releaser may begin at random or be guided by a sensory clue (a specific odor, for example), or by some previously acquired knowledge. The early ethologist Wallace Craig called the stage of the instinct-induced behavior—from the arousal of the instinctive need up to contact with the appropriate releaser—"*appetitive behavior*," while he named the stage after contacting the releaser the "*consummatory behavior*" (Craig, 1918, pp. 91–107).

The consummatory behavior (copulating or killing and eating the prey) is a true FMP fulfilling all the requirements of the traditional definition of instinct. The appetitive behavior, however, is a very different matter. Due to the unpredictable nature of a complex environment and other circumstances, the appetitive behavior cannot be preprogrammed in most of its aspects by heredity. It is a complex phenomenon containing diverse elements, some related to instinctive behavior, some not. Ethologist Niko Tinbergen formulated the problematic nature of the appetitive behavior in the following way: "Appetitive behaviour is a true purposive activity, offering all the problems of plasticity, adaptiveness, and of complex integration that baffle the scientist in his study of behavior as a whole. Appetitive behaviour is a conglomerate of many elements of very different order, of reflexes, of simple patterns like locomotion, of conditioned reactions, of 'insight' behaviour, and so on. As a result, it is a true challenge to objective science" (Tinbergen, 2003, p. 106).[3]

To render this abstract discussion more palpable, as well as to exemplify the complex interaction of different aspects of appetitive behavior with one another, it will be useful to quote here at length a concrete example Tinbergen used:

> [T]he hunting of a peregrine falcon usually begins with relatively random roaming around its hunting territory, visiting and exploring many different places miles apart. This first phase of appetitive behaviour may lead to different ways of catching prey, each dependent on special stimulation by a potential pray. It is continued until such a special stimulus situation is found: a flock of teal executing flight manoeuvers, a sick gull swimming apart from the flock, or even a running mouse. Each of these situations may cause the falcon to abandon its "random" searching. But what follows then is not yet a consummatory action, but appetitive behavior of a new, more specialized and more restricted kind. The flock of teal releases a series of sham attacks serving to isolate one or a few individuals from the main body of the flock. Only after this is achieved is the final swoop released, followed by capturing, killing, plucking, and eating, which is a relatively simple and stereotyped chain of consummatory acts. (Tinbergen, 2003, pp. 106–107)

It seems clear from this example that FMPs can also be encountered during the appetitive phase of instinctive behavior, interspersed with other elements of appetitive behavior previously mentioned. (Other examples of this kind of FMP are ritualized courting behavior in birds or sexual fighting in ungulates.) The consummatory act, however, consists almost exclusively of FMPs (with rare partial exceptions in highly evolved social animals and humans). However, in human behavior, such pure FMPs during the appetitive phase have disappeared completely, and they are also rare in such highly evolved social animals as the chimpanzees.

3) Any attempt at understanding the evolutionary continuity between animal and human behavior must take into account a distinct evolutionary trend with central importance, in my opinion—namely, the well-recognized trend from rigid, genetic preprogramming in simple animals toward an adaptively more flexible, more malleable

behavior in more evolved animals. This shift allows the modification, fine-tuning, and enrichment of fixed instinctive patterns or predispositions to an ever-increasing degree by experience and learning: "The degree of rigidity or plasticity of this [instinctive] behavior varies from the lower to the higher species in the evolutionary scale" (Fletcher, 1966, p. 18). Or: "The rigid patterns of innate behavior allowing only one route to an adaptive goal are replaced by the flexible patterns of behavior arising from drives and permitting a variety of adaptive strategies" (Boddy, 1978, p. 16).

Biologist and evolutionary theorist Ernst Mayr, in his often quoted computer analogy, describes two kinds of genetic programs, a "closed" and an "open" one. The closed genetic program is one "which does not allow appreciable modifications during the process of translation into the phenotype . . . closed because nothing can be inserted in it through experience. Such closed programs are widespread among the so called lower animals." An open genetic program is one that "allows for additional input during the life span of its owner." There is an evolutionary path that leads to "a *gradual opening up of the genetic program* permitting the incorporation of personally acquired information to an ever greater extent . . . there are two prerequisites for this to happen . . . [a] greater storage capacity . . . [that is] a larger central nervous system . . . [and] prolonged parental care. When the young of a species grows up under the guidance of their parents, they have a long period of opportunity to learn from them, to fill this open program with useful information on enemies, food, shelter, and other important components of their immediate environment" (Mayr, 1974, pp. 650–658, emphasis added).

What are the implications of these theoretical formulations on instincts—namely, the dichotomy of FMP and IRM; the separation of appetitive and consummatory phases of instinctive behavior; and the trend of behavioral evolution from mostly rigid, preprogrammed sequences toward more flexible and efficient modes that incorporate learning of progressively increasing complexity? How can such theoretical formulations aid us in our attempts to extrapolate the instinctive aspects of behavior from animals to humans?

Of course, we are dealing here with only the basic building blocks of human behavior, disregarding momentarily subsequent important developments induced by the increasing complexity of social organization, the cultural transmission of accumulated knowledge, and deeply rooted traditions, as well as the effect of a comprehensive and progressive relaxation of natural (including intragroup) selection pressures and the resultant diversification. However, we must examine these implications here, I think, because the effect of the incorporation of these later developments on human behavior will increase the complexity of our topic to a degree that could later prevent a clear exposition of its various intertwining aspects.

It seems clear that the evolution from a chain of FMPs and IRMs in simple animals toward incorporation of appetitive sequences of behavior (characteristic of animal behavior of increasing complexity) continued at an accelerating pace during human evolution and attained such overwhelming proportions that, as mentioned before, surviving traces of rigid FMPs and IRMs are hard to find in the behavior of mature humans. In newborns and young infants, we can still discern FMPs, such as signaling discomfort with crying; the differential smiling response to the human face, especially that of the mother; the grasp reflex, a relic of our descent from primates; etc. (For a more detailed discussion of this topic, see Eibl-Eibesfeldt, 2010, pp. 26–31.) In adult human behavior, on the other hand, apart from the behavior patterns that ensure the last step of vitally important exchanges of material substances with the environment (chewing, swallowing, drinking, copulating, which may be regarded as reflexes), pure FMPs and IRMs, as defined by ethologists, are almost nonexistent.

We may sometimes discern small involuntary moments that may be surviving traces of pure instinctive behavior, like small movements away from a frightening person or scene or a freezing reaction when perplexed or taken by surprise. But even if we accept these interpretations, it can surely be argued that the overwhelming majority of human behavior (including, of course, verbalization) cannot be interpreted as pure innate motor patterns or consummatory acts released toward circumscribed environmental agents "recognized" by a purely innate, inherited mechanism. To presume that human behavior contains only appetitive behavior and almost no consummatory behavior (in the form of FMPs) could be a tempting hypothesis, especially because appetitive behavior is understood as a much more complex form of behavior than pure innate motor patterns, involving learning, experimentation, and simple forms of problem-solving, according to Tinbergen's description (quoted in section 2.2).

However, such a hypothesis may lead to a deadlock. In the ethologists' formulation, the role of appetitive behavior is to bring the organism in contact with the releaser in order to enable the performance of the consummatory act. Unless appetitive behavior is the preparatory phase of some kind of consummatory behavior, it simply does not make evolutionary sense. Consequently, we would have to interpret the whole gamut of complex and varied human behavior, from an evolutionary point of view, as activities preparatory to eating, reproducing, and raising children (amid the constraints as well as advantages of social coexistence), which seems ridiculous.

Unquestionably, the appetitive behavior of the kind we have mentioned amounts to a considerable part of human activity. A few examples are: the effort to acquire an appropriate occupation, one that is in tune with one's innate traits, among other considerations; the preparation to master a daily routine that, if successfully accomplished, will assure the means of subsistence; the short- and long-term mating strategies in humans (which will be briefly discussed in the section on evolutionary psychology); the formation of coalitions in order to rise to a more dominant position, which may secure improved access to resources. Acquiring the relevant knowledge and developing the necessary skills to attain almost any desired goal in life have to be seen as having instinctive underpinnings of the appetitive type.

In addition, it will be proposed in the following that *the great flexibility of instinct-learning interaction in humans allows the channeling of aspects of instinctive behavior (energetic, intentional, and emotional ones, but not FMPs) into culturally induced behavioral avenues remote from common biological aims.* To better explain this, we will examine three related aspects of instinctive activity in humans.

1. It is clear that with the disappearance of the strictly preprogrammed FMPs and IRMs from the human behavioral repertoire, the most straightforward logical connection between observable behavioral phenomena and traditional evolutionary reasoning (which asserts that genetically transmitted behavioral dispositions are selected according to their fitness-enhancing ability) has been lost. Instead of clearly observable behavioral patterns, we have to conceptualize the effect of instincts on behavior in humans as being a kind of propensity, predisposition, striving, or urge—that is, an unobservable

phenomenon that can be sensed only by relying on our own sub-jective experience. Being unobservable, these categories of behavior cannot be approached directly by scientific methodology, but only *indirectly* through their secondary repercussions on observable be-havior and related physiological and neurophysiological changes. The impossibility of studying them by observational methods makes the impact of instincts on human beings unapproachable by tradi-tional scientific exploration.

Nevertheless, even in the absence of clear-cut FMPs and IRMs, it is evident that many human preoccupations are directed toward securing the means for biological existence. This is most evident in primitive societies, in which the greatest part of everyday activities serves to secure the means of subsistence and successful reproduc-tion; nor are these efforts absent in modern, technologically advanced societies. Frequently underlying more sophisticated, culture-related activities are strivings to achieve biological goals: We value biological survival for as long as possible, we care for our bodily integrity and health and promptly seek medical advice when in need of it, we work mostly to ensure we have enough daily life-sustaining resources, we are territorial (both regarding our private territory as well as that of the more extended communities of which we are a part), we compete for mates and for social hierarchy, we rear offspring till maturity and frequently beyond it, etc., without being explicitly taught to do these things or told why they are important, and without being coerced by the social milieu to do so. With respect to biological survival and reproduction, we are by far the most successful animal species on the planet. How could this feat be achieved without the participation of some kind of instinctive motivation?

2. If we accept the idea that we may have unobservable instinctive predis-positions, we can turn to the next question: How do these predisposi-tions assist us in our complex social coexistence—in our attempts to appease conflicting individual interests, such as the need for competi-tion but also for cooperation; the need to switch flexibly from one be-havioral tactic to another depending on the circumstances; the need to accumulate a large amount of socially and biologically useful cognitive material mostly transmitted from generation to generation; and so on?

As will be detailed in chapters 4, 5, and 7, skillful social behavior necessitates optimal placement of instinctive predispositions on the diffuse/differentiated scale. That is, to be useful in complex, quickly changing social circumstances, the instinctive predispositions have to be *blurred enough* in their capacity to induce overt behavior to allow the necessary degree of malleability. In other words, *a degree of uncertainty* regarding one's individual biological interest is needed in order to let it be guided to a certain extent by current *social* influences needed for good social adjustment—but not more and not less than that. As detailed in the following chapters, drives that are too *differentiated*, too narrowly goal oriented, interfere with good social adjustment by preventing the working out of flexible compromises between conflicting interests according to socially approved rules of behavior, leading to excessive clashes with other persons or social institutions. On the other hand, drives that are *too diffuse* are unable to guide people effectively amid the complexities of social life, due mainly to the inability to indicate and safeguard personal interests. They lead to a deficiency or inability to channel instinctive motives (their emotional, intentional, and energetic aspects) into socially as well as biologically adaptive activities. In subsequent chapters, I argue that this later disability constitutes the genetic predisposition of the schizophrenia spectrum of disorders.

We must hypothesize, however, that the *normative range* of *diffuseness/differentiation* of the instinctive predispositions in human populations has to be quite wide. At the more differentiated end of the spectrum, the drives induce excessive and narrow preoccupation with biologically relevant issues (food, sexuality, dominance/submissiveness, excessive accumulation of resources), while at the more diffuse end, they can be channeled into human activities far removed from biological needs—for example, a lifelong dedication to humanitarian, artistic, scientific, or religious activities. (When discussing the subject of displacement in section 5.2.1., clear evidence will show that even in more evolved animals both the energetic aspect of an instinctive striving as well as the releaser toward which it is directed may be displaced.)

3. Our third point concerns the question of how the appetitive phase of the instinctive activity can confer pleasure and satisfaction (the subjective experiential aspects of successful active instinctive activity) without a consummatory phase—that is, without contacting the biologically relevant releaser that ensures the behavior's

fitness-enhancing value. While eating or performing the sexual act, the source of satisfaction (in a biological sense) is evident. But what motivates and gives feelings of satisfaction to such activities as voluntary participation in Greenpeace or some other social or political organization, or dedicating a lifetime to being a policeman, a scientist, or a schoolteacher? In these instances, while no apparent dichotomy exists between the appetitive and consummatory phases of behavior, it is quite clear that the activities by themselves, at least in some people, possess strong emotional, intentional, and energetic charges, which cannot come from social learning but only from instinctive participation.

The answer to this question very probably lies in the ethologists' observation that, along with the intention to reach the adequate releaser, the appetitive phase of the instinctive behavior may possess its own separate motivation, accompanied by the relevant emotional, intentional, and energetic aspects. As Lorenz points out (1982, p. 135), "In dogs, hunting and killing are . . . independent of the motivation of eating: As everyone knows, it is impossible to wean a dog from its passion for hunting through abundant feeding. The consummatory act toward which the hunting appetite is striving is the killing, particularly the shaking of the prey, and not the devouring of it." It may be objected that the dog is a domesticated animal that is fed regularly and satisfactorily by its owner; therefore, its instinctively charged activities do not reflect their original, natural state. This objection is not valid in our case, however, since we modern humans in this respect are exclusively "domesticated animals," gratifying our daily need for food without special efforts. Moreover, Lorenz describes similar examples of appetitive behavior performed for its own sake in some bird species living in the wild (1982, pp. 135–136).

If we adhere to strict evolutionary reasoning, we have to presuppose that in natural conditions the motivation to perform appetitive behavior for its own sake can never override for longer periods the need for contacting the fitness-enhancing releaser. However, the examples just given strongly suggest the possibility that this evolutionary trajectory exists in nonhuman animals, too, wild or domesticated. It follows that humans, living in conditions that free them extensively from investing considerable amounts of energy and time in achieving basic life-sustaining resources, have the possibility of investing, sometimes massively, normatively diffuse instinctive motives in nonbiological, culture-determined activities that do not possess a consummatory phase.

Summarizing this discussion, we can propose that, while human instinctive drives lost their capacity to induce clearly discernible pre-programmed behavioral sequences toward genetically strictly specified releasers, they retained their capacity for:

1. *Directing behavior and social learning (in a dim, approximate way) toward fitness-enhancing avenues.*
2. *Activating the physiological mechanisms that prepare and assist the organism in discharging the respective instinctive behavior.* Sometimes this occurs in a way that was adaptive in the evolutionary past but may be superfluous or even harmful in present circumstances, as when, in a conflict situation, the body prepares for physical action (fight or flight), even when the expectation is for a verbal contest only.
3. *Conferring the intentional aspect, the energetic load needed for the proper intensity and length of the effort,* and providing the *subjective emotional experience* accompanying the instinctively fueled activity, even when the target of the activity is far remote from biological aims.

2.2.1. Instinct/Learning Interaction

Many of us have a tendency to associate learning with the experience of schooling or some other kind of formal education. This kind of learning, which is unique to human culture, primarily serves the need of the community to perpetuate certain cultural traditions and shared knowledge and values. Frequently, it is built on compulsory memorization of information that may be entirely unrelated to individual instinctive motivations. To be sure, educational achievements may contribute greatly to social adjustment. Being at school, for example, may teach discipline (while at the same time *suppressing* or *opposing* unrelated instinctive strivings), but apart whatever is relevant to one's profession or personal interests, the material that has to be memorized usually proves to be quite useless and eventually will be largely forgotten.

Because they are oversensitive to social pressures, many schizophrenia-predisposed "model" children, as well as certain individuals predisposed to anxiety disorders, can be easily induced to conform to this kind of compulsory learning. In general, people who suffer from mental disorders are, as a group, not different from the normative population in their ability to learn in this way, suggesting that mental disorders are not a consequence of decreased ability to learn culturally important material. However, this is not the kind of learning I have in mind when I refer to instinct/learning interaction.

Learning, *in the biological sense*, began much earlier in the process of evolution than institutionalized learning and is found in very simple, solitary animals. Biological learning is highly selective, unimaginable outside the domain of instinctive behavior, and is relevant indeed to the concerns addressed in this text.

Learning theories influenced by extremist behaviorist thinking (such as Watson's classical behaviorism or Skinner's operant conditioning) tend to downplay the importance of innate predispositions. They fail to acknowledge that the very nature of the measures employed in behaviorist research and therapy as praise or punishment, as positive or negative reinforcers, irrespective of whether they are applied to animals in a laboratory or to humans in a social environment—are innately determined. This is why we can never use electric shock or immersion into cold water as a reward or a desired food as punishment. Wallace Craig's innate "appetites" and "aversions" are an organic part of the biological nature of both human beings and animals. Put succinctly, *"All conditioning by reinforcement occurs within a context of appetitive behavior, at least in non-human species"* (Lorenz, 1982, p. 151, emphasis in original). Consequently, learning and inherited patterns or predispositions to behave (instincts), far from being antithetical (think of the nature/nurture controversy or Freudian psychoanalytic theory, which stresses the conflict between instincts and social prescriptions to behave), are inseparably intertwined or integrated in overt behavior. Niko Tinbergen (2003, p. 128) expresses this idea very clearly when he writes that: "all learning effects a change in the innate functions; *learning, therefore, is a phenomenon of ontogenetic growth of behavior superimposed on the innate patterns and their mechanisms* . . . learning is often predetermined by the innate constitution. Many animals inherit predispositions to learn special things, and these predispositions to learn therefore belong to the innate equipment (idem, p. 128, emphasis added; see also pp. 145–150 on "localized" and "preferential learning," as well as "critical periods of learning").

In humans, too, complex behavior patterns are built on simple innate reflexes, like locomotion, maintaining balance, grasping, and so on; our complex "gestalts" are built on innate predispositions to integrate various kinds of sensations into percepts and to automatically integrate crude percepts with relevant ontogenetically acquired material; our complex thoughts are built on innate predispositions for language acquisition (Pinker, 2007A).

Considering the many forms of learning, as well as disagreements about the nature of learning, it will be useful to discuss the modalities of "biological" learning in order of increasing complexity. These modalities are inconceivable independent of the domain of instinctive activity.

The simplest form of learning is *habituation*. Habituation means the extinction of an innate response (FMP) to a recurrent environmental stimulus that proves to be irrelevant to the innate behavior elicited by it (that is, the stimulus does not signal the presence of a releaser or of environmental danger). For example, the startle response to a sudden recurrent noise that is not followed by other signs of danger gradually subsides without affecting the same response to resembling stimuli that do signal danger (Lorenz, 1965, pp. 49–51). By fine-tuning the originally quite unselective innate response, this mechanism leads to a more energetically efficient, more differentiated response to environmental stimuli.

This kind of simple learning is universal in the animal kingdom, existing "in the lowest as well as the highest metazoa" (multi-cellular animals) (Lorenz, 1982, p. 265). Therefore, habituation is a simple learning mechanism that can function only in the framework of an innate, instinctive system.

Another very common learning mechanism is "*habit formation*" (idem, p. 272). This refers to a kind of learning that compensates for the oversimplified and unselective nature of the IRM by recognizing a biologically important environmental agent, like the individual parent or offspring. For example, the imprinting mechanism discovered by Konrad Lorenz induces a newly hatched, inexperienced graylag gosling to follow any moving object that emits an appropriate sound. This reaction was found to be nonspecific to a certain individual (a conspecific or a human person). The gosling recognizes only the general features of the species the individual belongs to. The individual itself (be it human or nonhuman) can be exchanged with another one of the same species without affecting the following response. However, in the next day or two, the gosling and its natural parent mutually learn to recognize each other *as individuals*, based on recognition of special facial features (idem, pp. 275–276). This refinement of the IRM toward greater selectivity by including additional distinguishing characteristics of the releaser is understood as a learning process.

The same is true in the case of humans. As René Spitz demonstrated, the smiling response of a human baby can be released at first by a very oversimplified dummy, but later only by the general features of a human face (while the formerly used dummy can elicit a fear response). However, at the age of five to six months, the smiling response becomes so selective that it can be elicited only by a familiar person's face, usually the mother's. At this age, the face of a stranger may elicit a fear response. It has to be stressed that *neither the fixed motor pattern nor the innate releasing mechanism is replaced by learned behavior.* The IRM is simply rendered more selective, enriched with details learned through association (idem, p. 273). This kind of learning is "extremely common" in higher animals: "IRMs which are *not* adaptively modified by experience, thereby having achieved a higher degree of selectivity, are not easy to find in higher vertebrates" (Lorenz, 1982, p. 274, emphasis in original).

It may be concluded from this discussion that instinctive behavior (both its motor pattern as well as the releasing mechanism) frequently needs, even in simpler animals, some learned additions, modifications, or honing in order to materialize or increase its inherent fitness-enhancing potential. This reasoning will become even more obvious as we proceed.

The next example represents a more complex interaction of the innate and learned aspects of behavior. It is learning through the consequences of the release of instinctive behavior. In the field of behavioristic psychology, this kind of learning is named *operant conditioning*. It occurs when the consequences of a certain behavior (success or failure to attain its goal, instinctive or partially learned) are fed back into the memory of the animal or human individual, thus rendering future similar behavior more effective or adaptive in similar circumstances (Skinner, 1976, pp. 51–79). K. Lorenz (1982) describes several forms of this kind of behavior in animals (pp. 283–314). Being generally known, they need not be detailed here. Instead, I want to discuss in some detail a related, but more complex, form of instinct/learning interaction in animals known as *exploratory behavior*.

In exploratory behavior, "the animal really tends to direct at the object of its curiosity practically all of the behavior patterns it has at its disposal" (Lorenz, 1982, pp. 325–326). The example Lorenz gives is illuminating:

> A young raven . . . confronted with an entirely new and unknown object treats it to a phyletically programmed sequence of behavior patterns whose functions are extremely varied. Caution being the better part of valor, the raven begins its exploration of an unknown object by treating it as if it were a dangerous predator, that is, something to be mobbed. The raven approaches the object edging sideways and even backward, crouched in preparation for flying off, delivers a very strong peck at it, and immediately, incontinently flees as fast as it can. If the object then follows in pursuit, the bird will try to get behind it and attack it from the rear. . . . If, instead, the object flees, the raven takes up the pursuit immediately, delivering peck after peck and proceeding, if possible, to the motor patterns for killing large prey. . . . If the object does not respond and proves to be "dead already," the raven will grip it with one foot and try to tear it to pieces. During this process the object may prove to be tasty, whereupon the raven will eat some of it and proceed to hide the remnant. If the object is of no use at all for any of these functions, it gradually loses its attraction and may then be used to sit upon, or its pieces may be used to cover and hide more interesting objects (Lorenz, 1982, p. 326).

In this way, the animal may accumulate a considerable amount of "latent knowledge" on relevant aspects of its environment (idem, pp. 327–328). This confers "the ability to adapt, through *individual* learning, to the most variegated biotopes. The raven can survive in the North African desert by behaving virtually like a vulture, soaring on thermal currents to discover carrion. It can live on islands in the northern seas, a parasite of bird colonies, eating eggs and young birds. . . . It can live in European forests, where carrion is scarce, by hunting insects and small vertebrates" (idem, p. 329, emphasis in original).

In this example, too, it is obvious that *learning is utterly dependent on innate behavior patterns* that give the animal both the "tools" with which to explore the unfamiliar object, as well as the guidelines for "interpreting" the results of these explorations—that is, what they "mean" for the animal's life-perpetuating intentions. In many respects, the more evolved social animal's selective learning about the environment, other conspecifics' behavior, and desired rules of conduct may involve essentially similar, albeit more complex, strategies. Similar behavioral mechanisms seem to be involved when small children explore their natural and social environment.

The general picture that begins to emerge from our discussion so far on the role of instincts in human behavior can be summarized by the views on human instincts of L. T. Hobhouse (an early theorist who antedates modern ethology): "Whilst believing that the permanent basis of human life is instinctive, Hobhouse conceives the instincts in man as being of the nature of innate promptings, cravings, or determining tendencies, which are specific in themselves but which do not comprise fixed inherited motor patterns of behavior.[4] The actual overt pattern of behaviour which is manifested in efforts to satisfy these promptings is not automatically given, to the same degree as in lower animal species, but depends largely upon the experience and intelligent control of the individual and also *upon the complex influences of the social traditions into which he is born and within which he lives*" (Fletcher, 1966, p. 46, emphasis in original).

2.3. The Subjective Experience of Instinctive Activity, and Communication Regarding This Experience: The Empathizing Mechanism

As argued previously (section 2.2), with the almost complete disappearance of rigid FMPs and IRMs from mature human behavior, the possibility of directly translating an innate instinctive motive into observable behavior (which reflects exclusively that instinctive motive without extensive acquired modifications) has been lost. Overt behavior became the final outcome of an interaction between instinctive promptings, felt subjectively as more or less circumscribed or diffuse intentions, propensities, etc., as well as (selectively) acquired, learned elements. Therefore, the possibility of directly deducing phylogenetically evolved roots from the overt, observable behavior, and of reliably differentiating that behavior from the ontogenetically induced elements, became more and more problematic. Since the main concern of the present work is the innate inherited predisposition to develop mental disorders, we cannot disregard this aspect, and must therefore search for ways that will enable us to identify these innate motivating elements of behavior as precisely as possible.

As mentioned at the beginning of this chapter, instinctive activity manifests itself in three domains:

1. In the domain of *somatic events* in the brain and the rest of the body.
2. In the overt, *observable behavior* (in a more-or-less concealed way).
3. In the domain of the *subjective experience*, mostly as emotions or intentions. In addition, we may have a subjective experience of some other aspects of the instinctive activity, such as the amount of energy it confers on the behavior, some of the physiological changes it induces (like blushing or muscle tension), or the quality or attractiveness of the object (the releaser) toward which the instinctive behavior in directed.

Each one of these domains in which the instinctive activity is apparent is approached by a different (more-or-less scientific) method. Each has assets and limitations, and no one of them can lead to an integrated, whole picture on the subject, as discussed in the section on methodological issues in psychiatry in chapter 1. Having no one clearly preferable way to approach instinctive activity in humans and more evolved animals, but three problematic ones, each illuminating a different aspect of it, I think, we cannot afford to disregard any of them. However, in order to avoid confusion, we have to take great care in how to integrate, and alternatively, when to keep apart, the data obtained by these different approaches.

The subject of this section is to clarify how we can obtain information on the subjective experience of instinctive activity of other minds, a knowledge of which may foster more precise communication in this domain with other conspecifics, in particular between mental health professionals.

Advocates of the psychodynamic approach to interpreting and treating mental disorders often encourage the therapist to enter the private, experiential world of the patient: "The therapist must have the plasticity to transpose himself into another strange and even alien view of the world" (Laing, 1960, p. 34). Carl Rogers, another existential psychiatrist, recommends that "the way of being with another person . . . means entering the private, perceptual world of the other and becoming thoroughly at home in it. . . . It means temporarily living in his/her life" (Rogers, 1980, p. 2155).

During my specialization in psychiatry, I often contemplated how exactly this feat can be accomplished. Is it meant literally or metaphorically? Does it denote a kind of indefinable psychological intent on the therapist's part, or more than that, a talent or special ability that only the few chosen ones possess? Or, alternatively, must it be built painstakingly during one's professional career? At any rate, these reflections sparked my thoughts on this subject, which are summarized in the following pages.

Subjective experience is defined in the domain of philosophy of mind as *"Direct or Immediate Knowledge . . .* not based on evidence or inference" (Kim, 2011, p. 18, emphasis in original). One of its main characteristics is *"Privacy, or First Person Privilege,"* meaning that "direct access to a mental event is enjoyed by a single subject, the person to whom the event is occurring" (idem, p. 19, emphasis in original). Strictly speaking, this is true for *any* subjective mental event, including simple sensations. Nobody has any direct means to "enter" the mind of another person. This is a problematic situation, because it follows that we cannot have any sure, scientifically verifiable knowledge on how other minds perceive the environment, their own bodies, and their own mental events. We have no interpersonally verifiable direct means that other minds even exist, nor do we even have any solid logical arguments as to whether the environment exists at all independently of any observer. This doctrine is named in philosophy "solipsism." It presupposes that the "first person perspective . . . stand[s] in various kinds of isolation from any other persons or external things that may exist" (R. Audi, 2005, p. 861). A philosophical standpoint based on solipsism's extreme implications was proposed by the Irish bishop and philosopher George Berkley, who maintained that "the only existing entities are finite and infinite perceivers each of which is a spirit or mental substance . . . physical objects exist if and only if they are perceived" (idem, p. 83).

We stipulated from the beginning, however, that our discussion here will unfold on the basis of evolutionary theory; that is, in Karl Popper's parlance, its chief "background knowledge" is Neo-Darwinism, which is taken (provisionally) for granted. On these grounds, we have to accept that organisms are inseparably interconnected with their environments, including the other organisms in their environments, in order to be able to fulfill life's main goals. They evolved sensory organs in order to provide themselves with necessary information about their environments as well as their own bodies and behaviors. Members of a species evolved in similar environments, among similar selective pressures and similar adaptive interactions with other conspecifics, and transmitted the same genes to successive generations. Diversification is constrained by more- or less-intense selective pressures. Therefore, we are justified in supposing that genetically closely related organisms acquire quite similar information on vital aspects of the environment with similarly evolved and inherited sensory organs, process this information in similar ways, and similarly use the received information in their quest for perpetuating life.

That is, we have to presuppose that sensory processing of environmental stimuli by organisms that are genetically closely related has to be quite similar, and that these organisms build behavioral tactics on this supposed similarity (even without being able to sense directly the subjective observational experience of other minds). The great success of scientific methodology, which is based on this presupposition regarding collaborating scientists, is convincing, albeit indirect, evidence of this similarity of sensory processing by closely related organisms.

However, as we move further away from simple information-processing on permanent and predictable aspects of the physical environment toward more fluid behavioral interactions with conspecifics, especially in more evolved social animals, diversification of subjective mental experiences and, consequently, uncertainty and imprecision in communication increases: "Evolved mechanisms that provide an interface between an organism and the physical world are fundamentally more constrained than are mechanisms that evolve to coordinate interactions among conspecifics, as in communication systems" (Barkow, Cosmides, & Tooby, 1995, p. 397). Sensory organs do not need to reflect the environment comprehensively and accurately so long as the information they convey remains useful to the organism for the perpetuation of its own special form of life.

From this discussion, we can conclude that basic sensory information, whether of an exteroceptive or proprioceptive nature (that is, sensing the environment external to the body or somatic events inside the body, respectively), is reasonably similar in healthy organisms closely related genetically. Nevertheless, between exteroceptive and proprioceptive sensations important differences do exist. Information about the external world in evolved animals and humans, at least of some sensory modalities (visual, auditory, and olfactory in animals), has a high resolution and supplies fine detail, while that coming from the body is much duller. Compare the wealth and precision of the information that vision, for example, can provide with that supplied by the sensation of pain. Considering the biological function of pain (to free the injured or diseased body part from strain), as compared to that of vision (supplying critical information on many relevant environmental events and objects), this difference is understandable.

However, the more important distinction from the point of view of the present discussion is that *exteroceptive sensory information can be interpersonally corroborated or validated while proprioception cannot.* The professional term for this corroboration, at least in a special area of scientific investigation, is *joint attention*: "The term joint attention denotes the process by which adults or other caretakers interact with a child to direct attention to a referent of a word (some object or activity) through eye contact, gaze, tone of voice, smiles, and so on. . . . A link is thereby established between some object or action and the sound-pattern or gestures that convey the word." (Incidentally, joint attention is not an exclusively human practice. It was found, too, in "chimpanzee mother-infant pairs starting at age three months" (Lieberman, 2006, pp. 368–369). Without question, this mechanism leads to the most precise verbal communication possible between humans (in addition to the communication in formal sciences of mathematics and logic). Joint attention (especially joint observation) can lead to the learning and sharing of words that represent exactly the same sensory qualities of objects, activities, or events for two or more observers. Communication based on this mechanism also constitutes one of the two mainstays of scientific activity: "The enterprise of science is founded on the hope that all rational beings who investigate and ponder on the same evidence, *derived ultimately from sense impressions* ('facts'), will be led to draw from this evidence the same conclusions" (Dobzhansky, 1962, p. 5, emphasis added). (The other mainstay of science is impartial, logical reasoning.)

While, as I have argued, observational information can be correlated relatively easily between different persons (at least concerning its sensory qualities), the process that leads from sense impressions to the perception of more complex entities, to abstraction and generalization, and to the formulation of empirical scientific regularities and theoretical constructs seems to be more problematic and fallible. In fact, observation itself may be fallible. Today it is unanimously accepted that the light spectrum is composed not of discrete colors but of a continuously changing range of wavelengths. The division of a gradually varying spectrum of light into discrete colors reflects the constraints of the color sensitive corpuscles in the retina; in addition, it is probably more useful for discerning discrete objects in the environment than sensing continuous gradations of the same sensory quality (a single color) would be. Besides, since the distortion is similar in all healthy conspecifics, it does not hamper exact communication about sensory percepts.

Formerly, we arrived at the conclusion that in our attempt to interpret the inherited aspects of behavior in general and the inherited aspects of mental disorders in particular, we cannot dispose of the information furnished by the subjective experience. Therefore, as an attempt to foster, no matter how modestly, the inclusion of this kind of information, at least in the behavioral and social sciences—including psychiatry, of course—I try in the following to analyze how people manipulate subjective experience in themselves in order to get information on the subjective experience of others. This analysis may improve the precision of communication with this kind of material. It is an important aspect of social communication in general and, I think, it may help the reader in grasping more precisely some of the arguments in the sections pertaining to unobservable mental phenomena.

As argued previously, nobody has direct access to the subjective experience of anyone else. For example, if I see a bird and show it to my grandchildren, I automatically presuppose (with good evolutionary reasons) that they will perceive exactly the same sensory qualities in that bird (color, shape, movements) as I do. Of course, the associations that enter into my integrated perception of the bird (prior memories, associated emotions, learned factual knowledge, aesthetic pleasure, etc.) may be considerably different from those of my grandchildren.

In trying to ascertain such unobservable mental experiences in another person (our main concern here being the subjective experience of the instinctive activity: intentions, emotions, subjective perceptions of physiological changes in the body, etc.), I evidently have to begin with *exteroceptive sensory activity* (observation). In this first step, I can see the respective person's *behavior* and some of his body's *physiological accompaniments*, like blushing, trembling etc., hear her vocalizations (including the tone and rhythm of her speech), and so on. Perhaps I recognize some *observable circumstances,* too, which may influence this person's observed behavior and subjective experience (suppose, for example, he is listening to some critical remarks of his superior).

The second step in trying to grasp another person's subjective experience—I cannot see any alternative—is to try to imagine *how I would have felt* had I behaved similarly, made similar vocalizations, shown similar physiological responses, and been exposed to similar circumstances. This mechanism is widely recognized in psychological, philosophical, and psychiatric literature, as well as in other areas, like fiction. It has been given various names by different authors. The great Scottish philosopher David Hume (1711–1776) gave it the name "sympathy." His conceptualization of this mechanism closely resembles the first two phases of the empathizing mechanism proposed in this work (Focquaert & Braeckman, 2011, pp. 241–242); Bertrand Russell, a well-known British philosopher, dubbed it "analogy" (Russell, 2002, pp. 667–669). More recent names for the same phenomenon are *empathy* and *mental simulation* (Bolton & Hill, 2007, p. 102).

Psychologist and anthropologist Robin Dunbar links this phenomenon to the emergence of self-awareness in humans: "Being self-aware provides us with the capacity to reflect on our internal mental states, and then to relate these internal states to the observable behavior of other individuals. From this comparison we can work out that other individuals also have mental worlds" (Dunbar, 1997, p. 91).

However, Darwin's observations on this phenomenon, which in this text will be dubbed "empathy" or the "empathizing mechanism," trace their origins to a stage of individual development when the "capacity to reflect on our internal mental states" seems absolutely impossible.

Relating an observation of his six-month-old son's behavior, he wrote: "When his nurse pretended to cry . . . I saw that his face instantly assumed a melancholy expression, with the corners of the mouth strongly depressed . . . an innate feeling must have told him that the pretended crying of his nurse expressed grief; and this through the *instinct of sympathy* excited grief in him" (Darwin, 1872/1965, p. 358, emphasis added; see also Eibl-Eibesfeldt, 2010, pp. 55, 462–463).[5]

More recently, primatologist Frans de Waal discussed the same phenomenon Darwin observed. He put the first expression of a rudimentary form of empathizing at an even earlier age: "Newborn babies cry in response to the sound of another baby's cries. It is not just a matter of noise sensitivity, as babies react more intensely to these sounds than to equally loud computer-simulated cries or animal calls. Their reaction is regarded as an expression of *emotional contagion*, and thought to provide the underpinnings of empathy.[6] Studies consistently demonstrate higher levels of this kind of emotional contagion in female than in male infants" (de Waal, 1996, p. 121, emphasis added; see also: Buss, 2008, p. 404).

The definition of *empathy* in *Campbell's Psychiatric Dictionary* is very similar to the one used by this text: "Empathy . . . [is] *intersubjective resonance*, a form of cognition that enables one to comprehend another person's subjective experiences from his or her own unique perspective [emphasis in the original]. . . . Empathy is not compassion, but an *affect attunement* is the first step on the empathic process of understanding and accepting the subjective validity of the other person's point of view" (Campbell, 2004, p. 219, emphasis added). (In our theoretical scheme, the empathizing or "affect attunement" is considered the second, not the first, stage of the empathizing process, the first being the observational phase.) It may be needless to remark at this stage of the discussion that the empathizing mechanism has to have *inherited* roots, which presupposes that it has some fitness-enhancing role.

The propensity for empathizing and the strength of the empathizing mechanism appear to vary according to several factors besides gender. One very important factor is the quality of the emotional relationship with the person empathized with. The stronger and more positive this emotional tie, the stronger (but not necessarily more precise) the empathizing tends to be. Since the strongest tie in this respect is the mother-infant bond, it may be that we have here a clue as to the evolutionary origins of the difference between the sexes in this respect. (For more on gender differences and the propensity to empathize, as well as its possible role in autism, see Focquaert & Braeckman, 2011, pp. 252–253).

Many years of close relationship with clients and a desire to grasp their mental states has led me to believe that the accuracy of the empathizing mechanism (though not necessarily its strength) can be improved by conscious effort, aided by an emotionally neutral and nonjudgmental approach. On the other hand, the more guarded or distant a relationship is, particularly when it is negative, condemnatory, or inimical, the more the strength and precision of the empathizing deteriorates.[7]

It can also be safely presumed that grasping another person's mental state through empathy will be more precise when the two persons have similar inherited emotional makeups, particularly when the empathizer has experienced relevant life events similar to those of the person empathized with. A therapist who has been through combat stress can find in himself an emotion closer to that of a soldier suffering posttraumatic stress than a therapist who has not gone through such an experience. In the same vein, a mother who has lost a child can reconstruct more faithfully in herself the feelings of another mother in similar circumstances than, say, a childless bachelor. In addition, we can hypothesize an innate diversity among individuals in general with regard to the inclination for introspection that consciously directs our attention to our own emotional or intentional states, an inclination that has to have an impact on the quality of empathizing. The terms *introversion* and *extroversion*—as used in psychology—express a diversified continuum, or more precisely in this sense, its extremes (*Concise Oxford English Dictionary*).

The discovery of "mirror neurons" in both macaques and humans may suggest that empathy, in a crude, primitive form, is evolutionarily older than humanity or even apes. "Mirror neurons discharge both when the individual performs a goal-directed action and when the individual observes another individual performing the same action. They therefore 'mirror' an observed action within the motor cortex of the observer. . . . Moreover, it is the *intention* of the action that is represented rather than the action itself. . . . Gallese has argued that the MNS [Mirror Neuron System] constitutes a neural basis for empathy" (Burns, 2011, p. 297, emphasis added; see also Gómez, 2004, pp. 258–261).

Is it possible that creatures so distant from humans as old-world monkeys are able to mentally reconstruct, through the mechanism of empathy, the (covert) intentions of conspecifics—or even of unrelated animals (like predators)—by observing their behavior? In fact, better anticipation or earlier recognition of another animal's intentions makes evolutionary sense, particularly when the other animal has aggressive intentions. This anticipation might occur through empathizing with the fear reactions of more experienced conspecifics or by comparing the motor patterns of a stalking predator, for example, with the empathizing animal's own behavior when it behaves as a predator.

The range of subjective mental phenomena that can be approached by empathy is wider than the unobservable aspects of instinctive activity, covering, for example, learned aspects of behavior as well as other thought contents. Allow me to summarize briefly here the main subjective elements of instinctive activity that *can* be sensed in another person through empathy, with some comments on their adaptive significance.

2.3.1. Intentions

In this context, *intention* means the subjective experience of an urge to act toward a certain environmental agent in a way that, at least in the phylogenetic past, possessed a fitness-enhancing quality. However, in the case of humans' more diffuse instinctive predispositions, this instinctive urge can be directed toward socially determined goals that may be remote from the original, life-sustaining biological ones. This line of thought is further detailed in the succeeding chapters.

Of course, I refer here only to the innate, instinctive foundations of what customarily is called intention. Simple intentions can be induced by a simple instinctive need, such as the intention to prepare a meal, and very complex intentions exist in which instinctive underpinnings interact with extensive acquired knowledge and forethought, like the intention to become a professional athlete or a physician.

2.3.2. Emotions and Moods

The way these terms are understood here needs some introduction, or more precisely, a certain account of the nature of subjective experience in general and of emotion in particular. Regarding these issues, however, we have no scientific explanations; therefore we have to be content with instructive metaphors.

Let's take the simplest instance of subjective experience, a sensation of some physical aspect of the environment. Scientific understanding defines light, for example, as a corpuscular or wave phenomenon (both models work well in spite of being irreconcilable). The light falling on the retina induces in its light-sensitive corpuscles chemical processes that are in turn transformed in the nerve fibers into electric impulses. On their way toward the cortical primary processing areas, these electric impulses are transformed several times into chemical processes, and vice versa, at the nerve synapses. However, at the primary visual cortex, these chemical and electric processes are somehow "translated" into the *mental phenomena of seeing*, say, a color. The nature of this translation from physico-chemical phenomena into the subjective experience of seeing remains a mystery.

In my opinion, the subjective experience of *emotion* is a similar translation of neurophysiological and neurochemical events into subjective mental experience, but instead of referring to a circumscribed stimulus of a physical nature, the experience of emotion reflects a more comprehensive appraisal of the individual's relationship with aspects of her natural/social environment: "Emotions can be argued as expressing the individual's awareness of his or her position in the world." And: "Emotions signal the relevance of events to concerns" (Frijda, 1994, pp. 112–113). And: "At their core, emotions . . . evaluate, assess, or appraise . . . they report, correctly or incorrectly, on how we are faring in the world." (Prinz, 2004, p. 44). Formulated somewhat differently: "Emotions provide information that functional systems are in a cost-deficit state (depression, frustration-anger, pain), will be in a cost-deficit state (anxiety), or are in a benefit-excess state (pleasure, power-elation-control)" (McGuire & Troisi, 1998, p. 117).

The reflection of this organism-environment interrelationship by emotions is a very unspecified, general, and crude one, and, of course, it is occurring at a preverbal level. At the most generalized level, the quality of the emotions may be divided into a pleasurable or unpleasant nature, reflecting a desirable or undesirable relationship between the individual and aspects of his environment. Both of these two all-encompassing categories can be subdivided into more discrete, circumscribed ones that may indicate more specific causative factors. This can be illustrated with the comprehensive category of unpleasantness: This very general indication, which could refer to the frustration of an active instinctive motive such as hunger, or to an environmental danger not related directly to discrete life-sustaining motives, such as the proximity of predators, can be better specified by more differentiated emotions. Take, for example, the emotion of fear. Many variables may influence a human's or animal's sense of security under given circumstances. On the individual side, an innate predisposition may accentuate the fear reaction's intensity or persistence. Furthermore, the amount of fear is influenced by how well the individual is acquainted with the nature of the respective threat, whether she has memories of similar situations in the past and of how they were resolved, whether he has at his disposal effective innate or acquired defensive tactics, how fit she is physically or mentally at the decisive moment, and so on.

On the side of the environmental variables are such factors as the nature and gravity of the threat, the characteristics of the terrain (especially if it favors escape or counterattack), whether other conspecifics on whom one can count for help are nearby, and so on. Animals and most humans in most circumstances are unable to analyze all these variables consciously, or even a portion of them. And in the case of humans, even when they are able to do so in optimal circumstances, usually no time exists for such an analysis. Environmental danger can strike quickly and unexpectedly. The calculation of the central nervous system (CNS) has to be made promptly and, as a rule, unconsciously. The mind informs its "owner" only of the final result of this computation in the form of an emotional tone (increased alertness, fear, anxiety, panic) and, at the same time, gives its instructions on how to react: "escape," "hold your place," "disregard," "defy in spite of being frightened." (See psychologist Robert Zajonc's data on this subject in Solomon, 2004, pp. 32–35.)

I am quite convinced that animals with an evolved CNS do experience these kinds of unspecified generalized subjective states, as do small children. I am reluctant to accept that such a central phenomenon in humans as the subjective experience of emotions emerged spontaneously from nowhere without rudimentary antecedents in other evolved animals. It somehow contradicts, to my mind, the spirit of evolutionary logic. It seems more appropriate to postulate that subjective experience in general, and nuanced emotions in particular, evolved in humans from a similar but more rudimentary phenomenon that was already present in evolved animals. I am even inclined to believe that the unique cognitive abilities of human beings evolved on the foundations of that crude, rudimentary, intrapsychic information about the relevant aspects of the organism-environment interrelationship that the emotional experience provides.

Emotions need not be prompt and short-lived but may reflect a more protracted, more stable interrelationship with elements of the environment—for example, a mother enjoying the sight of her sleeping baby or the blues we might feel when we are isolated from the outside world as a result of bad weather. Subjective experiences of this kind are customarily called "moods."

2.3.3. The Reflection of Instinct-Induced Physiological Changes in Subjective Experience

The famous James-Lange theory equates instinct-induced physiological changes in the body—more precisely the awareness of them in subjective experience—with the emotions themselves. This theory was refuted by W. B. Canon, who showed that emotional experience and emotional behavior continue to exist after the viscera are surgically or accidentally disconnected from the CNS; that the emotional response to visceral changes is diffuse and nonspecific; and that the emotional experience tends to appear faster in the framework of instinctive activity than the visceral responses (Gregory, 1987, p. 219; Mandler, 1984, pp. 20–24).

It seems to me, however, that the sensing of the functional state of the body has to contribute to the mind's evaluation of the individual's interrelationship with her environment, and therefore it does influence the subjective experience. The difference between feeling physically fit and prepared to face a challenge versus feeling considerably handicapped physically (as a result, for example, of a medical illness that prevents the instinct from preparing and activating the body properly) may dramatically affect one's chances of coping with a challenging or potentially dangerous situation, and in consequence has to be reflected in the subjective emotional experience. A self-confident, assertive attitude versus a tendency toward retreating, hiding, or submissiveness (accompanied by the subjective experience of fear) can make all the difference.

2.3.4. The Subjective Experience Reflecting the Biologically Adaptive Value or Significance of a Releaser

This point seems quite obvious. A beautiful post-pubertal woman arouses in men very different feelings or levels of attraction than a child, an old woman, or an already pregnant one (Buss, 2008, pp. 150–152). Or, during competition for dominance in a chimpanzee community, a strong male with a dominant predisposition is instinctively evaluated very differently (as a potential competitor) than a mature, strong male without aspiration for dominance (a possible ally), or an old, weak male (who is probably useless as an ally) (de Waal, 1982/2007).

2.3.5. The Subjective Experience of the Intensity (Energetic Charge) of an Instinctive Motive

Naturally, both overt behavior and the accompanying physiological processes are affected by the intensity of an instinctive striving at a given moment. In addition, we have a subjective experience of the instinctual striving that ranges from being barely perceptible at one extreme to being almost unconquerably intense at the other. This subjective information can be very important in humans when the instinctive motive is expressed in a social setting that puts restrictions on how, or how intensely, the respective striving may be expressed.

Before entering the discussion of the third step of the empathizing process, the corrective or compensatory mechanisms, I would like to quote English philosopher Gilbert Ryle, who summarizes succinctly the two steps of the mechanism we have already discussed (the observational and the empathizing stage proper) in the following way: "One person has no direct access of any sort to the events of the inner life of another. He cannot do better than *make problematic inferences from the observed behavior of the other person's body to the states of mind which, by analogy from his own conduct, he supposes to be signalized by that behavior.* . . . For the supposed arguments from bodily movements similar to their own to mental workings similar to their own would lack any possibility of observational corroboration" (Ryle, 1963, p. 16, emphasis added). It has to be clear by now that the degree of accuracy of this mechanism, more exactly its second phase—that is, the empathizing phase—depends first of all on the degree of similarity (or dissimilarity) of the participants' respective intrapsychic qualities (see also Thomas Nagel, 2002, p. 222).

2.3.6. Corrective or Compensatory Mechanisms

Discovering the (partially) hidden motives in another individual's be-havior can provide very useful information for a social animal about other conspecifics' possible attitudes and about actions they are inclined to take. This ability, which is related to empathy but has a wider scope in humans, including—in addition to instinctive predispositions—finding out the beliefs (including false ones), problem-solving strategies, decep-tive strategies, etc., of other minds, is called in cognitive science "theory of mind" abilities. It is believed some primates also possess them, in a less-developed form (Gomez, 2004, pp. 205–211, 226–237; Laland & Brown, 2011, pp. 200–201). It is hypothesized that good "theory of mind" abilities conferred fitness advantages during evolution, when hominids began to live in larger groups: "Reproductive success within relatively larger groups should have depended upon managing alliances and friendships with other members of the group; those individuals with theory of mind abilities should have been at a selective advantage at these tasks" (Mithen, 2007, p. 257).

It can also be hypothesized that since the previously mentioned sub-jective phenomena accompanying instinctive activity are the reflection of adaptive, evolutionary mechanisms subject to natural selection, they must be quite uniform across a population of animals with similar ge-netic endowment due to the strong selective pressures in natural en-vironments that constrain diversification. These subjective phenomena include the intention to act in a certain (fitness-enhancing) way, the emotions signaling the quality of the organism-environment interrela-tionship, the feelings about events induced in the body by the instinctive activity, and the emotional significance of the releaser.

On the other hand, in humans, who show great diversity in behavior and subjective experience due to their inhabiting self-constructed artifi-cial environments with considerably relaxed natural selection pressures, the mechanism of empathy, which is built on the expectation of simi-larities between the subjective experience of different persons, becomes more and more fallible. In other words, using our own emotions, inten-tions, etc., to sense that of others runs the risk of disregarding the true extent of human diversity and of interpreting others' emotions and in-tentions (and consequent beliefs, thoughts, and actions) as closely sim-ilar to our own (a phenomenon known in psychodynamic terminology as *projection*). In order to avoid this trap, or at least to reduce its biasing effect, a third step in the mechanism of empathizing is necessary. After observing the behavior and its eliciting circumstances (the first step), and after employing empathy (the second step), we can employ what may be called "*corrective or compensatory mechanisms.*" Their function is to assess the possible dissimilarities between one's subjective experience and that of the observed person's (despite any similarities in overt behavior and eliciting circumstances), and to compensate or correct accordingly. In the following, I enumerate some of these compensatory or corrective mental maneuvers of which I am aware:

1. The simplest compensatory mechanism is probably one that corrects for possible quantitative differences in the subjective experience. For example: I witnessed a sports accident in which an athlete broke several teeth. I may remember the experience of my last painful dental treatment and assume that the pain the athlete experienced had to be similar in its quality but much more intense. Another example: I hear the screams of a mother who was just informed that her child was injured in a road accident. My closest memory of similar circumstances and similar behavior on my part was of a road accident in which my beloved dog was killed. I may suppose that the subjective, emotional response of that mother is similar to mine in its quality but much stronger in its intensity. (For those who find this analogy offensive, let me add by way of clarification that in the context of this work the emotional bond with a family dog, kept for no utilitarian purposes but to be a loved companion, is interpreted as the relaxation and diversion of the IRM of the parental instinct [see K. Lorenz, 1982, pp. 164–165].)

2. Another, relatively simple, correcting or compensatory mechanism concerns facial and body features that are the result of inheritance or are otherwise acquired and convey the false impression of expressing emotions, intentions, or a whole array of instinctive configurations. Such features may consist of the relative proportions of the component parts of the face, the dimensions, form, and angle of the eyes and eyebrows, the distance between the eyes, the natural form, size, and curvature of the lips, the dimensions and relative proportions of the body parts, the natural body posture, and so on. These kinds of false impressions concerning possible intrapsychic contents can be

corrected relatively easily, as a rule, by better acquaintance with the respective person's behavior.

3. Frequently, we have to compensate for several parameters in which the observed person is different from ourselves because we suspect that these parameters may affect the emotional responses. They are: gender, age, cultural background, living circumstances, certain illnesses like Parkinson's disease, and so on.

4. The examples discussed up till now were built on impressions of isolated events or factors. However, most of us have longstanding or recurrent relationships with a considerable number of persons, which enables us to observe their behavior repeatedly, in diverse situations. Thus, we can see, over time, for example, how an emotional reaction is initiated, how it unfolds and clears, how its intensity changes with time, the extent to which another person expresses his emotions in his behavior in the first place, his preferred ways of expressing particular emotions, etc. In this case, individual observations followed by empathizing and corrective maneuvers over a protracted time period may add up and overlap to form a comprehensive picture of another person's subjective experience. This may happen during longstanding dynamic psychotherapy or longstanding relationships between family members, friends, etc., and may provide a truer, more comprehensive picture, I believe, of the observed person's emotional makeup.

5. Finally, what can be done when another person's behavior and verbalizations do not arouse in us any memory of emotions or intentions that can resonate with what we have observed? I refer here to experiences of persons who are in close contact with mental patients with bizarre psychopathology (schizophrenics in particular), such as mental health professionals or close family members. How can we make sense through the mechanism of empathy of such bizarre complaints as hearing voices inside one's head, or thought withdrawal and broadcasting (Schneiderian symptoms), or what is called in clinical psychiatry "incongruent affect"—that is, discordance between speech content and the accompanying affective expression?

It seems that the sheer impossibility of figuring out by means of empathy and compensatory mechanisms what may be going on in these patients' minds gives that strange, eerie impression that we are dealing with a fundamentally different world of mental experiences. Surely, if we have had no similar psychotic experiences ourselves, empathizing alone cannot help. However, I believe that these psychotic symptoms are the outcome of a long ontogenetic development during which a person with a dysfunctional instinctive configuration (instinctive drives that are too diffuse) tries to work out a fragile, unusual lifestyle amid more-or-less normative social circumstances. I think the approach to these patients' mental experiences needs, before empathizing, a highly abstract exercise in evolutionary reasoning. More precisely, we have to reconstruct in detail the ontogenetic development of these individuals before and after the outbreak of the psychotic disorder in order to try to understand the subjective aspects of the pathology of these patients. In this way, if I am right, the predisposition to the schizophrenia spectrum of disorders becomes a quantitatively different (instead of qualitatively distinct and thus incomprehensible) instinctive trait, whose subjective experiential aspect can be approximated, in principle, by the empathizing process. However, its third step needs to be much more complex than usual. I elaborate on this subject when discussing schizophrenia in sections 7.2.7. and 7.2.8.

A special case of empathy (particularly with regard to the compensatory and corrective mechanisms) is that concerning animal behavior.

This empathizing process in those dealing with animals arises, mostly unwittingly, as a reaction to those aspects of the animal behavior that remind us of analogous human behavior. Niko Tinbergen, in the quotation on the nature of appetitive behavior in animals (section 2.2), speaks about "purposive activity" and "insight," which naturally are unobservable phenomena, while Konrad Lorenz, in writing about the raven's exploratory behavior (section 2.2.1), ascribes to the bird such subjective experiences as "curiosity," "caution," or the loss of "attraction" if the investigated object proves to be useless. Additional examples are provided in section 5.2.3—for instance, attribution of feelings of dysphoria and dejection to young chimpanzees separated from their mother or depression in a widowed graylag goose.

The first two steps of the empathizing mechanism in this case seem to be quite similar to what we employ in the case of humans. Behavior that is partly similar to ours in partly similar circumstances arouses in us the appropriate intentional or emotional accompaniment, a subjective experience that we then attribute to the observed animal. If our dog devours its whole daily portion of food at once, we conclude that he was hungry. If a crocodile mother fiercely guards her eggs and later her young offspring, we conclude, as a rule without hesitation, that her actions are induced by parental feelings. (For Darwin's, Lorenz's, and especially Tinbergen's views on this subject, see Fletcher, 1966, pp. 268–271.)

The important difference between the mechanism of empathizing with animals as opposed to with humans has to do more with the corrective or compensatory mechanisms than it does with the nature of the empathizing itself. This may be the result of various degrees of difference between a human and the observed animal's body build, or the animal's ability or inability to generate movements resembling those of humans and, consequently, humanlike bodily and facial expressions. It is much easier to empathize with a chimpanzee mother caressing her newborn baby than it is with a whale or a bat mother's analogous behavior. Moreover, lack of a shared language and culture further taxes the empathizing mechanism.

Another special consideration concerning correcting mechanisms exists in the case of animals, one that is quite sophisticated in fact. It involves the degree of complexity of the part of the brain believed to be responsible for the generation of conscious emotional experience. While the behavior induced by parental instincts, for example, may resemble that of such widely different animals as crocodiles, birds, apes, and humans, we have to presuppose that those animals with less-developed limbic and neocortical areas have less clear-cut, conscious subjective experiences of emotion and even less developed self-awareness.

The readiness for and accuracy of empathy (I refer here to all of its components: observation, empathizing, and the corrective mechanisms) vary greatly from person to person and even within the same person in various mental states. In addition, its efficiency matures with age and experience. On the other hand, in a person possessed by a strong intention to act in a preconceived way or overwhelmed by a strong emotional state, the empathizing process will deteriorate. Others' behavior will be interpreted in a way that suits that dominant intentional or emotional state. In "functional" psychotic states, the last two phases of the empathizing process are seriously affected or lacking, while in organic psychosis all three phases, including the observational one, may be seriously compromised.

A more neutral emotional state, coupled with a genuine concern about another person's feelings, can improve the empathizing process. On the other hand, in methodical observation of behavior (human or animal) for scientific purposes, the observational phase will be accentuated, while the empathizing stage is purposefully checked or suppressed.[8]

During the professional career of a psychodynamic psychotherapist, all three aspects of the empathizing process can be purposefully improved. This improvement, I think, can be fostered considerably if the therapist becomes well acquainted with her own instinctive configuration (perhaps by undergoing psychotherapy herself), an awareness of where her instinctive configuration lies on the normative spectrum, and how it compares with that of the individuals she intends to understand and influence.

I cannot close this section without mentioning some instances of empathizing concerning normative persons. In spite of the great diversification of humans' instinctive constitution, which is the main subject of the present text, I suspect that the basic "building blocks" of this instinctive constitution, and consequently the subjective mental experience that goes along with it, have to be qualitatively quite similar across different individuals. Both the normative and pathologic forms of the diversification concern quantitative differences, disproportions, maladaptive intensities, or problems of integration between various emotional motives, or between these motives and other mental functions, such as the cognitive functions. *This hypothesized similarity in the basic constituents of the subjective experience of the instinctive activity has to be the reason for the frequently successful approximation during interpersonal transactions, with the aid of the empathizing mechanism, of others' mental states.*

I have often been puzzled by the uncanny ability of great fiction writers, stage performers, or movie actors to reconstruct in their writings or enactments a wide variety of characters (some of whom have instinctive configurations that are surely very different from their own). These descriptions or enactments elicit in readers or audiences an empathizing reaction. In the case of outstanding actors, the onlookers' feeling is that the enacting springs from the innermost being of the actor and is not merely the recitation of a memorized text accompanied by the imitation of the appropriate gestures. I wonder whether this impressive empathizing ability (both the actor's with the enacted character, as well as the receptive audience's) suggests the possibility that we humans share a wider and richer instinctive/emotional repertoire than the one we express in our everyday behavior. Might it be that, besides the dominant emotional and intentional attitudes with which we identify, we possess (in a weak and mostly suppressed form) emotional motives or instinctive promptings that cannot induce customary behavior but may nevertheless have a strong enough presence to allow the empathizing process to reconstruct another, widely different person's emotional, intentional configuration? Might it be that this ability contributed to the evolution of some forms of artistic creativity as well as to the enjoyment of its fruits by the audience? (For roughly similar thoughts, but in a wider context and in a more loosely associative style, see Summers & Crespy, 2013, p. 440.)

Notwithstanding, it has to be stressed that the fruits of the empathizing mechanism will remain always a kind of *approximation,* and communication with them can never achieve the level of precision that communication with percepts or through logical or mathematical reasoning possess.

On Additional Scientific Branches Relevant to the Present Theory

3.0. Introduction

This chapter discusses topics pertaining to various fields of research and theorizing: primatology, evolutionary psychology, evolutionary psychiatry, and theories of gene-culture coevolution—topics relevant to the present theory. It briefly looks at an example of the relaxation of selection pressures in the living conditions of bonobos as contrasted with those of chimpanzees and the comparative effects of that relaxation on the behavior and social organization of these two closely related ape species; it investigates how evolutionary psychology interprets normative human behavior's inherited underpinnings; and it examines evolutionary hypotheses on the genetics of mental disorders in the emerging field of evolutionary psychiatry. Certain gene culture coevolutionary and related theories are also discussed.

My intention here is to outline concisely the work done in these fields for nonprofessional readers in an attempt to endow them with a wider perspective on the scientific domain to which the present work applies. Some of the material presented here (like that discussed in the section on primatology) constitutes indirect supporting evidence for the ideas proposed in this text. Data and theoretical interpretations in other fields may be convergent, contradictory, or irrelevant to these ideas.

3.1. Primatology

The next step in attempting to bridge the gap between animal and human behavior, with an accent on its innate aspects, will be to take a closer look at some features of the behavior of our closest relatives, the apes.

The most well-known aspects of primate behavior are those complex cognitive abilities that can be experimentally demonstrated in the laboratory or in other controlled conditions. Both chimpanzees and bonobos possess the ability to communicate in sign language using a considerable number of words and to combine them spontaneously to produce rudimentary sentences (Gomez, 2004, pp. 277–278); solve concrete problems in controlled experiments (Lorenz, 1982, p. 240) and in captivity (de Waal, 1982/2007, pp. 61–63); recognize themselves in a mirror (Gomez, 2004, pp. 267–275); and use simple implements like a twig to "fish" (forage) for termites. But these research findings, while being indisputably important for understanding the evolution of cognitive abilities or the phenomenon of cultural transmission, are of limited importance for the purposes of the present work—the quest for understanding the evolutionary roots of mental disorders—and therefore will not be elaborated upon here. Instead, I will discuss some aspects of the social life of chimpanzees and bonobos that are more closely related to our concerns.

Even to one who, like me, is not a primatologist, *the great diversity in the social organization of anthropoid primates—which may signal highly developed behavioral flexibility and adaptability—is striking.* The Asian great apes are the *orangutan* and the *gibbon*. The least evolved ape, the *gibbon*—also called the "lesser ape"—lives in an exclusively arboreal habitat in permanent family groups consisting of the parents and their offspring (the *New Larousse Encyclopedia of Animal Life*, 1980, pp. 511–512). Orangutans lead solitary lives. Males have no socializing tendencies (idem, p. 513): "Full grown males are extremely intolerant of one another" (de Waal & Lanting, 1997, p. 16).

The African hominoid species are the *gorilla, chimpanzee*, and *bonobo*. Gorillas are largely terrestrial, and the group consists of one adult male, several adult females (the male's "harem"), and their immature offspring. "Once a male has established a harem, he tends to stay with it for life" (idem, p. 18). Males fight fiercely among themselves for the possession and retention of a harem.

As to the other two African hominoids, the chimpanzee and bonobo (our closest genetic relatives), the human lineage split off *before* the splitting between the chimpanzee and bonobo lineage took place; therefore, they are considered equidistant from humans on the evolutionary tree (de Waal & Lanting, 1997, p. 3). They are genetically very close to each other, both living in African forests; are about equally intelligent based on cognitive experiments; and resemble each other so much in physical appearance that early researchers could not tell them clearly apart and considered the bonobo a subtype of chimpanzee, the "pigmy chimpanzee" (idem, p. 7). However, these two species are strikingly different in their social organization and behavior (and on a closer look differ in some of the finer details of their anatomy, too).

The comparison of the way of life of these two species, closely related genetically to each other and to humans, seems to me important in the context of the present theory, because of interesting questions it raises about the interdependence of habitat, physical makeup, behavior, social organization, and—more specifically for our concerns—the consequences of the relaxation of natural selection pressures.

Before considering the differences, however, between the behavior of these two ape species, I would like to stress common characteristics (that are also observed in other apes), speaking from the presupposition that the instincts of apes (and to a lesser extent monkeys) are in a *more diffuse state* than that of other vertebrates, thus necessitating a greater contribution from learning for behavior patterns to become fully functional. Young female chimps, for example, are very interested in newborns, try to take them from the mother, and handle them in ways suggesting that maternal behavior is partially learned (de Waal, 1982/2007, pp. 65–66; Goodall, 2000, p.137). Even more impressive is *alloparenting*, that is, carrying and caring for infants by monkeys other than the biological mother, a phenomenon that probably reflects the diffuseness of the parenting instinct's innate releasing mechanism (Perry, 2011, pp. 229–232). And young chimpanzees reared in isolation and later reunited with their group are unable to engage in mature sexual behavior but can learn it from mature conspecifics.

Innate motor patterns during the *appetitive phase* of instinctive activity, like the ritualized courting behavior of birds or ritualized sexual fighting of ungulates and fish, are nonexistent in primates. On the other hand, culturally transmitted practices, like termite fishing (which exists in some chimpanzee populations but is unknown in others), are much more prevalent than in social animals lower on the evolutionary ladder (Ehrlich, 2000, pp. 70–71).

Now let's turn to an examination of the differences in behavior and social organization between the chimpanzee and bonobo.

Chimpanzee males are "coarse and hot tempered" (de Waal & Lanting, 1997, p. 9) and are prone to aggression, which escalates from displays of power and fierceness near the opponent to actual violence, though it is of a measured intensity. They dominate females, who are involved in the males' struggle for dominance by supporting one or the other party. The contestants themselves use sophisticated tactics in their efforts to rise to a dominant position, including the formation of "coalitions" with certain high-ranking males and with the female camp, as well as the use of intimidating techniques against supporters of the rival candidate. Fighting and reconciliation between rival males (and in general between quarrelling parties) alternate quickly even between the bitterest opponents contesting for dominance. The explanation for this quick alternation of aggression and reconciliation seems to be the need to balance intragroup competition for resources with the need for collaboration for the sake of group unity, especially during conflicts with groups of neighboring territories (Goodall, 2000, sections 5, 6, and 7).

The adult males form "patrol" groups, fight with intruders, collectively hunt smaller monkeys, and distribute their flesh among themselves (Goodall, 1988, pp. 198–200). When a male achieves dominant status, his behavior changes: He seems to consider himself above the group members, breaks off unwanted fights between them without taking anybody's part, or sides with the weaker party against the stronger, and breaks off attempts by lower-ranking males to mate with estrous females. The dominant male also tries to win the support of the adult female camp. The whole process of achieving, retaining, and practicing dominance is strikingly similar to processes in human populations but, of course, on a more instinctive, less premeditated level (de Waal 1982/2007; deWaal 2005, chapter 2).

Chimpanzees, like bonobos, live in "fission-fusion" groups; that is, depending on the abundance of food, the whole group may forage together or disperse. In chimpanzee groups, the males stay together while the females may travel alone or with their immature offspring. In bonobo societies, the female is accompanied by her male offspring, even after they reach adulthood.

With regard to food resources, in the case of the bonobo, "food occurs in concentrations large enough that multiple bonobos can forage together, typically in mixed parties" (de Waal, 2002, p. 56). The chimpanzee habitat, on the other hand, cannot provide food year-round for larger groups. The solitary travel of chimpanzee females with their off-spring is apparently not just a response to the scarcity of food resources but is also aimed at preventing male aggressiveness against very young offspring. Chimpanzee males are known to kill offspring of uncertain descent. In general, while bonobos "are sensitive, lively, and nervous . . . chimpanzees are coarse and hot-tempered." And: "Physical violence almost never occurs in bonobos, yet is common in chimpanzees" (de Waal & Lanting, 1997, p. 9).

The sexual act itself in chimpanzees is of very short duration, "less than a quarter of a minute," always in the same (ventro-dorsal) position, and, except occasionally in young females, without any apparent signs of sexual excitation as a rule on the face or in vocalizations of the participants (de Waal, 1982/2007, p. 154). Except for mating and masturbation in captivity, as far as I know, no other kinds of sexual practices have been observed in chimpanzees.

Male dominance does not exist in bonobo society. "Females occupy a central position in social life, . . . males . . . remain attached to the mother all their life, following her through the forest and depending on her for support against other males" (de Waal, 2002, pp. 57–58). "That bonobos lack formalized rituals of dominance and submission [so characteristic of chimpanzee society] tells us already how relatively unimportant status must be in their society" (de Waal & Lanting, 1997, p. 72).

In addition, one finds an "incredible individual diversity among bonobos. Like people, they differ widely from one another in intelligence, temperament, and behavior" (idem, p. 176). (To a lesser extent, this is true of chimpanzees and other evolved animals.) Another difference between these two ape species is that tool use has never been observed in bonobos, despite their high intelligence as revealed in laboratory studies, while it is common in chimpanzees.

Apart from the centrality of females and the lack of formalized dominance, male coalitions, and physical aggressiveness, the most peculiar aspect of bonobo society compared to that of chimpanzees (and for that matter of any other animal known to me) is their exaggerated and versatile sexual life, even by human standards: "Bonobos use sex for sex . . .

for appeasement . . . as a sign of affection . . . [and for resolving] tensions over food" (idem, pp. 99–100). Young females dispersing to neighboring groups in order to avoid inbreeding establish bonds with dominant females through sexual activity, thus securing their acceptance in the new group. Sexual contacts may occur—besides those between males and females—between males and males, females and females, males and juveniles, and females and juveniles. They use a variety of positions, including face-to-face mating, known to occur elsewhere only in humans. "Mouth to mouth kissing . . . often with excessive tongue to tongue contact," was also seen, as well as masturbation (in natural conditions, not only in captivity) (idem, p. 103). During some of the sexual practices, "the participants often bare their teeth and squeal in apparent pleasure" (de Waal, 2002, p. 53). Bonobos share with humans "a dramatically extended sexual receptivity . . . the chimpanzee female is receptive less than 5 percent of her adult life, whereas the bonobo female is so nearly half of the time" (de Waal & Lanting, 1997, pp. 106–107).

Various hypotheses, including extrapolations to humans, have been put forth to explain the exaggerated sexual life of bonobos. It has been linked to bartering sex for food or luring a male into a permanent relationship with a female, with resultant gains in securing food and protecting the offspring. An opposite argument exists as well, one that presumes that female promiscuity is meant to confuse paternity, thus conferring added protection for the offspring against male aggression (an assumption probably more justified in the case of chimpanzees). All these hypotheses seem to contain some partial truths but at the same time fail to account for the lifestyle as a whole of bonobos in their specific habitat, or for other differences between the bonobo and chimpanzee, such as the level of aggressiveness, the dominance relationship between the sexes, male immaturity and lack of bonding with other males in the bonobo, the bonobo's slenderer, humanlike physical build and posture, and "higher frequency of standing upright, shorter arms compared to legs, smaller skull, smaller teeth, [and] less size difference between sexes" (Klein & Takahata, 2002, p. 210; Ehrlich, 2000, p. 69).

The possible importance of *ecological factors* in leading to the differences between chimpanzees and bonobos is well recognized: "No doubt, ecological conditions were a key factor. Without large enough food concentrations . . . the entire [bonobo] scheme could never have worked" (de Waal & Lanting, 1997, p. 140).

Bonobo habitat today is restricted to the rain forests of equatorial Africa. Chimpanzees, on the other hand, due to the constriction of their territory (rain forests) during the glacial periods (in comparison with the interglacial ones) in the late Miocene through the Pliocene era (idem, p. 29), dispersed into much larger and less providing environments (idem, p. 10). "In the chimpanzee, evolutionary changes may have been prompted by a need to adapt to half-open habitats, such as woodlands. The bonobo, on the other hand, may never have left the protection of the rain forest" (de Waal, 2002, p. 42).

Assuming that the fundamental cause leading to the strikingly different evolutionary trajectories of the bonobo and chimpanzee lifestyle was a considerable relaxation in selection pressures regarding food resources in the case of the bonobo or, alternatively, the intensified selection pressures in the chimpanzee's new habitat, the following sequences of evolutionary events may be hypothesized: The need to protect a territory's food resources from competitors in the case of bonobos decreased substantially. In consequence the need for a hierarchically organized, cohesive male group, which is able to successfully fight rival groups, became superfluous. Aggressiveness in competition for survival (but not necessarily aggression in competition for mating) decreased. The pressure for dispersal of the group, that is the solitary travel of the females and their offspring in search for food in seasons scarce in resources, decreased in the case of the bonobo.

In addition, relaxation from constraints on feeding is conferred in the case of bonobos by improved ability, in comparison with chimpanzees, to feed on "terrestrial herbaceous vegetation, a common food source in their range." Consequently, "[f]ieldwork supports the assumption that the distribution and abundance of food is crucial to the contrasting lifestyles of the two *Pan* species" (idem, p. 64, emphasis in the original).

What would the repercussions of relaxed selection pressures (concerning food resources) be regarding reproductive activities? The study of behavior and physiology of domesticated animals may provide us with an important clue. Domesticated animals are released from selection pressures regarding feeding (even more than the bonobo), while being under strong artificial selection pressures for the traits desired by the breeders. In these circumstances, it was found that "resources formerly required for survival in nature are shifted to traits directly affecting fitness, such as reproduction, or to artificially selected traits. This theory is supported by the observation that in captive animals not selected for increased body size, *domestication has resulted in a rather consistent reduction in the size of body structures considered less important in captivity than in nature, while structures supporting reproduction have increased in size.*" This hypothesis is known as Beilharz's *"resource allocation theory"* (Price, 2002, p. 79; emphasis added). A similar phenomenon is considered in the next chapter concerning the RNA virus.

Some aspects of bonobo reproductive behavior and physiology do support Beilharz's hypothesis. The interval between births decreased (from about once in six years in chimpanzees to about once in four and a half years in bonobos), and the number of offspring per female increased slightly, too (de Waal & Lanting, 1997, pp. 106, 140, 190). The greatly extended sexual receptivity of the females as well as the dramatic increase in the instances and duration of sexual activity in both male and female bonobos can be interpreted as supporting Beilharz. Even the predominance of female interests (connected to rearing and guarding offspring) in bonobo society over predominantly male responsibilities can be seen as being in accordance with Beilharz's theory. In bonobo males, in comparison to chimpanzee society, efforts to secure a territory atrophied to striving to attain a higher rank under the mother's auspices.

However, other aspects of bonobos' sexual life cannot be explained by this line of reasoning. If we hypothesize a strengthening of sexual behavior in parallel with the weakening of selection pressures for survival, *why does no accentuated male competition exist for access to estrous females, accompanied by increased aggression* (as in gorillas)? An even greater problem with the resource allocation hypothesis is that it cannot explain the most conspicuous trait of bonobo sexuality, its great variety, which is mostly unconcerned with reproduction or even with attaining purely erotic pleasures. Sexual activity in bonobos crosses the boundaries between instinctive systems of very different nature. For example, it is aroused by the prospect of *eating*: "Feeding time . . . turned out to be the peak of sexual activity" (de Waal, 2002, p. 48); and *aggression* was deflated by sexual activity between members of two groups who had just met. In general: "Any object, not just food, that arouses the interest of more than one bonobo at a time tends to lead to sexual contact" (idem, p. 49).[1]

This aspect of bonobo sexuality, to my mind, can be explained only by assuming a considerable *diffuseness* (in contrast to differentiation or goal-orientedness) of the sexual instinct, as a result of a substantial *relaxation* of the selective constraints on the reproductive activity. This interpretation is not necessarily in disagreement with Beilharz's resource allocation theory. But while sexual activity became more intense, it also became less differentiated; that is, sexual arousal ceased to push the individual toward a circumscribed, preprogrammed action serving exclusively or largely reproductive aims. Instead, it became considerably *diversified*, both in regard to its *motor pattern* and, even more obviously, regarding the environmental agents and situations that *release* it. Some of these environmental agents and situations are remote from the reproductive aims. Some of these diverted pathways of sexual activity in bonobos acquired *secondarily* new adaptive or fitness-enhancing roles, like the deflation of aggressiveness between individuals belonging to different bonobo groups, or greater acceptance of a strange female into a new bonobo group. S. J. Gould and N. Eldredge named this kind of evolutionary diversion of a somatic or behavioral trait from its original function into a new, unrelated one, *exaptation.*

Other instances, like homosexual practices, masturbation, and sexual activity with sexually immature conspecifics may represent simply the "displacement" of sexual excitation (its FMP) toward biologically inappropriate releasers, which will be dealt with in section 5.2.1. All these practices (but not all the triggering conditions in the bonobo) are encountered in human behavior, too.

To put these arguments into a wider evolutionary context, it will be worth comparing the plasticity of sexual activity in the bonobo with the differentiated nature and rigidity of sexual behavior in other animals lower on the evolutionary scale that probably evolved amid more stringent selective conditions. K. Lorenz, discussing instinctive behavior patterns that cannot be conditioned by behavioral means, remarks: "Sexual action patterns, for instance, refuse to become conditioned to any kind of stimuli pertaining to other behavior systems—at least in nonhuman animals. It is impossible to teach a male pigeon, through food rewards, to utter his courtship coo, nor can female rats be conditioned through food or water rewards to assume the copulation posture, no matter how hungry and thirsty they may be" (K. Lorenz, 1982, p. 305). In contrast with rats, chimpanzee females may accept food from males in exchange for sexual consent, and in bonobos this seems to be the rule rather than the exception (deWaal, 2005, pp. 123–125).

It seems to me that, regarding the intermingling of different instinctive domains, bonobo sexual instincts are even more diffuse than those of humans. While humans may use a bewildering variety of sexual practices to obtain sexual pleasure in interaction with a wide assortment of releasers remote from the reproductive domain, I am not aware of instances in which excitement before feeding or reconciliation between rivals may lead to sexual arousal. Whether the relaxation of natural selection pressures in bonobo habitat may account by itself for the extreme diversification of bonobo sexuality is not clear to me.

A more general consequence of the relaxation of both natural and sexual selection pressures in bonobos may be their great diversity regarding intelligence, temperament, and behavior, mentioned formerly. However, bonobo society seems simpler, less structured, and less "task oriented" than that of chimpanzees, in spite of similar levels of intelligence. Bonobos did not disperse from their narrow original habitat of equatorial rain forest. Therefore, it seems to me that bonobos, for some reason, were unable to convert the diversification resulting from relaxation of selection pressures into fitness gains through tighter and more effective social organization, as did the chimpanzee, and especially the human, lineage.

3.2. Scientific Approaches to Human Behavior Influenced by Evolutionary Thinking

3.2.0. Introduction

Since the publication of Darwin's *The Descent of Man* and *The Expression of the Emotions in Man and Animals,* attempts to interpret human behavior using evolutionary reasoning have never ceased. The scientific branches relevant to our purposes are: Evolutionary Psychology, Evolutionary Psychiatry, and Gene-Culture Coevolution Related Theories. A detailed survey of these fields is beyond the scope of this text. However, characteristic examples in each area are relevant to the concerns of this work. (The ideas of Darwin's cousin Francis Galton on the inheritance of mental and behavioral traits, which served as the foundations of the ill-fated eugenics movement, will be mentioned in section 3.2.2.)

Ronald Fletcher's *Instinct in Man* (1966), a survey of the various disciplines that interpret human behavior in accordance with evolutionary thinking from the time of Darwin till the 1950s, is very informative on this subject. Laland and Brown's *Sense and Nonsense: Evolutionary Perspectives on Human Behaviour* (2011) discusses approaches to the same topic since the 1970s, in human sociobiology, human behavioral ecology, evolutionary psychology, and gene-culture coevolution theories. These two works are important sources for this section. Beginning in the late 1990s, several books written by psychiatrists attempted to interpret mental disorders according to evolutionary reasoning.

Although the present theory formally belongs in the category of gene-culture coevolution theories, its main aim is to *interpret dysfunctional human behavior as the result of the interaction of innate behavioral aspects with a restricted number of fundamental consequences of cultural evolution that increased ecological dominance and progressively decreased the intensity of natural selection pressures.* Existing gene-culture coevolutionary theories attempt to explain the evolution of culture in general, and the reciprocal interaction between genetic inheritance and cultural influences for normative, not dysfunctional, behavior, and examine the similarities between genetic and cultural transmission.

3.2.1. Evolutionary Psychology

The research methodology of evolutionary psychology is, in my opinion, closer to standard scientific methodology (discussed in the first chapter) than to the methodology of other approaches in psychology. Hypotheses are derived from a well-tested theory, an approach that, despite the shortcomings of Neo-Darwinism mentioned in chapter 1, still guides most scientific work in biology and related scientific fields. In consequence, evolutionary psychology allows us to draw similarities, parallels, and hypothetical evolutionary pathways between animal and human behavior that makes its somewhat more detailed investigations especially important for the purposes of this chapter. Hypotheses derived by evolutionary thinking are tested by controlled experiments and the results statistically evaluated. In addition, these hypotheses, when possible, are formulated in a way that allows their refutation (according to Karl Popper's rationale; see chapter 1) by systematic observation and experimentation. Furthermore, efforts are made to ascertain that behavioral categories thought to be evolved adaptive traits exist in ethnic groups widely different from one another, in order to exclude the possibility that the respective behavior pattern was induced exclusively by the peculiarities of a local culture or ecology.

However, while evolutionary psychology's findings seem important for understanding inherited aspects of normative behavior, they are useless in explaining the inherited underpinnings of mental disorders. The reason for this, I believe, is that in following traditional Darwinian thinking, these findings rely on the effect of directional selection pressures (instead of their relaxation) and target discrete instinctive mechanisms (instead of comprehensive ones). I will return to this point at the end of this section.

I will outline briefly some characteristic samples from the field of evolutionary psychology, a field that relies heavily in turn on the ideas of sociobiology (Laland & Brown, 2011, p. 105).

William Hamilton's *inclusive fitness theory*, published in 1964, is regarded by many as the "single most important theoretical revision of Darwin's theory of natural selection in this century" (Buss, 2008, p. 232)—alongside, of course, the synthesis between the Darwinian theory of evolution and developments in the field of genetics, both Mendelian and molecular. The inclusive fitness theory claims that an individual does not necessarily have to transmit his own genes to the next generation directly by producing offspring in order to influence the frequency of occurrence of those same genes in an interbreeding population. Instead, he can *indirectly* multiply his own genes in the population by promoting the survival and reproduction of blood relatives (who possess partially the same genes). This is true even in circumstances when the individual endangers his chances for survival and reproduction at the same time. The effectiveness of this strategy depends on the percentage of common genes with the respective blood relative (50 percent in the case of brothers and sisters, 25 percent in one's grandchildren, 12.5 percent in the case of first cousins, etc.).

This is the evolutionary rationale behind the close emotional relationship, solidarity, and mutual aid within families. It explains, too, the overwhelming role of extended families or clans in defending their own members or territory in pre-state societies (in which the state did not yet monopolize the right to use power in settling individual and group conflicts). It explains, too, further research findings in the behavioral and intrapsychic realm, such as the progressive decline, according to the degree of remoteness of the genetic relationship, of altruistic deeds and the willingness to help in life-threatening situations (Buss, 2008, pp. 238–240); emotional closeness as a function of genetic relatedness (idem, pp. 240–241); patterns of inheritance of material goods (overwhelmingly to blood relatives); investment of grandparents in grandchildren, and so on (idem, pp. 243–250).

Behavior patterns that can be explained in terms of the inclusive fitness theory exist in animals, too, such as the phenomenon of the "alarm call." The animal that emits the alarm call that warns the whole group of an approaching predator reduces its own chances of survival by giving away its hiding place or attracting the predator's attention to itself. Why is such innate behavior, in ground squirrels for example, or in several species of monkeys, not eliminated by natural selection? Alarm calling unquestionably increases the chances for survival of the surrounding conspecifics, which, being genetically related to the animal emitting the alarm call, may also possess the genes that induce or predispose for such behavior (Buss, 2008, pp. 236–237).

3.2.1.1. Reciprocal Altruism

Altruistic behavior is defined as a behavior pattern that decreases the fitness of the donator while at the same time increasing the fitness of the recipient. Reciprocal altruism refers to instances in which the participants are *not* genetically related. Therefore, it is a behavior pattern that defies conventional evolutionary logic. The theory that attempts to explain this kind of behavior in the animal kingdom in accordance with evolutionary reasoning is the *theory of reciprocal altruism among non-kin* advanced by biologist Robert Trivers in 1971 (Buss, 2008, p. 16). Reciprocal altruism involves the delivering of a biologically important resource (food, for example) to a conspecific (but not kin) with the expectation that the act will be reciprocated at some time in the future. The donated resource is not critical for the donor at the moment of its transfer but it may be so for the recipient. At the reciprocation of the deed, the desirability of the transferred resource for the participants may be reversed. (In fact, the accuracy of the word "altruism" in this instance has been questioned. See Mayr, 1991, p. 155.)

Reciprocal altruism is found in simple hunter-gatherer societies in addition to more evolved cultures, as well as in primates and even species as remote from humans as the vampire bat. For example, if an individual hunter in a primitive human society, in circumstances of scarce food resources, succeeds in killing a big animal that he and his family cannot consume in its entirety by themselves, it makes evolutionary sense for them to share it with other members of the group with the expectation of reciprocation sometime in the future (perhaps when the donor himself is in need of food). Such behavior may increase the chances of survival of both parties; therefore, its inherited predisposition may be transmitted genetically, in contradiction to the oversimplified traditional stereotype of the survival of the fittest "in the struggle for life."

Return of favors between acquaintances, friends, and even strangers is ubiquitous in human communities. Established social transactions of many kinds, such as buying, selling, providing various services in exchange for payment, and so on, may be seen as built on the instinctive foundations of reciprocal altruism. Because reciprocation may be delayed in time, sometimes for long periods, it is hypothesized that reciprocation led by natural selection to the evolution of special mental capabilities. Their function is to remember what one got from whom and what is expected in return, how to follow up and how to detect, avoid, or punish cheaters—that is, those who do not reciprocate or reciprocate inadequately (Buss, 2008, pp. 265–268; Barkow, Cosmides, & Tooby, 1995, pp. 169–170).

Reciprocation is well known in chimpanzees (de Waal, 1982/2007, pp. 200–203). What is even more impressive is that reciprocation was observed in vampire bats, an ancient mammal species with greatly reduced cognitive abilities compared to humans or apes. These creatures feed on the blood of larger animals like horses or cattle, and their survival is endangered if they cannot feed for more than three days. "Bats regularly regurgitate a portion of the blood they have sucked and give it to others in the bat colony, but not randomly. Instead, they give regurgitated blood to their friends, those from whom they have received blood in the past" (Buss, 2008, p. 269).

The evolutionary trajectory that created this kind of behavior can be imagined in the following way: The pattern of innate behavior, that is regurgitation of blood and its transfer to another bat, evolved in order to feed the young offspring (R. M. Nowak, 1999, p. 354), as an addition to lactation or after weaning, when the offspring is still unable to obtain the needed amount of blood independently. In a later stage of evolution, bats that used the same innate behavior pattern sporadically to aid genetically unrelated hungry adults (that may have emitted some sign of distress resembling that of a hungry offspring) and received the same service in turn when in need had a better chance to survive and transmit the respective gene(s) to the next generation. Thus, the IRM for this FMP widened to include certain other bats besides the immature offspring.

Reciprocal altruism (with an inherited basis) seems to be the foundation on which culturally determined standardized transactions between different persons or groups evolved (paying for different services or merchandise, for example). Such transactions are indispensable components of any social life, primitive or modern.

Our last example of human behavior that is shaped culturally on the foundations of instinctive predispositions consists of certain aspects of the relationship between the sexes as it emerges in evolutionary psychological thinking and research. This subject exemplifies beautifully the close relationship between behavior with innate underpinnings and subjective experience, as well as the relationships between behavior, physiology, and anatomy. Moreover, in keeping with evolutionary logic, the same adaptive instinctive motives (resembling behavior patterns) can frequently be demonstrated in both humans and animals.

One basic anatomic and physiologic difference underlying male versus female reproductive activity in the class of mammals is that insemination and conception take place inside the female body, which in turn leads to pregnancy, birth, lactation, and other forms of feeding, guarding (and possibly teaching) the offspring till maturity, with a lot of burden (or "cost" in evolutionary terminology) incurred by the female, with few exceptions in mammals. Developing further the theoretical implications of these facts, in 1972, the biologist Robert Trivers (whose theory on reciprocal altruism has been mentioned) advanced the *theory of differential parental investment,* which may explain many differences in the reproductive strategies of the sexes (Buss, 2008, p. 16). The investment of the male in transferring his genes to the next generation is very small compared to that of the female (one or several copulations), and keeping in mind the almost unlimited number of sperm cells at its disposition, a male can in principle inseminate a great number of females and thus sire a great number of offspring. In consequence, males compete fiercely between themselves (*intrasexual competition*) for access to that disproportionately great female investment in reproduction.

Since in all mammals this competition consists of some kind of exclusively physical contest (chimpanzees, bonobos, and humans being partial exceptions), selection—in this case sexual selection—led to an increased body size and power in males in comparison with females (sexual dimorphism). Moreover, the extent of the sexual dimorphism correlates well with the intensity of the competition and the degree of exclusivity the winner gains in terms of access to fertile females. The more intense the competition and the more exclusive the access of the winner to receptive females, the more serious the outcome of the competition will be for the reproductive success of the contesters. The winner takes all and the loser loses its reproductive chances altogether (till the next season); that is, the "variance in reproduction" increases (Buss, 2008, pp. 296–297). A few winners will inseminate most females in the colony. For example, in gorillas the winner acquires a harem that he guards intensely from other males. In consequence, the adult male gorilla's body size is about four times bigger than that of a female, and in orangutans the male is twice as big as the female (de Waal & Lanting, 1997, p. 16). In the more social and promiscuous chimpanzee and bonobo, as well as in human species, sexual dimorphism still exists but is much smaller.

With the increasing degree of access of multiple males to the same female in more promiscuous primates, another kind of sexual selection evolved, that of the "sperm competition." This means that in the case of copulation with several males, a greater volume of ejaculate containing a bigger number of sperm cells confers an advantage in the chances for fertilization. The number of sperm cells in the ejaculate is in turn correlated with the volume of the testicles. Gorilla males, by far the biggest of the apes with a body weight up to 180 kg (idem, p. 18), possess the smallest testicles among the ape species, while the promiscuous chimpanzee the biggest. These parameters—that is, sexual dimorphism and testicle size in humans, as compared to apes—gave birth to intriguing hypotheses about sexual behavior and the relationship between the sexes during human evolution. Evolutionary psychology hypothesizes, furthermore, that these evolved adaptive strategies in reproduction still powerfully influence contemporary humans' sexual behavior.

Females have no need to compete for access to the male's reproductive services as males do for the female's. In apes and humans, the interval between births is measured in years. While men can in principle sire hundreds of offspring by impregnating a great number of women in a relatively short period of time, women generally cannot give birth to more than twenty babies during their reproductive years. Therefore, female competition for male reproductive activities cannot increase their *reproductive variance*.

During human evolution, a fundamental change occurred in differential parental investment compared to that of other mammal species, including apes (except the gibbon). Men began to live in more stable, more committed relationships with one (or a restricted number of) women and began to invest increasingly in the successful upbringing of the offspring. This meant partly contributing the resources for survival (food, shelter), as well as guarding the mate and children from the dangers of the natural environment (predators, for example) and the possible aggression of other male conspecifics. It is hypothesized that at this stage women do begin to compete for males, but less for their value as fertilizers than for their value as potential providers of resources and protectors in dangerous circumstances.

In spite of the enormous cultural progress during human evolution, which radically changed the environment and lifestyle of modern humans, research in evolutionary psychology and related sciences has demonstrated that differential parental investment, male sexual competition in order to capitalize on the disproportionally greater female parental investment, and female competition for males able and willing to invest in rearing the offspring are still powerful inherited motives underlying the relationship between the sexes.

Women, for example, value those qualities in a man that predict willingness and ability to secure the needed resources for rearing the offspring. Those qualities include high social status, good financial prospects in the future, mental qualities like industriousness and ambition, dependability and stability, potential for love and commitment, and bravery. In addition, certain valued bodily qualities suggest good health and physical abilities, like an athletic build, tall stature, a symmetrical body and face, and a healthy physical appearance (Buss, 2008, pp. 110–129). Incidentally, the female gender's concern regarding male investment in parenting was found in other monogamous species quite distant on the evolutionary tree from humans. For example, the female African weaverbird chooses her mate according to its ability to build a nest to her satisfaction, which seems to be another example of convergent evolution (Buss, 2008, pp. 106–107).

Thus, human females, contrary to other mammals, do compete for their chosen male, but this competition is usually not of a physical nature. Rather, it relies on enhancing physical and mental qualities men value in a woman, most of which ultimately reflect or give the impression of good reproductive capabilities: youth, physical beauty, signs of good physical health, lack of signs betraying illness, lack of signs of previous sexual experience or pregnancy (Buss, 2008, pp. 139–169). Another strategy in the competition between women is the inclination to derogate rivals exactly on those items men prefer (physical appearance, sexual innocence, etc.) (Buss, 2008, pp. 167–168). Men's sexual competition, on the other hand, tends overall to be more physically aggressive, leading more frequently to lethal violence (Daly & Wilson, 1988, pp. 183–186). As expected on the basis of differential investment theory, men as a rule value casual sex more than women, demonstrate less emotional involvement in the overall relationship, seek sexual intercourse earlier, and are less choosy regarding casual sexual partners.

As a corollary of the partial conflict of interest between the sexes, both men and women can use deception. Each one can exaggerate, fake, or artificially create physical and mental characteristics expected to be of importance to the other sex. Men in their search for casual sex may fake longstanding commitment, while women may pretend sexual interest but withhold sexual access until thoroughly evaluating the qualities they desire in a mate.

Another important factor contributing to different strategies in the relationship between the sexes is the difference in the degree of certainty of biological parenthood. While the mother is always sure about her genetic contribution to the offspring, the father needs indirect evidence to be certain about paternity. The jealous guarding of the female by her mate from competing males, and males' checking of possible female promiscuous tendencies, is well known not only in humans but in apes, too (gorillas), as well as in monogamous birds (Barkow, Cosmides, & Tooby, 1995, pp. 292–297).

In humans, concealed ovulation and the woman's readiness to engage in a sexual relationship throughout her whole menstrual cycle further increases the paternal uncertainty. Paternal uncertainty (as well as the woman's aspiration for exclusivity concerning the man's resources) may lie at the foundations of the social institution of marriage, which makes clear "who was mated with whom . . . thus potentially reducing conflict within male coalitions" (Buss, 2008, p. 156). Even with this assurance, as a rule, mothers provide more parental care than fathers. In evolutionary terms, "males suffer tremendous costs by channeling their resources to other men's descendants" (idem, p. 201).

These differences in behavior patterns, paralleling the different physiological mechanisms of reproduction between the sexes, as well as their reflection in the subjective mental experience (emotions, intentions), are amply supported by statistically significant evidence in evolutionary psychological literature. The same is true of other domains in which behavior was subjected to selective forces during human evolution, such as the detection of dangerous animals or situations, the complexities of the parent-offspring relationship, language acquisition, and even the extent of esthetic pleasure in certain landscapes (which has been correlated with the possible fitness-enhancing value of certain environmental configurations during human prehistory (Barkow, Cosmides, & Tooby, 1995, pp. 555–575).

The converging arguments from these widely different fields of enquiry—animal as well as human behavior, anatomy, physiology, and subjective experience—coupled with a scientific methodology of research as rigorous as possible in each individual field, impressed me as strongly convincing (destroying at the same time my formerly held romantic beliefs in unselfish, disinterested love, friendship, self-sacrifice, etc.). However, one field of human behavior, that of the disturbed, dysfunctional one, so ubiquitous in human populations, was conspicuously absent, not only in the literature concerning evolutionary psychology but, to my knowledge, in other evolutionary approaches to human behavior, too (sociobiology, gene culture coevolutionary theories, etc.). For example, David Buss's 2008 book, *Evolutionary Psychology: The New Science of Mind* (one of the more comprehensive accounts of the field of evolutionary psychology—used as the main source for this short account), dedicated less than one and a half pages out of 423 to dysfunctional behavior, without mentioning clinical disease categories. In attempting to understand why this conspicuous omission happened in a scientific field that held the promise of shedding light on the origins of mental disorders, several possible explanations have occurred to me.

The central one, which disregards the effects of the relaxation of selection pressures during human evolution and the resultant excessive diversification of behavior, the thesis that is central to this work, was recognized by David Buss himself: "Most research and theory in evolutionary psychology has focused on species-typical psychological mechanisms. . . . Individual differences, in contrast, have been relatively neglected. . . . Evolutionary biologists, as a general rule, have tended to focus on species-typical adaptations ignoring individual differences except in their role of providing the raw materials on which natural selection operates. Individual differences, particularity those that are heritable, are often relegated to secondary status because they are thought to originate primarily through non-selection forces such as random mutations. . . . Genetic differences are sometimes viewed as 'noise' or 'genetic junk' maintained within a population precisely because they are presumed to be *unrelated* to the core of the evolutionary process: adaptation and natural selection" (Buss, 2008, p. 408, emphasis in the original. See also: Gangestad & Simpson, 2007, p. 417).

In other words, as already mentioned, the research in evolutionary psychology is built on and interpreted according to hypotheses concordant with traditional evolutionary thinking—that is, adaptation through directional selection. For example, arguing according to the logic of the inclusive fitness hypothesis—that one favors genetic relatives over unrelated individuals—evolutionary psychologists Daly and Wilson state: "Substitute parents are less likely than natural parents to experience the emotional rewards that make the costs of parenthood tolerable . . . only 53 percent of stepfathers and 25 percent of stepmothers could claim to have 'parental feelings' toward their stepchildren and still fewer to 'love' them" (1988, pp. 83–84). That "only" 53 percent and 25 percent, while disappointing from a humanitarian perspective, in the context of the logic of natural selection is significant. Adoption of genetically unrelated offspring is nonexistent, to my knowledge, even in highly evolved social animals. The only line of reasoning I can imagine that may explain this behavior among human beings is the diversification of the innate releasing mechanism of parental behavior and its related feelings as a result of relaxation of selection pressures in order to include other people's children as targets and recipients of parental feelings.[2]

The other possible reason of the nonrecognition of evolutionary processes in bringing about mental disorders in humans has to do with the reliance on discrete instead of comprehensive instinctive mechanisms. This theme will be discussed in the following, and again in some detail in the section on evolutionary psychiatry.

3.2.1.2. The Genome Lag Hypothesis

A line of reasoning frequently employed to explain dysfunctional human behavior on evolutionary psychological grounds (generally without specifying clinical disorder categories) consists of the following: The specifically human instinctive predispositions evolved during the Pleistocene period—that is, from about 1.6 million till about 10,000 years ago. The environment during the Pleistocene period is dubbed in evolutionary psychological literature the *environment of evolutionary adaptedness* (EEA), in which humans lived in hunter-gatherer groups of about 50 to 150 persons. Evolutionary psychological reasoning hypothesizes that the diverse innate predispositions to behave in specifically human ways evolved in this period.

In the last 10,000 years, with the appearance of agriculture, permanent settlements, food surpluses and the accumulation of other goods, social stratification etc., and with the explosion of the human population, cultural evolution progressed at a quick pace, one that wasn't matched by a corresponding genetic evolution that could adjust innate behavioral predispositions to the adaptive requirements of the changed social environment. Appropriate examples are the retention of innate fears from ancient natural dangers like snakes, spiders, and high or closed places, which induce genetically predisposed avoidant behaviors that, if unduly strong, are diagnosed as simple phobias. No corresponding innate fears exist from the much more frequent and serious dangers of the modern environment, like electricity, automobiles, firearms, and so on, because directional selection pressures could not act in these instances. This line of reasoning, called *the genome lag hypothesis*, is also employed in evolutionary psychiatry, in order to explain psychopathology.

The whole idea of "adaptation to the modern environment," or, for that matter, the claim of maladaptation to the modern environment because of discrete inflexible behavioral predispositions that we have been dragging along from the Pleistocene era, is such a wide, overinclusive generalization that it will be difficult to build meaningful scientific exploratory and explanatory strategies on it. The environmental conditions in which, say, a police officer on her beat in an overcrowded metropolitan area has to function successfully requires immensely different mental, behavioral, and physical abilities than that of, say, a clerk working in a small office. The same can be said of a successful salesperson versus a research scientist, or of a good politician versus a talented poet, and so on.

It seems clear that most people, with widely diversified mental makeups, succeed in finding the appropriate social circumstances in which they can function properly, earn a living, and bring up children. Otherwise it is impossible to explain the rapid population growth worldwide.

It is true that a minority, for various reasons, cannot function adequately in any of the social niches at their disposition. But it is equally true that we can easily find individuals adapted extraordinarily well to a particular modern social niche. Think, for example, of those businesspeople who, rising from a poor socioeconomic background, succeed in building a financial empire; of famous or infamous military and political leaders (Alexander the Great, Napoleon, Hitler, Stalin, etc.) who changed the history of nations or even continents and the fate of hundreds of thousands or even millions of people; of great scientists who brought about unprecedented theoretical understanding or technological advances in their field of expertise. In modern societies, *great individual diversity is balanced by great diversity of social niches* (not only in occupational fields but also in family size and structure, leisure activities, social life, etc.). With improving social mobility, one can frequently find the niche that best suits one's innate and acquired abilities and minimizes the costs of the deficiencies or limitations of one's physical and mental constitution. The ability to cooperate and compete across this great diversity according to the rules of social coexistence is what makes, in my opinion, cultural achievements possible. "The flexible nature of our learning and culture allows us to survive and flourish in a broad range of environments, including those that are heavily niche-constructed by modern humans" (Odling-Smee et al., 2003, p. 368).

Another reason, besides disregarding diversification, that makes the real discoveries of evolutionary psychology unhelpful in fostering the understanding of mental disorders according to evolutionary logic, is its focus on discrete, specific instinctive mechanisms adapted to a circumscribed field of activity—that is, "domain-specific mental modules." As noted in the introductory section, *most mental disorders, and especially the most widespread and deleterious ones, are the result of the diversification of comprehensive instinctive mechanisms that affect several or most discrete instinctive predispositions* together, instead of a single one. Circumscribed eating, sexual, or phobic disorders that do not affect other instinctive domains constitute the minority of conditions in a general psychiatric practice in comparison with affective, schizophrenic, paranoid, anxiety, and personality disorders. Underlying these disorders are dysfunctions in multiple discrete instinctive domains.

In fact, the same reservation that is raised in the case of evolutionary psychology can be expressed with respect to *human ethology*. Human ethology deals with the *biological underpinnings of human behavior*. A comprehensive treatment of this topic can be found in Irenäus Eibl-Eibesfeldt's *Human Ethology*, 2010. Its more than eight hundred pages contain an impressive wealth of data and theorizing on this topic (some of which was used to support claims made in this text), yet not one single word on mental disorders. It seems to me that without the systematic exploitation for explanatory purposes of the mechanism of relaxation of the inherited aspects of human behavior from natural selection pressures and the resultant behavioral diversity, and without the recognition of comprehensive inherited mechanisms that are capable of influencing a wide variety of specific instinctive domains all at the same time, the chances of finding lawful theoretical connections between animal behavior, normative human behavior, and mental disorders are very small indeed.

3.2.2. Evolutionary Psychiatry

The search for evolutionary explanations of mental disorders began early in the history of psychiatry. In 1845, German psychiatrist W. Griesinger, defying the prevailing view that mental disorders represented moral failure and punishment by God, argued that they are "disorders of the brain" and that deterioration of cognitive functioning forms a continuum with mental health (Brüne, 2008, p. 2). And "many psychiatrists of the second half of the nineteenth century and early twentieth century saw strong implications of evolutionary theory for psychiatry. This was in part based on the growing acceptance of philosophical monism and abandonment of mind-body dualism . . . the confusion of natural selection with steady progress led to the view that mental illnesses were the result of abolishment of the forces of natural selection" (idem, p. 3).

In 1920, Kraepelin wrote: "If we . . . try to fit the expressions of insanity to the individual stages of personality development . . . it will be necessary to trace back manifestations of our psychological life to their roots in the psyche of the child, [and] of primitive man and animals. In this way we can discover to what extent certain illnesses reflect a recrudescence of emotions hitherto concealed in our individual and phylogenetic developmental history" (Beer, 1992, p. 529). In the same period, a large increase in the number of psychiatric inpatients was observed, a phenomenon that "led psychiatrists to conclude that the abolition of natural selection had induced a degeneration of the population," an idea that was in accord with Kraepelin's opinion "that mental illnesses were the result of a domestication-induced degeneration" (Brüne, 2008, p. 4).

The concern with the relaxation or disappearance of natural selection and resultant possible degeneration of the innate behavioral foundations of human behavior, initiated by Darwin's cousin Francis Galton's work on this subject, combined with a poor understanding of the mode of inheritance of mental and behavioral traits, led to the ill-fated eugenics movement. In its quest for improving the genetic qualities of human populations, eugenics recommended the introduction of drastic measures, including sterilization of the mentally and morally deficient. Compulsory sterilization became a legal policy in several European countries and the United States in the 1920s and 1930s and later in Japan as well (Lynn, 2001, p. 34). The ensuing public protest in liberal, humanitarian regimes, while of course quite understandable, nevertheless gave this complex subject (namely, how the social/cultural environment influences natural selection in human populations) a strong negative ideological and emotional charge that, in my opinion, ultimately hindered impartial scientific inquiry (Mayr, 1982, pp. 623–624). A contemporary trend in evolutionary sciences, called "niche construction," attempts to fill exactly this gap in our understanding—that is, how the changes by animals and humans effected on their environment may influence natural selection.

The growing interest in the second half of the twentieth century in the role natural selection plays in normative human behavior led to the evolutionary approaches already discussed—sociobiology and evolutionary psychology, as well as to human ethology, human behavioral ecology, and the gene-culture coevolutionary theories. It also led to renewed attempts to understand certain aspects of mental disorders according to evolutionary principles. Beginning in 1996 (to the best of my knowledge), several works attempted to integrate data from clinical psychiatry, research, and psychotherapy with an evolutionary perspective. I will mention three of them, which constitute the main sources of the present short account on this subject: Anthony Stevens and John Price's *Evolutionary Psychiatry* (1996), Michael McGuire and Alfonso Troisi's *Darwinian Psychiatry* (1998), and Martin Brüne's *Textbook of Evolutionary Psychiatry* (2008).

Evolutionary psychiatry builds hypotheses and interpretations about the interrelationship between disordered mental functioning and evolutionary considerations on several bodies of knowledge (facts and theories). The most important of these, in my understanding, are:

1. The *psychoanalytic and psychodynamic theory* of mental functioning, or more specifically, of the unconscious instinctive motivations underlying overt behavior. This approach is especially prominent in the work of Antony Stevens and John Price.
2. The scientific achievements of *biological psychiatry*—that is, the evidence of close interdependence between localized brain events (electrical, chemical, etc.) on the one hand, and behavior and subjective experience on the other. While brain anatomy, brain development, and physiological functioning can be explained only in terms of biological reasoning, the close interdependence between them and mental and behavioral manifestations fosters the interpretation of the latter, too, according to evolutionary principles. Some of the specific topics in this respect are:
   - Correlations between drug-induced neurochemical events in the brain, on one hand, and mental events, both pathological and normative, on the other.
   - Behavioral and experiential consequences of localized brain stimulation or lesion due to organic pathology (like temporal lobe epilepsy), experiments during brain surgery (Calvin & Ojemann, 1994), or experiments on laboratory animals.
   - Correlations between measurable functional states of circumscribed brain areas (measured by functional neuroimaging techniques) and mental events, normative or pathological.
3. The theoretical formulations of the biologically adaptive foundations of normative human behavior by *sociobiology, evolutionary psychology*, and related disciplines.

4. The insights of *human ethology*, the application of comparative animal ethology to human behavioral research, but with a wider scope that also includes socially induced behavior with evolutionary connotations (Eibl-Eibesfeldt, 2010, p. 4).

5. Most importantly, perhaps, the findings of *psychiatric genetics*, which strongly support a considerable degree of inheritance for mental disorders (Plomin, De Fries, McClearn, & McGuffin, 2008), as well as the findings of *psychiatric epidemiology*, which also frequently support the aggregation of mental disorders in biological relatives. (It seems needless to point out that the presence of genetic inheritance is an unquestionable indication of the involvement of phylogenetic evolutionary processes. We have no other scientific framework with which we can inquire and interpret the mechanism of gene transmission from one generation to the next.) The most salient examples are the worldwide occurrence and comparable prevalence of several major psychiatric disorders across populations with different cultures, different levels of economic and technologic development, living in different ecological conditions; the differential prevalence of mental disorders according to sex and maturational stage; the strikingly modest effect of the "shared environment" in the causation of most common forms of psychopathology, a finding which speaks against cultural causation (Plomin et al., 2008, pp. 306–309).

In the following, I describe concisely some common *evolutionary mechanisms* that can potentially lead to dysfunctional behavior, as hypothesized by evolutionary psychiatric literature. It is not my intention to give an exhaustive account on this topic, only to illustrate the characteristic themes and ways of reasoning:

3.2.2.1. The Genome Lag Hypothesis

The genome lag hypothesis (see also section 3.2.1.2) was originally suggested by English psychiatrist and psychoanalyst John Bowlby (Brüne, 2008, pp. 26–27), and was taken over by mainstream evolutionary psychological theorizing. It maintains that specifically human instinctive predispositions developed in the Pleistocene period (from approximately 1,6 million years ago till the advent of agriculture and permanent settlements about 10,000 years ago). During this period, humans lived in small, kin-based hunter-gatherer groups.

After the advent of agriculture, it is thought, albeit not unanimously (see Odling-Smee et al., 2003, p. 368), that culture progressed much more quickly than natural selection and that this asynchrony led to a situation in which modern humans have to tackle the requirements of a complex social environment with behavioral strategies built on instinctive predispositions that were adaptive in the environmental circumstances of the Pleistocene era. Examples supporting this reasoning are the specific fears of various natural dangers and lack of instinctive fears of dangers of the modern environment as mentioned previously (Brüne, 2008, p. 229). Another example is *xenophobia* (fear of strangers), which has an undeniable genetic base. It appears in human infants around the age of six months and is hypothesized to be an adaptive reaction to the infanticidal potential of strange males (known to exist in several animal species, too, including some closely related to humans) (Brüne, 2008, pp. 77–78). Xenophobia in mature humans means fear of (and aggressiveness toward) strange people, particularly toward representatives of other races, ethnic groups, or religions. In the form of tribalism, it had a fitness-enhancing role in our evolutionary past and no doubt continues to flourish in our age of globalization (Wilson, 1998, pp. 253–254).

Stevens and Price illustrate the deleterious effects of a drastic change from a natural to an artificial, man-made environment with the case of baboons transferred together from several different troops to a zoo. The ensuing vicious battles for dominance led to fatalities. Another example concerns a group of hunter-gatherers from Uganda transferred to an agricultural settlement: "They rapidly became demoralized, depressed, anxious and ill, and behaved with psychopathic indifference to their children and their spouses" (Stevens & Price, 2000, pp. 30–31). However, these examples reflect the consequences of a *sudden* environmental change, while the human environment changed much more gradually during the last 10,000 years and even more so during the Pleistocene era, allowing for at least partial adaptation to the changing environmental conditions through mainly genetic but also acquired means.

Stevens and Price ask some embarrassing questions regarding the genome lag hypothesis: "How is it that so many of us remain disorder-free? And how may we explain that psychiatric disorders exist and are recognized in hunter-gatherer societies still living in environments that closely approximate the EEA [Environment of Evolutionary Adaptedness in the Pleistocene period]" (Stevens & Price, 2000, p. 39). That second question hints at an *earlier origin of mental disorders than the end of the Pleistocene*. This is in accordance with the present theory, which traces back the beginnings of the ecological dominance, RfNSPs, and the resultant diversification of instinctive predispositions to the beginning of the Pleistocene era, not to its end. The first stone tools, markers for ecological dominance, appeared 2.5 million years ago, and the first evidence of the use of fire 1.6 million years ago (Buss, 2008, p. 21).

My view of the genome lag hypothesis is the following: While we certainly possess instinctive predispositions selected during our evolutionary past (I refer not only to instincts acquired during human evolution but instincts we share with other mammals, with birds, and even reptiles), due to the relaxation of selection pressures, they became considerably diversified in their strength, degree of differentiation, interrelatedness, and other parameters—a matter that is discussed in chapters 5 and 7. They are only predispositions, which by themselves can induce overt behavior only after being integrated with acquired knowledge. On the other hand, as I have emphasized, the "modern environment," composed of very different "social niches," cannot be regarded as a unitary entity. Therefore, it cannot be hypothesized that only one or a very restricted number of ways lead to successful adaptation. The great diversity of both the individual, innate predispositions, as well as the environmental niches from which one may choose (particularly in modern egalitarian societies), boils down to the question professional career counselors, for example, ask: What occupation and working environment from among those that are available suit the individual's innate and acquired mental and physical capabilities (Brown, & Associates, 2002, pp. 3, 154–156, 260–261, 379)? The idea of a uniform natural environment even during the Pleistocene period, with uniform natural selective pressures, must be questioned, given the diverse habitats early humans occupied after the two major migrations out of Africa, as well as the great climatic changes during the ice ages and interglacial periods.

3.2.2.2. A Hypothesis in Evolutionary Psychiatric Literature That Recognizes the Pathogenic Potential of the RfNSPs

When McGuire and Troisi (1998) enumerated possible evolutionary explanations of mental disorders, they included one they call "minimal selection pressure": "Conditions such as dyslexia, which might go unnoticed in nonliterate societies, or conditions that appear after the critical reproductive years, such as postmenopausal depression, late-life depression in males, and Alzheimer's disease, might be explained in this way" (30). Deleterious genes that manifest their effect only *after* the reproductive period is over (and consequently have been already transmitted to the next generation) indeed escape the weeding-out effect of natural selection. To my mind, this is one of the more convincing explanatory hypotheses of the exclusively old-age disorders, physical or mental, with a hereditary base. Survival into old age, however, is itself a consequence of the shielding effect of the modern, technologically advanced social environment against natural selective pressures.

As to the genes or gene configuration underlying dyslexia in nonliterate societies, in my opinion they have to be qualified as *neutral* gene recombinations or mutations. Having no positive or negative effects on adaptation, they are transmitted to the next generation by genetic drift.

3.2.2.3. Dimensional Instead of Categorical Conceptualization of Mental Disorders

One form of natural selection, *the stabilizing or normalizing one*, acts on traits induced by multiple genes (that is, by *polygenic inheritance*). In this kind of inheritance (discussed in more detail in 4.1.1.2), the involved genes' summative effect does not lead to a qualitative change but to continuously varying gradations of the same trait. Natural selection's role in this instance is to keep to a minimum deviation from the desirable optimum. Deviation in any direction from the optimum will reduce fitness (Bell, 2008, p. 189). For example, it was discovered that birds with a longer or shorter wingspan than the optimum tend to die more often during migration than birds with an average wing length (Stevens & Price, 2000, p. 40).

Inherited traits concerning behavior are exclusively of the polygenic type. In humans, too: "Adaptive strategies such as flight, fight, freeze, withdrawal, dominance, submission, and attachment will be *normally distributed* through every population. Most individuals will be placed in the middle of the distribution curve and will be reasonably well adapted to prevailing environmental conditions. Exaggerated or inadequate endowment with these strategies might account for psychopathology among individuals placed toward the tails of the distribution of adaptive traits. . . . It is theoretically possible that tendencies to anxiety, depression, sensitivity to adverse events, to criticism, and so on, could be similarly distributed. This approach is linked to the *critical threshold* concept: those too sensitive to criticism go under; those too insensitive take too little care; those with an *anxiety neurosis* will suffer the symptoms of arousal in response to a minimal stimulus, while those with an *anti-social personality disorder* will experience no symptoms of arousal when confronted with a major threat" (Stevens & Price, 2000, p. 40, emphasis in the original). In the same vein: "Every single sign or symptom has its functional counterpart as part of a set of evolved psychological mechanisms. Accordingly, dysfunctions (or psychopathological signs and symptoms, as we may call them) can be described as *extremes of variation of normal adaptive mechanisms*" (Brüne, 2008, p. 90, emphasis added).

Such views are very close to the standpoint of the present theory, with the stipulation that we mean exclusively the *inherited predisposition to dysfunctional behavior* (in normative environmental conditions) and not the full-blown clinical disorder.

In fact, *all* the comprehensive instinctive-mechanism categories and subcategories that, as this text proposes, underlie mental disorders possess a graded continuum with a normative range at the middle and dysfunctional-behavior-inducing potential at the extremes.

3.2.2.4. Balancing Pleiotropic Effects

Pleiotropy means that a single gene determines or contributes to multiple traits in the phenotype. In certain kinds of pleiotropy, some of these multiple effects may be deleterious to the individual organism, while other effects are fitness-enhancing. In heterozygote carriers of recessive genetic disorders, only the desirable trait is expressed in the phenotype, while in homozygous individuals it is the maladaptive trait that is expressed. The classic example from medicine is the gene that in its heterozygote form confers increased protection from malaria, while in homozygous form causes sickle cell anemia. The concept of balancing pleiotropic effect was used in evolutionary psychiatry to build explanations about some mental disorders—those that lead to considerably lowered fitness through maladaptive behavior but, contrary to evolutionary expectations, are not eradicated by natural selection from the gene pool (Brüne, 2008, p. 38).

In accordance with this reasoning, the genetic predisposition to schizophrenia was hypothesized to be a "a trade-off of human language acquisition or creativity," or, alternatively, conferring "reduced risk of cancer in relatives of schizophrenic patients" (idem, p. 193).

And: "Perhaps the clearest example of pleiotropy would be that of greater fecundity among females during reproductive years (the adaptive trait) coupled with late-life (post-reproductive) vulnerability to conditions such as depression. . . . In this example, late-life conditions would have a reduced chance of being selected against, while increased fecundity would ensure the pleiotropic gene(s) will be present in subsequent generations. Similar interpretations have been offered for unipolar depression and bipolar disorder, where a greater-than-chance occurrence of highly creative persons (the adaptive trait) . . . has been reported" (McGuire & Troisi, 1998, pp. 29–30). Other experts in evolutionary theory have doubts about the fitness-enhancing potential of creativity (Cochran & Harpending, 2010, pp. 126–172). Pleiotropy is discussed in more detail in section 4.1.1.1.

3.2.2.5. Evolutionary Explanations Regarding Circumscribed Mental Disorders or Groups of Disorders

In this section, I summarize the explanatory hypotheses offered by evolutionary psychiatry concerning discrete clinical disease entities and, when possible, relate them to the views espoused in the present theory.

3.2.2.5.1. Anxiety Disorders

"Anxiety is thought to have evolved as a system warning . . . that high-priority biological goals may be jeopardized" (McGuire & Troisi, 1998, p. 106). In Martin Brüne's definition, anxiety disorders "reflect exaggerated responses to internal and external signals of perceived danger or threat. The autonomic part of the anxious response pattern prepares the organism for one of several behavioral options to terminate the anxiety-eliciting situation, namely flight, immobility, submission, or aggression" (Brüne, 2008, p. 228).

While the present theory is by and large in agreement with these views, it makes some further distinctions. In the framework of this text, *anxiety is seen as the subjective experiential aspect of only a certain kind of instinctive reaction in the face of perceived danger. The term* anxiety *will denote here the subjective experience in situations in which the behavioral pattern activated is inadequately differentiated genetically—it is diffuse, in the present work's terminology—and thus incapable of inducing, or serving as the instinctive foundations for, an effective behavioral reaction* that has a good chance of coping with the external danger that elicited it. The subjective experience of a differentiated and effective avoidant response to threat or danger will be called here a *fear reaction* or, in milder cases, a *focused reactive mental arousal*. In humans, the reaction may be physical (flight, for example), but it is more frequently verbal, mostly reflecting acquired knowledge of how to cope with the respective situation (which, as a rule, is of a social nature). Even in this latter case, in my opinion, the reaction is built on innate, instinctive arousal manifesting itself frequently in small involuntary moments or gestures (for example, away from the source of threat), and, in physiological arousal preparing the body for physical activity, betraying the reaction's instinctive foundations. This subject is elaborated upon in the sections on "reactive behavior" (5.3.3 and 5.4.2.1).

Evolutionary psychiatry interprets individual anxiety disorders as specific instances of the fear or anxiety reaction. A *panic attack*, for example, is "seen as the extreme version of a preparatory set of physiological changes typical of immediate flight or escape behaviors" (Brüne, 2008, p. 232); *posttraumatic stress disorder* on a physiological level represents "a hyperactivity of the alarm system," which, coupled with "persistent feelings of impending danger," precludes the return to the pre-traumatic appreciation of the level of dangerousness of the environment (idem, p. 234).

These views are convergent with those proposed in this text, as becomes clear when the respective clinical disorders are discussed in detail in chapter 7.

3.2.2.5.2. Personality Disorders

Evolutionary psychiatry also views personality disorders mainly according to a dimensional view rather than as discrete categorical entities; that is, they "are conceptualized as extremes of normal variation of strategies, which are pursued in a rigid, inflexible, or excessive way" (Brüne, 2008, p. 261). Persons with antisocial personality disorder for example are hypothesized to use the adaptive strategy of nonreciprocation or "free riding." In a population consisting of cooperative and reciprocating people, the strategy of nonreciprocation may pay off, particularly if the risk of detection is low and the nonreciprocator remains socially mobile (Stevens & Price, 2000, pp. 88–89).

While the argument for the existence of such a strategy certainly seems convincing, at least in subgroups of antisocial psychopaths, it addresses only one possible symptom of this disorder, overlooking other characteristics—for example, the good "theory of mind" abilities of non-violent psychopaths, which detects the most suitable victims; the lack of anxiety and remorse; an excessive predisposition, in some individuals, to aggressive coping strategies; and, above all, the wider context of personality disorders in general. In order to illustrate how varied, even contradictory, the clinical signs of personality disorders may be, I will quote here some characteristics of obsessive personality disorder that are the complete opposite of the behavioral strategy in antisocial personality disorder:

> Perfectionism which interferes with task completion. Excessive devotion to work and productivity to the exclusion of leisure activities and friendships. . . . Over consciousness, scrupulousness, and inflexibility about matters of morality, ethics, or values. (Svrakic & Cloninger, 2005, p. 2086)

This is not a criticism of evolutionary reasoning or of psychiatric nosology; these two personality disorders are accurately described by current diagnostic systems. My argument is only for the precedence of defining first the common characteristics of the more encompassing nosological entity of personality disorders in general (and for that matter, that of anxiety disorders, too) from the point of view of evolutionary theory *before* addressing the particular characteristics of each individual clinical category. This seems to me to be an effective way to avoid drowning in a sea of unconnected observational data.

Anxiety disorders and personality disorders each possess specific, well-known characteristics that distinguish these two comprehensive groups of pathology from each other. The distinguishing factors are, in fact, well illustrated by the earlier quotation from Stevens and Price regarding the respective reactions of these disorders to stressful events (section 3.2.2.3).

Anxiety disorders are, as we have seen, dysfunctional strategies to deal with perceived environmental danger. The individual perceives the danger as more intense, more overwhelming, more paralyzing (regarding efficient adaptive behavior) than would a normative individual.

The innate behavioral predisposition is predominantly the avoidant flight reaction (not freezing or counteraggression). The subjective experience is that of an intrapsychic tension and suffering induced by an agent *outside* the individual; that is, it is ego-dystonic.

On the other hand, personality disorders represent predominantly *"self-expressing" behavioral strategies at the expense of optimal social adaptation*, strategies that overstrain the immediate social environment, or at least strike surrounding people as strange and unusual forms of behavior, as in schizoid or schizotypal personality disorders. The individual feels that her behavior originates from her own natural inclinations to achieve biological and social goals, or at least to attain mental peace and delivery from the strain of social pressures, with relatively little regard for the critical and antagonistic reactions her behavior may trigger in surrounding people; that is, it is excessively ego-syntonic. In short, *the pressure for expressing internal (active) instinctive motives overpowers the suppressing, altering effect of the undesirable or inimical reactions of the social environment.*

To summarize the evolutionary reasoning that may account for the personality disorder/anxiety disorder dichotomy, I must first enumerate the three most complex instinctively fueled behavior patterns that this text addresses (detailed in sections 5.3.3 and 5.4).

1. Behavior fueled by *active instincts*, that is, instinctive motives serving directly self-centered life-sustaining and reproductive aims.
2. Behavior fueled by *self-defending, reactive, instinctive behavioral tactics* that are employed when an environmental disturbance or danger is sensed. These behavioral expressions (of which the basic ones are freezing, avoidance or flight, and counteraggression) are present in both solitary and social animals.
3. Behavior induced by the need to *alter* the preceding two kinds of behavioral predispositions in order to submit or conform to the imperatives of social coexistence. This kind of behavior is present only in evolved social animals and has attained a high degree of complexity and sophistication in humans.

A balanced interrelationship, adaptive alternation, or correct intermingling of these three comprehensive, instinctively underpinned behavioral strategies is critically important for fitness in evolved nonhuman social animals. It merits mentioning that both Freud and K. Lorenz hypothesized the existence of a higher-order brain center that fulfills the above or similar tasks. Freud called it "ego," while Lorenz named it "the superior command locus" (Lorenz, 1982, pp. 204–210).

In this text, it is argued that a relaxation of natural as well as intragroup selection pressures in humans affected the strict, fitness-enhancing relationship between these three comprehensive behavioral categories, leading to various imbalances between them, which, if pronounced enough, may be categorized as mental disorders. The predominance of active instinctive motives (and sometimes of well-differentiated individual reactive ones, as well) over submission to, compliance with, or internalization of the rules of social coexistence constitutes in our theoretical framework the innate predisposition for personality disorders.

On the other hand, the predominance of high levels of anxiety (that is, diffuse reactive arousal), or, alternatively, an unusually strong inclination to submit or conform to social pressure, underlies most kinds of anxiety disorders. Both excessive levels of anxiety and an inclination to conform excessively tend to suppress the expression of active as well as egocentric, self-defensive reactive motives to a more than desirable degree. This line of reasoning will be detailed in chapters 5 and 7.

3.2.2.5.3. Depression and Bipolar Disorders

Depression is a good example, in my opinion, of how syndromes that resemble each other descriptively can be brought about by widely different evolutionary mechanisms. In fact, the literature of evolutionary psychiatry recognizes all three instinctive mechanisms related to depression described in this text, though without clearly linking the right evolutionary mechanism to the correct depressive pathology.

One of the hypothesized explanations for depressive states is related to frustration of life-sustaining or reproductive instinctive drives: "Depression is thought to have evolved as a somatic indicator that biological goals have not been or are not being achieved" (McGuire & Troisi, 1998, p. 109). The subjective, emotional state is a "severe inhibition of both consummatory and appetitive pleasures" (idem, p. 108). In the present theory, the proposed evolutionary mechanism underlying this kind of depression is one of the three basic instinctive responses to frustration of life-sustaining or reproductive goals (instincts of the active kind), that is *dysphoria.* This kind of depression conveys the impression of intensi-fied instinctive activity manifesting itself as psychomotor agitation ex-pressing intense mental pain. The innate aspect of this state is seen in its purest form in young offspring suffering from biological deprivation (for example, hunger or cold) or (in humans and more evolved animals) suffering from being separated from the mother. These situations show clearly the instinctive roots of this mental state as well as its original fitness-enhancing function, which is to mobilize help from the appro-priate mature conspecific, as a rule the mother. In the present work, this kind of depression is named *dysphoric depression,* and discussed in more detail in section 5.2.3.

Another, in my opinion, very different kind of adaptive evolutionary mechanism leading to depressive states concerns "physiological slowing" (idem, p. 156) in order to "conserve energy" (idem, p. 160). This kind of depression (which in the official psychiatric nomenclature is not separated from the previous one) is interpreted in the framework of the present theory as originally representing adaptation to seasonal variations in the intensity of life-sustaining and reproductive instinctive activities. Winter depression may be considered the paradigmatic example of this kind of depression.

Furthermore, I suggest that the relaxation of natural selection pres-sures in humans and resultant diversification also produced other, related forms of retarded depression, like that encountered in bipolar disorders or post-psychotic depression. (I enlarge on this topic in section 5.1.) That both retarded depression and bipolar disorder represent strong fluctu-ations in the intensities of active instinctive drives is well recognized in evolutionary psychiatric literature (Stevens & Price, 2000, p. 65):

> From the evolutionary standpoint . . . it would seem that low mood and high mood evolved as means of adaptation to alterations in one's resource holding power and chances of reproductive success . . . if success is in one's grasp, then it is best to pull out all the stops and go for bust. This is the essence of the depressive and manic adaptations.

It seems clear that at Earth's higher latitudes, the main determinants of fluctuations in the abundance of life-sustaining resources (light, tem-perature, the availability of vegetation and prey animals, etc.) are the seasonal cycles, which last several months and thus parallel well the average duration of major affective episodes. A related interpretation in evolutionary psychiatric literature is that manic and retarded depressive states represent dominant and submissive behaviors, respectively. Based on ethological analysis, Brüne argues that: "although pathologically exaggerated in depression, these nonverbal behaviors aim at reducing aggression of others . . . and avoiding harm by displaying deescalating appeasement strategies in situations of (perceived) defeat or inferiority" (Brüne, 2008, p. 211). And: "In contrast to depression, mania represents the pathologically extreme variation of dominance behavior and competition-enhancing strategy at the behavioral, emotional and cognitive level" (idem, p. 214).

While Brüne as well as Stevens and Price recognized, rightly in my opinion, that depression and mania in bipolar disorders represent the extremes of ebbs and tides in active instinctive intensity, what is (again) lacking here is the indication of the natural selective mechanism involved. We are talking about episodes of psychopathology lasting as a rule several months. Some of them, like winter depression and spring hypomania or the seasonal bipolar disorders, follow closely seasonal periodicity. Moreover, light therapy ameliorates winter depression. (The length of daily light entrains the circannual clock.) These are strong hints, in my opinion, of the possibility that these two phenomena, depression and mania on one hand and seasonal variations in environmental resources on the other, must be closely related.

Relaxation of the original natural selection pressures (by such shielding effects of the modern environment as abundance of food in winter, heating appliances, and artificial lighting during the dark hours of the day), and the resultant diversification, may explain additional forms of fluctuating affective psychopathology that do not follow the alternation of seasons. It may also explain why the majority of humans (with genetic ties to those populations who, during human evolution, lived at high latitudes, where seasonal fluctuations were prominent) lost this adaptation that adjusted the fluctuation of instinctive intensity to seasonal variations in the amount of natural resources and instead adapted in this respect to the conditions of the modern human environment, which requires stable, all-year-round functioning.

The last explanatory hypothesis regarding depression in accordance with evolutionary reasoning is probably the oldest one: "In 1936, Lewis suggested that depression is a way of eliciting help from others" (McGuire

& Troisi, 1988, p. 160). This hypothesis, in the present theoretical framework, seems to be related to *the relaxation of strong intragroup selection pressures, that is, to a loosening of the requirement to function optimally on a permanent basis* (so essential in prehistoric hunter-gatherer groups during their quest for survival), as well as from the requirement not to be a burden on these groups' meager resources. The relaxation of these kinds of intragroup selection pressures may at first have relieved selection against the inherited predisposition for mild and short depressive episodes of the retarded or dysphoric kind. The relaxation of these selection pressures increased progressively with technological and related cultural advances.

However, this evolutionary mechanism is not specific to depression but acts in the case of any kind of temporary handicap in functioning of a physical or mental nature, with or without an inherited base. This seems true especially if it is accompanied by expressions of considerable subjective suffering, which may resemble those found in immature offspring, touching thus on the surrounding individuals' (diversified) parental or nepotistic instincts.

McGuire and Troisi mention all three evolutionary mechanisms involved in depression but without distinguishing between the widely different nature of their phylogenesis: "The symptoms of depression may warn a person that past or ongoing strategies have failed [first mechanism, leading to dysphoria]; physiological slowing and social withdrawal may remove a person from high-cost, low benefit social interactions [second mechanism, retarded depression as a phylogenetic response for meager environmental resources of seasonal nature]; and signaling one's state to others may initiate others' help without requiring long-term payback [third mechanism—that is relaxation of intragroup natural selection pressures] (McGuire & Troisi, 1998, p. 156).

These three evolutionary mechanisms are discussed in more detail in sections 4.6., 5.1., and 5.2.

3.2.2.5.4. Schizophrenia

The great paradox posed by schizophrenia to traditional evolutionary reasoning is well recognized in evolutionary psychiatric literature: "Schizophrenia presents a major challenge to the explanatory powers of evolutionary psychiatry. How can the bizarre qualities so typical of the disorder be derived from underlying characteristics to be found in normal biology and psychology? How is it that the genetic predisposition that results in schizophrenia in some people can result in adaptive traits in others? For adaptive the predisposition must be, otherwise it could no longer be with us. Far from eliminating it, natural selection has fixed it as an enduring component of the human genome" (Stevens & Price, 2000, p. 146).

Since the evolutionary foundations of schizophrenia proposed in this work have no convergent threads of reasoning with any of the hypotheses in the field of evolutionary psychiatry, its elaboration is postponed until sections 5.3, 7.2.7, and 7.2.8.1.

3.2.3. Additional Theoretic Approaches to Human Evolution: Cultural Evolution, Gene-Culture Coevolution, and Niche Construction

Because the theory proposed in this text belongs formally to the category of gene-culture coevolutionary and related theories, it is useful to give a short account of this topic, drawn primarily from Laland and Brown's comprehensive treatment of the approaches to (normative) human behavior and culture based on evolutionary reasoning, *Sense and Nonsense: Evolutionary Perspectives on Human Behavior* (2011).

In spite of the overlap between these three classes of theories—cultural evolution, gene-culture coevolution, and niche construction—all of which deal with the interrelationship or interaction between natural selection and the effect of the changes brought about by organisms on their environment, each has particular characteristics that may be formulated in the following way:

1. *Theories on cultural evolution* deal chiefly with the question of how cultures possibly evolved; the similarities, parallels, and differences between genetic and cultural evolution, and the ways relevant information is transmitted, preserving useful, adaptive information and selecting out useless or maladaptive information.

2. The main focus of *gene-culture coevolution* theories is on ways that the cultural process influences or alters natural selection in human populations and, conversely, how the long phylogenetic evolution of humanity (our concern here is chiefly with the

innate behavioral predispositions) constrained certain directions of cultural development while catalyzing others.

3. The *niche construction* theory discusses in a wider context the ways environmental changes are brought about by living organisms, from the simplest to the human, and how this activity affects and changes the character of natural selection pressures on the same kinds of organisms, as well as on unrelated ones. Of course, for our purposes, the most important aspect of this theory concerns the effects of human culture on the physical, biological, and social constituents of the human environment, and how these effects in turn change the natural selective pressures on human populations.

Despite the fact that the present theory belongs to the aforementioned classes of theories (particularly those mentioned under points 2 and 3), their relevance to the present work is restricted for at least two reasons. First, these theories deal with the culture-gene interrelationship in the sphere of *normative* human anatomy, physiology, and behavior, not with dysfunctional or pathologic behavior, the main focus of this work. Second, the above theories focus on positive, directional selection pressures and disregard or downplay the importance of the relaxation from these selective pressures, which is the main evolutionary mechanism on which this work concentrates. A partial exception to this is the niche construction theory.

Nevertheless, several lines of reasoning in the gene-culture interrelationship, as well as niche construction theories, do converge with or support the theory proposed here. I refer first of all to the almost universal agreement that the changes organisms (mature phenotypes) effect on their environment alter in certain circumstances the selection pressures on subsequent generations of conspecifics.

3.2.3.1. Theories on Cultural Evolution

Theories of cultural evolution presume that culture evolves in a manner more or less similar to natural evolution; that cultural elements come in units, like genes; that they can be transmitted according to principles more or less similar to those by which genes are transmitted in the course of natural selection; and that their retention and perpetuation depends on the selection of the fittest or most adaptive cultural traits. Moreover, human cultural achievements evolve from simple toward more complex ones, like some forms of life in nature. Social Darwinism, in this respect, can be seen as an early version of that group of theories.

British biologist Richard Dawkins advanced the most extreme theory of the similarity between natural and cultural evolution. (He also proposed the most extreme gene-centered theory of biological evolution in the same famous and controversial book, *The Selfish Gene*.) Dawkins uses the term *memes* to refer to basic cultural elements that in his opinion are transmitted much in the same way as genes: "Just as genes propagate themselves in the gene pool by leaping from body to body via sperms or eggs, so memes propagate themselves in the meme pool by leaping from brain to brain via a process which, in the broad sense, can be called imitation. If a scientist hears, or reads about, a good idea, he passes it on to his colleagues and students. He mentions it in his articles and his lectures: If the idea catches on, it can be said to propagate itself, spreading from brain to brain" (Dawkins, 1989, p. 192). Other proposed similarities between genes and memes are: "variation, heredity, and differential fitness, the three characteristics necessary for evolution" (Laland & Brown, 2011, p. 140). Several more parallels between cultural and biological evolution exist that are not relevant here. The interested reader can consult Laland and Brown's *Sense and Nonsense* (2011, pp. 145–148).

In spite of the seeming similarities between genetic and cultural transmission, Dawkins's theory of memes was severely criticized on a number of grounds (idem, pp. 157–164): Cultural information does not comes in neat, standardized packages, as the meme concept implies; what is transmitted is acquired knowledge, not inherited traits (in this respect his model resembles Lamarckian inheritance, long discredited in biology); the direction of propagation of genetic inheritance is exclusively from parents to offspring, while cultural information can be transmitted in any direction between people, generations, groups of people, and even, in a way, disconnected from individuals through the media, books, etc. However, the most important reservation concerning the claim that memes resemble genes, in my understanding, is that genes are transmitted directly, obligatorily from one generation to the next (that is, the recipient of the genes cannot do anything about it), while *cultural information, arriving to the individual as external stimuli to the senses, is reconstructed and reinterpreted by each individual brain.*

But if that is so, how can we explain that cultural traditions (like religions, forms of government, local ethnic traditions, aspects of scientific practice, etc.) frequently survive seemingly unchanged for long periods of time? The most evident answer is blind acceptance or imitation, of course. But what should we choose to imitate from the multitude of often contradictory information and other stimuli of the modern cultural environment? Proposed answers to this question are: "Copy the most successful individuals," "Copy in proportion to the demonstrator's payoff," and "Conform to the majority" (Laland & Brown, 2011, p. 148).

Charles Lumsden and Edward Wilson (1981) proposed a theory partially resembling that of Dawkins in their book *Genes, Mind and Culture*. They called the units of cultural material transmitted *culture-genes*. However, in their theory, besides the *parallels* between cultural and genetic transmission, their *interaction* was emphasized as well: "Culture is generated and shaped by biological imperatives, while biological traits are simultaneously altered by genetic evolution in response to cultural innovation" (Lumsden & Wilson, 1981, p. 1). Furthermore: "Genetic biases will evolve that affect what cultural information is adopted . . . weak genetic biases can be amplified by conformity of behavior . . . and . . . culture can both slow down and speed up the rate of genetic change" (commented by Laland & Brown, 2011, p. 167). What are these specific "genetic biases" affecting social behavior in Edward Wilson's opinion? They include: "division of labor between sexes, bonding between kin, incest avoidance, other forms of ethical behavior, suspicion of strangers, tribalism, dominance orders within groups, male dominance overall, and territorial aggression over limiting resources. *Although people have free will and the choice to turn in many directions, the channels of their psychological development are nevertheless . . . cut more deeply by the genes in certain directions than in others*" (Wilson, 1994, pp. 332–333, emphasis added). That last sentence is a felicitous metaphor, to my taste, for the way instincts are supposed to influence human behavior—that is, as an inclination, predisposition, or aspiration, not as obligatory behavior patterns.

3.2.3.2. Gene-Culture Coevolution Theories

Culture and biology are never completely dissociated: "Cultural learning is always reliant on biologically evolved knowledge-gaining structures" (Laland & Brown, 2011, p. 166). This is true even when the cultural phenomenon seems to be completely irreconcilable with or contradictory to any biological rationale, like altruistic suicide (the Japanese kamikaze pilots of the Second World War, for example), or the celibacy of priests in certain religions.

Therefore, the central concern of gene-culture coevolutionary theories is the way natural selection and cultural phenomena interact. In the following, I mention several examples of how genetic inheritance can constrain culture and how culture can change natural selection (and, in consequence, gene transmission).

The human genome project discovered evidence of quite recent genetic evolution in certain physical traits through directional selection. The most frequently mentioned ones are:

1. *Changes in several parts of the digestive system's anatomy and physiology after the introduction of agriculture some 10,000 years ago*, which produced new kinds of food (like milk and its processed products) and better ways of food preparation. Hard evidence exists, for example, that tolerance to lactose, which is lost in most humans and mammals in general after weaning, is preserved throughout the whole life span in a high percentage of the European population due to a particular mutation that is only a few thousand years old (Cochran & Harpending, 2010, p. 22; Laland & Brown, 2011, pp. 177–179).

2. *Adaptive changes in the immune system due to a directional selective response to the appearance of devastating contagious diseases.* These epidemics were the consequence of a chain of causes and effects: The introduction of agriculture led to a rapid increase in food resources and, in consequence, to the size and density of the population, which led inevitably and unavoidably to crowded permanent settlements (Cochran & Harpending, 2010, pp. 85–91).

3. That cultural progress has an effect on the selection of *new mental traits* is less well substantiated, but, in my opinion, some quite convincing arguments support this claim. It is reasonable to presume that successful functioning in an agricultural setting needed considerably different personality traits than those suited for good survival and reproductive prospects in a hunting-gathering lifestyle—for example, *"the ability to defer gratification*

for long periods of time. This was a practical requirement for farmers, since they had to save a portion of their crop for seed and some of their domesticated animals for breeding stock. This wasn't easy. Food was often shortest just before sowing" (Cochran & Harpending, 2010, p. 114, emphasis added).

Other personality traits whose spread in the population was facilitated by the agricultural way of life are: "patience, self-control, and the ability to look to long-term benefits instead of to short-term satisfaction"; miserliness in order to accumulate property, which increases "an individual's fitness or that of his children and relatives"—traits not required for a foraging way of life. Another such trait is the propensity for working hard on a permanent basis, as opposed to the hunting-gathering lifestyle, which requires "bursts of strength in war and hunting." This last presupposition also has genetic support. The differential prevalence of genes that induce the production of a specific protein needed for the fast-twitch muscles versus those genes that increase aerobic efficiency of the muscle activity and confer endurance changed in favor of the latter (Cochran & Harpending, 2010, pp. 116–118). Industriousness and hard work on a permanent basis may be considered adaptive for farmers, while periodic "laziness" (after short-lived bursts of intense activity) was adaptive for hunter-gatherers, since it conserves energy (idem, pp. 114–117). The relevance of the newly evolved personality traits for pathological developments (when they become excessive and interfere with good social adjustment) is discussed in chapter 7, in the section on obsessive-compulsive personality disorder.

The second example that *culture-induced natural selection* brought about regarding changes in heritable personality traits is *selection for increased ability to cooperate*. By "culture-induced natural selection" I mean the transmission of genes from generation to generation according to the rationale of natural selection—enhanced fitness in a given environment, as we have seen in the above examples—but as a result of cultural alterations in that environment. It will be sharply distinguished from *cultural selective practices with an impact on gene transmission*, which do not follow the rationale of natural selection. This subject will be detailed in section 4.6.

Cooperative behavior cannot be imagined without inherited foundations; in a simple form, it is already widespread in social insects.

Moreover, it has to be hypothesized that from the dawn of human evolution, a strong selective pressure for intragroup cooperation of progressively increasing complexity has been present. Cooperative behavior lies at the foundations of human culture. Learning and practicing cultural traditions necessitates a cooperative form of interaction between individuals. Cooperative behavior between genetically unrelated individuals (sometimes possessing widely different mental and physical abilities) made possible cooperative hunting, food sharing, and exchange of merchandise in prehistory that evolved gradually into such basic constituents of contemporary, technologically advanced societies as "large scale organizations and nation states of the Western world" (Laland & Brown, 2011, p. 179).

More recent theories on gene-culture coevolution attempt to explain the evolution of *human reciprocity* (which comprises cooperation, norm-abiding behavior, and *"altruistic punishment,"* that is, a willingness to punish transgressors, even if it incurs an exaggerated cost to the punisher) by a kind of *group (instead of individual) selection*: "The cultural groups that exhibited strong reciprocity are argued to have outcompeted those cultural groups that did not exhibit this behavior, resulting in selection for genes underlying prosocial behavior, and eventually leading to a universal trait of prosociality being exhibited across the whole species" (idem, p. 179). However, in biology, *intergroup* selection is considered a weak form of selection, easily overpowered by selection for individual rather than group fitness (Williams, 1974, pp. 92–96) as well as by intermingling between competing groups (Laland & Brown, 2011, p. 180), and weakened by individuals with nonreciprocating behavioral tactics. Moreover, success in intergroup competition or hostility seems strongly dependent on the intensity and quality of the preceding *intragroup* selection for cooperation. As a consequence, instead of group selection along the rationale of genetic inheritance, Richerson and Boyd proposed an alternative form of selection named *"cultural group selection."* For the details of these two different approaches to group selection and their respective merits and problems, see Laland & Brown, 2011, pp. 179–183.

3.2.3.3. Niche Construction

Niche construction is a quite recent trend in evolutionary theorizing that studies the natural selection–altering effects of those organismic activities that bring about lasting changes in the environment. (In the case of humans, niche-construction theories strongly overlap with gene-culture coevolution theories.) The great importance of this mechanism is not appreciated properly in traditional evolutionary theorizing, which accentuates a unidirectional kind of selection in which the environments' parameters are seen as more or less rigidly given and the organisms have to accommodate to them (Odling-Smee et al., 2003, pp. 1–32). Put more succinctly: "Organisms are assumed to adapt to their environments, but environments are not assumed to 'adapt' to their organisms" (idem, p. 239).

In the more extreme, gene-centered, view of evolution, the individual organisms (phenotypes) are considered the passive ephemeral vehicles for the transmission of the (almost) immortal genes—or, more precisely, of the information they contain (Dawkins, *The Selfish Gene*). However, it is clear that selection for survival and reproduction operates on and through the whole phenotype (not only the genotype), which is the product of an interaction between genotype, epigenetic processes that influence the expression of the genes in the phenotype, and a continuous bidirectional interaction between the organism and its environment during maturation and adult life, when reproduction becomes possible.

Niche construction theory adds an additional dimension to this already complex mechanism. Its underlying concept is that, in addition to being the recipients of environmental selection pressures, organisms actively choose the environment suited best to their genetic makeup and may change certain elements of that environment. *If those changes are far-reaching and enduring enough, they in turn may change that environment's selection pressures.* Moreover, these changed selection pressures are "inherited" (by a mechanism different from genetic inheritance) by later generations of the same species, or may influence other, genetically unrelated, living organisms, a process that, if all-encompassing enough, may be able to change the direction of some evolutionary trajectories. This reasoning led to the view of *dual inheritance*. Dual inheritance means that organisms inherit from former generations not only their genes but also the environmental changes effected by previous generations, which may in turn alter that environment's selection pressures. This second form of inheritance, particularly in humans, may be the result of activities based partially on acquired knowledge; that is, it is Lamarckian in nature. Acquired knowledge cannot be transmitted through genes to the next generation, but in the case of humans' outstanding scientific and technological achievements, for example; or cultural traditions that affect the choice of a mate or the average number of children in a family; or the way in which the physical environment is built, rebuilt, and perfected by countless former generations (as in a city or town, for example)—all of which unquestionably alter natural (and sexual) selection—is transmitted to successive generations by the mechanism of cultural inheritance.

The effect of niche construction activities of even very simple organisms may have been so profound that it changed the direction of some evolutionary pathways on a global scale. The following example may illustrate this point: "When photosynthesis first evolved in bacteria, particularly in cyanobacteria, a novel form of oxygen production was created. The contribution of these ancestral organisms to the earth's 21 percent oxygen atmosphere . . . highly likely . . . modified natural selection pressures, [which,] stemming from the earth's changed atmosphere, played an enormous role in subsequent biological evolution. For example, many organisms have evolved a capacity for aerobic respiration, and they also evolved other mechanisms . . . that protect cells against oxidation" (Odling-Smee et al., 2003, p. 12).

While the renewed interest in the subject of niche construction is quite recent, Darwin was already aware of the basic idea and one of its central implications—that complex organisms can evolve only on the foundation of simpler ones: "Darwin realized that in general the contingencies favored diversification: The very existence of one group of organisms created new niches for another group (as in the simple case of plants creating *ipso facto* the possibility of animals) and so on indefinitely. The level of complexity of organisms was thus a function of the level of complexity of the organic environment" (Howard, 1982, p. 82).

The human-made artificial environment thus has to be seen as a special case of niche construction. A centrally important question concerning the present theory is how this altered environment changed the natural selection pressures on human populations. According to the niche construction theory, it affected selection in two opposed ways.

The first way is similar to directional selection in nature: "Some acts of niche construction, known as *inceptive* niche construction, initiate a change in an environmental factor and lead to changes in allele frequency" (Laland & Brown, 2011, p. 172, emphasis in the original), which leads to a new trait or the modification of an existing trait. The examples given previously about how agriculture changed diet, which in turn modified functional and anatomical aspects of the digestive apparatus; how permanent, overcrowded settlements led to the spread of contagious diseases, which in turn altered the immune system; and how agricultural adaptation required a changed lifestyle, which selected for new inherited personality traits, belong to this category.

The second alternative means of changing selection pressures by niche construction is called *counteractive niche construction*. This category is especially relevant for the theory proposed in this work: "For example, when human beings light fires or put on more clothes in response to living in cold temperature regions, the temperature changes actually experienced are dampened relative to the external environment, and consequently the selection on the genes is weak [relaxed, in the terminology of this text]. . . . These activities, known as *counteractive* niche construction, oppose or nullify the effects of environmental change and function to protect organisms from shifts away from environmental states to which they are adapted" (Laland & Brown, 2011, p. 172, emphasis in the original).

Might this be a hint about how to approach a puzzling recurrent finding—namely, that humans have considerably *less* genetic variations than our closest animal relatives? "Even though chimps, gorillas, and humans diverged from one another about seven million years ago and are all consequently the same age, humans appear to be depauperate in genetic variation compared to our closest relatives" (Marks, 2002, pp. 86–87). This hypothesis, concerning the protective human environment's counteractive effect on genetic adaptation to varied natural selective pressures, is further discussed in the next chapter.

A consequence of this counteractive effect on allele frequency is related even more closely to the present theory: "One prediction from this cultural mitigation of selection is that one might expect more (of what would otherwise be) deleterious alleles in the humane gene pool than we would in the absence of cultural activities, such as genes associated with short-sightedness that are counteracted by spectacles." Therefore: "Hominid populations may have become increasingly divorced from local ecological pressures" (Laland & Brown, 2011, p. 172). And again: "It is widely recognized that culture can also shield low-fitness genetic variants from selection. . . . For instance, improved levels of health care and sanitation are examples of culturally mediated counteractive niche construction that mitigate selection against individuals with some gene-related disorders or susceptibilities, who may survive and reproduce in the modified environment" (Odling-Smee et al., 2003, p. 354).

The great geneticist Theodosius Dobzhansky and his coauthors formulated this mechanism succinctly many years earlier: "While all organisms adapt to their environments by changing their genes, man alone adapts mainly, though not exclusively, by creating the environments that suit his genes" (Dobzhansky, Ayala, & Stebbing, 1977, p. 452).

This idea, however, has even earlier origins. It was known to Darwin and Alfred Wallace: "Mr. Wallace . . . argues that man, after he had partially acquired those intellectual and moral faculties which distinguish him from the lower animals, would have been but little liable to bodily modifications through natural selection. . . . For man is enabled through his mental faculties 'to keep with an unchanged body in harmony with the changing universe.' He has great power of adapting his habits to new conditions of life. He invents weapons, tools, and various stratagems to produce food and to defend himself. When he migrates into a colder climate he uses clothes, builds sheds, and makes fires" (Darwin, 1871/1998, pp. 131–132).

The present theory is obviously built on similar reasoning concerning human evolution—almost. The above quotations imply that, *as a result of the human cultural environment's shielding effects against new natural selection pressures, the genetic code adapted to the former environment remains unchanged.* This implication is most evident in the quotation from Laland and Brown: "*counteractive* niche construction . . . protect organisms from shifts away from the environmental states to which they are adapted." The quotations from Darwin and Dobzhansky hint in the same direction.

As mentioned above, the problem with these opinions is the presupposition that when changed environmental selection pressures are mitigated or nullified, the preexisting genetic adaptations of the concerned traits remain unchanged. However, this is not the case. *The genetic foundations of a trait (somatic or behavioral) need continuous maintenance, which prevents their disintegration, atrophy, disappearance, or excessive diversification*: "Any system will degenerate to the extent to which there is a relaxation of selection pressures for its maintenance" (Williams, 1974, p. 266). The main evolutionary mechanism of this process is well summarized by Ernst Mayr when he states, "Diversification is the most characteristic attribute of evolution," and "As soon as the selection pressure is relaxed, as in the case of the eyes of cave animals, individuals with imperfect structures will no longer be eliminated" (Mayr, 1991, pp. 62, 115–116).

The present theory argues that some of the above consequences of the RfNSPs, particularly diversification, were exploited during human evolution to build specifically human behavioral adaptations, and that the extremes of these diversified continuums led to dysfunctional forms of behavior. These topics are enlarged upon in the following chapters.

4

Relaxation of
Natural Selection Pressures

4.1. Introduction and Basic Principles of
Genetic Transmission

I have to admit that this chapter is built on more incomplete knowledge of its topic than previous chapters. While I am a nonprofessional regarding the subject matter of chapters 1 through 3, I at least had at my disposal books written by specialists who integrated the respective domains to create an overall picture that was accessible to nonprofessionals. However, to my knowledge, no such integrated works exist on the consequences of the relaxation from natural selective pressures (RfNSPs) in the living world. The reason for this situation seems to be that directional selection (the main mechanism through which new adaptive traits arise), as well as purifying and stabilizing selection (discussed in the following), are of more interest to professionals than the mechanism through which existing traits atrophy, disintegrate, or are otherwise altered when they are no longer useful. While this bias is understandable, and probably has no serious undesirable consequences with regard to the study of most animals, in the case of humans, the disregard for this mechanism, in my opinion, misses a crucially important aspect of human evolution.

Lacking integrated works on the subject of this chapter, I had to gather the necessary information from various sources whose main subject wasn't the RfNSPs: genetics, evolutionary theory, animal domestication, primatology, human biodiversity, etc., potentially missing or misinterpreting important points or not seeing the whole picture clearly.

In spite of these inadequacies, this chapter nevertheless seems indispensable to the preliminary part of a work, one of whose two main themes is the RfNSPs during human evolution. It attempts to put this subject matter into the wider perspective of RfNSPs in the living world and clarify their unique consequences in the case of humans.

The main topics to be discussed are the following:

Various kinds of genetic mechanisms exist, some affected by natural selection, others not. An example of the latter type is *internal selection*, which ensures smooth cooperation between the constituent parts of an organism. It begins to act before the maturing organism is submitted to natural selection pressures and remains mostly independent of them, or may even oppose the effects of natural selection. Other examples are *domestication,* along with other forms of artificial selection; and certain social traditions and selective practices that influence gene transmission from one generation to the next. A short discussion of the general principles and different kinds of genetic transmission seems desirable in order to provide the nonprofessional with the necessary background knowledge, before turning in the next chapter to topics more directly relevant to the concerns of this work.

RfNSPs are not the only source of genetic diversity, and not even the most important. Other sources are sexual (as opposed to asexual) reproduction, outbreeding, and random drift of neutral mutations. (On the other hand, migration and intermingling of populations, inbreeding, and incest *reduce* the extent of diversification.) Apart from accentuating diversification, RfNSPs have other consequences that are addressed briefly in this chapter.

Diversification itself takes different forms. The type relevant to the concerns of the present work is diversification in the form of *graded continuums,* which is the outcome of a specific form of genetic mechanism, the *polygenic,* as well as of a specific form of natural selection, the *stabilizing* or *normalizing one.* This topic is also discussed briefly. Then, in order to further widen the perspective of this chapter and to substantiate the idea that some consequences of the RfNSPs are universal in the living world, the effect of RfNSPs on the simplest known living organism, the RNA virus, is discussed.

The last and longest section of this chapter represents an effort on my part to enumerate and discuss briefly all the various selective mechanisms playing a possible part in human evolution: those relevant to the present as well as those that took place in the distant past; those consisting of relaxation of selective pressures as well as those leading to their intensification; those complying with the rationale of natural selection as well as those originating in social practices with an effect on gene transmission. I argue that this last mechanism has different outcomes than natural selection acting in the wild. My aim here is to illustrate the great diversity and complexity of the factors influencing genetic inheritance in humans. This discussion will close the preliminary part of this text.

4.1.1. Basic Principles of Genetic Transmission

The theory of evolution concerning sexually reproducing organisms, in its modern, neo-Darwinian form, rests on the idea that the *recombination of chromosomes during meiosis* (germ-cell formation) leads to "the mixing of the paternal and maternal genes [of each participant in a reproductive act] ... [which] provides a great abundance of new genotypes (far more than mutation) on which natural selection can act" (Mayr, 1982, p. 768). Some alleles (alternative forms of a gene) lead to adaptively desirable effects, while others to deleterious ones. Still others are neutral in this respect. In addition, one allele is often dominant over the other, the recessive one, which determines in heterozygotes which alternative will be manifest in the phenotype. Deleterious *dominant* alleles are selected out rapidly in an interbreeding population, while deleterious *recessive* ones, due to heterozygosis, may survive in the genome for long periods of time.

Darwin himself had no knowledge of the above genetic mechanisms and believed instead in the existence of two different kinds of inheritance: the blending inheritance, which means that the genetic traits of the two parents blend and the child inherits a kind of average, as well as Lamarckian inheritance, which argues in favor of the inheritance of characteristics acquired during the life span of the individual. Both of the above hypotheses were proved to be erroneous. The theory of Jean-Baptiste Lamarck (1744–1829) was refuted because acquired characteristics cannot change the genome of the germ cells. (In fact, the information flow from DNA toward the somatic protein synthesis in any cell is a strictly unidirectional process, Mayr, 1991, p. 120.) And Gregor Mendel (1822–1884), who established the science of genetics, rejected blending inheritance. Given these circumstances, it seems an outstanding achievement in theoretical science that Darwin's evolutionary theory survived and is accepted by most biologists today (with the reservations and additions discussed in section 1.3).

Mendel laid down the two basic laws of genetics. The first states that the two genes of the same trait, each inherited from one of the parents, do not blend but remain separate (the law of segregation). The second law states that genes for separate traits are assorted independently from one another in the germ cells—that is, the inheritance of a gene for one trait is not related to the inheritance of any other trait (the law of independent assortment) (Plomin et al., 2008, pp. 4–18).

During the twentieth century, geneticists discovered exceptions and supplements to Mendel's laws. For example, mutation in the germ cells themselves can induce new traits in the offspring not possessed by the parents (idem, p. 22), and genes sitting close to one another on the same chromosome may be transmitted together (in contradiction to the law of independent assortment). More important to our subject are two other additions to Mendel's laws, gene pleiotropy and polygenic inheritance.

4.1.1.1. Gene Pleiotropy

Gene pleiotropy means that a gene may have multiple effects at multiple loci on the phenotype, affecting possibly different traits of the body and behavior: "Hereditary diseases in man and higher animals are often complicated 'syndromes,' composed of changes in many body parts, organ systems, and physiological functions. A mutation in the rat caused thickened ribs, narrowed lumen of the trachea, emphysema of the lungs, hypertrophy of the heart, blocked nostrils, blunt snout, and low viability. Gruenberg (1938) found that the whole syndrome stems from a single primary change, an anomaly of the cartilage" (Dobzhansky, 1970, pp. 61–62). And: "Particularly interesting are pleiotropic syndromes that combine morphological traits with specific changes in behavior . . . the mutant yellow in *Drosophila melanogaster* changes the body color and also the courtship pattern" (idem, p. 63, emphasis in the original). The explanation of pleiotropism therefore seems to be, in light of the first example, that the *genes define a certain chemical process* in the body that *may have multiple secondary effects on observable phenotypic traits*: "A 'dogma' of molecular genetics is that each structural gene specifies one and only one polypeptide chain in a protein. On the molecular level, then, we would find no 'genuine pleiotropism.'" However: "As the traits studied are further and further removed from the primary gene action, the possibilities of epistatic interactions of different genes, as well as modifications due to environmental influence, increase" (idem, p. 64). Epistatic interactions are interactions between different genes; for example, one gene may suppress the effect of other genes (Mayr, 1982, p. 792), which ultimately determines their phenotypic effect.

This last statement has an important implication for psychiatric genetic research. This research tries to find correlations between discrete genes, on one hand, and the clustering of full-blown clinical disorders in blood relatives, on the other hand. However, all the nosological entities in psychiatry are the outcome of intense or longstanding interactions between genetic and ontogenetic influences. Therefore, the increased unpredictability when the studied trait is remote from the primary gene action (as the result of multiple gene interactions or ontogenetic influences) takes on ominous significance in this context. This subject is further discussed in sections 7.2.8 and 8.3.

4.1.1.2. Multiple Gene Inheritance

The other important addition to Mendel's laws from the point of view of the present text is *multiple gene inheritance* leading to polygenic traits. Mendel (probably not entirely by chance; see Mayr, 1982, p. 712) studied the traits of plants induced by variants of a single gene. Nevertheless, it became clear that many body traits and all the behavioral and mental ones are induced or influenced by multiple genes. Each one of these genes is inherited according to Mendel's laws, but their final effect has to be *additive* or *complementary*. These traits will not manifest themselves in an either/or fashion, but if diversified will induce *continuous quantitative gradations* of the respective trait in a population (which can indeed mimic blending inheritance) (Plomin et al., 2008, pp. 32–34); "Most human aptitudes form graded continua and are conditioned by several genes reinforcing each other's action" (Dobzhansky, 1973, p. 47). The genetic aspects of general cognitive ability, emotional makeup, and predisposition to mental disorders, in fact all the behavioral and mental manifestations, are such polygenic traits (Plomin, et al., 2008, pp. 34–37).

4.2. Different Kinds of Natural Selection

4.2.1. Directional, Purifying, and Stabilizing Selection

Darwin, in his epoch-making book *The Origin of Species by Means of Natural Selection, or the Preservation of Favored Races in the Struggle for Life,* meant by "natural selection" first of all *directional selection.* That is, any inherited variation that confers an advantage in survival or reproduction (increases fitness) amid specific environmental circumstances tends to be preserved and perpetuated by natural selection. Its simplest form is the qualitative one determined by a single gene that affects a single trait, a rare phenomenon. More widespread, especially in the behavioral sphere, are those polygenic traits that may induce quantitative variations around an optimum for a specific adaptive purpose. The sort of natural selection that deals with these kind of traits is called stabilizing or normalizing selection (Bell, 2008, pp. 189–190).

The section on evolutionary psychiatry used migrating birds as an example of this kind of selection. It was found that birds with longer or shorter wingspans than the optimum tend to die more often during migration. In fact, this example comprises not only directional selection but also its opposite, *purifying selection,* which means the elimination of adaptively undesirable variants of a trait. The alleles inducing optimal wing length tend to be preserved and perpetuated (directional selection), while the variations influencing wing length toward both extremes of the spectrum tend to be selected against (purifying selection).

If stabilizing selective pressures are strong, it can be predicted that they will "reduce both phenotypic and genetic variance" (Bell, 2008, pp. 189–190). Conversely, if stabilizing selection weakens, variance has to increase, leading to accentuated diversity of a quantitative nature (that is, a wider, gradated continuum away from the optimum in both directions). *This is the genetic mechanism that induces human behavioral diversity whose extremes constitute the innate predisposition to mental disorders.* Edward Wilson formulates this point—that is, that advanced technology releases heredity from the influence of natural selection and thus induces increased variability, by stating that "people are using scientific knowledge to gain conscious control over their heredity. . . . The evolutionary effect will be to relax stabilizing selection at an increasing rate and thereby increase the genetic variability of humanity as a whole" (Wilson, 1998, p. 275).

4.2.2. Additional Forms of Natural Selection

4.2.2.1. Internal Selection

A special form of purifying selection called *internal* or *developmental selection* begins to select out deleterious mutations long before the organism attains the maturational phase at which it can be subjected (without parental assistance) to natural selective pressures. This refers to the "selection of mutants . . . in accordance with their compatibility with the internal coordination of the organism" (Whyte, 1965, p. vii). And: "*The conditions of biological organization restrict to a finite discrete spectrum the possible avenues of evolutionary change from a given starting-point*" (idem, p. 22, emphasis in the original). In other words, any genetic variation that has a chance of being transmitted to the next generation first of all has to be of a nature that does not significantly disrupt the smooth, coordinated activity of neighboring or functionally interconnected structures.

Thus, internal selection, to a considerable extent, is independent of the selective pressures existing outside the organism; it may even act in an opposite direction to selective environmental effects. For example, if a new genetic variation has the desirable phenotypic effect of coping better with a certain environmental difficulty but at the same time interferes with the coordinated activity of other interrelated genes, the respective mutation very probably will be selected out. This mechanism has to have a powerful effect on the final outcome of the RfNSPs. Years ago, I frequently contemplated the curious fact that modern human populations, living for thousands of years in conditions of considerably relaxed natural selection pressures, did not diverge much more in mental and physical features and abilities from those (today rare) populations living in isolated places (disconnected from the effects of civilization), in natural conditions not especially different from those of our hunter-gatherer ancestors of the Pleistocene era—apparent evidence that internal selection can operate independently of external selection pressures and relaxations.

Ernst Mayr argues in favor of dual causation of evolutionary trends. Changes in the nature of external selection pressures push toward correlated changes in the phenotype. "On the other hand . . . the *internal cohesion of the genotype* . . . places severe constraints on the morphological changes that are possible" (Mayr, 1982, pp. 530–531, emphasis added).

4.2.2.2. Diversifying Selection

Diversifying selection may arise when a species has to adapt to varied environmental conditions:

> Individuals of a species who live together in the same locality . . . may have a variety of environmental opportunities: An insect may feed on different plants, a plant may grow in different soils, people may obtain their livelihood by hunting or fishing or agriculture. It may be that one genetic endowment enables its possessor to deal equally efficiently with all these environments. If so, a genetic uniformity will prevail. But if one genotype "agrees" better with one environmental factor and another genotype with another, then a genetic diversity may be preferable. Under these conditions a population which contains several adaptive genotypes will be better off than a genetically uniform one, and natural selection will favor diversity, polymorphism, the presence of genotypes suited for different aspects (ecological niches) of the available environments. (Dobzhansky, 1962, p. 222)

The human social environment, particularly in modern, technologically advanced societies, is extremely complex, varied, and diversified. Its members, as a rule, have the opportunity to choose from among various social niches the one that (among other considerations) best suits their genetic endowment. However, the possibility that human genetic diversity was created by different directional selection pressures of these diverse social environmental niches seems improbable.

In this regard, Theodosius Dobzhansky discussed the Indian caste system, which he considered "the grandest . . . apparently unsuccessful genetic experiment ever performed on human populations" (idem, p. 234). The Indian castes and sub-castes were rigidly separated reproductively by the prevailing social rules; the members of a specific caste or sub-caste lived in comparable living conditions and practiced a restricted range of occupations. This social system lasted for more than two thousand years (idem, p. 237). Nevertheless, Dobzhansky found that it did not induce directional selection, leading to subpopulations with different genotypes:

"The caste system . . . attempted to breed varieties of men genetically specialized in the performance of different functions. To all appearances, such a specialization has not been achieved" (idem, p. 258). Therefore, it seems quite clear that human behavioral diversity is the result of relaxation of natural selection pressures (of the stabilizing or normalizing kind) on genetic variability, rather than the result of multiple kinds of directional selection. The match between the diversified genetic makeup and environmental requirements is made mostly by *social mobility,* through which an individual chooses or partly creates the environment that suits best her genetic makeup (or alternatively is chosen by social selective mechanisms such as admission interviews or examinations).

Further mechanisms exist through which organisms, even simple ones, can cope with varied environmental conditions. "Versatility" is the ability of an organism to maintain growth over a broad range of conditions (Bell, 2008, p. 267). "Plasticity," on the other hand, means that "the same genotype will be able to express different phenotypes in different environments. . . . A more plastic genotype *has a more variable developmental program* that enables it to exhibit greater stability with respect to components of fitness over a broader range of environments" (Bell, 2008, p. 267, emphasis added).

That last concept seems especially relevant to humans. The human brain has to be considered the most plastic organ in the living world. Being extremely immature at birth (compared to that of other highly evolved animals as well as to other organs of the body), during maturation it undergoes an extensive reorganization. In this period, a great number of neurons and existing synapses are lost, and a great number of new synapses are formed. The most obvious reason for this great plasticity (nonexistent in any other organ or tissue in the body) seems to be the need to adapt the behavior of the maturing organism to its complex environmental—chiefly cultural and social—conditions (Kalverboer & Gramsbergen, pp. 6, 7, 34, 51, 118, 834).[1]

Another consideration in evolutionary theory that may interest us here concerns "the balance between generalization and specialization that should evolve in populations that live in a heterogeneous environment. . . . [G]enotypes that have very high relative fitness in an environment are mediocre or even inferior in others" (Bell, 2008, p. 268). These genotypes lead to phenotypes that are "specialists." On the other hand, some experiments on single organisms suggest that in certain conditions "generalists will evolve in variable environments" (idem, p. 301). Some environmental conditions "should select for versatile generalists capable of mounting a rapid physiological response to environmental variation" (idem, p. 302).

I do not dare to extrapolate to humans from experimental results and evolutionary reasoning concerning such simple organisms as fruit flies and bacteria. Nevertheless, these considerations suggest that the need to develop evolved mechanisms that grapple with environmental variability and heterogeneity, rather than being a specifically human attribute, appeared early in the evolution of life.

Other kinds of natural selection are relevant to human evolution, such as *rudimentation* or *regressive evolution*: "Adaptation to new environments may decrease the importance of some organs or functions that were vital in past environments; such organs and functions may then become vestigial and disappear . . . many internal parasites have lost the alimentary canal and sense organs, such as eyes, that were doubtless present in their free-living ancestors" (Dobzhansky, 1970, pp. 405–406). A relevant example of such regressive evolution with regard to our main topic is the disappearance of fixed instinctive motor patterns from humans' behavioral repertoire.

Evolutionary processes also exist that lead to increased diversity without involving relaxation of natural selection, such as the drift of neutral mutations from one generation to the next: "Given many genetic variants that are selectively neutral or very nearly so, their frequencies in populations will drift at random until some of them are lost and others reach fixation" (idem, p. 259). Taking in account that clearly advantageous mutations will spread rapidly in a population and thus become fixated (that is, they become an organic part of that species' genome), while unequivocally deleterious mutations are rapidly eliminated, most mutations existing in the genome will be neutral. However, the condition of a mutation being considered advantageous, neutral, or deleterious is more complex than it seems at first glance. These attributes of the mutations are not intrinsic qualities but are determined by the existing selective pressures and relaxations in a given environment: "Hairiness or lack of hair, skin pigmentation, or visual acuity may have been important for survival in the past; at present these traits may be nearly neutral or may affect fitness via differential mating success rather than differential survival" (Dobzhansky, 1970, p. 296). Indeed, it seems to me that in the varied social conditions (containing diverse social niches), in which natural selection is considerably relaxed, it may frequently be difficult to categorize mutations as clearly advantageous or deleterious, particularly in the behavioral and mental sphere. (The same consideration applies to genetic diversification due to recombination of homologous chromosomes during gamete formation, discussed in 4.3.)

4.2.2.3. Artificial Selection and Domestication

Artificial selection of domesticated animals (as well as of wild and laboratory animals) is a complex situation from the perspective of natural selection. These animals are subjected to strong directional selective pressures for the traits desired by the breeder and at the same time are released from most of the selective pressure that existed in their natural habitat. They are provided with shelter, food, protection from predation and diseases, etc., and the mating animals are selected by the breeder (rather than by sexual competition), again in accordance with the desired traits.

What further complicates the picture is gene pleiotropy and polygenic inheritance. That is, breeding for a selected trait may lead to unexpected changes in other areas of functioning or bodily attributes. For example: "Continuous selection for a calm temperament in foxes [cultivated for their fur] resulted in negative effects on maternal behavior and neurological problems . . . physiological and behavioral problems increased in each generation. In fact, some of the tamest foxes . . . cannibalized their pups" (Grandin & Deesing, 1998, p. 21).

In spite of these confusing factors, in view of the rarity of significantly relaxed selective pressures in the environments of wild animals, the findings of artificial selection are used in the following discussion to clarify or illustrate the effects of relaxed selective conditions on morphology, physiology, and behavior. Beilharz's resource allocation theory, based on some of the effects of domestication, has already been cited in the section on primatology (section 3.1) to account for bonobos' exaggerated sexual life.[2]

4.3. Genetic Diversity and Relaxation from Natural Selective Pressures (RfNSPs)

Genetic diversity, as mentioned in the introductory section, is one of the three indispensable requirements of evolution in its traditional

Darwinian sense (the other two being the selection of variants with adaptive significance and their propagation to successive generations). RfNSPs are only one of several causes leading to genetic diversity in a Mendelian (or interbreeding) population, and with the exception of their role in human populations, not the most important one; nor is that cause indispensable. The most important source of genetic variability in a sexually reproducing population is the *recombination (crossing over) of homologous chromosomes* during meiosis, that is, during gamete formation (Mayr, 1982, pp. 537, 588; see also Dobzhansky, 1970, pp. 90–92). Other causes of genetic diversity are *reassortment of the parental chromosomes* during conception; *mutations in the gametes; outbreeding*, or the avoidance of reproduction between genetically closely related individuals; and the *drift* or *"random walk"* of neutral mutations. The augmented variability caused by the reassortment of parental chromosomes during sexual reproduction, when compared with uniparental reproduction, and in consequence the improved adaptability to changing environmental conditions, is thought to be the chief reason for the very existence of sexual reproduction, despite its much greater costs compared with asexual reproduction (Mayr, 1982, p. 599).[3]

On the other hand, certain factors decrease diversity, like *inbreeding* (the opposite of outbreeding), *intensification of stabilizing selective pressures*, as we have seen, or *intermingling of genetically different populations* (during migration or military conquest, for example).

Although evolutionary theory downplays its importance, relaxation from selection pressures and its consequences are an integral part of the evolutionary mechanism. Many selection pressures vary periodically, or in an irregular way, in their intensity. These fluctuations can concern one trait or a multitude of traits, short-lived or longstanding and affecting many generations, and can be local or widespread. The availability of feeding resources, as well as the nature of physical conditions (heat, light, humidity) oscillate diurnally or with the seasons; the prey-predator ratio may change, leading to increased or lessened selection pressures on certain traits in both predators and prey animals; habitats can change as a result of a natural catastrophe; or the climate may change locally or on a global scale (as during glacial and interglacial periods), etc. If the change in the intensity of selection pressures is short-lived, it cannot lead to changes in gene frequencies in a population; if the oscillations in selection pressures are regular, like seasonal and diurnal variations, the organisms can develop innate adaptive strategies (migration, hibernation, synchronization of the reproductive cycle with seasonal changes, the sleep-wakefulness cycle); if a new selection pressure or relaxation is longstanding (relative to the duration of the life cycle of an organism), it can lead to considerable changes in the gene pool of a population.

4.4. The Effects of Relaxation from Natural Selective Pressures

The relaxation from natural selection pressures may have various consequences besides diversification, the area on which the present work concentrates. A short discussion may help place our more restricted topic in a wider perspective.

4.4.1. Gradual Degeneration, Atrophying, and Possible Disappearance

The most obvious effect of the RfNSPs on an organ or system is its gradual degeneration, atrophying, and possible disappearance. To reiterate, "Any system will degenerate to the extent to which there is a relaxation of selection pressures for its maintenance" (Williams, 1974, p. 266). This has to be the case, especially when the respective organ's or system's maintenance involves an energetic cost or some other kind of cost to the organism.

4.4.2. Atavism

We must be careful in concluding that the disappearance of a phenotypic trait means invariably the deletion of the genetic machinery underlying it. It may happen that only the genes that *activate* the genetic mechanism responsible for the development of the respective phenotypic trait are switched off. This is the case in the phenomenon of *atavism*, those rare instances in which a long-disappeared trait of a remote ancestor sporadically reappears in an individual or in several genetically closely related individuals. The reappearance of dense hair over the whole body and face, characteristic of our apelike ancestors (generalized congenital hypertrichosis), is an example of such a phenomenon (Klein & Takahata, 2002, pp. 203–204).

4.4.3. Resource Allocation Shift

As mentioned in the discussion on the bonobos (section 3.1), R. G. Beilharz proposed a theory according to which domesticated animals in relaxed selective conditions for survival withdraw resources from organismic functions released from the respective selective pressures and reallocate them to reproductive functions, or traits under intense artificial selection (Price, 2002, p. 79). And: "Haase (2000) compared a number of reproductive parameters in male wolves . . . and dogs . . . living under similar conditions in captivity. Wolves exhibited maximal levels of testosterone and 5-α-dihydrotestosterone in the peripheral plasma and maximal testicular weights during the winter months (which is their natural breeding season), whereas the dogs exhibited relatively high levels of these parameters during all seasons. Has the loss of reproductive seasonality in the dog been due to relaxed selection, since they are normally protected from harsh climatic factors and their basic needs are provided by human caretakers? Has natural selection in captivity favored dogs that breed more than once a year?" (Price, 2002, pp. 69–70).

4.4.4. Changes in Selection Pressures during Adaptation to a New Environment

This point is especially important to our concerns. During the transition from one kind of adaptation to another as a response to changing environmental conditions (for example, the transition of certain mammals from a terrestrial to aquatic or aerial existence, as in the case of cetaceans or bats), the *selective pressures relevant to the former environment relax while at the same time the new environment's selective pressures intensify.* No other order of the relevant events can be imagined.

This principle can be illustrated with the evolution of the human hand and hand-eye coordination. Hands have to be relaxed from the selection pressures of primates' arboreal locomotion (in fact from any kind of selective pressures for locomotion) to undergo directional selection for fine coordination suitable for object manipulation (use of tools, weapons, etc.). The diversification or atrophying of structures that became useless as a result of changing environmental circumstances seems to be a precondition for upcoming new directional selection pressures to be maximally effective.

This order of events also has to be true for other behavioral domains more relevant to our discussion: "The obvious trend in the evolution of behavior in the direction of greater plasticity and increasing influences of learning and insight has to be regarded at least as much as a *consequence of reduction and disintegration of innate fixed patterns* as of higher development of those functions which, in the individual's life, affect adaptive modification of behavior" (K. Lorenz, 1965, p. 5, emphasis added).

It must also be hypothesized that some diversified forms of the former adaptations may have proven useful as building blocks for the new adaptations. For example, the dexterity and high-level hand-eye coordination needed for arboreal existence may have been retained because it was also needed for the new function of the hands, object manipulation.

The same reasoning will be employed in the present theory when it is proposed that normative human behavior is built on preexisting instinctive motives of animal behavior. However, *these instinctive motives in humans have to be diffuse enough or undifferentiated enough to enable the necessary far-reaching modification of behavior by learning.* On the other hand, they have to remain differentiated enough to preserve the ability to guide learned behavior in an approximate manner toward biologically important goals. This argument is developed further in the discussion of the predisposition to the schizophrenia spectrum of disorders.

4.4.5. Exaptation

Exaptation should be mentioned here because it is proposed as an alternative or, more exactly, as a supplement to classical evolutionary theory, to explain the evolution of certain human traits, particularly language. In addition, this concept overlaps with the subject discussed above (section 4.4.4).

It is argued that not all the "biological forms have evolved as functional adaptations to local conditions" (Barkow, Cosmides, & Tooby, 1995, p. 393). Exaptations are defined as "characters, evolved for other usages (or for no function at all), and later 'coopted' for their current role" (idem, p. 393). However, evolutionarily old structures employed for new functions by changing environmental pressures are in fact predicted by traditional evolutionary principles. Therefore, the demarcation between exaptation and changing adaptation as a result of changing selection pressures becomes blurred: "Legs in terrestrial animals are also exaptations rather than adaptations, since they evolved from lobed fins, which were used for swimming rather than for walking" (idem, p. 394).

4.4.6. Unique Consequences of RfNSPs during Human Evolution

In human evolution, as mentioned in the introductory section, relaxation from natural selection pressures and consequent diversification has special significance. In evolving human populations, accentuated diversification of mental and physical characteristics was apparently *not* only a preliminary stage that enabled new directional selection pressures resulting from environmental change to be more efficient; rather, the *diversification itself (which progressively increased with ecological dominance) had a critically important role in the evolution and structuring of human populations and the emergence of culture.* Without efficient cooperation between individuals widely different in their mental and physical abilities, human societies, with their hierarchical structure; their technological, scientific, and artistic achievements; their wide range of social institutions; and so on, would be simply unimaginable. The authors whose opinion on the subject is known to me unanimously recognize human diversity—physical and mental—without exploiting the great explanatory potential of that diversity with regard to subsequent human cultural evolution.

It is also recognized that diversity *within* human groups (polymorphism) is always more impressive than differences *between* groups (polytypism), and consequently: "within group variation (polymorphism) indeed held the key to understanding biological diversity in the human species" (Marks, 1995, p. 101). Body height, strength, body proportions; fine coordination and degree of hairiness; the considerable variability of the human brain's weight, macroscopic appearance, and microscopic structure are but a small sample of variable physical characteristics. Memorizing ability, personality traits, and emotional constitution, as well as the uneven distribution of special talents, capabilities, and mental weaknesses, are examples of variable mental traits.

Shortly after the publication of Darwin's *The Descent of Man*, scientists also recognized the great quantitative diversification of instinctive predispositions to behave, the extremes of which are reflected most clearly in mental disorders. But instead of a balanced, objective evaluation of this diversification's assets and drawbacks, distinguished scientists and philosophers saw diversification and the loosening of highly differentiated instinctive predispositions of nonhuman animals as having undesirable, even catastrophic significance for humankind.[4] The German philosopher Friedrich Nietzsche (1844–1900), for example, compared modern man to an "ailing animal" suggesting that "ignoring and denying our instincts has made us weak, sickly, and even mad" (Adriaens & De Block, 2011, pp. 23–24). See also Emil Kraepelin's and other psychiatrists' opinions on this subject (section 3.2.2).

Mental disorders, particularly schizophrenia, "this most human of maladies," were seen, too, as a consequence of the above evolutionary development, and it was believed (including by Freud) that studying mental disorders can help elucidate aspects of normative behavior: "Mental illnesses show in an exaggerated and magnified way tendencies and mechanisms that constitute our humanity" (idem, p. 24). (The view that mental disorders can help elucidate aspects of normative behavior, and that schizophrenia is "the most human of maladies"—in the sense that it is intimately related to unique aspects of human evolution—is also embraced by the present work.)

It is intriguing to see in these views the different combinations of genuine scientific insight, emotionally laden exaggerations, one-sidedness, and unfounded beliefs.

4.5. Relaxation from Natural Selective Pressures and the RNA Virus

The ideas suggested in the following are strongly hypothetical. They propose parallelisms and analogies between the consequences of RfNSPs in higher animals, including humans (where we have mostly precarious inferences based on indirect evidence), on one hand, and on the other, the effects of the RfNSPs on the simplest known living organism on earth, the RNA virus, where we have sound laboratory findings. I suspect that entering upon this subject I may come close to suffering the consequences of not heeding the nuclear physicist Richard Feynman's warning quoted in the first chapter: "In talking about the impact of ideas in one field on ideas in another field, one is always apt to make a fool of oneself" (see section 1.2).

Nevertheless, the importance of what we may learn from the following example concerning the RNA virus seems to me to warrant the risk. While the analogies that will be drawn here are not directly relevant to our narrower subject matter, mental disorders, they may suggest that some of the effects of the RfNSPs are universal in the living world.

The RNA virus in question is called Q-Beta, which infects bacterial cells. "The Q-Beta RNA encodes a number of proteins, including one that specifically catalyses the replication of Q-Beta RNA. . . . The host cell provides the rest of the apparatus for producing the protein, and a supply of raw material from which the new Q-Beta genomes can be constructed. . . . The viral genome encodes several kinds of protein that are used to transmit itself from one host cell to the other" (Bell, 2008, p. 1). In addition, the virus possesses several biological mechanisms whose function is to overcome the defenses of the host cell.

In the laboratory experiments described here, the virus was cultured in a solution that provided all the chemical elements it needed for replication. The experiment was designed to study the general principles of natural selection, and its conclusion was that selection's ultimate concern is the *rate of replication*. This "is the only attribute that is selected directly" (idem, pp. 1–19). As was expected, replication errors (mutations) appeared at an increased rate and the "populations of self-replicating RNA molecules . . . [became] *markedly diverse* . . . the same principle applies to the replication of DNA, although this is a more rigorously controlled process" (idem, p. 3, emphasis added).

The following describes more specific consequences of the virus being completely relieved of its usual environmental selection pressures, which arise from "parasitizing complex and hostile bacterial cells" (idem, p. 4). It was found that in these conditions, that is, in an experimental solution that provided all its needs for replication, the virus rapidly lost all the now-unnecessary machinery designed to cope with its natural environment (the bacterial cell). During this process, the chain of the RNA, which originally comprised 3,300 nucleotides, began to shrink rapidly: "After 70 transfers, the evolved variant . . . was only 550 nucleotides in length" (idem, p. 6). The virus "shifted" all its fitness-enhancing resources toward more rapid replication—that is, the more rapidly replicating populations replaced the slower replicating ones. The explanation proposed for the shrinking of the RNA sequence was also related to the rate of replication, which became quicker as the RNA chain became shorter. What eventually remained in the laboratory solution was "not a random fragment . . . but rather a minimal sequence that supports efficient replication" (idem, pp. 6–7). Of course, "the ability to infect bacteria is almost completely lost after four or five transfers. . . . Selection will cause progress, but it is a constrained and restricted sort of progress, with advance in one direction being associated with regress in others" (idem, p. 7). Biologist Brian Goodwin described a similar experiment, concerning the replication of the DNA (instead of the RNA) molecule, which arrived at similar results and theoretical implications (Goodwin, 1996, pp. 35–36).

Any extrapolations we can draw from this experiment to higher organisms of course have to be made very cautiously. We are dealing here with the simplest organism on earth, which is not only less complex than unicellular organisms but one whose status as a living organism can even be questioned. It does not possess metabolism (in the sense of exchange of materials with the environment), and it cannot replicate itself autonomously without the aid of the host cell's replicating apparatus and supply of needed raw material. (See the definition of the phenomenon of life according to Michel Morange, discussed in section 0.2.4; see also Morange, 2008, pp. 36–38; and Medawar & Medawar, 1977, pp. 8–9, 89.)

Of course, the RNA virus possesses no behavior in the sense this word is used by the behavioral sciences, and therefore it has no direct relevance to the main subject of this text (human dysfunctional behavior). Nevertheless, drawing analogies from these experimental results to higher organisms is tempting. The relaxation of selection pressures led to greater diversification than existed in natural conditions. In addition, *the RfNSPs led to the rapid loss of coping mechanisms* no longer necessary for survival and reproduction. It is intriguing to ask whether the more slender and fragile body of humans; their almost complete hairlessness; the more sensitive, choosy digestive apparatus, which cannot cope with many kinds of unprepared food; the loss of any effective defense mechanism against predators of the evolutionary past, and in general against many natural stressful conditions without the protective shield of human technology and culture; human beings' great physical and behavioral diversity, etc., are essentially different evolutionary developments from those of the cultivated RNA virus, or do they differ only in their degree of complexity?

The observed transfer of resources of the RNA virus from coping with environmental difficulties toward increased rate of replication recalls, as we have mentioned, Beilharz's resource allocation shift theory concerning domesticated animals (section 3.1 and in the present chapter, section 4.4.3). The once-a-year reproductive cycle of many mammals in the wild (synchronized with the seasons) increased to two or more per year in our cats and dogs; the excessive sexual life of the bonobo, which lives in relatively relaxed environmental conditions, was discussed in the previous chapter. And what about humans in this respect? Humans have been dubbed the most sexually active animals on earth (perhaps excluding the bonobos). Our developed technology, resourcefulness, and liberal social organization enable us to enjoy the pleasures of sexual activity while eschewing such undesirable or harmful consequences as sexually transmitted diseases, unwanted pregnancies, etc. Sexual desire and pleasure can be defined, in accordance with the definition of instincts in general in this text, as the subjective reflection of the instinctive urge to reproduce while enacting the appropriate FMP common to all mammals (and to some animals of other classes). In most cases, the great differences in the aim of sexual activity in humans versus other animals (pleasure instead of reproduction) result from the impact of technological achievements and cultural influences, not from the nature of the innate biological mechanism involved.

Finally, the considerably shortened RNA sequence of the cultivated Q-Beta virus relieved from the selection pressures of its natural environment invites another risky analogy with humans. At the beginning of the genome project, it was expected that the human genome would comprise far more genes (around 100,000) than were actually found (fewer than 25,000 protein coding genes, about equal with that of the mouse). In light of this, it is tempting to conclude that becoming human in a genetic sense meant the loss of genes no longer needed for survival, and successful reproduction at a rate comparable to the gaining of genes that underlay new, specifically human inherited traits (see also J. Marks in section 3.2.3.3).

That last idea can be illustrated by an intriguing paradox concerning the increase of brain size during human evolution. While the human brain tripled its volume (from 450 to 1,350 cubic centimeters) in comparison with that of the chimpanzee since the hominid line's divergence about 6 million years ago, this increase came to an end 200,000 years ago, after which time no further increase in brain size took place (Mithen, 1966, p. 12). This paleological finding leads to the following questions:

1. What selection pressures led to the immense brain expansion in the period between 4 million and 200,000 years ago (idem, p. 12) taking into account that in this period there existed no developed technology, no science, no arts, and no complex social organization, or developed symbolic language, cultural achievements we tend to attribute to increased brain capabilities?

2. Why did the growth of brain volume come to a stop *before* all the abovementioned cultural developments appeared or began to accelerate, attaining gigantic proportions after the introduction of agriculture, the emergence of city-states, the Industrial Revolution, the extensive use of electronic appliances, and so on? How, in a monistic framework of reasoning, can we account for the discrepancy between the impressive increase in mental abilities needed for adapting to these developments and the lack of a corresponding quantitative increase of its material substrate?

In fact, these questions have already been asked: "early cultures remained amazingly constant during a period of substantial growth in brain *size*," and conversely: "cultural changes took place at astonishing speeds with no significant change in the physical appearance of people or in the characteristics of their brain that can be divined from fossil sculls" (Ehrlich, 2000, p. 106, emphasis in the original).

Obviously, providing definitive answers to these questions exceeds by far my knowledge and intellectual abilities. However, it is intriguing to speculate that, up until the advent of Homo sapiens, our prehistoric ancestors' ecological dominance was minimal, which means that *no considerable relaxation of natural selection pressures* existed, and, consequently, most of the behavioral and somatic adaptations to the natural environment were retained (see Wilson, 1978, pp. 83–84). On the other hand, living in cohesive groups necessitated the evolution of complex mental and behavioral abilities needed for successful social coexistence, competition, and cooperation, which surely required additional brain volume (for details, see: idem, pp. 84–88). Indeed, an impressive correlation exists in anthropoid primates and some other social mammals between brain volume (more exactly neocortex size, which amounts to 80 percent of the total brain volume in humans), social-group size, and the ability for social manipulation (Dunbar, 1997, pp. 62–69, 94, 111–112).

However, with the advent of Homo sapiens, some 150,000–200,000 years ago, the ecological dominance increased substantially, leading to more accentuated relaxation of natural selection pressures and consequently the loss or weakening of corresponding adaptive traits. (We can presume these evolutionary events, for example, from Homo sapiens' more slender build, more sophisticated stone tools and other artifacts, possession of trading networks, and replacement of the Neanderthal population [idem, p. 117].) Neanderthals' greater adaptation to natural selective pressures compared to anatomically modern humans' can be deduced from their body's increased surface to volume ratio, a somatic adaptation to cold environment, and, more interestingly, from comparing the brain configuration of these two hominid species. Overall, Neanderthals possessed brains at least as large as those of modern humans, but the ratios of the cortical regions differ. Cortical areas dealing with visual functions and body maintenance and control were relatively large, while the regions dealing with higher cognitive functions, like the frontal lobes, were proportionally smaller (Pearce, Stinger, & Dunbar, 2013; see also Ehrlich, 2000, pp. 101–104).

We can presume that as cultural achievements progressed from this time forward, they posed increased demands on the mental (primarily cognitive) functioning and its underlying material substrate. However, at the same time, the ecological dominance, as well as the impositions to behave in socially prescribed (that is, predictable) ways, also increased, thus reducing progressively the need for the mental and behavioral abilities that are meant to deal with strong natural and social selective forces. The division of labor increased, as did the assortment of social niches, too, compartmentalizing the tasks of the individual brain. This in turn led to a quite constant daily routine adapted to the individual's restricted social/physical niche. Consequently, it may be hypothesized that instead of further increase in brain size, the anatomically modern human brain *reallocated* its resources in accordance with the changing intensities of the diverse selection pressures coming from the natural and social environment. (For examples of the high degree of cortical plasticity—the brain's ability to transfer functions from one area to another, or to change the function of an area in accordance with ontogenetic events—see Lieberman, 2006, pp. 204–207.) The diversity of the brain volume in modern humans is impressive, too (most probably as a result of the relaxation of selection pressures), having a range of 1,200–1,700 cc (Ehrlich, 2000, p. 92).

It has to be stressed, however, that the relationship between an organism's level of complexity and the size of its genome is a more complex issue than a simple parallelism (see Maynard Smith & Szathmáry, 1995, pp. 5–6).

4.6. An Overview of the Selective Mechanisms (Natural and Social) that Influence Gene Transmission in Humans

In the following, I try to compile a list of mechanisms that influence the transmission of genes from one generation to the next in human populations. I begin with natural selection pressures and relaxations, past and present, and continue with social selective practices influencing gene transmission. The list is partial and the discussion of each point sketchy, and I do not investigate how these mechanisms interact with each other in human populations with different cultures and ecologies. Nevertheless, I hope to sketch out a tentative picture of the complex subject of human genetic "evolution" at this point in the text, which marks the end of the preliminary chapters, before entering into the main topic—the genetic predispositions of discrete mental disorders.

4.6.1. Relaxation of Natural Selection Pressures

As we have stated, the effect of natural selection pressures on the genetic inheritance (both directional as well as purifying) decreased in human populations as a result of the shielding effect of the man-made artificial environment. Of course, I do not mean here the basic body structures and functions (organs, systems, tissues), which continue to be determined quite rigidly by inheritance. The decreased effect of selection pressures becomes apparent, instead, in less critical areas whose diversification can be contained by the shielding effects of the cultural environment or by various, more specialized social niches. These less critical areas include variations in body height, weight, body proportions, physical power and stamina, level of motor coordination, extent of acuity of sensory organs, hairiness, predispositions to various medical diseases and somatic dysfunctions. Of course, internal selection (which ensures the coordinated functioning of different body parts and organs and begins to act prior to the stage at which the mature organism becomes subject to environmental selective pressures) continues to act, irrespective of the human environment's shielding effects.

The relaxation of natural selection pressures—those that were active mostly during the evolutionary past—on certain body structures and functions does not mean that new, natural selection pressures ceased to appear altogether in humans. Directional selection pressures have been identified in human populations of relatively recent origin. Immunity to contagious diseases that resulted from crowding and tolerance to new diets (lactose tolerance), both of which occurred at the transition to an agricultural way of life, are effects of such recent directional selection. Nevertheless, it seems to me that these evolutionary events are relatively rare in comparison with the multitude of examples in which the social environment shields the human body from noxious environmental influences.

However, this diversifying effect of the relaxation from natural selection pressures is much more evident in the field of innate predispositions *to behave* (instincts) than in the case of bodily traits (Barkow, Cosmides, & Tooby, 1995, p. 397). In fact, the same hypothesis was advanced in the case of domesticated animals: "Behavioral traits are more subject to relaxed selection than associated traits at lower levels of biological organization" (Price, 2002, p. 65). The diversification in humans greatly surpasses that observed even in highly evolved social animals in such areas as general intelligence, memorizing ability, personality traits, innate predisposition to react with anxiety or aggressiveness of inappropriate intensity or duration, sociability, and so on.

4.6.2. Relaxation of Sexual Selection Pressures

Natural sexual selection pressures of the kind observed in other mammals and lower animals also decreased considerably in the human social environment. As discussed in the section on evolutionary psychology, long-lasting pair formation in humans, a rare phenomenon in other mammals, shifted the preference of females from those body and behavioral characteristics valued in general in animals (body size, physical power, bravery, ornaments) toward traits that indicate willingness and ability for longstanding investment in a relationship and offspring. Interestingly, this preference can be found in birds, too. A lot of data exists showing that female birds value the tenacity of courting, the investment in nest building carried out by the male, and offerings of gifts like delicious food, suggesting convergent evolution between birds and humans in this respect, possibly as a result of similar selective pressures (see chapter 3, section 3.2.1.2).

In the case of women, however, evolutionary psychological research produced some surprising experimental findings that suggest the *partial preservation of characteristic mammalian sexual preferences*, findings that led to the so called "Good Genes Hypothesis." It was found that some women, living in a good, long-term relationship with a man (who presumably possesses behavioral traits that predict good abilities for long-term investment in the relationship) may, at the peak of their menstrual cycles—at ovulation—prefer extramarital short-term sexual relationships with men of markedly different characteristics than those possessed by their long-term partners. These characteristics are: masculinity, tallness, muscular build, and intrasexual competitiveness—"qualities that indicate *social dominance*" (Buss, 2008, pp. 190–191, emphasis added).

Human sexual activities are without question the most diversified in the animal kingdom. It is true that, in bonobos, sexual activity can be triggered by a wider range of circumstances than humans, some of which, like eating or reconciliation, pertain to an entirely different instinctive domain. Nevertheless, in humans, the variety of patterns of sexual activity and the assortment of releasers capable of leading to sexual excitation is greater by far than it is for any other members of the animal kingdom.

The wide variety of disorders of the sexual drive (paraphilias) that exists in humans has no counterpart in other animals. The most extreme example of a releaser capable of awakening sexual feelings in humans, to my knowledge, is encountered in a condition known as "object love" or "object sexuality." In this condition, the releaser is completely different from anything related, even distantly, to the human body, to part of it, or to an accessory that comes in contact with it, or even to the body of a domesticated animal whose genitalia are accessible to human sexual activity. In articles available on the internet, I found descriptions on "falling in love" (and sometimes even "marrying") such objects as the Eiffel Tower, the former Berlin Wall, engines, computers, etc.

Mate selection of the natural kind in humans is strongly interfered with by cultural practices and traditions, sometimes to such an extent that the choice of whom to marry and have children with is not in the hands of the couple at all, as in the "minor marriage" custom of some traditional cultures (Wilson, 1998, p. 175), where marriage decisions are made by fathers or some other authority figure. This topic, however, belongs to *social selective practices*, rather than natural selection, and will be discussed in section 4.6.6.

4.6.3. Relaxation of Intragroup Natural Selection Pressures

It must be presumed that intragroup purifying natural selection pressures stemming from intolerance toward handicapped or deviant members of groups were strong evolutionary forces during human prehistory. Such individuals, as a result of defective inheritance or other reasons, would have been unable to function properly in the group's framework.[5]

The earliest evidence of relaxation of intragroup selection pressures known to me is that of the Neanderthals surviving with serious injuries (as can be inferred from healed bone fractures of Neanderthal skeletons) surely as a result of help received from other group members (Glynn, 2000, p. 62).

The opposite evolutionary force, directional selection, will be referred to in this text as *secondary intragroup directional natural selection pressures*, meaning that inherited traits enabling or improving social coexistence and cooperation are positively selected. They are "secondary" because in the framework of this text it is hypothesized that the necessary precondition for this kind of selection to occur is a prior relaxation of natural selection pressures *external* to the group and consequent diversification of the respective inherited behavioral inclinations. This kind of selection is discussed in section 4.6.5.

It is also important to clarify here that these selective pressures are called "natural" since, in spite of being the effect of social coexistence, they act on gene transmission in accordance with evolutionary principles, contrary to some "social selective practices" that act on gene transmission in a way unrelated to or even in opposition to conventional evolutionary logic, the subject matter of section 4.6.6. Such social selective practices might include the "minor marriage" custom, for example, or being found fit for an elite combat unit.

A further clarification concerns *intragroup* as against *intergroup* selection. While in humans, intragroup selection—both in its initial intensification and subsequent relaxation—was a strong evolutionary force, intergroup natural selection pressures (that is, the differential survival and reproduction of the better functioning group) must be a weak evolutionary mechanism in humans. As mentioned previously, individual selection overpowers group selection (Williams, 1974, pp. 82–124); migration, interbreeding between members of different groups, and the custom of exogamy (marrying outside one's clan or tribe) dilutes intergroup selection; and, perhaps most importantly, the effective, in-concert functioning of group members, which eventually secures the success of the group as a whole, depends heavily on prior *intragroup* selection.

Returning to our original subject in this section, we have to conclude that at later stages of human evolution, due to increasing technological sophistication, growing supplies of life-sustaining resources, increasing standards of living, and humanitarian ideologies, the intensity of the *intragroup purifying selection pressures* gradually decreased, gaining really "unnatural" proportions in modern democracies.

The relaxation of purifying intragroup selection pressures in technologically advanced democracies is very impressive indeed. Massive aid to survive as long as possible and to assure improved quality of life is provided to a wide range of dysfunctional and disabled individuals. This includes not only the temporarily handicapped but the chronically ill, the old, the mentally ill, the crippled, and so on, incurring a heavy economic burden on the community, and frequently also a psychological one on the caretakers. That the budget of modern states can permit this burden without becoming unduly overstrained is certainly not the only issue here. I suspect that a long evolutionary history of mutual aid and identification with the plight of other more unfortunate members of the narrower community as a result of specific selection pressures or relaxations concerning certain mental traits predisposed us to such maladaptive efforts (maladaptive from a biological, not humanitarian, point of view), even when they incur a considerable economic and psychological burden. Whether this mental mechanism is related to the relaxation of instinctive predispositions that originally evolved for purposes of parenting and inclusive fitness concerning genetic relatives (section 3.2.1)—or to be more precise, the widening of the IRM leading to the inclusion of nonrelatives, too—I cannot say. At any rate, the "emotional contagion" of babies when hearing expressions of distress, mentioned in section 2.3, may be a good indication that *innate, instinctive* predispositions (empathizing and sympathizing) are involved here. The whole subject of relaxation of intragroup selection pressures in modern societies is an emotionally and ideologically charged one, posing an additional difficulty for impartial scientific analysis.

The following passage from Ernst Mayr summarizes well our discussion up to this point and introduces our forthcoming subjects: the consequences of the diversification of human mental and physical abilities; the emergence of varied social niches; and how social selective practices may influence the transmission of genes from one generation to the next.

> Could the now existing human species evolve as a whole into a "better" new species? . . . To be sure, there is abundant genetic variation within the human genotype to serve as material for appropriate selection, but modern conditions are very different from the time when some populations of *Homo erectus* evolved into *Homo sapiens*. At that time, our species consisted of small troops, in each of which there was strong natural selection with a premium on those characteristics that eventually resulted in *Homo sapiens*. Furthermore, as in most social animals, there was undoubtedly strong group selection [intragroup selection for cooperation, to my understanding]. Modern humans, by contrast, constitute a mass society and there is no indication of any natural selection for superior genotypes. (Mayr, 2001, p. 261)

4.6.4. Diversification of Physical and Mental Characteristics Resultant on Relaxation of Natural, Sexual, and Intragroup Selective Pressures and Consequent Diversification of Social Niches

The relaxation of these three kinds of selective pressures, as repeatedly argued in this work, led in the case of polygenically inherited physical, and especially mental, traits to a degree of diversification unprecedented in the animal kingdom. The relaxation of normalizing (or stabilizing) selection produced a graded continuum of variability of physical traits, and even more of mental traits, around an optimum. Since each trait is inherited more-or-less separately from the others, their different assortment in individuals means an almost infinite extent of variability. If we combine this variability with the widely different social niches a person can choose from, we can appreciate the degree of uniqueness of each individual path of life.

It is reasonable to presume that this diversification proceeded at a pace matched by the increasing ecological dominance of humans, efficiency of cooperation within groups, and increasing complexity of social organization, leading to the emergence of increasingly varied social niches that were in turn able to contain and profit from most of these diversified abilities. Among persons with similar talents or abilities, possibly working in similar social niches (in a university environment, for example, or in the past, in the same guild), a useful exchange of ideas takes place, and perhaps a constructive competitiveness, which further enhances those specific abilities.

Different social niches and the achievements of individuals with different abilities may become interconnected according to the interests of the wider community. For example, a new understanding by a gifted theoretical scientist may be further developed by other, more practically minded scientists into a prototype of an instrument or appliance that may solve a commonplace difficulty in everyday living. If the final product seems to be useful for the wider community, individuals with still different abilities (in this case marketing skills) will attempt to advertise the product's assets and sell it to an increasing number of consumers. Frequently, the lonely inventor possesses superior abstract thinking and imagination, coupled with less than average practical sense and socializing abilities. The businessperson, on the other hand, seems to be endowed with high social intelligence and an ability to manipulate other persons (but perhaps has a less-developed talent for independent creative thinking than the inventor), exemplifying the need for cooperation between individuals with widely different mental abilities for a successful cultural or technological achievement.

Geneticist Theodosius Dobzhansky summarizes this subject well: "It is possible and indeed probable, that occupational differences between human populations usher in some correlated genetic differences" (Dobzhansky, 1962, p. 250). And: "When the environment is highly diversified, as it is in civilized societies, all these genotypes may find suitable opportunities. The countless occupations, callings, and pursuits that advanced human societies offer (corresponding to ecological niches on a biological level) are enough to accommodate human polymorphism" (idem, p. 223).

I found it quite difficult to translate the abstract concept of "social niches"—that is, those "occupations, callings, and pursuits" that correspond "to ecological niches on a biological level" into concrete examples. At first, I tried to define these niches as social institutions like universities, hospitals, and armies, or occupations that require one to have some particular innate abilities to ensure success. However, I found this approach inadequate. Teachers and doctors experience the environment of an institution they work in very differently from those employed in administration or maintenance: Clerks, technicians, and janitors are submitted to different kinds of occupational pressures and rewards. Likewise, a surgeon's working environment is very different from that of a psychiatrist. Even within the psychiatric profession, different niches exist—biological, clinical, and psychodynamic psychiatry; hospital or outpatient clinic; research, education, or administrative tasks—each requiring somewhat different innate predispositions for their optimal execution. Even inside each one of these sub-niches, one can develop a unique working style.

Psychologist John L. Holland built his theory of vocational counseling on the requirement of a personality-environment match.[6] Holland presumes that, based on their personality, persons tend to gravitate toward the environment that will assure them the most satisfaction, as well as proficiency in their vocational career (Brown, & Associates, 2002, pp. 375–385).

Another complicating factor is that other incentives besides personality-environment match exist in the choice of a profession. Such a choice may be powerfully influenced by the supply and demand of the market, the social status and financial reward of various occupations, family tradition regarding an occupation, the influence of an authority figure, current fads, financial means to pay for an expensive education, etc.

Being concerned in this work mainly with the inherited predispositions to certain behaviors, I will mention here only some personality traits with a known inherited component, as well as occupations that seem better- or less-suited respectively for these traits. Probably the most pertinent example is the case of inherited special talents (artistic talents, for example) and the appropriate occupations that may exploit them. Someone with a submissive, self-effacing personality can exploit this trait by working as a secretary or waiter, for example, while self-assured, expansive personalities with a strong social presence can perform better as a superior, a manager, or a field officer. High social presence, practical intelligence, and realistic flexibility in personal relationships may be very useful for politicians or businesspeople, or those in other occupations requiring manipulative interactions with various kinds of people, while asocial personality traits, an inability for close cooperation with others, and low theory-of-mind abilities are better suited for a solitary occupational milieu. Persons in whom stress and possible dangers elicit strong and longstanding anxiety need a social/occupational environment that shields against such stresses, while a fearless person may flourish or even actively seek out occupations or pursuits involving dangerous situations and enjoy overcoming them.

Other traits with a possible innate element that may influence occupational orientation are: high versus low capacity to support social disapproval, high general activity levels versus easy tiring, levels of motor coordination, the ability or inability to express emotions adequately in a social setting, and so on.

It must be noted that the individual is not the only factor that decides the fit between inherited predispositions and social niches. The individual not only chooses but is also chosen, accepted into, or rejected from different social or occupational settings. This can happen at the preparatory stage of acquiring a profession (by taking an entrance examination, for example) or during its practice. I suspect that a key factor in the superior achievements of modern, socially highly mobile, egalitarian democracies in many domains needing the cooperation of persons with diverse abilities (science, technology, economics, commerce, military power, etc.) when compared with more traditional, rigidly stratified societies is the much higher efficiency of this bidirectional selective process, which ensures a better inherited traits/occupational environment match. (In chapter 3 of his 2012 book *Games Primates Play*, evolutionary biologist Dario Maestripieri illustrates impressively how in present-day Italy, as well as in many other countries, nepotism and cronyism interfere with the efficiency of this selective process with destructive consequences for cultural achievement.)

The desirability of the social niche/inherited-predispositions-and-abilities match also seems to apply to other areas of activity besides work, such as the type of marriage relationship one is suitable for (if at all), the optimal number of children, type of leisure activities, whether an urban, suburban, or rural environment is most suitable, and so on.

In opposition to the eugenics movement, which saw in this wide diversification of cognitive, emotional, and physical abilities (resulting from RfNSPs) a serious danger to subsequent human biological evolution, geneticist Theodosius Dobzhansky warmly welcomed genetic diversity: "Human genetic diversity is not a misfortune or a defect of human nature. It is a treasure with which the evolutionary process has endowed the human species. . . . Any human society from the most primitive to the most complex (the latter more than the former), needs a diversity of men adapted and trained for a diversity of functions" (Dobzhansky, 1973, p. 44).

While I fully identify with this view, the following chapters of this text will reveal that this accentuated diversification of genetic predispositions to behave has a dark side, too, that of predisposing for dysfunctional behavior at a level of severity that can no longer be contained by the diversified social niches at one's disposal.

4.6.5. Directional Intragroup Natural Selection Dependent on and Secondary to Antedating Relaxation of Natural Selection Pressures External to the Group Life

I would like to introduce this section with quotations concerning *ecological dominance*. Mark Flinn and Richard Alexander define ecological dominance as "the relative lack of selection from extrinsic causes compared with the relative importance of selection from interactions with conspecifics" (2007, p. 253).

David Geary comments on the same topic as follows: "Humans are *ecologically dominant*, and once dominance was achieved, there was an important shift such that the competing interests of other individuals and coalitions of other people became the central pressure that influenced the evolution of brain and cognition in humans" (2007, p. 306, emphasis in the original).

As we have seen before, an evolutionary event resulting from a far-reaching and long-lasting environmental change frequently brings about relaxation of the former environment's selective pressures, leading to a diversification of the traits released from these selective pressures. The new selective pressures of the changed environment may make use of some of these diversified elements of the old structures or traits, selecting from among them usable elements to build new adaptive structures and functions while the unusable elements as a rule atrophy, degenerate, or disappear altogether. As already mentioned, this sequence of events was probably at work at the dawn of the evolution of the terrestrial animals, when the fins of fish were transformed into legs, and many millions of years later, when the reverse event took place—that is, when the cetaceans (whales, dolphins) evolved from terrestrial mammals. In that later case, for instance, many features of the terrestrial skeleton and respiratory organ (inconvenient in an aquatic environment) were retained.

It seems to me logical to presume that during early human evolution the order of events was partly similar with regard *to physical traits.* For example, the relaxation of selection pressures for an arboreal existence released the limbs from those respective selective pressures, leading to their diversification. These diversified structures came under the new selective pressures of more specifically human ways of life, leading, for example, to erect posture, bipedal locomotion, and adaptation of the hands and arms for finer and more varied object manipulation, while the legs became the organ of terrestrial locomotion, losing the prehensility of the big toes, and so on.[7]

The same events seem to occur in *nonhuman* animals with respect to behavioral characteristics. The preprogrammed, rigid instincts (IRMs and FMPs), adapted to simpler and more predictable environmental conditions than human beings occupy, when relaxed from these constraints as a result of environmental change, may become more diverse and less differentiated. The new environmental pressures may use some of these diversified behavioral traits to build the new adaptive behavior (that is, altered FMPs and IRMs) in a way quite similar to that of the bodily traits. On the other hand, the learning abilities of animals need less adjustment in these circumstances, being evolved in the first place to cope with unpredictably changing or more complex elements of the environment.

Of course, the evolution of characteristically human behavior and mental abilities cannot be accounted for in a similar way. It is true that the decisive change thought to initiate human evolution was a change of the natural environment, probably from African rain forests to savannas. Nevertheless, this change occurred in an already highly developed social ape, with hands already quite well adapted to object manipulation and highly developed learning abilities, including, it seems, an innate mental potential to acquire rudimentary symbolic language. (This conclusion assumes that the evolution of the extant ape species, which we use for comparison, was less dramatic than that of humans and did not evolve from an ancestor with considerably inferior mental and behavioral abilities than modern apes; Gomez, 2004, de Waal, 2005, p. 144.) The instincts of that ancestor *were already more diffuse*, that is, less differentiated into specific FMPs. It possessed no ritualized courting patterns, like many birds, and no ritualized fights between males evolved specifically for intrasexual competition. When a young chimpanzee is separated from its group of conspecifics at an early age and later reunited with it, its social and sexual behavior will be inadequate, needing a period of learning in order to redress its lack of innate ability. It has an extended period of maturation for perfecting through learning these as well as other kinds of behavioral predispositions. (The release of the bonobos' sexual behavior from strictly reproductive aims, and the bonobos' and chimpanzees' great diversity with regard to mental characteristics like personality, sociability, and intelligence, were mentioned in section 3.1.)

Dobzhansky described this same evolutionary trajectory of instinctive behavior: "Beach (1948) found that male rats reared in complete isolation copulate normally when offered receptive females, and females reared in isolation react normally to the approaches of males. This is not so with the chimpanzee. . . . Sexual behavior is quite variable in different individuals; individuals raised without an opportunity to observe the mating behavior of others do not copulate when first meeting a receptive partner. However, if left with a suitable mate for a few months they eventually do so. In man . . . [t]he overt manifestations of . . . libido are, however, largely culturally determined and hence learned; even the consummatory sex act is not wholly instinctive" (Dobzhansky, 1962, p. 205).

Therefore, the seemingly unique path of human behavioral and mental evolution cannot be seen as entirely distinct from primate evolution in general. I cannot explain, of course, what specific environmental conditions, selective pressures, or intrinsic possibilities of the genetic material led (after quite a number of blind alleys) to the appearance of anatomically modern humans or why our ancestors did not "choose" the evolutionary path of the chimpanzee or bonobo. What seems striking, however, is the great difference between the *relative proportions* of the different kinds of selective pressures and relaxations (natural versus social) that shaped human evolution in comparison with that of the apes.

As we have seen before, the selection pressures of the natural environment progressively decreased in humans, something that cannot be said about orangutan, gorilla, chimpanzee (or even bonobo) evolution. At the same time, *the intragroup selection pressures, that is the requirement to behave in a way that enhances the function of the group as a coordinated unit and does not disrupt its cohesion but, if possible, enhances it, gained supremacy,* becoming the main shaping force for what has to be seen as innate predispositions for specifically human behavioral and mental qualities.

It seems reasonable to presume that, in Pleistocene conditions, attracting group aggression to oneself as a result of deviant behavior, risking being expelled from the group, or even risking more than fleeting group disapproval was a serious blow to survival and reproductive aspirations. *Since most of these selective pressures originated from quickly changing, complex social interactions, this state of affairs discouraged the evolution of new, rigidly differentiated instinctive behavior patterns* (that is, specific FMPs and IRMs common in animal adaptation to changed environment). Instead, it seems that *the influence of these kinds of pressures on the instinctive predispositions was to keep them in an optimally diffuse and diversified state, experienced subjectively as dim intentions, predispositions, and preferences* that needed the selective learning of socially provided information in order to lead to mature human behavior. Therefore, the importance of progressively more advanced forms of learning increased without, of course, becoming dissociated from or entirely replacing basic biological aspirations.

Almost all the specifically human instinctive predispositions (hypothesized to evolve during the Pleistocene era) discussed in evolutionary psychology and related scientific fields are the result of such selective pressures originating from social coexistence.

Some of these social-life-induced instinctive predispositions seem to be directly related to coping with the problems posed by the diversified mental and physical characteristics of conspecifics in a group setting. Cooperation between group members, even if they are different in physical and mental endowments and are not closely related genetically, seem to me to be one important predisposition in this context. Cooperation seems to necessitate the evolution of "theory of mind" abilities—that is, inferring in others such intrapsychic contents as convictions, intentions, emotions, thoughts, loyalties, etc., that are not expressed explicitly in behavior and verbalizations, as well as the acquisition of *face-saving and tension-reducing tactics* during social interactions. Reciprocation or reciprocal altruism (section 3.2.1.1) is one of these forms of cooperation that evolved progressively into highly complex transactions like the redistribution systems of states (Harris, 1989, pp. 358–390).

Social norms, customs, laws, etc., evolved (among other incentives) in order to secure the settling of conflicts between individuals (including between those with widely different physical and mental characteristics), or groups of individuals, in ways that do not disrupt seriously social coexistence and cooperation. Controlled, attenuated, rule-bound competition (intrasexual, or for a desired working place or social role, etc.) serves the same purpose.

Other inherited predispositions resulting from social pressures are less dependent on diversification and seem to be influenced more directly by the requirements of social coexistence, with no regard to whether the involved individuals are more or less different from each other. Examples of such inherited predispositions include conformity to cultural norms and traditions; the subordination of individual interests to those of the group to a certain extent, particularly during intergroup conflicts; the evolution of greatly improved verbal communicative ability (in its inherited aspect, Pinker, 2007); the tendency to imitate the behavior of the majority or that of the successful (Richerson & Boyd, 2005, pp. 120–126); and xenophobia (fear of strange persons outside one's own group or culture).

However, at a later stage of human evolution, intragroup selection pressures began to relax progressively (as detailed in section 4.6.3). The main reasons for this seem to be that the formerly mentioned ability for social coexistence and cooperation were already highly developed and widespread in human populations (that is, their inherited aspects were fixated in the human genome). In addition, the vastly increased ecological dominance and the abundance of life-sustaining resources attenuated the uncompromising need for optimal group functioning in securing survival and reproduction.

4.6.6. Social Selective Practices and Other Cultural Traditions That Influence Gene Transfer from One Generation to the Next

With the increased diversification of inherited physical and mental traits and predispositions, with the progressively improving mental and behavioral abilities that welded human groups into more and more efficient functional units, as well as with the resultant multiplication of "social niches" that are supposed to contain and profit (when possible) from this diversification, there appeared *social selective practices* in order to ensure a better fit between these social niches (occupations, social roles, etc.), on one hand, and individual abilities—both inherited and learned, on the other. These social selective practices *may affect, sometimes powerfully, gene transfer between generations, but at the same time they frequently do not comply with traditional evolutionary principles.* Evolutionary logic presupposes a strong causal interdependence between success in survival; success in competing with conspecifics for resources, including access to fertile females; ascent on the scale of dominance hierarchy (when applicable), which ultimately secures higher rates of reproduction; and adequate parenting skills (when applicable), which means increased chances for the offspring to attain reproductive maturity in order to transfer genes to the next generation. Failure at any one of the links of this chain precludes or seriously compromises success in the following ones.

This strong interrelatedness between successive links of the above evolutionary adaptive chain frequently disintegrates in the case of social selective practices or other social/cultural traditions; that is, *success in one link of this chain does not necessarily predict increased chances for success in the following one(s), and conversely, failure at an earlier link does not necessarily predict failure at a later one.* For example, failure to ascend in the social hierarchy or secure more resources does not necessarily predict decreased rates of reproduction. Being found fit for important social functions or roles and the successful accomplishment of their requirements, which may mean ascendance in social hierarchy to a more dominant position, does not necessarily predict improved rates for survival and reproduction. Being found fit for military service in a prestigious combat unit may seriously lower the chances of survival (particularly in wartime); being found suitable for the priesthood in religions with obligatory celibacy for priests precludes reproduction; passing an entrance examination to a university, which means good prospects for acquiring a well-paid and well-respected occupation, enhances the chances for a higher standard of living or rise in the social hierarchy, but not necessarily of having more children.

As mentioned in the section on sexual selection, in many traditional societies, the choice of a mate is not in the hands of the future couple at all but is decided by the father or some other authority, and the important considerations are not good innate physical and mental traits in the prospective mate (good genes) but how much material profit the bride's family will gain. Or, the match may be determined by cultural tradition or serve the promotion of important social alliances, etc. In modern societies, it was found that the correlation between spouses is much higher for similar levels of education than for physical characteristics like height and weight or mental characteristics with an inherited base like personality traits (Plomin et al., 2008, p. 160).

Sometimes the reproductive rate of an entire nation is forcefully affected by ideological incentives or by the official policy of a totalitarian regime. In modern China (until 2015), only one child per family was permitted. In Rumania, under the dictatorship of Ceauşescu, contraception and artificial abortion were strongly opposed. Many religions (including branches of Christianity, Judaism, and Islam) categorically prohibit the use of contraception or artificial abortion, or even the use of sex for pleasure instead of reproduction.

4.6.6.1. Dissociation between Ascent in Social Hierarchy or High Standard of Living and Reproductive Rate

Probably the most striking divergence from the interdependent natural selective chain in modern societies is the dissociation between the first part of the chain (access to resources, high social status, high living standard, success in the social or professional sphere) and reproductive rate. The consequence of this dissociation is that those in the affluent social stratum have fewer children than those who are poor; and more intelligent or educated individuals have fewer children than the less intelligent or less educated. This phenomenon, which has been linked to historical change, is known as the "demographic transition." There was a period in post-industrial societies "in which dramatic changes in fertility and mortality have occurred along with a rise in living standards . . . rich families reduce their fertility earlier, and often more markedly, than the rest of the population" (Laland & Brown, 2011, p. 89).

The biologist and anthropologist Jonathan Marks argues that humans, in fact, passed through two (instead of one) demographic transitions, both strongly connected to a "relationship between economics and reproduction" (Marks, 1995, p. 204). The first occurred with the introduction of agriculture, when the low population densities of hunter-gatherers (maintained by meager life-sustaining resources, high offspring mortality rates, and extension of the period of lactation with its anti-ovulatory effect) rapidly increased. The second "was the time when the economic trend that began with industrialism and culminated with modernization began to take hold, and to transform once again the structure of the population" (idem, p. 206). Both demographic transitions involved a drastic change in the women's social role, which had repercussions on the reproductive rate. (For the different explanatory hypotheses of this phenomena, see Laland & Brown, 2011, pp. 88–91, and Jonathan Marks, 1995, pp. 204–206.) Cultural and psychological incentives thought to "conflict with the biological imperative to reproduce" are: the improvement and comfort of the parents' life, when more children are perceived as an economic and psychological burden; ideological reasons like the "moral authority of asceticism"; and sexuality's new function in the human species—in addition to reproduction—of enhancing emotional intimacy and bonding (Marks, 1995, pp. 196–197).

4.6.6.2. Dissociation between the Biological Parents' Parenting Abilities and the Offspring's Survival, Social Success, and Reproductive Rate

Contrary to other animals, in which parenting instincts are narrowly directed toward the biological progeny, in humans—with social progress, increased abundance of life-sustaining resources, and the relaxation of the IRM with regard to parenting instincts—a dissociation took place in this domain, too. The dissociation occurred between the biological parents' ability to bring up their own children and the actual success of these children in the domains of survival, occupational/social status, and rate of reproduction.

In the offspring of evolved nonhuman animals, a biologically desirable outcome is strongly dependent on the biological parents' parenting abilities. In the case of humans with a serious disability that interferes with bringing up their children, this role may be taken over by biological relatives (a practice known in chimpanzees, too), biologically unrelated adoptive parents, or an orphanage or foster home. Even in the case of well-functioning parents, social institutions regularly take over extensive aspects of the children's care and education from an early age, thus increasing considerably the likelihood of success in the previously mentioned parameters.

It has to be stressed, however, that in spite of the departures from the natural chain leading to reproductive success in highly evolved social mammals, the human species remains in this respect by far the most successful. It seems that the benefits of human culture and social organization greatly surpass their drawbacks in supporting successful reproduction.

4.7. Conclusions

The preceding overview of the similarities and differences in the selective mechanisms of inheritance between humans and nonhuman animals illustrate how complex, varied, and unique the mechanism of gene transfer is in human populations (even before considering interactions between them and the possible influence of genetic engineering), and, as a result, how unpredictable future human evolution can be. However, it seems to me that some tentative predictions can be attempted. Before doing so, however, I would like to put the effects of relaxation from natural selection pressures into proper proportion.

Since the main theme of this work is the RfNSPs and some of its consequences, it may be conveying an inflated picture of the magnitude of diversification and the trouble it produces. Therefore, it must be stressed that, from a wider, biological point of view, the diversification of *physical* characteristics, with all its importance and concomitant problems, is ultimately quite modest. Most people function acceptably in their social niche in spite of various acute and chronic medical conditions, such as obesity, diabetes, hemorrhoids, backache, etc., that either have a genetic component or result directly from unique aspects of human evolution (like bipedal locomotion or overeating). In most cases, such disorders are compensated for or corrected satisfactorily by modern medicine. As a result, in modern societies the average life expectancy is around eighty years, much greater than it is in simple societies with primitive technology and a lifestyle resembling that of our ancestors. One of the mechanisms that seems to protect humans from physical diversification of ominous proportions, in spite of considerable relaxation of natural selection pressures, is the "internal selection" (mentioned in section 4.2.2.1), which ensures the "selection of mutants . . . in accordance with their compatibility with the internal coordination of the organism" (Whyte, 1965, p. vii). Or, as Stephen Gould points out: "All organisms evolve as complex and interconnected wholes, not as loose alliances of separate parts, each independently optimized by natural selection" (Gould, 2007, p. 464). And again: "The individual [genotype] as a whole is the target of selection . . . [that] sets severe limits for the response of the phenotype to selection" (Mayr, 1982, p. 590).[8] The deleterious impact of the diversification in the mental and behavioral sphere, however, seems much more impressive than it is with regard to somatic traits. It was found, for example, in 1994, that the "estimated lifetime prevalence of [all] mental disorders in the United States approaches 50 percent" (McGuire & Troisi, 1998, p. ix; see also pp. 17, 159, and 182). Another study from 2005 confirmed this estimate and added the finding that one out of three persons in the United States suffered from a mental condition during the last year (Plomin et al., 2008, p. 194). If we add to this figure dysfunctional cognitive or emotional states that do not affect occupational or social functioning to an extent warranting a psychiatric diagnosis (as mentioned in the introductory section) we can surely argue that there are very few persons entirely free of dysfunctional behavior or of disordered mental states during their lifetimes. This topic is detailed in chapter 7.

However, the proportion of grave mental disorders that lead to a lifelong inability to function adequately in the occupational and social sphere is relatively small in developed countries and even smaller in the underdeveloped ones (section 7.2.7.2). Therefore, the impact of grave mental disorders on the functioning of entire communities is relatively small. I am not aware of any instance in which an entire community was seriously stressed economically or psychologically or in daily functioning as a result of an overwhelming prevalence of grave mental disorders, or that a modern state was hindered in its technological or cultural progress because of the same reason.

It can be argued, I think, that while humans progressively lose adaptive traits important for coping with a *natural* environment, rendering their chances of survival in these conditions questionable without technological and social support, they became progressively better adapted to their *social* milieu. At the same time, the social milieu itself seems to become, through some complex, bidirectional feedback mechanisms, more and more responsive to its members' individual needs. Think, for example, how easily the use of computers and of related digital appliances has been mastered during the life span of just one generation and how in turn they have changed almost every aspect of modern life, mostly for the better.

It seems appropriate to assume that, barring some unexpected—or expected—global disaster, technology will continue to progress, leading to further ecological dominance and compensating even more for human physical and mental frailties, which means further relaxation of natural selective pressures on gene transmission and a widening range of diversification that should be contained by even more varied, more specialized future social niches. Hopefully, the social selective mechanisms that are supposed to match individual mental and physical diversity with the appropriate social niches will improve, too.

On the undesirable side of the above scenario, it seems that, at the same time, the prevalence of dysfunctional behavior patterns will increase in technologically advanced democracies. We already have research findings in this respect, in particular concerning depression (Healy, 1999, pp. 251–252; Plomin et al., 2008, p. 208; Tsuang & Tohen, 2002, p. 404). Perhaps carefully constructed, increasingly specialized future social niches with increased responsiveness to inherited individual drawbacks and more considerate social selective practices will be able to compensate better for these disabilities.

The following chapter deals with comprehensive instinctive mechanisms whose diversification in humans as a result of RfNSPs, as it is hypothesized in this text, underlies the inherited predispositions for common mental disorders.

5

Four Comprehensive Instinctive Mechanisms and the Relevance of Their Excessive Diversification to Human Dysfunctional Behavior

"I am pretty sure that most of the great mysteries,
if they are in some sense to be solved, will be solved
within an evolutionary context."

—PAUL EHRLICH, 2000, p.306

5.0. Introduction

As mentioned in the introductory section, the present work grew out of a clinical observation—namely, that the overwhelming majority of mental disorders I encountered in general psychiatric practice affect a wide range of specific instinctive mechanisms rather than a single one. Of course, there are mental disorders that affect primarily or exclusively a circumscribed instinctive sphere, such as sexual activity, feeding, or a fear reaction to a specific agent (simple phobia). Nevertheless, in general psychiatry, these disorders are encountered rarely in comparison with those that affect at least several specific instinctive mechanisms at the same time, including more generalized anxiety disorders (neuroses), personality disorders, affective disorders, and schizophrenia. The non-recognition of the part played by these comprehensive mechanisms in generating dysfunctional behavior precludes, in my view, any successful attempt at a theoretical understanding of the etiology (more precisely, the inherited aspect) of these disorders. This claim can be illustrated with Freudian psychoanalytic theory's lack of success in interpreting mental conditions as the outcome of dysfunctions of the sexual drive, or with evolutionary psychology's inability to explain the evolutionary roots of mental disorders, in spite of its considerable success, in my opinion, in elucidating the role of circumscribed instinctive motives in the evolution of normative human behavior.

The present chapter contains the description and discussion of four such *comprehensive instinctive mechanisms* whose relaxation from natural selection pressures (RfNSPs), and consequent excessive diversification, as hypothesized in this theory, may account for the innate predisposition to major mental disorder categories. Three of them are relatively simple. They are:

1. *Seasonal fluctuations in the intensity of overall instinctive activity* as an adaptation to seasonally changing amounts of life-supporting environmental resources (temperature, amount of daylight, food availability).

2. The *three basic kinds of transformations* that active (that is, directly life-supporting) instinctive motivations undergo when their discharge toward a suitable environmental agent (the releaser) is frustrated.

3. *The evolutionary trajectory of drives from differentiated to diffuse.* This trajectory underlies the variable complexity and flexibility of animal behavior. The rigidly preprogrammed behavior patterns of simple animals (unchangeable by learning or personal experience) progressively became predispositions and inclinations that are more and more tentative as we ascend the evolutionary ladder. The loosening of fixed patterns allowed the flexible modification of behavior by experience and learning acquired during ontogeny, a process that attained its culmination in humans. This process ensures more accurate adaptation to more extensive, more varied, more complex, and more rapidly changing physical, biological, and especially social environments.

4. The fourth comprehensive instinctive mechanism influencing a wide range of circumscribed instinctive motivations is much more complex than the first three. It refers to *the dichotomy, interrelationship, and interaction between two already highly complex behavioral categories, active and reactive behavior.* Active behavior unites under a

single concept all the instinctively based activities that directly serve the goals or needs of the individual life, its sustenance, maturation, and reproduction, and when appropriate, the efforts made toward increasing the offsprings' chances of attaining reproductive maturity (parenting). Reactive behavior refers here to all those instinctively based reactions whose aim is to protect the organism from harmful or dangerous environmental influences (like evading a predator). Excluded here are evolved behavioral patterns that are the result of long-standing, highly predictable harmful environmental influences, like migration or hibernation, as well as harmful external influences that frustrate one single active instinct in progress, mentioned in point 2.

The optimal interrelationship between these last two behavioral complexes has great fitness-enhancing value; therefore, in nonhuman animals it is tightly controlled by natural selection. *The RfNSPs in human societies, on the other hand, lead to extensive diversification concerning all four instinctive mechanisms, with the potential for psychopathological developments at the extremes of the resultant diversified continuums.*

While the adaptive significance of the four comprehensive instinctive mechanisms may seem quite straightforward, reliably detecting their effect in a particular observable behavioral sequence in humans is by no means straightforward or easy. In fact, we are facing here a big methodological dilemma. On one hand, it seems clear that *the effects of vague, tentative behavioral predispositions cannot be observed reliably in overt human behavior* and even less can they be measured quantitatively. Being unobservable entities, they cannot be approached directly by scientific experimentation and inductive generalizations (see chapter 1, section 1.1, on scientific theory). This is in contrast to the external stimuli and the behavioral output (the final result of the central nervous system's computations), which are *observable* entities.

However, in the study of behavior, it is generally very important to *address first the innate behavioral incentives* before trying to interpret the effects of the external stimuli on the final behavioral output: "Any investigator of the ontogeny of animal behaviour is well advised to begin with the time-honored procedure of first searching for whatever may be blueprinted by heredity . . . in any system whose function consists of the interaction of many parts, the least changeable are the best point of departure, because their properties appear *most often as causes and least often as effects* in the interaction that is to be studied. An even better reason is that phylogenetically adapted structure represents, in the ontogeny of behaviour, *the indispensable prerequisite for guiding modification [through learning] in the generally improbable direction of adaptedness*" (Lorenz, 1965, pp. 34–35, emphasis added).

Extrapolating this line of reasoning to humans, it seems hopeless to build an adequate theory on the cultural milieu's influence on individual human behavior without trying to clarify first the role of innate predispositions in selecting and processing these cultural influences and evaluating their impact on the biological needs of the individual organism. In pursuing such clarification, the only way out from the impasse caused by the unsuitability of human instinctive predispositions to direct scientific experimentation seems to be the hypothetico-deductive theory, detailed in the first chapter, some of whose implications do lend themselves to scientific investigation.

5.1. Seasonal Fluctuations in the Intensity of Instinctive Activity in the Animal Kingdom; Their Alteration and Diversification by the Artificial, Human-Made Environment; and Diversification's Relevance to Dysfunctional Behavior

For psychiatrists working in a psychiatric emergency room, the admission of patients suffering from an acute psychotic episode is a commonplace occurrence. Most of the time, the onset of the episode cannot be linked to a specific life event or circumstance or to some medical condition like drug intoxication or some other kind of organic causation. Frequently, the individual has had similar episodes in the past and has been diagnosed as suffering from one of the major psychiatric disease categories with a fluctuating course: bipolar or unipolar affective disorder, schizophrenia, or schizoaffective disorder. Frequently, the person seems to be in an emotional turmoil, is agitated, and shows pressure of speech. Or the patient may seem withdrawn and disconnected from the surroundings. (Sections 6.2.1 and 6.2.2 discuss how intense instinctive activities can lead either to excitation or inhibition of the overt behavior.)

Just as the onset of the episode cannot be linked to a circumscribed life event, the cause of its clearing up—as a rule, several months later—also remains a mystery. (This phenomenon is more evident without antipsychotic drug therapy.) Sometimes, the acute psychotic episode is followed by a retarded depressive phase called (in the case of nonaffective mental disorders) post-psychotic depression. In bipolar affective disorder, the retarded depression following a manic episode is considered an organic part of the disease.

For quite a long time I assumed that the cause of these acute episodes has to be a longstanding suppression of strong emotions (or, in the context of our discussion here, the subjective experiential aspect of instinctive activity) in these often sensitive and socially maladjusted persons. The accumulation and eventual unrestrained eruption of these emotions into overt behavior leads to these acute episodes. (This so-called "hydraulic model" of instinctive activity was hypothesized both by Freud as well as by Konrad Lorenz.) Being expressed (acted out), the accumulated instinctive energetic charge slowly loses its strength until it subsides completely. (As I am an introverted person who cannot express his emotions easily during everyday interactions, this explanation seemed to me quite convincing, intuitively.) However, I began to doubt the validity of this hypothesis in the case of acute psychotic episodes.

When one suppresses emotions in frustrating environmental conditions, it is usually a single or a very restricted range of emotions that are involved. When this emotion is ultimately discharged, the behavior is directed toward a circumscribed environmental agent, the cause of the frustration, or it is channeled along some appropriate avenue of displacement; for example, rebellious feelings against a domineering superior may be discharged as anger toward a subordinate.

As a rule, however, in acute psychotic episodes, *a wide range* of instinctive activities are involved; in mania, for example, sexuality, aggressiveness, dominance, and the overall intensity of the behavior strengthen, while a decrease occurs in fearfulness, anxiety, skepticism, cautious foresight, and attention to detail and to useful but undesirable kinds of information, along with the amount of sleep. Furthermore, in acute psychotic episodes, the emotional behavior is not directed toward one circumscribed environmental agent (the cause of the frustration or its displacement object). In acute paranoid schizophrenia, for example, the target of the aggression or fear is as a rule a remote or abstract entity like an intelligence service, the police, or a superhuman being, not a well-defined real person, situation, or event; this is not to mention catatonic or hebephrenic agitation in which the emotional arousal may lack any concrete or imagined target.

My second problem with the hypothesis that fluctuations of circumscribed frustrated instinctive motives are the cause of acute psychotic episodes concerns the *duration* of the "suppressed," as well as "cathartic" periods. While the cathartic period or acute episode may last for months—the average duration of mania, for example, before effective drug treatment was available, was six months (Tsuang & Tohen, 2002, p. 454)—the remissions may last for years. This timescale seems to me too long for a mechanism that involves ventilating circumscribed emotional frustrations. Nevertheless, the changes discrete active instinctive motives go through during frustration are considered an important topic for the purposes of the present text and are detailed in section 5.2.

My third problem with the hydraulic model, and the factor that was decisive in its abandonment as a hypothetical explanation for acute mental disorder episodes, is that certain fluctuating conditions, like winter depression or other forms of seasonal affective disorders (psychotic or not), were evidently related to, or synchronized with, seasonal periodicity. The six-month average duration of a manic episode, mentioned above, corresponds well with the duration of seasonal fluctuations in the amount of environmental resources at higher latitudes. I began to suspect that acute mental disease episodes desynchronized from seasonal rhythms are not entirely different entities from those synchronized with seasonal rhythms but are the result of the relaxation of natural selection pressures on the appropriate brain apparatus, with a resulting diversification and disintegration of an originally adaptive mechanism in the human-built, artificial environment. This environment significantly alters the triggers that synchronize the fluctuations in the intensity of the overall instinctive activity with the abundance of the seasonally changing life-sustaining resources (above all, the amount of daily light), and obviates this evolutionary adaptation by providing constant, year-round resources needed for the life processes, such as food, extended daylight, and optimal ambient temperature.

While the frustration model of instinctive intentions was intuitively appealing, seasonal fluctuations of instinctive intensities as the cause of the inherited predisposition for several episodic mental disorders seems to me much better founded in evolutionary logic and better supported by verifiable facts. Before entering the subject of seasonal behavioral fluctuations, however, we need to take a short detour into the nature of psychiatric nosology and the relationship between the descriptive aspects of clinical mental disorder categories and their genetic underpinnings.

The psychiatric disease entities are defined in the DSM-III classification system (*Diagnostic and Statistical Manual of Mental Disorders*, 3rd ed.) as admittedly atheoretical, descriptive categories: "DSM-III is generally atheoretical with regard to etiology. . . . DSM-III attempts to describe comprehensively what the manifestations of the mental disorders are and only rarely attempts to account for how the disturbances come about. . . . That general approach can be said to be descriptive in that the definitions of the disorders by and large consist of descriptions of the clinical features of the disorders" (Kaplan, Freedman, & Sadock, 1980, p. 1053). This approach has not changed in principle in the decades that have elapsed since the publication of DSM-III.

In other words, discrete dysfunctional behavior patterns encountered regularly in clinical practice (and severe enough to considerably disturb daily functioning) are labeled as mental disorder categories. In addition to descriptive features, these clinical descriptions also contain some *general regularities* as to the typical age of onset of the disorders, their course, the manner in and gravity with which they affect social and occupational functioning, and their possible reaction to different treatment modalities, again as empirically observable regularities (or laws). The "mapping" of the observable "territory" or subject matter of an *empirical* science (that is, all scientific branches except mathematics and logic), its categorization and systematization, has to be the first step in any empirical scientific enterprise. Systematization means: "bringing ordinary things under ever wider and more abstract concepts" (Quinton, 1973, p. 284). This activity is the precondition for the discovery of empirical laws or regularities previously mentioned (that is, time of onset, characteristic symptom complexes, course, prevalence, etc.).

The assumption of distinct or unique patterns of heritability for each one of these empirically defined disease categories is, however, an entirely different matter, because it brings into the discussion theoretical and causal considerations besides empirically observable regularities. If inheritance can distinguish between whole clinical disorder categories, it means that, in accordance with evolutionary logic, such clinical disorders must have separate, phylogenetically evolved adaptive mechanisms as their foundation. They thus cease to be exclusively descriptive categories anymore. Instead, we expect them to express some specific inherited adaptive trait, adaptive in the present or during some period in the evolutionary past, which involves phylogenetic causality, at least as a *predisposition* to develop the respective disorder. Therefore, we have to ask here to what extent inheritance can distinguish between categories of clinical mental disorders.

The answer to this question is not a straightforward yes or no but instead a "mixture" of yes and no elements. In fact, it was discovered that almost any kind of behavioral trait subjected to genetic research possessed an inherited aspect. However, heritability rarely surpasses 50 percent, and the remaining 50 percent of causal factors have to be non-genetic (Plomin et al., 2008, p. 89). In the field of mental disorders, the situation is even more complex. Certain clinical categories seem to be quite well separated from one another on genetic grounds while others are not. For example, most studies I know of found that schizophrenia and bipolar disorder each had quite distinct inheritances. On the other hand, while the comprehensive disease category called "schizophrenia" was found heritable, its subtypes (catatonic, paranoid, disorganized) were not (or at least not by the same genetic mechanism as the comprehensive category of schizophrenia). That is, "although schizophrenia runs in families, the particular subtype does not" (Plomin et al., 2008, p. 203). A somewhat similar problem is found with respect to good versus bad outcome schizophrenia. The bad outcome schizophrenia, containing mostly negative or passive symptoms, seems to have a more pronounced heritability than the good outcome schizophrenia, which contains predominantly positive or active symptoms (idem, p. 204).

Problems concerning genetic inheritance have been found not only *below* the inclusive category of schizophrenia but *above* it, too. While schizophrenia belongs to the wider category of psychotic disorders, additional mental conditions found with greater frequency in schizophrenic families (schizophrenia spectrum disorders) belong partly to the inclusive category of *personality disorders*: schizoid, schizotypal, and paranoid personality disorders.

Some personality disorders, as well as obsessive-compulsive disorder, substance abuse, etc., may co-occur with the schizophrenic syndrome. Therefore, the most parsimonious hypothesis (Occam's razor) has to be the coexistence of a range of different genetic predispositions in the same patient (instead of presupposing that each clinical category has a different, disease-specific genetic background).

In summary, it seems that mental disorder categories do contain genetic predispositions *specific* for the respective disorder, but, overall, the overt clinical picture usually reflects not just that genetic predisposition but additional ones as well, along with non-genetic influences. These considerations are detailed in chapter 7.

A peculiar outgrowth of the dissatisfaction with this loose interrelationship between the clinical pictures and their possible underlying genetic foundations is the search for "endophenotypes." Candidates for endophenotypes consist of circumscribed, measurable behavioral or somatic traits that are better correlated with the supposed genetic underpinnings of the respective mental disorder than the whole clinical picture, and at the same time aggregate in the patient's "healthy" biological relatives. Endophenotypes are "measurable components . . . along the pathway between disease and . . . genotype. . . . Endophenotypes represent simpler clues to genetic underpinning than the disease syndrome itself . . . to be most useful, endophenotypes for psychiatric disorders must meet certain criteria, including association with a candidate gene or gene region, heritability that is inferred from relative risk for the disorder in relatives, and disease association parameters" (Gottesman & Gould, 2003, pp. 636–645). Examples of endophenotypes given in this article are: "difficulty in filtering information from multiple sources" and difficulties with "working memory" and "executive cognition" in schizophrenia.

It seems to me that the central concepts of this text (like seasonal variations of instinctive intensities, the subject matter of this section) may fulfill some of these functions of the endophenotype. These innate behavioral incentives—predispositions—do not manifest directly in overt behavior but always in combination with learned elements and lie along the "pathway" between 1) the distant phylogeny of behavior in the animal kingdom; 2) its modification by more recent characteristically human evolution (first of all RfNSPs); and 3) the observable behavioral outcome (the clinical syndrome or disorder) resulting from the interaction of diverse inherited predispositions with the natural and social environment during ontogeny.

I have no alternative but to continue to use the traditional disease categories in this text, since these categories are the basis of the shared language among professionals in human behavioral sciences. However, it has to be stressed that the instinctive mechanisms discussed in this and the following sections of this chapter, while they are differentially relevant to different clinical categories, do not exclusively pertain, as a rule, to one specific disease category or another. Seasonal variations in the strength of the overall instinctive activity, for example, are reflected in a pure, almost unaltered form in winter depression. The same seasonal variations in instinctive activity, which may attain unusual degrees of intensity as a result of RfNSPs and, as a result of far-reaching alterations of the natural triggering clues in the artificial human environment, may become desynchronized from the actual seasonal rhythm, underlie different forms of bipolar and unipolar affective disorders lacking seasonal synchronization. The same mechanism is proposed here to account for the episodic and affective aspects of schizoaffective disorders, in which it is hypothesized that two different instinctive mechanisms are active: seasonal fluctuations in instinctive intensity, as we have just seen, as well as more than normatively diffuse instinctive strivings (section 5.3), which constitutes the genetic predisposition for the schizophrenia spectrum of disorders.

It is understandable that the reader may find the conclusions sketched premature and even unwarranted. My purpose up to this point has been only to illustrate the specific ways the comprehensive categories of instinctive predispositions presented in this work are relevant to traditional clinical disease categories. The bulk of the reasoning, as well as the evidence on which my views are based, have yet to be detailed. The only additional point I wish to make in advance of that further exposition is that—should the above line of reasoning prove to be valid—*the comprehensive instinctive mechanism that is predominant in the respective clinical picture will determine as a rule the main clinical diagnosis (affective disorder or schizophrenia respectively), as well as the main inherited foundation* of that clinical disorder.

———

Much of the following discussion on the *seasonal periodicity* of animal and early human behavior at higher latitudes of the earth is based on data provided by Foster and Kreitzman in their 2009 book, *Seasons of Life, The Biological Rhythms That Enable Living Things to Thrive and Survive.*

The fact that the axis of the earth's rotation is oblique in relation to the plane of its orbit around the sun leads to seasonal changes in the amount of daily light, temperature, and humidity on almost the whole surface of the earth (Foster & Kreitzman, 2009, pp. 11–16). These variations of the environment's physical properties affect *in a cyclic and predictable manner* the growth of vegetation, available quantities of water, and so on, which have a decisive influence on the whole food chain depending on it.

Living organisms have to adapt to these regular fluctuations. In animals, the two major forms of behavioral adaptation in this regard are seasonal migrations toward more providing environments and the reduction (to different degrees) of the intensity of physiological processes and behavior in those seasons that are scarce in life-sustaining resources (dormancy, torpor, and true hibernation). Many animal species migrate seasonally from cold areas to warm ones, or from dry areas toward more humid ones with more abundant vegetation. These adaptations affect not only the feeding patterns of these animals but also their reproductive cycles, and in fact they ensure the spread of animals to areas on the globe that only seasonally possess the needed life-supporting resources.

Seasonal reduction of the intensity of behavioral and physiological processes in animals takes several forms: *Diapause* is "a state of low metabolic activity, usually in insects, associated with increased resistance to environmental extremes and altered or reduced behavioral activity" (idem, p. 245). More relevant to our subject is the periodic reduction of the intensity of physiological and behavioral activities at higher latitudes during the winter months, and, conversely, the "explosion" of life-sustaining activities (feeding, reproduction, parenting) in the warmer periods of the year with increased daylight and food availability. In higher animals this reduction in the intensity of physiological and behavioral activities has several forms and degrees: *Torpor* is a "short-term state of decreased physiological activity, usually characterized by reduced body temperature . . . and rate of metabolism." *Dormancy* is defined as a period in which "growth, development and physical activity are almost completely suspended temporarily"; while *hibernation* is: "an adaptive winter physiology of sustained inactivity and metabolic depression, characterized by a profound lowering of body temperature, breathing, and metabolic rate" (idem, p. 96). Hibernation is interrupted by periodic arousal when the animals "urinate and defecate, move about and change their position" (idem, p. 99).

The example of the female black bears' winter torpor will illustrate how intricate the adaptation of life processes is to seasonal fluctuations. (Bears, contrary to popular belief, do not hibernate.) "Mating in black bears occurs in early summer, before most berries and nuts ripen. Implantation of the fertilized eggs is, however, delayed until November, so that birth in the dens occurs in January. . . . The cubs are fed on milk, and although the mother is in winter torpor, she is still alert to her cubs' needs, responding to vocal demands for warmth, comfort, and suckling. By delaying implantation and timing birth in the den to midwinter, the cubs spend less time confined in the den" (idem, p. 100).

Animals in the period preceding hibernation "typically feed intensively to build up their fat stores. Some . . . fill up on carbohydrate- and lipid-rich foods" (idem, p. 99). This must also be true in the case of dormancy and torpor, though in a less intense form. (This aspect of the adaptation to seasonal changes will be of interest for us when we later explore the evolutionary roots of some unusual features of winter depression, namely overeating, weight gain, and craving for carbohydrates.)

The actual seasonal variations in the amount of light, temperature, and abundance of food do not directly induce the abovementioned physiological and behavioral changes. Ground squirrels, for example, kept in laboratory conditions of constant temperature and daylight, "continued to hibernate about once a year and each hibernation episode was preceded by a marked increase in body weight and food consumption. . . . Even golden-mantled ground squirrels that had been born in captivity and had never experienced a natural photoperiod still had a circannual rhythm" (idem, p. 101).

Therefore, it is clear that these animals had evolved phylogenetically an *internal biological rhythm* ("circannual clock"—based on the year) that had become independent of the actual external circumstances. However, it was found that this circannual clock was not synchronized precisely to seasonal changes. The role of seasonal variations (the drop in daily light hours and temperature) is understood to provide signals that *entrain* and thus correct that imprecise internal biological clock (idem, pp. 101–104).

In the following, we touch on the topic of our human ancestors' migration out of Africa toward higher latitudes as well as possible related phylogenetic developments: "Our earliest ancestors came out of the equatorial forests and left a mildly seasonal environment for one in which they had to battle continually with the cyclic changes between hot and cold, wet and dry, calm and stormy. *Those changes were engines of natural selection*" (idem, p. 149, emphasis added).

The last (second) migration out of Africa happened 50,000–60,000 years ago, enough time for evolution to build specific adaptations for seasonal changes. As they gradually migrated toward higher latitudes, replacing the Neanderthal populations, these early humans came under the influence of more extreme seasonal variations. The adaptation to these variations physiologically, behaviorally, and culturally became an indispensable requirement for survival.

However, we have to be careful when we try to draw parallels between the ways these early human populations adapted to increasingly harsh seasonal fluctuations and the way that lower animals did so. These nomad hunter-gatherer groups were organized, possessing a primitive but efficient technology for hunting, fishing, cooking, and preparing efficient shelters. The taming of fire and the use of animal fur and hide for clothing protected them from the worst extremes of low temperature and fewer daylight hours. Therefore, adaptations like dormancy, torpor, and hibernation, which contain prominent physiological and metabolic changes, beside behavioral ones, were irrelevant in their case.

On the other hand, they had no efficient ways of storing food for longer periods. Even present-day isolated agricultural communities with undeveloped technology have difficulty in this respect. The scarcity of wild animals and gatherable plant food to eat as well as the long nights and short days (which could be extended only marginally by the fire inside the shelters) surely forced them to reduce their daily activity considerably during the winter months. Being a lifesaving behavioral adaptation, recurring predictably in a cyclical way for tens of thousands of years, it surely led to some kind of phylogenetic adaptation. I think it can also be hypothesized that the underlying brain apparatus that induces this behavioral (and neurophysiological) adaptation in humans is similar to that in animals. We have seen that in animals the degree of reduction of physiological processes (growth, metabolism, etc.), as well as of behavioral intensity, ranges in its strength from torpor to hibernation. The characteristic form and intensity of the adaptation have to be an optimal compromise between the severity of external conditions, the phylogenetically acquired body traits that shield the animal from them (fur and fat layers, body volume-to-surface ratio, etc.), as well as the adaptive oscillations in the intensity of physiological processes and behavioral output.

Humans migrating to higher latitudes had bodily characteristics adapted to the central African climate and surely were inadequately equipped somatically for the climate of northern regions. But what they lacked in physical characteristics was more than compensated for by cultural and technological means. These early humans replaced the Neanderthal populations, who were better adapted to a cold climate, with a better body mass to surface ratio. For these early migrators toward northern regions, the adaptation to the local climate must have been primarily an optimal compromise between technological and cultural shielding, on one hand, and the reduction of behavioral, and, consequently, physiologic activity, on the other.

For this later presupposition—that early modern humans who lived at higher latitudes evolved phylogenetically seasonal fluctuations in the amount of instinctive energy channeled into physical and mental activities—we have very good clinical evidence in the form of winter depression (or seasonal affective disorder) and the therapeutic effect on depression of increased amounts of artificial light, which will be detailed later in this section.

The rapid increase in the ecological dominance of humans, the progress of technology whose achievements became available to increasing segments of the population and shielded them more and more successfully from the effects of seasonal changes (mainly in the last 10,000 years and accelerating in modern times), drastically reduced the need to adapt to the change of seasons with appropriate changes in the amount of physical and mental activity. The modern environment in technologically advanced, liberal nations reduced to a minimum—and in some respects wiped out completely—the effects of seasonally induced changes in environmental resources (light, temperature, and food availability). From our point of view, the consequence of this development was *a substantial relaxation* of selective pressures on the respective behavioral adaptations, leading to their increased diversification. Moreover, modern social environments in most instances not only rendered the cyclical changes in behavioral intensities superfluous but put a premium on year-round stable functioning in the occupational, social, parenting, etc., spheres. Fluctuating intensities of innately determined (that is, instinct-induced) behavior, on the other hand (synchronized or not with the seasons), became more and more disadvantageous. When these fluctuations are strong enough to interfere considerably with fitness-enhancing or adaptive behavior, they can receive the label of a psychiatric disorder.

If this is so, the difficult question that may be asked is whether, in these circumstances, the modern human environment may lead to *directional selection pressures for year-round stable functioning* (as well as to purifying selection against considerable seasonal variations) even in regions where the seasonal shifts are prominent. Statistically, in certain populations living at high latitudes, the prevalence of subclinical Winter Depression is up to 25 percent (Foster & Kreitzman, 2009, p. 207), and the clinical form (Seasonal Affective Disorder) varies between 3.6 percent (Iceland) and 16.5 percent (Siberia) (idem, pp. 210–211); while the average prevalence in the general population is 4–9 percent (Provencio, 2005, p. 169). While these figures in higher latitudes seem high, the other side of the equation is that at least 70 percent of the population in these regions is free from seasonal affective fluctuations. Theoretically, in these regions, in certain social and economic conditions, year-round stable functioning may promote increased rates of survival, increased rates of reproduction, and increased rates of offspring survival until reproductive maturity. On the other hand, the modern technologically advanced physical environment, as well as the tendency of human communities to protect their (temporarily) disabled members, led to relaxation of the natural and intragroup selective pressures in this respect, resulting in increased diversification. Lastly, immigration of populations adapted to warmer, less seasonally fluctuating conditions may dilute, that is, reduce, the overall frequency of genes predisposing to seasonal fluctuations in behavioral intensity.

The interaction between these four kinds of selective mechanisms: 1) relaxation of natural selection pressures and resultant diversification as a result of the protective effects of human environments; 2) possible directional selective pressures for stable social and occupational functioning; 3) relaxation of intragroup selection pressures; and 4) intermingling of populations (more precisely, their genes) living at different latitudes as a result of migration, is a complex situation beyond the scope of the book. However, of the four comprehensive instinctive mechanisms, I have presented first the seasonal fluctuations of overall instinctive intensity, because it illustrates in the most obvious manner the following chain of evolutionary events: well-demarcated natural selective forces; evolutionary adaptations to them by lower animals; adaptation to them by prehistoric human populations; subsequent relaxation of these selective pressures as a result of increasing ecological dominance by humans and consequent diversification; and the emergence of genetically predisposed forms of dysfunctional behavior. This chain of evolutionary events exemplifies well what is meant in this text by "gene-culture coevolution."

In addition, it seems to me that the relevant behavioral and mental conditions—functional as well as dysfunctional—can be arranged into another chain, interconnected by the same evolutionary logic. This chain begins with conditions in which cyclical changes in the instinctive intensities may still be adaptive (or useful or desirable); proceeds through pathological conditions in which these fluctuations, preserved in their original (formerly adaptive) form and synchronized with the seasons, became dysfunctional in modern social settings fulfilling the requirements of the "genome lag" hypothesis (sections 3.2.1.2 and 3.2.2.1); to mental-disorder categories in which the increasing relaxation of selection pressures produced forms of dysfunctional behavior whose relationship to the original phylogenetic adaptation becomes less and less obvious. The discussion of these behavior patterns is our next topic.

The few instances known to me where fluctuations in instinctive intensities, synchronized with the seasons or not, may be adaptive in contemporaneous societies derive, as a rule, not from scientifically verified data; they are mostly anecdotal or conjectural in nature or based on personal observation. Nevertheless, for the sake of comprehensiveness, I thought, it would be worthwhile to mention the following examples here:

1. Certain isolated human communities with underdeveloped technology living in higher latitudes or mountainous regions where the shielding effect of modern culture and technology remained very partial: The members of these communities are not well protected against the hardships of winter months, the amount of physical and mental effort needed for survival fluctuates greatly with the seasons, and food resources in winter months are restricted. In these conditions, I think, moderate fluctuations in behavioral intensity synchronized with the seasons definitively seems adaptive.

2. The summer versus winter occupations of certain European watchmakers: "Long before production lines come into existence, a cottage

industry existed in Switzerland, France, and Germany, where poor farmers, after trying to make a living from infertile soil during the summer months, would spend long winter days and evenings producing, cog by cog, the forerunners of today's [mechanical] wristwatches" (*Watches, The Collector's Corner*, Grange Books, 1999, p. 9). To my mind, the great difference between the kind and amount of physical and mental effort required of these farmers in summer versus winter is striking indeed.

3. Reports exist of the high prevalence of emotional fluctuations or even (hypo-) manic and lethargic depressive episodes of clinical severity in creative personalities and those in leading positions. Sometimes hypomanic episodes, sometimes depressive ones, and sometimes both are thought to augment creativity. Here are some relevant quotations: "The cycle of rest and creation of the artist resembles that of manic depression, and manic depression may occur more frequently in creative people. Andersen (1978) found that, in a group of successful writers from the University of Iowa, 60 percent had personality disorders, usually cyclothymic personality, and 67 percent have experienced affective disorder (15 percent bipolar, 54 percent unipolar)" (Colp, 1980, p. 3114). And: "Persons with hyperthymic temperament and soft bipolar conditions in general possess assets that permit them to assume leadership roles in business, the professions, civic life, and politics. . . . Notable artistic achievements are found among those with soft bipolar disorders, especially cyclothymic disorders . . . depression might provide insights into the human condition, and the activation associated with hypomania helps in producing the artistic work" (Akiskal, 2005a, p. 1639).

Of course, the fluctuation of the instinctive intensities is only one of several requirements for creative work, some others being talent, the right kind and level of intelligence, a deep commitment to the creative field, etc. Nevertheless, the right intensities of the instinctive energies during creative periods can be an indispensable component for achieving completion of the artistic product. The "soft" sub-depressive episodes (of the "lethargic" type discussed in the following section) may promote in persons with the appropriate talents periods of silent, unbiased, self-absorbed reflection on the topic of their interest and assure the patience needed for the emergence of the appropriate intuitions, without the urgency to promptly convert instinctive tensions into overt behavior in order to achieve concrete, goal-directed results in a limited timescale. Hypomanic periods, on the other hand, may provide the energies and self-confidence needed to finalize the creative product and to make it available to the targeted audience (see also Brüne, 2008, p. 215).

Several years ago, I treated an American bipolar patient. He reported (and the details were confirmed by his current wife) hypomanic episodes during which he had significant success in business enterprises, earned a lot of money, and got married. During the subsequent depressive episodes, he lost his earnings and divorced at the request of the wife. This sequence of events was repeated at least three times. Earning a lot of money in business and being attractive to women seems to me to be the epitome of success in certain social circles.

5.1.1. Winter Depression

In the official psychiatric classification (DSM-5), winter depression is not classified as a distinct clinical disorder but as Major Depressive Disorder (with a seasonal pattern). However, in the context of the present text, winter depression has characteristics that link it in a straightforward manner to the evolutionary adaptation hypothesized to exist in early human ancestors (while at the same time separating it sharply from another common form of clinical depression, detailed in section 5.2.3).

Human beings' progressively increasing ecological dominance alters former behavioral adaptations to different degrees in different individuals. However, in the case of winter depression, this alteration is minimal. The syndrome's mild or subclinical form (affecting up to 25 percent of the population, females more than males, and younger persons more than older ones) is very widespread in northern regions where the winter is harsh and longstanding, and its status as a mental disorder is contentious (Foster & Kreitzman, 2009, pp. 206–208; DSM-5, p. 188). The crew members of polar expeditions described this phenomenon repeatedly, and it may be particularly severe in creative persons (Foster & Kreitzman, 2009, pp. 201–202). The *subjective feeling* in these states was dubbed as "lethargy . . . languor brought out by winter darkness" (idem, pp. 199 and 202); "winter blues." From an evolutionary point of view, the condition was succinctly summarized by Foster and Kreitzman in the following way: "Many of us get a little 'down' in winter, and in the past this may have been a useful protective measure as people slowed down, did less, used up less energy and so managed the harsh conditions of winter" (idem, p. 217).

Winter depression in its full-blown clinical form is characterized by two sets of symptoms. One set is shared with the more inclusive category of *retarded* depression (in contrast to the *agitated* form of depression that will be dealt with in the next section); this first group of symptoms includes fatigue, loss of energy, low mood, reduced sexual desire, and reduced daily functioning in the vocational, social, and familial domains.

The second set, which is quite specific to winter depression and atypical in other forms of depression, includes an increase rather than decrease in the amount of sleep; increased food consumption (rather than lack of appetite), in particular of carbohydrates (like sweets); and weight gain (Provencio, 2005, p. 169; Foster & Kreitzman, p. 213). Winter depression sets in with the decreasing amount of daylight at the end of autumn or beginning of winter, and remits with the advent of spring, when winter depression can switch into a mild hypomanic episode.

The incidence of winter depression is high in the general population (4–9 percent) with preponderance in women, and "as much as 20 percent of the population may have subsyndromal features" (Provencio, 2005, p. 169). The prevalence of winter depression seems to increase with higher latitudes and decreases or disappears altogether nearer to equatorial regions, although exceptions to this regularity are not rare (Foster & Kreitzman, 2009, pp. 210–211; DSM-5, p. 154). The heritability of this disorder was found to be higher at higher latitudes (British Columbia, Canada) (Plomin et al., 2008, p. 218).

Another clinical fact, which further strengthens the hypothesis that winter depression is closely related to one form of evolutionary adaptation to seasonal variations in the animal kingdom is its positive response to "light therapy." The depressive symptoms improve considerably or resolve entirely and the condition may even switch into a mild hypomanic episode (Akiskal, 2005b, p. 1638). Light therapy increases the amount of experienced daylight by adding artificial illumination to the shortened amount of natural daytime light in winter months, bringing it up to amounts characteristic of the warmer seasons. The best predictor of a good therapeutic response (which is claimed to be around 85 percent, much higher than that of the antidepressant drug therapy) is the presence of typical signs of winter depression: increased amount of sleep, overeating, weight gain, and especially craving for carbohydrates (Foster & Kreitzman, 2009, p. 211). The amount of daily light is probably the most important factor, which entails the internal circannual clock in hibernating animals (idem, p. 103). It should be mentioned here that retarded depressive episodes *desynchronized* from the seasons may also be improved by light therapy, suggesting that they are induced by the same (diversified) phylogenetic mechanism as winter depression.

It would be interesting to know how the two specific symptoms of winter depression—increased amount of sleep as well as overeating, especially with high-caloric content food—are related to the mechanism of adjustment to seasonal variations in life-sustaining resources.

As to the *excess of sleep*, it is tempting to interpret it as part of a general inhibition of the overall neurophysiological and mental activity in dormancy, torpor, and hibernation. The word *dormant* in the case of animals in fact means "in or as if in a deep sleep" (*Concise Oxford English Dictionary*). Comparative electro-encephalographic recording might support or alternatively disprove the hypothesis that a neurophysiologic resemblance exists between the reduced central-nervous-system functioning in animals in these states and humans' excessive sleep during winter depression.

The more interesting symptom cluster, however, seems to be overeating (especially with carbohydrates), and weight gain. Evidently, animals during torpor, dormancy, or hibernation cannot eat at all, making this symptom cluster in winter depression more enigmatic.

However, it is well known that the same animals during the period *preceding* hibernation (and milder related states) "typically feed intensively . . . to build up their fat stores. Some, such as the edible dormouse . . . fill up on carbohydrate-and lipid-rich foods." The success of this phase is vital for the survival of the hibernation period: "Most of the deaths [during hibernation] are thought to result from the [premature] burning-up of food reserves" (Foster & Kreitzman, 2009, p. 99). The simplest evolutionary hypothesis that may account for this symptom complex in winter depression, I would conjecture, is that humans overeat during the period of winter depression because:

1. *No specific innate triggering mechanism* exists in the animal world, which stops overeating before hibernation or related states. (In fact, no need exists in nature for such a mechanism, since food disappears anyway with the advent of winter.)
2. *The human-made unnatural environment does provide a continuous food supply* (and other life-supporting resources) during the winter months. In animals, on the other hand, it is safe to presume that the gradually shortening daylight, increasing scarcity of available food, inconvenience of low temperatures, and activation of the instinctive mechanisms concerned with preparatory activities preceding hibernation (like finding or preparing an appropriate den) are enough to prompt a cessation of eating even without a specific innate triggering mechanism.

An alternative hypothesis could be that one or more of the above changes in the natural environment (decreasing amounts of daily light, temperature, and available food) fulfill the function of triggering the cessation of overeating. Since modern human environments compensate for these shortages, overeating continues. Research findings about the possible presence of overeating in the period *preceding* the appearance of lethargy in winter depression might support this hypothesis, while its absence would weaken it. In addition, it is well known that a proportion of lethargic patients suffer from lack of appetite instead of overeating and insomnia instead of excess sleep. This topic is addressed in section 7.2.1.

5.1.2. Additional (Unipolar and Bipolar) Affective Disorders

In the case of winter depression, the relaxation of normalizing selection and consequent diversification played almost no role. Perhaps the diverse degrees of severity of the condition can be attributed to it. Winter depression can be seen as a "soft" version of the slowing down of behavioral activity in animals in the winter months without the characteristic physiological changes, such as considerably reduced metabolism and body temperature, which are probably obviated by the protective effect of the human environment. The accompanying emotion, that is, a certain kind of depressed (more precisely, apathetic) mood, which is detailed later, represents *the subjective, emotional experience of the weakened instinctive activity*. On the other hand, human culture and its technological achievements are hypothesized to be responsible for the persistence of overeating during the depressive episode.

However, in the categories of mental disorders that are discussed in the following sections, the considerable relaxation of normalizing selection and consequent broad range of diversification, even disintegration, of the respective (originally adaptive) behavioral patterns is hypothesized to play an increasingly prominent role. In order to be methodical, I discuss the diversification of the three main components of this phylogenetic mechanism separately, in spite of the fact that in actual clinical cases we frequently see a mixture of them. These components are:

1. *Relaxation of the normalizing selection pressures on the intensity and duration* of the seasonally fluctuating instinctive activity, leading to its diversification and to increasing dysfunctionality at the extremes of the resultant diversified continuum (manic and depressive states of clinical proportions).
2. The *desynchronization from the seasonal rhythm* of these fluctuations of instinctive activity, probably as a result of the alteration in the human environment of the natural triggers that entrain the circannual biological clock (duration of daylight and possibly change in temperature) in the artificial human-made environment.
3. As a result of a more complex phylogenetic mechanism, *clinical syndromes appear in which the permanent (lifelong), overall, energetic level of daily activities is suboptimal or excessive*. These mental conditions are categorized, as a rule, as temperamental or personality traits (hyperthymic or dysthymic). Their close relationship to bipolar disorders is indicated by the frequent presence of short manic, hypomanic, or depressive episodes. These episodes are the *opposite* of the respective enduring temperamental trait (that is, hyperthymic persons suffer from depressive episodes and dysthymic persons from manic or hypomanic episodes).

In the following sections, these three evolutionary mechanisms are discussed in some detail.

5.1.2.1. Relaxation of the Normalizing Selection Pressures on the Intensity and Duration of the Seasonally Fluctuating Instinctive Activity

Regarding the fluctuations in the intensity and duration of the overall instinctive activity in accordance with seasons, we must hypothesize that in nature these parameters are precisely tailored by normalizing selection to the abundance (or scarcity) of the resources or other relevant parameters. If, for example, the female black bear's winter torpor while tending for her cubs (section 5.1) had been allowed by natural selection to become considerably deeper or lighter, or shorter or longer, than the optimal, it probably would have led to there being no surviving cubs to raise.

In humans, on the other hand, relaxation of selection pressures and consequent diversification of such parameters led to various forms of dysfunctional behavior, without as a rule affecting survival and reproduction. Considerably stronger or weaker overall instinctive activity than the optimal can lead to the well-known manic, depressive, or bipolar episodes. If the winter depressive episodes become unduly intense, interfering considerably with normal functioning, we will diagnose a unipolar depressive disorder of the retarded type; if both the winter depressive as well as the "hypomanic" episodes of warmer seasons become excessive in intensity or duration, the diagnosis will be bipolar affective disorder.

Mental health professionals have noted the resemblance between seasonal fluctuations in animal behavior and bipolar disorder: "As Lange observed . . . many animals in their normal adaptations to changes in the physical environment exhibit behavioral and physical adjustments that are similar in kind and degree to those exhibited by manic and depressive patients. For example, at different times of the year animals eat more or less, gain or lose weight, become active or lethargic, sleep more or less, become more or less interested in sex, and interact more or less with their physical and social environment" (Wehr & Rosenthal, 1989, p. 836).

It is worth mentioning that Kraepelin's original interpretation of the depressive and manic pathology stressed mental and physical slowing or acceleration: "For Kraepelin, the core pathology of clinical depression consisted of lowered mood and slowed (retarded) physical and mental processes. In mania, by contrast, the mood was elated, and both physical and mental activity were accelerated" (Akiskal, 2005b, p. 1564).

5.1.2.2. The Desynchronization of Manic and Depressive Episodes from Seasonal Periodicity

The temporal relationship of acute affective episodes (of unipolar depression and bipolar disorder) to the periodicity of seasons is very variable. Sometimes a correct synchronization exists with the appropriate season; in other cases, no interrelationship exists at all, and in still other cases a partial or odd synchronization exists. Sometimes, a chronological regularity exists that does not respect the circannual cycle (that is, affective episodes occur once every several years or several times in the same year). In still other cases an *inverse* synchronization to that of the adaptively correct one exists, such that manic episodes may appear in winter months and depressive ones in summer. Otherwise, all these varieties of affective episodes are quite indistinguishable clinically from each other. As mentioned already, the relationship of these episodes to the seasons does not influence the actual clinical diagnosis of the disorder in the current psychiatric nosology.

The following example illustrates the sometimes odd relationship of affective disorders to seasons. For an extended period, I followed a patient who exhibited characteristic hypomanic episodes once every two years. These episodes began in autumn (in a Mediterranean climate), lasted throughout the winter, and resolved after about five months in spring. This chronological regularity continued for at least ten years. Between the acute episodes, the patient was slightly (sub-clinically) depressive; by this I mean he showed retarded type of depression.

The desynchronization of the fluctuations in instinctive intensity from the actual cycle of seasons in animals, as well as humans, has to be a more complex phenomenon than it seems at first glance. In some cases, it seems to be simply the result of the lack or alteration of the usual natural triggers, more exactly the entraining effect of the circannual clock mentioned earlier. "Cyclic recurrences . . . can be independent of seasonal cycles for example when animals are removed from their natural environment" (Wehr & Rosenthal, 1989, p. 836). In this case, the relaxation of natural selection pressures evidently plays no role at all, since the desynchronization happens in the same animal, not in successive generations. In domestic animals, too, seasonal events like the reproductive cycle or shedding the coat are desynchronized from the seasons: "All domestic animals have lost their seasonal patterns of reproduction and tend to breed at any time of the year" (Heird & Deesing, 1998, p. 207).

Human sexual activity and the timing of birth also seem to be quite independent of seasonal cycles. However, statistical regularities between seasons and birth rates are "typical of agricultural populations at higher latitudes," and were even observed in populations living in modern conditions (Foster & Kreitzman, 2009, pp. 151–154).

It seems very difficult to reconcile instances of desynchronization as a result of altered environmental circumstances with the laboratory observations of ground squirrels mentioned earlier (section 5.1) that continued their annual hibernation under constant laboratory conditions without being exposed to any relevant environmental triggers. Perhaps the *typical hibernation* cycle is "entrenched" more deeply in the genetic apparatus, and thus less susceptible to alteration by changing environmental conditions than the milder forms like torpor or dormancy.

Other questions more relevant to our subject but for which no ready answers exist include: What triggers affective disorder episodes in humans who are desynchronized from seasonal variations, when it seems clear that the usual natural triggers (light, temperature, lack of food) are evidently not candidates for this role? How does it happen that in some cases a strange periodicity (as in the case described earlier of hypomanic episodes that occurred once every two years, with an inverse relationship to the seasons) can occur while in other cases no chronological regularities or seasonal synchronizing is apparent? What exactly happened to the mechanism of the circannual biological clock and its interrelationship to triggering (entraining) factors as a result of the progressively decreasing natural selection pressures throughout more recent human evolution?

It is not clear whether the sensitivity to natural triggering factors that activate the seasonal clock decreased and diversified; whether, as a consequence of this diversification, some new triggers related to technological/economical/cultural influences emerged; or whether in populations living at high latitudes this diversification was less pronounced, manifesting itself in increased prevalence of winter depression; but it does seem clear that *the whole pattern of originally adaptive behavior to seasonal changes underwent progressive diversification, even disintegration.*

In a more optimistic vein, it seems that the chances that scientific (specifically, chronobiologic) research may discover the brain mechanism underlying seasonal behavioral changes are good. Many aspects of the related circadian (diurnal) clock mechanism are known: "We now have a good idea of where that clock is, what it is made of and how it works. We have a broad understanding of the molecular processes that enable a near-twenty-four-hour rhythm in the rise and fall in abundance of various proteins to be sustained and entrained to the daily solar cycle by the light signal of dawn and dusk" (Foster & Kreitzman, 2009, p. 37). The success of a similar project in the domain of the circannual mechanism may represent an important step toward more effective treatment, or even prevention, of a common kind of affective disorder, which in turn may alleviate a lot of human suffering.

5.1.3. The Permanent, Overall Energetic Level of Behavior and Its Relationship to Episodic Fluctuations in Behavioral Intensity

Personality traits refer to certain enduring aspects of human behavior with an innate foundation. Often, they are already discernible in childhood, "crystallize" around adolescence, and remain more or less unchanged for the rest of the life span. They are at the foundations of basic adaptive or maladaptive patterns of behavior in social settings, influencing powerfully such vital aspects of everyday functioning as coping with stressful situations, patterns of relationship with the immediate social environment (love, the character of social relations, the handling of interpersonal conflicts, the amount of competitiveness, etc.), as well as the attitude toward work. The topic of personality and its disorders is further considered in section 7.2.4.

A basic element on which personality is built is the *constant, overall energetic level* fueling behavior. Evidently, this is an innate aspect of behavior, since it cannot be learned, albeit its way of utilization (from optimally useful to utterly wasteful or self-damaging) depends on other mental mechanisms, including intelligence and learning. (Psychoactive drugs—antipsychotics, antidepressants, psychostimulants—can powerfully influence this aspect of behavior, as discussed in the next chapter.) Animals have a more or less constant overall intensity of the energetic level of behavior, which is a biological variable. What follows is a short discussion of its adaptive significance in nonhuman animals.

In animals, this basic level of behavioral intensity—that is, the speed with which an animal moves as a rule during the waking state, the proportion of activity versus rest, the proportion of time it remains motionless as opposed to actively following life-sustaining aims, and so on—differs broadly between different animal species. Some animals are inclined to sustain high activity levels during most of their waking hours (like rats, many small birds, the mongoose, *New Larousse Encyclopedia*, 1981, p. 565), while others (sloths, hippos, chameleons, etc.) move slowly, rest a lot during the day, or stay motionless for other reasons. Another example is the striking dissimilarity in the overall intensity of behavior between two monkey species belonging to the same family and inhabiting the same South American tropical forests: the capuchins and the howlers. While capuchins "are rambunctious and feisty all day long," the howlers "spend about 80 percent of the daylight hours sleeping, and all of the night as well" (Perry, 2011, p. 17).

The basic activity level of animals represents an evolutionary adaptation, and as such, according to evolutionary logic, has to be regulated tightly by natural selection. This optimal level of activity depends on a host of factors relevant to adaptation to a special environmental niche, and frequently it constitutes an optimal trade-off or balance between conflicting adaptive interests. For example, it is influenced by body volume and weight (as a rule, the bigger the body, the slower the animal tends to move); by the energy content of consumed food and the balance between energy intake and expenditure (howler monkeys, for example, are leaf- and fruit-eating, while the very active capuchins' diet is much more diverse and includes fruit, insects, and meat [Perry, 2011, pp. 52–66]); in the case of predators, by the balance between remaining motionless and inconspicuous as long as possible in order to get closer to prey animals versus adopting a more active searching and chasing strategy; and so on. Seasonal variations in resources modulate this basic general level of activity.[1]

In the case of humans, it is quite evident that this aspect of behavior is considerably relaxed from normalizing selective pressures, leading to a wide range of interpersonal diversity. It is not difficult to find individuals with high overall energetic levels, capable of working intensely (frequently under stressful conditions) for many hours a day. With the same ease, one can identify persons with meager energetic resources and easy fatigability: "Normal energy levels vary considerably. . . . Some people fatigue easily and are perceived by themselves and others as having 'weak constitutions,' whereas others appear to have almost boundless energy and much less need for sleep" (Yager & Gitlin, 2005, p. 972).

In my experience, the level of overall behavioral intensity does not differentiate individuals with mental disorders as a group from the rest of the population, nor is it a characteristic of specific mental disorder categories, with the exception of a few instances detailed below. However, some factors, like longstanding preventive psychoactive medication and the great diversity of the general level of activity in normative as well as psychiatric patients, make it difficult to be sure about more subtle regularities in this respect.

Returning to the subject of this section, that is, the fluctuations in instinctive intensities in mental disorders, it must be stressed that those with both unipolar and bipolar disorders should have in principle two kinds of energetic levels, the normative one seen in periods of remission, as well as dysfunctional levels during episodes of highs and lows. Yet, according to the literature on this subject, as well as my professional experience, many affective patients lack entirely the normative periods of behavioral intensity. Instead, one of the two polarities, the high or the low, becomes (at a milder intensity) the lifelong, constant energy level of behavior, representing thus an aspect of the personality makeup or temperament of the individual. The high version of this permanent level of instinctive intensities, characterized by an enduring state of mild hypomania, is called in psychiatry *Hyperthymic Personality Disorder*. The behavior of these individuals is characterized by "unusual energy, ebullience, confidence, intensity" (Yager & Gitlin, 2005, p. 992). Although the term continues to be used by mental health professionals, this condition has received no officially recognized mental disorder status in the classifications of mental disorders (DSM-III, DSM-IV-TR, and DSM-5).

The opposite state, that is, longstanding low energy levels, is called in psychiatry *Dysthymic Personality Disorder*. Unlike Hyperthymic Personality Disorder, Dysthymic Personality Disorder is recognized by the DSM-IV-R classification as a distinct mental disorder. Its characteristics are: "Poor appetite or overeating . . . Insomnia or hypersomnia . . . Low energy or fatigue . . . Low self-esteem . . . Poor concentration or difficulty in making decisions . . . Feelings of hopelessness." Besides, these "patients complain that they have always been depressed . . . most cases are of early onset beginning in childhood or adolescence" (Akiskal, 2005b, p. 1629). (DSM-5 classifies the same disorder as one form of depression under the denomination of "Persistent Depressive Disorder [Dysthymia]"—DSM-5, p. 168.) The clinical resemblance of this condition to episodes of the retarded form of depression or to winter depression—except for its permanent (not fluctuating) nature—is striking.

The third kind of disorder characterized by abnormal intensities of overall behavior in the long run is the most peculiar. It consists of fluctuations between high and low energy levels. This disorder is called "cyclothymia." It may be worthwhile to quote here a more detailed description of the condition, since it combines socially adaptive as well as maladaptive features:

> During adulthood these patients may be very adaptive, particularity during the hypomanic or elated phase, when their optimism, social gregariousness, good humor, high drive, and ambition may make them successful in business, professional life, public service or academic pursuits. . . . During the depressed phase the patient may be underactive, have difficulty in concentrating, be underresponsive, and perform at lower levels than normal. Conversely, during the periods of elation or hypomania, the patient may behave inappropriately socially, with excessive sexual behavior, poor management of funds, and poor judgment in family, business, and social activities. (Klerman, 1980, p. 1337)

Furthermore, and this may be the most interesting point for our purposes, hyperthymic and dysthymic disorders predispose the individual to acute affective episodes of a nature opposite to that of the longstanding temperamental aspect—that is, as mentioned earlier, persons with dysthymic temperament develop acute hypomanic or manic episodes, while those with hyperthymic temperament may suffer from acute episodes of retarded depression. Hagop Akiskal, an authority in the field of affective disorders, argues (correctly, according to my clinical experience) that this situation is more the rule than the exception in bipolar disorders: "New systematic clinical observations have revealed that bipolar II disorder (characterized predominantly by depressive attacks) arises more often from a hyperthymic or cyclothymic baseline, whereas bipolar I disorder (defined by manic attacks) not uncommonly arises from the substrate of a depressive temperament. . . . *Bipolarity* is conventionally defined by the alternation of manic (or hypomanic) and depressive episodes [emphasis in the original]. The foregoing data on temperaments suggest that *a more fundamental characteristic of bipolarity is the reversal of temperament into its 'opposite' episode*" (Akiskal, 2005b, p. 1639, emphasis in the last sentence is mine).

The hypothetical conclusions arrived at in this section up to this point can be summarized as follows: The more or less permanent, overall intensity of behavior in the animal kingdom, which varies according to species, represents the optimal trade-off between the life-perpetuating interests of the organism and the environmental resources, as well as between the deleterious influences of the environmental niche the organism inhabits and the defensive strategies against them. This overall energy level of behavior is kept at a narrow range of variability around an optimum by normalizing natural selection. In those environments in which living conditions vary considerably according to seasonal changes, this genetically determined biological mechanism came under the modulating effect of an additional innate mechanism, which regulates the intensity of the behavior (and of physiologic processes) according to the periodically and predictably changing amounts of environmental resources. This mechanism's functioning is also kept at a narrow range of variability by normalizing selection.

In humans, as a result of the changed conditions of the culturally determined environment, both mechanisms—that is, the one that determines the constant energy level of behavior, as well as the one that modulates the former mechanism according to seasonal fluctuations in the abundance of resources—were relaxed progressively from the strict control of normalizing selection. This led to considerable diversity, in both the enduring basic intensities of behavior, as well as in its adaptive fluctuations tailored to the effects of seasonality. Moreover, as a possible result of the interaction between these two diversified mechanisms, in some individuals *the basic level of behavioral intensity somehow changed permanently toward a lower or higher intensity than the normative range.*

As a matter of fact, it may be hypothesized alternatively that the hyperthymic and dysthymic constitutions represent the extreme ends of the diversification of the first mechanism, that is, the permanent energy level of behavior alone. But we have a clinical regularity that seems incongruous with this inference, namely that dysthymic and hyper-thymic personality disorders are frequently accompanied by acute episodes of the opposite affective polarity.

This close relationship between acute disease episodes and certain personality or temperamental traits is another example (besides that mentioned already in the case of schizophrenia spectrum of disorders) that suggests that the sharp distinction in psychiatric classification between mental disorders characterized by alternation of acute disease episodes and remissions, on one hand, and (permanent) personality disorders, on the other, may *not* be paralleled by an equally sharp distinction in their underlying hereditary foundations. This line of thought is further developed in section 7.2.5.

5.1.4. Schizoaffective Disorder and Schizophrenia

It would be utterly premature to attempt here the discussion of the more characteristically "schizophrenic" aspects of these disorders. In order to do so, it is important to first introduce the instinctual diffuse/differentiated scale (the subject of the third section in this chapter), as well as other, yet undiscussed, topics. Therefore, here I intend to detail only the *episodic nature* (or lack of it) of these disorders, as a continuation of the subject matter of this section—namely, the fluctuations of instinctive intensities. The basic suggestion here will be that different aspects of the disorders are induced by different (originally adaptive) inherited biological mechanisms. In view of the descriptive and atheoretical nature of psychiatric classification, we have no logical or factual grounds for insisting on a single specific, inherited mechanism underlying *all* the descriptively (or intuitively) coherent-looking aspects of a nosological category. It will be proposed, therefore, that the causative biological mechanism for the *episodic* aspect (and that aspect alone) of schizoaffective disorders, as well as of the episodic form of schizophrenia, is basically the *same* biological mechanism responsible for the fluctuating intensities of behavior in more characteristically affective disorders. The characteristically schizophrenic symptoms of these disorders, on the other hand, are induced by a different biological mechanism, which will be discussed later. After all, the acute episodes of schizoaffective and schizophrenic illnesses are of comparable duration to those of purely affective disorders and respond to the same psychoactive drugs (a topic that is detailed in the next chapter). Besides, indications exist that the affective and episodic aspects of schizophrenia may have a different kind of inheritance than the more characteristically schizophrenic ones.

5.1.4.1. Schizoaffective Disorder

Schizoaffective disorder is a clinically multifarious mental illness in which the symptomatology of a "full mood syndrome," as well as of "core schizophrenic" pathology, occur together in various combinations (Fennig, Fochtmann, & Carlson, 2005, p. 1533). The presupposition that brought me to include this disorder here was, as previously mentioned, that its episodic course, and even more characteristically *the episodes with typical affective symptomatology, have the same underlying biological mechanism that is responsible for the typical unipolar and bipolar affective disorders, while the more characteristic schizophrenic symptoms have an entirely different biological mechanism as their foundation* (which is discussed later).

For this purpose, the most relevant subtypes of the disorder are one in which affective symptoms predominate over the typical schizophrenic symptoms, and another in which typical schizophrenic symptoms predominate over the affective ones. The course of the disorder and its potential to lead to progressive deterioration in social and occupational functioning was found to lie between that of the two major psychiatric disorders that co-occur in this condition.

The data on the genetics of schizoaffective disorder indeed suggest that two different genetic mechanisms co-occur (in accordance with Mendel's law of independent assortment): "The risk of schizoaffective disorder may be increased among individuals who have a first-degree relative with schizophrenia, bipolar disorder, or schizoaffective disorder" (DSM-5, p. 108). And more impressively: "In a number of well-designed family studies, relatives of patients with the affective subtype of schizoaffective disorder . . . were found to have increased rates of mood disorders in comparison with normal control families but not increased rates of schizophrenia. . . . In contrast, patients with the schizophrenic subtype of schizoaffective disorder showed an increased prevalence of schizophrenia in their relatives" (Fennig, Fochtmann, & Carlson, 2005, p. 1533).

How the co-occurrences of the genetic predisposition for these two major psychiatric disorders may influence each other is entirely unknown.

For example, we do not know whether fluctuations in the instinctive intensities can further disorganize an otherwise mild schizophrenic symptomatology, or how a mild predisposition to the schizophrenia spectrum of disorders may alter the expression of considerable fluctuations in the energetic charge of behavior. Nor, to the best of my knowledge, do we know how the predisposition to one of these disorders may be capable of inducing or imitating specific symptoms of the other. However, some factors that may play a role in this puzzle that will be discussed later in more detail can be mentioned in advance.

As mentioned earlier (when discussing the concept of endophenotype, section 5.1), the basic genetic predisposition to mental disorders is not expressed directly in the full-blown clinical picture, and in consequence symptoms that at a descriptive level seem to be inseparable aspects of the disorder may have different causations, hereditary or ontogenetic. For example, it's not certain at all that the *psychotic* symptoms in schizophrenia are a direct manifestation of the genetic predisposition underlying schizophrenia (in spite of schizophrenia being considered the epitome of psychotic disorders). The spectrum disorders of schizophrenia, which are genetically related to it, contain personality disorders (that is, disorders without psychotic symptoms). Another consideration that can create confusion is that, from an evolutionary point of view, we see two entirely different types of depression; the first is discussed in this section, and the second is detailed in the next. Both can be present in schizophrenic symptomatology, but they have very different biological mechanisms as their foundation. Moreover, as detailed later, affective symptoms characteristic of schizophrenia can easily be confounded with affective symptoms of mood disorders (the flat affect or emotional detachment found in schizophrenia, for example, may be confounded with the retarded, lethargic form of depression). Schizophrenia itself has two partly different forms, an episodic and a progressively deteriorating one, with seemingly different inheritance. Until these and other considerations are assimilated into the psychiatric research design, and unless the search for the genetic roots of mental disorders differentiates between clinical symptoms that are straightforward expressions of the biological predisposition and symptoms that are secondary, ontogenetically induced developments, the interpretation of the resultant experimental findings will remain problematic.

5.1.4.2. Schizophrenia

As mentioned, I deal here only with the fluctuating aspect of the disorders that contain typical schizophrenic symptomatology—that is, the alternation of acute-illness episodes with periods of remission. The course of schizophrenia in this respect is very variable. While it is considerably altered by the universal use of long-term preventive antipsychotic medication (and most likely as well by its long-term socio-environmental management), we still seem to see in this disorder all the transitional forms between the "good prognosis schizophrenia," with infrequent acute episodes and remissions with relatively good social and occupational functioning, and the "bad prognosis schizophrenia," with a chronic, unremitting course of progressive deterioration.

In the 1980s, psychiatrist Timothy Crow, drawing attention to the different nature of these two forms of schizophrenia, called them type I and type II schizophrenia (Crow, 1985, pp. 471–486). This distinction is still relevant to modern psychiatric genetics: "Type I schizophrenia, which has a better prognosis and a better response to drugs, involves active symptoms such as hallucinations. Type II schizophrenia, which is more severe, has a poorer prognosis and passive symptoms such as withdrawal and lack of emotion. Type II schizophrenia appears to be more heritable than Type I" (Plomin et al., 2008, p. 204). The differential response of these two types of schizophrenia to drugs as well as the difference in their heritability is important for the following discussion.

The assumption advanced in the present theory is that the episodic, good-prognosis type of schizophrenia (Type I), like the schizoaffective disorder, contains two different underlying genetic mechanisms, one that induces its episodic nature and another that is responsible for the characteristic schizophrenic symptomatology. The episodic nature of the disorder seems to be induced by the same biological mechanism that induces the episodic nature of both typical affective, as well as schizoaffective, disorders. Nevertheless, the more conspicuous, more malign schizophrenic symptoms (a topic that is enlarged on in section section 5.3.) may *conceal or disguise* the characteristic affective symptomatology. The intensified emotional motives in an acute schizophrenic attack are expressed in a bizarre, typically schizophrenic way, in contrast to the characteristic (good prognosis) manic behavior that can be understood as exaggerated intensities of normative emotional and intentional motives and thus are accessible to the empathizing efforts of normative persons. It has to be mentioned, too, that sometimes an acute schizophrenic psychotic episode is followed by a period of "post-psychotic depression," which closely resembles the retarded, lethargic form of depression, characteristic of bipolar mood disorder and winter depression. (However, it is sometimes quite difficult to disentangle the lethargic depressive element from negative schizophrenic symptomatology [like "emotional blunting"] and from the effects of antipsychotic therapy.) On the other hand, the bad prognosis schizophrenia is a purer, more exclusive expression of the typical schizophrenic genetic predisposition, undisturbed by the predisposition to considerable fluctuations in instinctive intensities.

In fact, we can construct a list of mental disorders and their subtypes in which the weight of the fluctuations in the intensity of instinctive activity gradually decreases while the presence of the typical schizophrenic inheritance becomes gradually more prominent, as follows:

1. Typical, episodic mood disorders (no schizophrenic inheritance or pathology).
2. Affective subtype of schizoaffective disorder.
3. Schizophrenic subtype of schizoaffective disorder.
4. Episodic (good prognosis) form of schizophrenia.
5. Unremitting or progressively deteriorating (bad prognosis) form of schizophrenia (only the typical schizophrenic genetic predisposition is present, and no episodic fluctuations in the intensity of instinctive activity are discernible in the clinical picture).

This conception of the genetic underpinnings of these disorders seems to me more in accordance with Occam's parsimony principle ("What can be done with fever assumptions is done in vain with more," Wilson, 1998, p. 53) than trying to hypothesize a discrete, specific inherited mechanism for each.

5.2. The Three Universal Responses of Active Instinctive Drives to Frustration in Evolved Nonhuman Animals and Humans: Displacement and Vacuum Behavior, Aggression, and Dysphoria

As the reader may remember from the introductory section, the kinds of instincts whose function it is to deal *directly* with issues concerning the sustention and reproduction of the individual life by acting upon the appropriate environmental agents are called in this text *active* instincts. They are sharply separated from *reactive* instincts, whose role is to induce behavior that protects the organism from influences deleterious or dangerous to the life processes coming from the environment (physical, biotic, or social). This section examines what happens to active instinctive behavior or predispositions when an obstacle arises between the active instinctive aim (which induces the appropriate appetitive and consummatory behavior) and the releaser that is designed to fulfill or gratify it.

In fact, the previous section dealt with the same subject, or, more precisely, with a special form of it. In that case, seasonal fluctuations in the abundance of life-supporting resources (the obstacle in the way of stable active instinctive gratification) comprehensively affected a wide range of environmental resources, thus frustrating several basic active instinctive (and physiologic) motives at the same time (feeding, reproductive, and parental activity, thermoregulation, etc.). Furthermore, the regular (seasonal) *rhythmicity* of these fluctuations enabled natural selection to build specific behavioral and physiological mechanisms to cope with them (migration, hibernation, etc.).

The frustrating factors or conditions that are dealt with in this section are of a very different kind. Their effect on active instinctive intentions operate within a far more restricted range, and they have no diurnal, seasonal, or any other kind of temporal regularity. They affect directly, as a rule, only *one single active instinctive aim* at a time. I have in mind such factors as temporal shortage or lack of food due to various randomly occurring causes in an individual organism's environment; or the event in which another conspecific competing for the same food resource or sexual access blocks the direct route toward the desired releaser; or the death of an immature offspring, frustrating the continuation of parenting activities; and so on. The instinctive response to these kinds of frustrating events frequently is *quick,* contrary to the response to seasonal variations, when the behavioral response unfolds *slowly.*

In the introductory section of this work, I stressed that I will deal only with instinctive mechanisms with a *comprehensive* effect on multiple individual instinctive aims. It may seem that, in raising the present subject, I contradict myself. However, in the present section, although the three behavioral transformations act each time on *a single,* circumscribed active instinctive motive, *they occur uniformly—that is, they transform the innate behavioral expression in the same way, even in the case of very different, circumscribed instinctive motives.* The aggressive behavioral response to frustration, for example, may be quite similar, whether it occurs in the framework of competition for food resources or sexual aggression between competing males or when a mother protects her immature offspring from dangers coming from other animals. Although the behavior patterns may differ somewhat (for example, the aggressive behavior of a mother defending her offspring compared to ritualized intrasexual competition of males), even in these instances the *aggressive nature* of both kinds of behavior is evident. In humans in whom these three responses to frustration are diversified and frequently unbalanced, the preponderance of one form over another may determine the overall "style" of behavior in its entirety (aggressive or dysphoric personality traits for example).

Besides the three forms of frustrated active behavior, other kinds of behavior patterns exist that can be interpreted as responses to frustration of active instinctive aims—above all, the *submissive response.* However, because the relatively simple submissive responses of animals evolved in the complex human environment into highly sophisticated behavior patterns like *conformity* to social rules of conduct or *internalization* of social values, this subject is discussed in the fourth section of this chapter. In addition, humans have other behavioral and mental mechanisms employed to deal with instinctive frustration, like deception or sublimation. These behavior patterns, needing highly developed cognitive abilities (along with the innate predisposition), do not pertain to the relatively simple and universal behavioral categories dealt with in the present section.

Another reason why I discuss these three kinds of responses together and exclude other related ones is that they frequently appear in human behavior in a more or less regular *succession,* as a response to the same event frustrating the same circumscribed instinctive aim. The following example will illustrate the three forms of behavioral transformations in frustrating conditions that I have in mind, as well as their successive appearance in response to the same circumstance frustrating the same instinctive motive.

During *mourning* as a response to the demise of a person with whom one had a strong positive emotional relationship (a child, a mate, a good friend, a parent), all three responses to frustration can appear in succession. As a rule, the first response is *denial* of the loss, which enables the continuation of the appropriate, instinctively induced relatedness even in absence of the deceased person. Nothing in her room is changed (if she was a family member), she frequently reappears in the mourning person's dreams or imagination as living and engaging in his everyday activities, or in vivid memories of especially prized moments in the relationship. We have called these kinds of responses to instinctive frustration *vacuum behavior*, which means the discharge of the original instinctive behavior (accompanied by the usual pleasurable emotion) in spite of the absence of the respective releaser. This kind of response is well known in animals, too. In fact, the term was coined by ethologist Konrad Lorenz.

A related response, which in my opinion differs only in degree from vacuum behavior, will be called in this text *displacement* response. The common factor between these two forms of behavior is the unavailability of the appropriate releaser for the respective active instinctive aim. The difference, on the other hand, is that while vacuum behavior is discharged *in vacuo*, that is, in the total absence of any kind of releaser, displacement behavior is discharged toward an inappropriate releaser. Because the releaser is different from the original object toward which evolution has built the respective instinctive mechanism, the ensuing behavior's adaptive value in natural conditions is lost or greatly reduced. This is even more obvious in the case of vacuum behavior.

Various kinds of displacement behavior in tamed animals are related in Konrad Lorenz's book *King Solomon's Ring* (1961). An especially lovely example is that of a tamed jackdaw male that "fell in love" with Lorenz, and as a gesture of courtship tried to feed him minced worm. When he refused to cooperate (that is to open his mouth), the bird, as a less desirable alternative, tried to push the worms into his ears (pp. 135–136).

The second phase or form of mourning, frequently succeeding the phase of vacuum behavior (denial), consists of *aggression*. The mourning individual may abuse or attack in imagination or in reality, verbally or (rarely) physically, persons or institutions he believes to be responsible for the death of the loved person. The temporary transformation of a life-sustaining instinctive motive or behavior into aggression when frustrated seems to be the most frequent response to active instinctive frustration in the animal kingdom (at least in the vertebrate subphylum, animals whose behavior I am more familiar with). Its adaptive role, in my opinion, seems clear. It constitutes *an attempt to remove the obstacle that is in the way of active instinctive gratification by contacting the appropriate releaser.*

The third form or stage of mourning, in the context of our discussion, is the *dysphoric form* or phase. This form's subjective experience has an intensely painful, distressful quality. In an approximate order of growing intensity, it manifests itself in sighing, moaning, complaining, crying, wailing, screaming, blaming oneself for the loss, agitation, complaints about the worthlessness of life. In my opinion, its pathological correlate—agitated depression or melancholia, which manifests itself in depressive mood (of an acutely painful not lethargic quality), self-accusations sometimes of a delusional nature, suicidal thinking and acts, etc.—while comprehensively suppressing other (adaptive) daily activities, differs only in *degree,* not qualitatively, from normative responses to significant emotional loss or frustration.

Dysphoric behavior, while being less frequent among mature animals in the animal kingdom than displacement or vacuum behavior, and certainly less frequent than aggression, is by no means unknown, and instances of it can easily be found in ethological literature. I will bring some examples later in this section.

In the following, I give a more detailed description of the three responses to frustration, as well as try to analyze their adaptive value from an evolutionary point of view. In addition, I discuss the possible effects of the relaxation of selection pressures on these instinctive mechanisms in humans, their resulting diversification in behavior, and the possible significance of this diversification for psychopathology.

5.2.1. Displacement and Vacuum Behavior

I would like to begin this section with an example of this phenomenon in animals. Early ethologist Wallace Craig "was the first to describe this phenomenon in the motor patterns of courtship of the blond ring dove. . . . A male of this species, after having been isolated from conspecifics for some time, was ready to court a domestic pigeon, something he had refused earlier. After isolation had been continued even longer, the bird was ready to court a human hand; still later, he directed his courtship activities toward the rear corner of his box where the convergence of the three edges offered at least a point of fixation" (Lorenz, 1982, p. 124). The same phenomenon can be observed in the case of the feeding and nest-building instincts, too (for detailed descriptions, see Lorenz, 1982, p. 127).

Displacement phenomenon, and especially vacuum behavior, were seen by Lorenz and other ethologists as proof of the *autonomous internal causation of behavior*, which disproves the stimulus response theory (espoused by the behavioristic school in psychology). It was proposed instead that activities needed for the sustenance and reproduction of life do not depend on the appropriate stimulation from the environment but are generated by (and serve the fulfillment of) the internal, physiological needs of the organism (which often cyclically wax and wane). If these phylogenetically built instinctive motor patterns cannot contact the appropriate releaser, a "damming up' of accumulated readiness" to act occurs (Lorenz, 1982, p. 124) (reminiscent of Freud's "hydraulic model" of instinctive behavior), which will be eventually discharged in the form of displacement or vacuum behavior. This phenomenon is very widespread in the animal kingdom: "Spontaneity . . . not clearly related to changes in the exteroceptive stimulation . . . is a commonplace in animals from protozoans . . . and sea anemones . . . to mammals" (Hinde, 1970, p. 311).

As the last section of this chapter shows, this theorem has only partial validity in our theoretical framework. It is valid in the case of active behavior only whose aim is to interact with, alter, or incorporate environmental agents according to the *internal* needs of the organism. Reactive behavior, which is built to avoid or neutralize randomly occurring deleterious *external* influences, lacks spontaneity or rhythmicity. (It would be gravely damaging, for example, for a prey animal's fitness to periodically interrupt feeding or some other active instinctive activity and spontaneously flee when no approaching predator is present; or, alternatively, not to flee because this instinctive activity—to use an analogy to a characteristic trait of active behavior—is in its refractory period.)

The "closing time phenomenon," identified by evolutionary psychology, may represent an example of mild displacement behavior in humans. It consists of "shifts in judgments of attractiveness over the course of an evening in singles bars. . . . As closing time approached, men viewed women as increasingly attractive . . . [which may] be attributable to a psychological mechanism that is sensitive to decreasing opportunities over the course of the evening for casual sex" (Buss, 2008, p. 180). (The analogy between this phenomenon and the behavior of Wallace Craig's isolated blond ring dove, which in the absence of a female conspecific was willing to court a domestic pigeon, seems compelling.)

For those clients of the singles bars who missed the opportunity for casual sex before closing time, a remaining alternative to sexual gratification would be lonely masturbation. It has to be noted, however, that masturbation, in spite of being done in the total absence of a real releaser, cannot be considered typical vacuum behavior. The person involved in masturbation—aided by dexterity, advanced technology, and human inventiveness—frequently reconstructs artificially many of the relevant external stimuli existing before and during the sexual act performed with an adequate releaser: tactile auto-stimulation, the use of a sex doll generating tactile and visual stimuli, visual and auditory stimuli by watching pornographic material, and so on.

An interesting instance of displacement (in this case concerning the parental instinct) was described as a cultural phenomenon by anthropologist Ruth Benedict: "Among certain of the Central Algonkian Indians . . . [u]pon the death of a child a similar child was put into his place" and adopted—often a captive brought back from a raid (for details, see Benedict, 1968, pp. 184–185).

The question that has to be asked when considering the ubiquitous nature of displacement and vacuum behavior in the animal world is: Why has natural selection preserved these instinctive activities or predispositions in view of their obvious lack of fitness value? To the contrary, these behaviors may be deleterious by using up precious energy or making the animal more conspicuous to predators.

In humans and highly evolved social mammals, like bonobos or chimpanzees, we can in fact find adaptive value in some displacement activities. As mentioned in the section on primatology, bonobos use sexual activities for improving social coexistence. A young female that has recently arrived in a new bonobo group can facilitate her acceptance into that female-dominated society by performing sexual acts with an older, high-status female. Similarly, individual or group tensions (for example, during feeding, or when two strange bonobo groups meet) can be alleviated by sexual activities.

The aggressive transformation of active instinctual intentions is the topic of the next section. However, it should be mentioned here that the displacement of an aggressive response from the conspecific that elicited it in the first place toward another, less dangerous, one may have adaptive, even lifesaving, value. This tactic is known in chimpanzee societies—for example, when aggressive intentions elicited by a strong dominant male in a male with a lower social status is often displaced toward the weaker, less dangerous, females. (Displacement of aggression, in this case, is in fact a more complex mechanism, reflecting the interaction of active aggressive instinctive motives with fear, a topic pertaining to the fourth comprehensive instinctive mechanism discussed in this chapter.)

Nevertheless, all these adaptively useful examples of displacement of instinctive activities cannot explain the same phenomenon in less evolved animals lacking a complex social network, as in the birds mentioned earlier. It seems, rather, that these examples of displacement are *secondary evolutionary developments*, superimposed on or superseding the original or more basic aim of the instinctive transformation. A more profound, more comprehensive explanation may have to do with the deleterious effects of the damming up of the tensions or energies or "readiness to act" that fuel a circumscribed active instinctive motivation, and that motivation alone, the *action specific excitation,* to use Konrad Lorenz's term (Lorenz, 1982, p. 110). This action-specific excitation, when blocked from discharge (from performing the innate motor pattern on the relevant releaser), may attain high levels of intensity that may suppress or interfere with other important life-sustaining or danger-evading instinctive activities.

As everyone knows, strong emotions may disturb optimal cognitive functioning—that is, the ability to see in a balanced way the situations one has to deal with. A strong active instinctive motivation leads to selective perception of the environment—to being more sensitive to the clues of the looked-for releaser and less sensitive to other important but unrelated environmental stimuli, such as those coming from an approaching predator or from social disapproval. Using a loose human analogy, it resembles an infatuated person's misguided or inadequate behavior in complex social situations. The discharge of mental tensions connected to a frustrated active instinct by displacement or vacuum behavior may prevent or alleviate this situation. This interpretation of the adaptive value of displacement and vacuum activity was proposed by the best ethologists: "Tinbergen and others have suggested that displacement activities may serve as a safety valve discharging superfluous motivation. . . . This interpretation is probably correct for those cases in which a sudden cessation of external stimulation leaves the animal with an excess of specific excitation which has to be got rid of in some manner" (Lorenz, 1982, p. 252).

If this line of reasoning is valid, it can be hypothesized that normalizing natural selection pressures have to regulate tightly the action specific excitation's intensity (its threshold) at the moment when the striving to interact with the fitness-enhancing releaser is given up and the respective motor pattern is displaced or discharged "in vacuo." If this threshold becomes too low, if the discharge of the innate motor pattern is allowed too easily to quit the striving toward the fitness-enhancing releaser and be diverted toward an unsuitable object, the adaptive value of the respective instinctively underpinned behavior is prematurely lost. On the other hand, if natural selection allows this threshold to rise above optimal levels, and the animal persists for too long—particularly, when the chances of attaining the adequate releaser are small, or the continuation of a stubborn striving may become too costly or dangerous—the chances of survival and reproduction may be jeopardized.

It should be noted that renowned ethologist Nikolaas Tinbergen used the concept of displacement differently than the way just described. By displacement, he meant the "sparking over of the surplus motivation" of a frustrated instinctive activity to another, irrelevant instinctive domain (for example, from the domain of reproductive activity to that of feeding or caring for body surface—activities less problematic, as a rule, from an interactional point of view). In consequence, the fixed motor pattern(s) (FMP) of these new domains are used in order to discharge the accumulated surplus tensions belonging to the original instinctive motive (Tinbergen, 2003, pp. 113–119). These kinds of "sparking over" activities appear, as a rule, when a conflict is present between two antagonistic drives or when a strong instinctive arousal is blocked from discharge through its original FMP. For example, preening in several bird species serves as an outlet of sexual arousal (idem, p. 115). Tinbergen argued that similar instinctive mechanisms may be present in humans, too. "The general occurrence of scratching behind one's ear in conflict situations almost certainly has an innate basis . . . it is striking that displacement scratching can be observed regularly in primates" (idem, p. 210).

While this phenomenon bears little direct relevance to our narrower subject matter in view of the paucity of FMPs in human behavior, it has, in my opinion, a very important theoretical implication. It suggests that Lorenz's "action specific excitation," or energy, in fact is not strictly "action specific" after all, since it may pass, or "spark," over into an unrelated instinctive domain. While in animals this rechanneling involves another instinctive domain, it makes more feasible the presupposition that in humans (and probably in evolved social animals) the same "action specific" instinctive energies *can be rechanneled into culturally created activities or avenues.* Combining this realization with Craig's and Lorenz's interpretation of the concept of displacement, which suggests *the possibility of redirecting instinctive strivings toward unrelated, naturally maladaptive releasers,* further substantiates the tenability of the assumption that different aspects of the instinctive activity—energetic, intentional, subjective experiential (but not FMPs)—can be rechanneled in humans into, and integrated with, culturally created (that is, learned) ways of behaving. Furthermore, this phenomenon is not an exclusively human trait, since its simpler evolutionary antecedents can be observed in the behavior of lower animals. The reader should also consider in this context the examples of instinct-learning interaction described in the second chapter—such as habituation, habit formation, exploratory behavior, and so on—which also suggest of the possibility of integrating learned material with instinctive behavior.

The next question relevant to the central topic of this text is the following: What happens when this tight control of natural selection regarding when to quit the striving toward the appropriate releaser (and discharge the accumulated "action-specific-excitation" in a maladaptive way) becomes relaxed in the protective, culture-made human environment? How does the ensuing quantitative diversification on both sides of the optimum threshold affect human behavior, and what significant implications does this diversity have for psychopathology?

However, before addressing this question, we have to acknowledge that the extrapolation of animal displacement to human behavior is fraught with considerable difficulties. In animals, at least in the more characteristic instances, a particular displacement behavior seems quite simple: A well-defined FMP (a courtship pattern, for example) is not directed toward its appropriate, biologically adaptive releaser (a receptive female conspecific), thus losing its fitness-enhancing value. In humans, however, living in a complex social environment, the identification of a displacement or vacuum behavior is frequently highly problematic. This difficulty arises for several reasons:

1. As mentioned repeatedly before and as detailed in the next section, the human instinctive predispositions became more and more blurred or diffuse, becoming incapable of inducing pure fixed motor patterns. Therefore, FMPs disappeared completely from the appetitive phase of instinctive activity, being clearly recognizable only at the final phase of the consummatory act (chewing, swallowing food, coitus proper). As a result, attempts at the extrapolation to humans of displacement activities in animals in Tinbergen's sense (that is, as the "sparking over" of an "action specific excitation" from one instinctive domain to another) becomes in most cases impracticable.

2. If the above kind of extrapolation to humans is impracticable, what, then, is the situation with respect to the displacement of an instinctive striving toward an inadequate releaser or its discharge without any releaser being present? In parallel with the increasing diffuseness of instinctive motives, and the instinctively underpinned behavior's great susceptibility to modification by learning, in humans the natural releasers of an instinctive domain also became diversified and, to a great extent, influenced by cultural factors. However, they are apparently modified to a lesser extent than FMPs. In human behavior, as a rule, we can still differentiate between the proper releasers belonging to, say, the sexual sphere, versus those belonging to the feeding drive or parenting domain, in spite of their diversification and sophistication.

3. The third category of difficulties in extrapolating displacement or vacuum behavior from the animal kingdom to human behavior

stems from the fact that an adaptive releaser or avenue in a natural environment for a circumscribed instinctive activity may be strongly dysfunctional in a social environment and vice versa. For example: In a natural environment the adoption of the offspring of a different species is strongly maladaptive—in fact so much so that I am not aware of such instances in nonhuman animals. However, the same practice may seem desirable in humans in certain circumstances. If a lonely person or an older couple, for example, keeps a pet animal, a dog or a cat, in order to satisfy parenting motives and alleviate feelings of loneliness or worthlessness, the behavior may be considered desirable or useful.

Furthermore, in humans, intentional or emotionally loaded behavior (hinting at its instinctive underpinnings) may be channeled into culturally determined activities that have no counterpart or adaptive value in the natural world. For example, should we regard cheering a favorite athlete or team, volunteering to promote a political or social program, and watching motion pictures or reading fiction for the pleasure of identifying with the plight of an imaginary hero and so on as undesirable displacements of our natural instinctive strivings, or, conversely, as desirable adaptations to a cultural environment? After a critical remark from our superior, is the displacement of resultant angry feelings toward a colleague who is lower in the hierarchy dysfunctional, adaptive, or something in-between?

In spite of these difficulties, in my opinion, we still may discern in most everyday activities fitness-enhancing instinctive motives analogous to those seen in the animal world (expressed frequently in covert, culturally modified forms) directed toward adaptively appropriate releasers paralleling the natural ones. In the majority of men, sexual attraction arises toward healthy, post-pubertal women, not toward a child or an old woman, or a pregnant or physically ill one. We care for our body surface in an effective way, do what is required to live as long as possible, try to acquire and accumulate resources needed for physical well-being parallel to those in nature (food, shelter, devices for keeping optimal temperature in our immediate physical environment, etc.). The inefficient or self-damaging sidetracking of such strivings is still frequently considered dysfunctional or pathologic (as in the paraphilic displacements of the sexual drive).

If we accept that the concept of displacement and vacuum behavior in animals can be extrapolated to human behavior in spite of the difficulties, our next question has to be: How does relaxation from natural selection pressures (RfNSPs) in human societies influence that critical threshold that determines when the individual must abandon the effort to attain the appropriate releaser and accept a less than desirable alternative, or express the instinctive motive in vacuo? Keeping in mind the diversity and flexibility of normative human behavior, that threshold has to be much wider than it is in animals (particularly the nonsocial and simpler ones). How can we quantify the normative level of persistence when, for example, courting a discriminating woman with "high mate value" (in the parlance of evolutionary psychology)? How long should a suitor insist and compete with other suitors before quitting and contenting himself with a less desirable but more easily attainable partner to be considered still normative? Where are the borders beyond which the degree of insistence is considered abnormally, maladaptively strong or weak?

The wide range of the parameters in normative behavior can be illustrated with individuals' varying degrees of insistence on sexual access. Some persons give up easily and prefer to indulge readily in masturbation instead of insisting on the real sexual relationship; this may occur even when same-sex competition is not an issue, the prospects of consent of the other party are good, and the act is not condemned by cultural norms. At the other extreme of the normative range, we may find persons who endanger a social status or job as prestigious as prime minister or president by insisting on illicit, socially condemned (or even legally prohibited) sexual affairs. (Right now, as I am writing these lines, a former Israeli president is sitting in jail for rape.)

Cultural customs and prohibitions play an important role in generating displaced or in vacuo behavior. They influence forcefully which instinctive motivations are allowed to be expressed directly, in what form and circumstances, and toward what kind of releasers, and when they have to be suppressed or expressed in a concealed, displaced, or roundabout way, particularity in the realm of sexuality and aggression.

5.2.1.1. Displacement and Vacuum Behavior in Mental Disorders

Displacement and vacuum behavior are not the exclusive features of one mental disorder or group of them but are encountered in most mental conditions. In fact, the difference between displaced forms of instinctive expression in pathologic versus normative behavior is mostly quantitative, not qualitative. Moreover, the other comprehensive instinctive mechanisms discussed in this chapter (fluctuations in overall instinctive intensity, the differentiated-diffuse scale of instincts, and the nature of active/reactive behavior interactions) may play a causative role in the generation of displacement or vacuum behavior. I discuss this consideration in some detail in the following sections.

5.2.1.1.1. The Role of Fluctuations in the Overall Intensity of Instinctive Activity in Inducing Displacement or Vacuum Behavior

High intensities of active instinctive pressure push the individual to convert strong intrapsychic tensions into overt active behavior quickly and in a way that is as free from intrapsychic or interpersonal constraints as possible. As we have seen when describing the cyclothymic personality, more exactly its hypomanic states (which suppresses normative circumspection), these persons may express socially inappropriate behavior in domains like sexuality, interpersonal transactions, management of funds, and so on. Both Konrad Lorenz and Niko Tinbergen stressed the importance of the increased *intensity* of instinctive tensions in displacement and vacuum activity (Lorenz, 1982, p. 249).

However, we have to be careful in our conclusions regarding the nature of the resultant pathologic behavior. At the beginning of hypomanic or manic episodes, active instinctive drives are in fact expressed in an unaltered form directed toward suitable releasers. The individual's problems arise mostly from the lack of heeding the respective social constraints and prohibitions. Therefore, the resultant behavior is maladaptive not from a biological but from a social coexistence perspective, and thus resembles antisocial behavior.

However, with the intensification of a manic episode, the interconnection between the instinctive motive and the appropriate releaser progressively loosens. The individual exhibits an inability to focus the attention (and consequently the behavioral response) on a particular object or topic (which functions as an appropriate releaser). Instead, the attention shifts frequently from one object to another, leading to *distractibility*, a characteristic clinical sign of the manic state. These frequent shifts in attention can be induced by irrelevant stimuli and may lead to quite random responses, including what is called in clinical psychiatry "clang associations" ("association based on similarity of sound, without regard for differences in meaning," Campbell 2004). Evidently, at this stage, the directing of clear-cut instinctive aims toward appropriate releasers becomes impossible.

Slightly less than normatively intense active instinctive activity (mild lethargic depression in the terminology of this text), on the other hand, has a very different effect on mental functioning and behavior. Contrary to the manic state, it may *improve and widen*, in appropriately gifted persons, the unbiased grasp (that is, unbiased by strong intentions, personal interests, or emotions) of physical, cultural, or intrapsychic realities. This state, by sparing the individual the efforts to seek continuously the discharge of relatively narrowly goal-oriented activities in a continuous, demanding interaction with the immediate social environment, allows longstanding contemplation, which, coupled with the right mental abilities or talents, may lead to original, creative achievements. This observation was made already in the times of the ancient Greeks: "The melancholic temperament . . . was described as lethargic, sullen, and given to brooding and contemplation. . . . A long tradition, dating back to Aristotle . . . attributed creative qualities to the otherwise tortured melancholic temperament in such fields as philosophy [which in this period was not separated from the domain of science], the arts, poetry, and politics" (Akiskal, 2005a, p. 1562; see also: Brüne, 2008, p. 215). Whether this kind of transposition of (low level) active instinctive energies from biologically relevant goals toward creative mental activity can be considered displacement in the ethological sense seems to me an impossible question, reflecting the great distance between the ethological versus cultural understanding of human adaptedness.

5.2.1.1.2. Relevance of the Increased Diffuseness of Human Instinctive Predispositions to Displacement and Vacuum Behavior

It seems obvious that if we hypothesize that human behavior is built on significantly more diffuse, more tentative instinctive predispositions than animal behavior, in order to be readily modifiable by learning, the motor patterns of the resulting overt behavior, as well as the releasers toward which it is directed, will be significantly more varied and frequently widely different from those of animals in a natural environment.

Therefore, the question whether a certain sequence of human behavior represents displacement, vacuum behavior, or adaptation depends on whether we refer to:

- Adaptations to achieve *biological* goals in a *natural environment*
- Strivings to achieve individual goals *that parallel biological ones* but in the context of the specific constraints and vastly improved opportunities of the *social environment*
- Behavior that can be fitness-reducing concerning the individual's biological needs but highly prized by the respective community (becoming a Japanese kamikaze pilot, or a celibate priest, for example)

As a general statement, we can therefore say that *human instinctive predispositions must be diffuse enough to allow behavior's flexible modification by social influences, thus ensuring social adaptation, but at the same time, differentiated enough to preserve the ability to guide the behavior amid complex, changing, inconsistent, or contradictory social influences toward fitness-enhancing (in the biological sense) individual goals.*

The hypothesis of this text is that more than optimally diffuse drives that lose their capacity for the attainment of biological goals through social adaptation underlie the predisposition to the schizophrenia spectrum of disorders, while more than optimally differentiated drives may predispose to antisocial personality traits, as I detail in sections 7.2.4.1 and 7.2.7.

5.2.1.1.3. The Relevance of Active/Reactive Behavior Interrelationship to Displacement and Vacuum Behavior

Section 5.4 details the active/reactive behavior interrelationship and its diversification in humans, while in chapter 7 we discuss its pathological consequences. Here I intend to mention only one central aspect of this relationship. Because of the possibility that harmful environmental influences may strike an organism unexpectedly and quickly (a falling rock or a predator's attack, for example), while organismic needs (hunger, sexual drive) arise and intensify comparatively slowly, as a rule, and their fulfillment can be temporarily postponed, reactive behavior almost always takes priority, initially, over active behavior. When a human or nonhuman animal perceives a potentially dangerous situation, she instantly suppresses any active preoccupation she is engaged in at the moment and does not return to it until the danger is over.

In the animal kingdom, flexible transitions between active and reactive behaviors (including precise pace and timing) are highly fitness-enhancing and therefore are tightly controlled by natural selection. The relaxed selection pressures in the human environment led to a wide diversification in this respect. In persons predisposed to, or already suffering from, *anxiety disorders* (neurosis), as hypothesized in this theory, reactive behavior (its avoidant or flight aspect) is employed more readily than normative behavior, it is more intense than optimal, and its dissolution, after the potentially harmful situation is over, takes more time. This means that the recapturing of the active preoccupations is considerably delayed. Consequently, among the normatively stressful conditions of social coexistence, active motives may remain suppressed for long periods during which they may intensify and ultimately reappear in consciousness or behavior in a displaced or otherwise dysfunctional form. Examples of relevant mental conditions include some of the obsessions and compulsions in the framework of obsessive-compulsive disorder, or posttraumatic disorder. In this latter condition, the individual is unable to recapture the more balanced active/reactive behavior interrelationship that existed before the traumatic situation. That is, the mental and physiological apparatus continue to react to the traumatic situation for a long time after it is over. The active drives suppressed for long periods may eventually surface in "angry outbursts (with little or no provocation)," or as "reckless or self-destructive behavior" (DSM-5, 272). The active instinctive motives in this instance are not only clearly displaced or expressed in vacuo but also show aggressive or dysphoric forms of frustration.

5.2.1.1.4. Paraphilias (Formerly Sexual Deviations)

Paraphilias are clear instances of displacement of the expression of the sexual instinct from a biological and, depending on the culture, a social point of view as well. They include fetishism, masochism, sadism, exhibitionism, pedophilia, zoophilia (sexual attraction toward animals), necrophilia (sexual attraction toward corpses), and so on. In some of the paraphilias, the *releaser* of the sexual behavior is biologically (and frequently also culturally) maladaptive, as in pedophilia, zoophilia, and necrophilia. In other paraphilias, the way chosen to awaken or intensify the sexual arousal (that is, the sexual behavior's appetitive phase) is inappropriate, as in fetishism, voyeurism, sadism, masochism, and exhibitionism. The aroused sexual drive in these cases may be discharged in a biologically normative way (intercourse), or through masturbation (vacuum behavior).

5.2.1.1.5. Homosexuality

Homosexuality seems to have a complex causative mechanism. The complexities are not only socio-cultural in nature but biological, too: in animals, "Homosexual behavior is common . . . from insects to mammals, but finds its fullest expression as an alternative to heterosexuality in the most intelligent primates, including rhesus macaques, baboons, and chimpanzees. In these animals the behavior is a manifestation *of true bisexuality latent within the brain.* Males are capable of adopting a full female posture and of being mounted by other males, while females occasionally mount other females" (Wilson, 1978, pp. 143–144, emphasis added).

The social attitude to homosexuality varies widely among cultures. In the official psychiatric classifications, homosexuality is no longer considered a sexual deviation. In light of the complexities of homosexuality, both from a biological as well as from a cultural perspective, its relatedness to displacement behavior will not be detailed further in this text.

5.2.2. The Aggressive Form of Active Instincts Frustration

This section deals with *individual, instinctive,* and not *premeditated* or *organized* group aggression, since these two forms of aggression in humans are quite separate topics. Soldiers in face-to-face combat may be angry, fearful, excited, emotionally detached, and so on. However, a soldier who launches a missile in armed conflict acting on the orders of his superiors is definitely not aggressive in the sense meant in this text.

Some famous students of behavior have viewed aggression as a *separate active instinctive drive.* Freud conceptualized aggression as "a drive that constantly seeks release" (Wilson, 1978, p. 101). This interpretation of aggression is similar to this text's concept of *unfrustrated active instinct.* Konrad Lorenz in his book *On Aggression* sees the aggressive instinct in a somewhat similar way. He believes in the "spontaneity" or autonomous internal generation of aggressiveness (1982, pp. 132–133), a description that resembles the concept of unfrustrated active instinctive activity too in our terminology.

In this text, as the title of the present section implies, aggression *is not viewed as a separate active instinctive entity, but as an adaptive tactic, through which different active instincts (feeding, sexual, territorial, parenting) try to overcome, remove, or neutralize an obstacle in the way of the instinct's gratification*—more precisely, an obstacle in the way of the active instinctive behavior's "intention" to contact, to interact with, its proper adaptive releaser. If the aggression is successful in removing the obstacle, its intensity subsides or ceases altogether, and the animal reverts to its former unfrustrated active behavior (foraging, sexual activity, etc.), showing that the aggressive act was not the *ultimate goal* of the behavior but a means that enabled the fulfillment of an organismic need among certain unfavorable environmental circumstances. This interpretation of aggression has been known in ethology for a long time: "A hindrance to goal-directed behavior (frustration) elicits an aggressive response to help overcome the obstacle, and in this context *aggression serves other motivations.* According to J. Dollard et al. (1939), one need not presume the existence of an independent aggressive drive, since such behavior is entirely reactive. The reaction pattern of responding to frustrations . . . with aggression is innate" (Eibl-Eibesfeldt, 2010, p. 366, emphasis added).

While this formulation is very close to the position adopted in this text, the three forms of transformation of active instinctive behavior in frustrating conditions are viewed as closely related, or part of the domain of *active behavior,* because, in spite of being elicited by a disturbing environmental agent, they appear exclusively in the framework of an active behavioral intent in progress. This is in contrast to *reactive behavior,* which—in the terminology of this text—is viewed as a reaction to dangerous or disturbing environmental stimuli *not related directly* to a specific endogenous striving of the organism unfolding at the time of a specific environmental interference. One tactic of the reactive behavior also manifests itself as an aggressive reaction. The difference and interrelation between aggression in the framework of reactive behavior and aggression in the framework of active behavior is important in the context of this work and is discussed in section 5.4.0.

E. Wilson's opinion on aggression is essentially similar to Dollard's: "There is no evidence that a widespread unitary aggressive instinct exists"

(Wilson, 1978, p. 103). Furthermore, if aggression were a specific active instinct, like feeding, sexuality, or nurturing offspring, I would expect it to possess an autonomously generated energetic aspect, or "action-specific excitation" that waxes and wanes periodically, even in the absence of externally originating frustrations, challenges, etc., which is not the case.

It is good evolutionary reasoning to presume that the ease with which an aggressive response is elicited in certain nonhuman animals—its intensity and duration, the quickness of its escalation from mildly threatening to life-and-death struggle, the promptness with which the animal returns to the original active instinctive preoccupation after the obstacle is removed, etc.—varies according to many factors. These include the species, the nature of the obstacle, the intensity of the active instinctive motivation underlying the behavior before the frustration, and so on. Furthermore, the aggressive response, discharged correctly, has strong fitness-enhancing effects, whereas when it goes astray it may have disastrous consequences for the same animal. Therefore, it is reasonable to presume that *its attributes (intensity, duration, and so on) are tightly regulated by normalizing selective pressures.* As a simple example, take the guarding of a carcass by a predator, in this case a lion, when challenged by another predator, a hungry hyena. I watched this scene repeatedly in one of *National Geographic's* nature documentaries. When the hyena approaches the eating lion, the lion suspends eating, chases away the intruder, and in rare instances, may even kill it. The lion then returns to the carcass and continues eating. It can be confidently presumed that the intensity and duration of the lion's aggressive response and the timing of its return to the carcass to continue eating are tailored by natural selection quite precisely in order to optimally fulfill its adaptive goal. If the aggressive act is weaker than what is required, it may lose its deterrent effect, inviting further challenges by the hyena or other predators that may be watching the interaction. Conversely, if the aggressive response is too fierce and longstanding—if the lion, for example, chases the hyena for a longer than optimal distance—the prey may be eaten or stolen by other hungry animals. In other words, the intensity and duration of the aggressive response, as well as the quickness with which the unfrustrated active behavior before the challenge is recaptured after the obstacle is removed, contribute greatly to (and even determine) the adaptive value of the aggressive act, and consequently have to be checked tightly by natural selection. This line of reasoning can be exemplified by comparing the previous example with the occasion when several lions (of the same pride) feed on the same carcass. The intensity of the aggressiveness against the conspecifics is considerably milder than it is against an opponent of another species and consists as a rule of angry growling and small threatening movements.

Jane Goodall, the best known early researcher of chimpanzee behavior in the wild, described how quickly the fierce fighting for dominance in male chimpanzees turned into mutual grooming after one of them showed signs of surrender (Goodall, 1988, pp. 115–117). Witnessing this kind of behavior, Goodall comments: "It is one of the most striking aspects of chimpanzee society that creatures who can so quickly become roused to frenzies of excitement and aggression can for the most part maintain such relaxed and friendly relations with each other" (idem, p. 117). She observed a similar incident between a dominant male and an old female: "How was it possible for her to enjoy such a relaxed interaction with Mike so soon? The secret perhaps lies in the fact that, although a male chimpanzee is quick to threaten or attack a subordinate, he is usually equally quick to calm his victim with a touch, a pat on the back, an embrace of reassurance" (idem, p. 118). It seems that, unlike humans, nonhuman social animals cannot afford the luxury of feeling offended or bearing a grudge for a longer period of time than is needed to attain the instinctive goal.

Similar behavior was described in a captive chimpanzee group by primatologist Frans de Waal in *Chimpanzee Politics*. Like Goodall, he stressed the presence of quick alternations between aggression and reconciliation. De Waal interprets this phenomenon as the result of the need of these highly social animals to balance two opposing adaptive goals— the need to pursue conflicting individual interests on one hand and the need to preserve group cohesion on the other (de Waal 1982/2007, pp. 27–29). It may be hypothesized that this balance of opposing adaptive necessities is also controlled tightly around an optimum by normalizing selective pressures.

Evidently, a serious imbalance between these two adaptive requirements invites trouble in highly evolved nonhuman social animals. If the tendency for reconciliation, for a conflict-free relationship, is too strong, the animal's deterring ability when pursuing vital individual interests (competing for food, sexual access, etc.) will suffer. If, on the other hand, the animal strives too strongly to attain individual instinctive goals in an "inconsiderate" way toward others in the group (behaves like a psychopath, I would say), it risks group aggression or ostracizing.

After this brief sketch of the adaptive value of the aggressive behavior in nonhuman animals, and its tight control by natural selection, we have to ask our usual question: What happened to human individual aggressiveness among the relaxed normalizing selection pressures of the culture-made, diversified environment, an environment in which various social niches are able to contain a wide range of aggressive predispositions, aggressive predispositions can be hidden or disguised to a certain extent by cognitive foresight, and even the disruptive intensities of innate aggressive predispositions do not necessarily negatively influence the genetic transmission for this maladaptive trait?

It should not be too difficult to convince the reader that, in human populations, particularly in modern societies, *a wide range of diversity exists with regard to the attributes of the aggressive response*, which cannot be checked by normalizing selection. Instead, persons with different aggressive predispositions tend to gravitate toward the social niches (occupations, subcultures, etc.) that can best contain or even profit from the constellation of their respective aggressive response.

Individuals in whom strong but flexible aggressive responses can be easily aroused (that is, they are capable of flexible oscillation between the aggressive response and the original active instinctive aim) may gravitate toward occupations that require this ability. Examples are law enforcement personnel, who have frequent confrontations with potential lawbreakers, or military personnel, who must function satisfactorily amid dangerous physical encounters. Persons with low levels of aggressive response, on the other hand, may gravitate toward such professions as office work, the medical or other helping professions, waiting on tables in a restaurant, etc.—jobs that help other people or provide a service without requiring an aggressive attitude.

Nevertheless, in the normative range of diversified aggressive predispositions, an observation traceable back to the ancient Greeks should be noted—namely, that certain individuals react with aggression (mostly verbal) more promptly and intensely than the average person, even to minor nuisances or challenges. The Greeks called this predisposition "choleric" temperament.

With regard to psychopathology, aggression is an important aspect of many mental disorders, such as the aggressive form of a manic episode;

acute psychotic disorder; antisocial or borderline personality disorder; impulse control disorder; sexual sadism; and so on. It is needless to stress that for no mental disorder category is the conspicuousness of the aggressive reaction the only or even the defining characteristic of the diagnosis. In each of these categories, the quality of the aggressive response is only one aspect of the whole clinical picture. However, in one category of common mental disorders a peculiarity of the aggressive response is the central pathological mechanism.

In *paranoid disorders*, whether we are speaking of a delusional disorder (persecutory type), paranoid personality, or paranoid ideation of subclinical proportions, the peculiar innate predisposition, to my mind, is *an inability to revert to the original, unfrustrated instinctive attitude until long after the aggression-eliciting situation has been resolved in one way or another.* This disability amid the ubiquitous conflicts of interests in a social setting leads to the erroneous belief in a longstanding or permanent inimical interrelationship between the individual and the surrounding people and social institutions. This subject is detailed in the section on paranoia in chapter 7.

5.2.3. The Dysphoric Form of Active-Instinct Frustration

As described in the stages of mourning (section 5.2), dysphoria represents an especially painful, distressful transformation of the originally pleasurable active instinctive motive in frustrating conditions. The overt behavior of dysphoria consists of expressing mental suffering of varying intensities—sighing, crying, screaming, up to severe agitation. The accompanying subjective mental state and related thought content vary from mild distress to grave mental agony and despair, unrealistic self-accusations, and suicidal ideation or intentions. The discharge of the instinctive tension lessens its intensity (as well as the subjective suffering), enabling the recurrence (at least partially and temporarily) of other, unfrustrated, more pleasurable instinctive strivings. In the event the frustrating condition continues, the same fluctuations may appear again and again.

An extreme example of this kind of frustration is the realization of impending inevitable death (in the case of a terminal illness or death sentence, for example). In this instance, in spite of life in its entirety being endangered and not simply one circumscribed active instinctive motive, the ensuing forms of frustration are the same. According to Kübler-Ross, the situation is dealt with by the following mental mechanisms: denial, anger, bargaining, depression, and acceptance (Kübler-Ross, 1969, chapters 2–7). Nevertheless, the succession of these phases is seldom regular and is not specific to the realization of impending death but consists of "general reactions found in many situations of loss" (Weisman, 1980, p. 2045). And: "Seldom does any dying patient follow a regular series of responses that can be clearly identified. Some patients deny, then accept, deny again. Some become angry, querulous, depressed, serene, anxious, depressed, over and over again" (idem, p. 2045).

Returning to the topic of this section, it has to be stressed that dysphoric behavior is encountered in mature animals, too, albeit in a less clear-cut and less widespread form than displacement or aggression. Many years ago, I witnessed a scene that left me deeply dejected for some time. A jackal mother was watching the play of several of her older cubs when suddenly one of them was shot dead by a careless hunter. The jackal mother's gruesome howls, which instantly followed the kill, reminded me of the screaming of another mother, a human one. As part of my duties as a resident in psychiatry, I was present when she was told about the death of her son.

Lorenz describes the response of a graylag goose to the disappearance of its mate after the bond between them was firmly established, a response that bears partial resemblance to that of humans in distress: "The first response to the disappearance consists of the anxious attempt to find him or her again. The goose moves about restlessly by day and night, flying great distances . . . uttering all the time the penetrating trisyllabic long-distance call. . . . From the moment a goose realizes that the partner is missing it loses all courage and flees even from the youngest and weakest." In spite of the prominent signs of fear and loss of self-confidence, Lorenz concludes that "all the objectively observable characteristics of the goose's behavior on losing its mate are roughly identical with those accompanying human grief" (1974, pp. 178–179).

The next example relates to chimpanzee group life. It exemplifies, besides dysphoria, several other behaviors discussed in this section, such as the quick, flexible transitions between anger, despair, and the regaining of the original active instinctive motive, as well as the possible socially adaptive role of a display of despair. It contains an especially impressive attempt on the part of primatologist de Waal to grasp animals' subjective experiences through empathy; and it also provides hints as to the phylogenetic origins of the motor pattern of dysphoria (which are detailed later). The quotation refers to the closing episodes of a dominance takeover in a captive chimpanzee group from an older, experienced, but physically inferior, leader named Yeroen.

De Waal writes:

[A] phenomenon . . . which I have come to interpret as the beginning of the end, is the losing party's temper tantrums. . . . With an unerring sense of drama he would let himself drop out of a tree like a rotten apple and roll around on the ground screaming and kicking. . . . These hysterical outbursts gave an impression of scarcely suppressed despair and abjectness. When he had regained some of his selfcomposure, Yeroen would run yelping to the females, throw himself down on the ground a few meters away, and stretch out both hands to them. This was not a begging gesture but a beseeching gesture, beseeching them for their support. If the females refused to help, or even went out of their way to avoid him, Yeroen would once again break down and have a tantrum.

His reaction was very different if the females did offer their support. Then he would leap up immediately, embrace them, and turn on his rival with the females close behind him (de Waal 1982/2007, pp. 98–99).

The question that has to be asked here is, what is the adaptive significance of this kind of dysphoric response to an active instinctive motive's frustration in mature animals? In the case of evolved social animals like the chimpanzees and ourselves, we can easily find an adaptive motive— the mobilization of some kind of help from surrounding conspecifics. In human societies, in certain kinds of emotional loss like mourning, this help is even ritualized and institutionalized. However, dysphoric responses can be seen in animals with much less developed social life. Examples include birds, the jackal mother mentioned earlier, and other instances that will be discussed below. In these cases, a mature animal cannot receive effective social help from conspecifics.

Moreover, dysphoric behavior may be strongly maladaptive. The psychomotor agitation it induces is very energy intensive and exhausting; its noisiness makes the animal conspicuous and more exposed to predation; it suppresses other (fitness-enhancing) instinctive activities, maybe even more than aggression or displacement behavior. We have to hypothesize again, as in the case of displacement and vacuum behavior (and frequently in displaced aggressive behavior, too), that the primary function of dysphoric behavior in mature animals is *to discharge accumulated active instinctive tensions that have a disturbing effect on balanced mental functioning*. Note that Yeroen, the challenged dominant chimpanzee, "regained some of his self-composure" after the temper tantrum was over. The importance of this balanced mental functioning will become even more conspicuous after discussing the active/reactive behavior interrelationship at the end of the present chapter.

Taking in account dysphoric behavior's highly maladaptive nature in mature animals, natural selection has to check it tightly—that is, it must be rendered as short, subdued, and as mild as is optimally adaptive for the species in its natural habitat. Extreme examples in which the dysphoric response to the loss of offspring (as we know it in humans) is strongly suppressed by natural selection include lions and langur monkeys. In these animals, the pride or group is frequently taken over by a new dominant male. S. B. Hrdy's description of languor monkey behavior (1977), to my best knowledge, seems applicable to both species. In Laland and Brown's formulation: "males will be selected to eliminate unweaned infants that were not their own, as females would then ovulate sooner than if the infant had lived and continued to suckle. The willingness of female langur monkeys to mate with infanticidal males was viewed by Hrdy as an adaptive strategy on the part of females in response to the high turnover of males in the group. The mother of an infant would make some attempts to prevent such attacks, yet would often mate with the male that had just killed her offspring" (Laland & Brown, 2011, p. 70).

I am aware of another even more extreme instance of suppression of the dysphoric transformation of the parental instinct in animals. A female mouse will devour her newborn offspring before fleeing when its hiding place has been discovered by a predator. The evolutionary explanation of this cannibalistic behavior is that the mouse needed that precious protein, which, in this extreme situation, would have been wasted anyway.

It may seem hardly believable, but I found an example of an adaptation in a human population living in extremely difficult environmental conditions that, while it is not such a brutal blow to the parental instinct as in the story of the langur monkeys, is a comparably brutal insult to the emotional tie between wife and husband: "Among the Eskimo, when one man has killed another, the family of the man who has been murdered may take the murderer to replace the loss within the group. The murderer then becomes the husband of the woman who has been widowed by his act. This is an emphasis upon restitution that ignores all other aspects of the situation—those which seem to us the only important ones" (Benedict, 1968, p. 184).

The evolutionary origin of the motor pattern characteristic of a dysphoric reaction (and surely its accompanying emotional state) seems quite obvious. The overt behavior in dysphoria, at least the basic, innate aspects least affected by ontogeny (crying, agitation, and other physical expressions of helplessness and mental suffering), closely resembles the behavior of immature, dependent offspring when instinctive needs are frustrated, or body functions deranged (as in the case of hunger, disturbances in thermostasis, physical injury or painful disease, separation from the parenting figure, and so on). In these cases, the dysphoric behavior has a clear, adaptive (life-sustaining, lifesaving) value. It mobilizes parental aid in order to redress the harmful or frustrating situation. In more intrapsychic terms, dysphoric behavior expresses pain and suffering, which induces in the parent an instinctively founded urge to discharge the desired, helpful, or guarding activity (lactation, for example) when practicable. In more complex situations, as in more serious bodily injury, expressions of suffering on the child's part induce through empathy similar distressful emotions in the human parenting figure, which induce her to search for more sophisticated help, such as seeking the advice of an expert.

Taking into account this line of reasoning, it is not surprising therefore that mature individuals—animals, as well as humans—when suffering physically or mentally, may resort to the same or similar innate behavioral patterns as the immature, dependent offspring. Of course, mature humans, or highly intelligent social animals (like the old chimpanzee challenged in his dominant status), acquire throughout ontogeny a vast amount of learned material superimposed on a basic instinctive "layer." While this sophistication superimposed on instinctive predispositions is utterly important for certain fields of behavioral sciences, it is less relevant to the concerns of this text. What is more important here is that, in most mature nonhuman animals, the adaptive value of the dysphoric behavior is lost; no parenting figure is around that can be induced to behave helpfully in frustrating conditions. And even if they were, innate parenting behavior is of little help in the suffering of *mature* animals. Therefore, it can be hypothesized quite confidently in this case *that natural and intragroup selection pressures will select against those individuals predisposed to express unduly strong or longstanding dysphoric behavior at the expense of more adaptive strategies,* as we have seen in the examples of langur monkeys, lions, and mice.

In the case of humans, however, particularly in those living in technologically advanced democracies, the situation is vastly different. The natural and intragroup selection pressures against expressing dysphoric behavior are considerably relaxed. Moreover, in many instances, surrounding people may be willing and able to provide effective, or at least emotionally soothing, help. The close family circle and friends are, as a rule, positively empathetic and helpful in these circumstances. Furthermore, as was mentioned previously, in certain cases of loss, such as mourning or temporary unemployment, modern societies provide ritualized or institutionalized emotional or material help. I do not intend to argue on behalf of positive intragroup natural selection pressures for expressing dysphoria in human societies, but certainly *the relaxation of selection pressures led to an accentuated diversification of the readiness to express dysphoria*—that is, the *threshold* at which an original adaptive instinctive striving transforms into dysphoric behavior has a wide range of graded continuum. Furthermore, the *intensity* of the dysphoric reaction is also diversified, which influences the extent to which it can suppress the expression of other, more useful instinctive activities unaffected by the frustrating condition. And finally, the *length* of the dysphoric reaction is also diversified. Therefore, the dysphoric response (or, more precisely, the innate predisposition to it) can be strong and short, subdued but protracted, elicited more or less easily, or an expression of many in-between combinations. In these circumstances, it can be hypothesized that a dysphoric reaction that has become too intense, too protracted, or significantly suppresses other instinctive motivations may lead to pathological mental states.

Before mentioning the forms of psychopathology related to the dysphoric transformation of active instincts, however, we have to more fully discuss an inadequacy (especially from an evolutionary point of view) in the psychiatric classification of depressive disorders, a topic that was mentioned in section 3.2.2.5.3.

When discussing the seasonal fluctuations of instinctive intensities and their diversification, we encountered a form of depression whose prototype is winter depression. Underlying this kind of depression are *low*

intensities of active instinctive activity, which represents a phylogenetically evolved adaptive response to the seasonal scarcity of environmental
resources. In addition, as we have seen, the dysphoric transformation
of active instinctive strivings, when unduly intense and protracted, can
lead also to depressive pathology, but of an entirely different kind. Psychiatric classificatory systems do not differentiate between these two
forms of depression. Both kinds are categorized as Major Depressive
Episode. (The kind of depression involving low instinctive intensities, if
longstanding [more than two years], receives a different diagnosis, that
of Dystimic Disorder [DSM-IV-TR, 2000, pp. 168–169 and 173–179].)
Considering that psychiatric classification is atheoretical and descriptive, this approach may be at least partially understandable, although in
my experience, and as I will detail below, conspicuous differences can be
found in the clinical picture, too (as well as the subjective experience),
between these two forms of depression. Indeed, in a text like the present
one, based on evolutionary reasoning, the differences between these two
forms of depression seem striking.

For the sake of brevity, I refer to these two forms of depression as
"lethargic" and "dysphoric" depression. The basic cause of lethargic depression is *low intensities of a wide variety* of active instinctive drives.
In dysphoric depression, on the other hand, the fundamental underlying cause is the accentuated dysphoric transformation of (usually)
one single active instinctive motive. Because the frustration blocks the
access to the appropriate releaser, the drive's "action-specific excitation"
intensifies, which in turn *energizes* the behavior instead of *inhibiting* it
(as in the case of lethargic depression). Accordingly, the evolutionary
adaptive aim of these two forms of depression is widely different, even
contradictory. In lethargic depression, the original adaptive scope is to
survive a longer period of scarce environmental resources by slowing
down behavioral and physiological processes; in dysphoric depression,
the behavior's original adaptive scope is powerfully signaling pain and
suffering (physical or mental) with the expectation of getting external
help in order to redress the harmful influence here and now. (At least
this was its original adaptive role in the immature offspring.) Therefore,
rather than suppressing, it intensifies the overall level of activity.

Lumping together these two different adaptive mechanisms (more
exactly, their clinical manifestations), as is currently practiced in clinical psychiatry, may considerably confuse research efforts that seek to

discover their respective somatic substrates in the brain (which have to be different), their genetics, as well as the effect of antidepressant drugs on the depressive symptomatology and related brain chemistry. This failure to distinguish between these two different kinds of pathology can be the cause of the low efficacy of antidepressants found in randomized clinical trials. (I will elaborate on this subject in the next chapter).

Genetic and epidemiologic research data illustrate the confusion caused by not distinguishing between these two forms of depression. My presumption will be that most episodes of depression in the framework of bipolar disorder are of the *lethargic* type, while a considerable amount of unipolar depression is of the *dysphoric* type: "Relatives of unipolar probands are not at increased risk for bipolar depression . . . but relatives of bipolar probands are at increased risk for unipolar depression" (Plomin et al., 2008, p. 209).

In the case of bipolar depression, twin studies "suggest extremely high heritabilities of liability," and "conventional diagnostic rules assume that an individual either has unipolar or bipolar disorder and bipolar disorder trumps unipolar disorder. However, in this twin study, this diagnostic assumption was relaxed and a genetic correlation of 0.65 was found between depression and mania, which supports the model. However, 70 percent of the genetic variance on mania was independent of depression, which does not support the model. A model that explicitly tested the assumption that bipolar disorder is a more extreme form of unipolar depression was rejected, but so too was a model in which the two disorders were assumed to be genetically distinct" (idem, p. 212).

As we have seen, mania or depression in bipolarity may be of clinical or subclinical intensity (which does not change its heritability), or one of the polarities may be a personality trait while the other is episodic in nature. Moreover, the two genetic predispositions (that is, for lethargic and dysphoric depression) are independently heritable; therefore, both can co-occur in the same individual (as in dysphoric mania). I cannot imagine how a significant breakthrough in the genetic research on depression can be achieved so long as the research criteria for diagnosis remains a more strictly formulated variant of clinical disease criteria of a descriptive nature, instead of relying on those clinical signs that are the closest reflection of the basic evolved adaptive mechanism.

The idea that depressive pathology can be divided into two different clinical entities is not new. It was first proposed about a thousand years ago. Persian philosopher-scientist-physicist Avicenna (980–1037) "speculated that a special form of melancholia . . . was 'coupled with inertia, lack of movement, and quiet' . . . Avicenna further observed that the mixture of anger and restlessness in melancholia indicated that the disease was manic in nature [that is, induced by high energetic levels] and the appearance of such signs and symptoms along with violence heralded the transition from melancholia to mania" (Akiskal, 2005a, p. 1562).

In fact, the confusion that results from failing to clinically separate *the low level of intensity* of the instinctive activity underlying one kind of depressive behavior from the predisposition to react preferentially with the *dysphoric form of transformation of active instincts*, due to ever present frustrations, has plagued psychiatric nosology from its beginnings: "For Kraepelin, the core pathology of clinical depression consisted of lowered and slowed (retarded) physical and mental processes. In mania by contrast, the mood was elated, and both physical and mental activity were accelerated" (Akiskal, 2005b, p. 1564). This view of the bipolar affective disorders has remained influential up to the present day: "The clinical features of mania are generally the opposite of those of depression. Thus, instead of *lowered mood*, activity, and self-esteem, there is *elevated mood*, a rush of ideas, psycho-motor acceleration and *grandiosity*" (idem, p. 1620, emphasis added). Nevertheless, on the following page of the same textbook, where the affective state in mania is described in more stringent clinical detail, a different picture emerges, one that is closer to the viewpoint of this text: "The prevailing positive mood in mania is not stable, and momentary crying or bursting into tears is common . . . when crossed, patients can become extremely irritable and hostile. Thus, lability and irritable hostility are as much features of the manic mood as is elation. In mixed manic states they dominate the clinical picture, giving rise to what is now termed *dysphoric mania*" (idem, p. 1621, emphasis in the original). Dysphoric or "mixed" mania is described as "characterized by dysphorically excited moods, irritability, anger, panic attacks, pressured speech, agitation, suicidal ideation, severe insomnia, grandiosity, and hypersexuality, as well as persecutory delusions and confusion" (idem, p. 1636). This clinical description seems like a mixture of, or quick alternations between, all three forms of frustration of the active instincts described in this section, brought about by excessive instinctive energies and a low threshold concerning the transitions between these frustrated forms, leading to considerable behavioral disorganization.

Therefore, it has to be stressed that *the intensity of the psychomotor activity by itself does not determine its emotional quality*. Low intensities of psychomotor activity, as for example in the *refractory state* after sexual or feeding needs are completely satisfied, or the total exhaustion of physical and mental energies after a successful ascent to a high mountain peak, need not be painful or depressing; in fact, they are as a rule pleasurable. In the same vein, *high* intensities of instinctive activity are not felt always as euphoria. In their dysphoric form of frustration, they can be extremely painful. *The intensity of the instinctive activity can determine the intensity of the respective form of its transformation during frustrating conditions but not its kind.* A manic patient (high intensities of instinctive activity) may be mostly euphoric when left alone but may become highly aggressive or dysphoric during enforced hospitalization or when coerced to do something against her will.

For the sake of clarity, I would like to summarize here concisely the emotional accompaniment—that is, the subjective emotional experience—of each one of these three forms of frustration, despite their having been mentioned already in a casual way. I also include the instance in which the original active instinctive motive resists these transformations, which usually constitute the most desirable alternative.

When an active instinctive motive pursues a goal (tries to contact its appropriate releaser) and *resists transformations amid frustrating conditions*, or when these transformations are mild and short-lived and the individual regains quickly and flexibly the original, unfrustrated instinctive motive toward the desirable goal, the basic emotional state consists mostly of a *pleasurable expectation*. This can be exemplified by the emotions one feels while courting a desirable woman, or preparing a delicious meal, or working on a self-initiated professional project with good chances to complete it successfully. Individuals with these kinds of instinctive attributes were seen by the ancient Greeks as possessing a "sanguine" temperament or personality.

In the case of *displacement or vacuum behavior*, as we have seen, the original instinctive motive (its motor pattern) is enacted in an unchanged form but directed toward an improper releaser or discharged without any releaser being present. Probably as a result of the instinctive behavior *preserving its original form*, the accompanying emotional state is *pleasurable* in spite of being incapable of achieving its ultimate adaptive goal. As I have noted earlier in a different context, Konrad Lorenz has gone so far as to claim that the main motivation of an animal in some instances is the enactment of the appetitive phase of an instinctive striving, not the achievement of the consummatory phase: "In dogs, hunting and killing are . . . independent of the motivation of eating: As everyone knows, it is impossible to wean a dog from its *passion* for hunting through abundant feeding" (Lorenz, 1982, p. 135, emphasis added).

During the *aggressive* transformation of the original active instinctive activity, its accompanying pleasurable mental experience is lost. Nevertheless, the person (and this is probably true for nonhuman animals as well) feels powerful and confident in possessing the behavioral means to redress the problematic situation. Aggression suppresses physical pain, anxieties, hesitation, and other psychic sensibilities like empathy, self-pity, and self-accusations in order to assist the individual in carrying out uninhibitedly the aggressive act.

In the *dysphoric* form of frustration, the individual has no behavioral means of effectively tackling the frustrating object or situation. For some reason (which differs in different circumstances and instinctive constellations), he cannot "disconnect" himself from the unavailable releaser in order to express the instinctive motive in a displaced way or in vacuo (while preserving its pleasurable subjective experience), nor can he fight aggressively in order to try to remove the frustrating obstacle. The instinctive motivation continues to strive toward its original releaser or goal (a missing child, for example, in the case of frustrated parental instincts), and in its absence, the "action-specific excitation" accumulates, leading to an intensifying feeling of unpleasant mental and physical tension, pain, or agony. This increasing instinctive tension suppresses other instinctive motives (hunger or sexual drive, for example), which may have a proper available releaser at hand, and whose enactment could bring a pleasurable mental experience. The only means at the sufferer's disposal to obtain some temporary and incomplete relief is incessantly expressing that painful internal tension (felt, as a rule, as both physical and mental) at various levels of sophistication, from verbal expression to body language, while using the inherited motor and physiological patterns of a suffering, dependent child.

This emotional state is clearly different from the subjective experience in lethargic depression. In this latter state, one experiences loss of energy, languor, apathy, psychomotor slowing, cognitive emptiness, fatigue, a *reduction* in the intensity of different instinctive motivations (like sexuality and territoriality, though not as a rule hunger), as well as an increased propensity to feel defenseless and anxious in disturbing or stressful situations. As we have seen before, this emotional state, induced by low instinctive intensities, does not necessarily have to be especially painful. It seems to me that the painful quality is induced *secondarily when the consequences of inactivity* become unduly strong or protracted. The affected person becomes unable to function adequately in the role of parent, husband, or wife, unable to tackle occupational and social duties, and so on. If an otherwise industrious person with normative moral standards becomes unable to continue working, if a mother sees that her children are not looked after properly, etc., this situation may provoke anxiety or feelings of inadequacy or guilt.

In my experience, aggressiveness or suicidal and homicidal intentions are not encountered in this state (in contrast to dysphoric depression), although the individual may complain of the futility, emptiness, and meaninglessness of life. Epidemiological data that link the suicide rate to the seasons of the year support this reasoning. Suicides (particularly those that are violent) peak in May and June in the Northern Hemisphere and in November and December in the Southern Hemisphere. This time of the year in those persons inclined genetically to seasonal variations in instinctive intensities is characterized by increased (rather than decreased) active instinctive activity, which, according to our theoretical scheme, suits the dysphoric form of depression. Conversely, researchers have documented a drop in suicide rates in the winter months of about 20 percent (Foster & Kreitzman, 2009, pp. 221–224).

Increased rates of suicide at the beginning of antidepressant therapy are also hypothesized to result from increased "willpower," while the thought content is still under the influence of the depressive mood. Antidepressant drug effects in this context are discussed in the next chapter.

Naturally, the predisposition to dysphoric versus lethargic depression are two independent variables—that is, the presence of one of them does not exclude the presence of the other. Therefore, the presence of lethargic depression does not preclude the inclination to respond readily with a more intense, more protracted dysphoric state to painful or frustrating circumstances. However, it must be presumed that the low overall instinctive activity in this case will *reduce* the intensity of the dysphoria, too. (This subject is discussed further in section 7.2.1.) Conversely, a person with a predisposition to respond readily with the dysphoric transformation of the active instinctive motives may also suffer from periods of intense overall instinctive activity (mania) during which the dysphoric state may become especially severe.

5.2.3.1. Dysphoria in Mental Disorders

This topic has already been partly addressed. As in the case of the other forms of active instinctive frustration, dysphoric symptoms may appear in a wide variety of mental disorders. In general, it may be concluded that the intensity (and in consequence the pathological nature) of the dysphoria depends mainly on the following three factors:

1. The overall intensity of the instinctive energies due to causes unrelated to active instinctive frustration (as in mania, abuse of psychostimulants, or the antidepressant drug effect).
2. The extent of the innate predisposition to respond overwhelmingly or exclusively to frustrating conditions with dysphoria.
3. The severity, multitude, and protractedness of the frustrating circumstances. This aspect is especially prominent in the late-life or postmenopausal form of agitated depression. (This subject is discussed briefly in section 7.2.1.)

For the way dysphoria is expressed in schizophrenia, see section 7.2.8.9.1.

Our next topic is the *flexible transitions and overlap* between the three forms of active instinctive frustration. In addition, we look at the transitions between the instinctive motives' original (unfrustrated) and its frustrated forms.

The story of the old chimpanzee Yeroen whose dominant status was challenged by a younger and stronger chimpanzee (section 5.2.3) exemplifies several aspects of these transitions and overlapping of forms of frustration. The phenomenon of the temper tantrum itself, so familiar in human children as well as in chimpanzees (in young offspring and adolescents during weaning), contains elements of both aggression and dysphoria. The individual does not bear his suffering alone, as a solitary plight. Temper tantrums happen regularly in the presence of others. The impression received while witnessing a temper tantrum is that the observer is forcefully drawn into the situation, compelled to empathize with the suffering, to feel the same pain, and to feel himself guilty for it, as well as for any possible physical injury its continuation may cause. He feels forced to provide some kind of help, or alternatively to yield to the demand from the child that induced the temper tantrum in the first place. The temper tantrum itself may contain dysphoric and aggressive elements in different proportions. At the dysphoric end of the scale, it contains only the loud expression of physical and mental pain, as well as intentions of self-harm, while at the aggressive end it may be overtly threatening. The following quotation is a description of tantrum behavior in adults who as children were used to obtaining their frustrated emotional goals in this way: "When frustrated or threatened, they may act like bullies, glare, snarl, yell, shout, intimidate, pout and sulk, and sometimes be physically violent" (Yager & Gitlin, 2005, p. 995).

Returning to our chimpanzee, Yeroen, we have seen that when the females consented to support him "he would leap up immediately, embrace them"—that is, he returns quickly to the usual (unfrustrated) behavior of a dominant male toward his supporters—". . . and turns on his rival with the females close behind him" (de Waal, 2007, p. 99). When support was successfully achieved with the aid of the dysphoric behavior, he quickly regained his aggressive intentions against the primary frustrating source.

I suspect that in those (fortunately rare) cases of homicide followed by suicide, the underlying emotional motivation has to be a blend of strong aggressive and dysphoric mood, which for some reason (intrapsychic, environmental, or both) could not find a less tragic outlet. This, sadly, may be what is occurring when a parent, usually the mother, kills her children and commits or attempts suicide; or when, in the case of an old couple living in poverty, with one or both of them suffering from serious physical illnesses, the husband kills his wife and commits suicide; or when a young man enters his former school, shoots (randomly) at pupils and teachers, and eventually kills himself, and so on.

Human behavior, which is so richly nuanced, can show other transitional states, too, between the forms of frustration or between the unfrustrated and frustrated forms. Teasing and mocking between peers contains joyful as well as aggressive elements; arrogance signals at the same time inflated self-esteem and scorn toward others. Sarcasm, irony, satire, and a cynical attitude are additional, more complex, examples.

We do know of instances in which two or all three forms of the in-stinctive frustration appear in a more or less regular succession: As we have seen, *mourning*, as well as the mental response to the *awareness of impending death*, frequently contain these three responses to frustra-tion appearing in succession or interspaced with other mental states. We have seen, too, that in the course of a manic attack all three forms of frustration can appear in an irregular sequence or in a mixed form, partly as a reaction to the conditions of enforced hospitalization if that has occurred.

Carlson and Goodwin, however, describe a different regularity during an untreated manic attack (including its phase of resolution), which may be relevant to our discussion. In their scheme, first the euphoria dominates the clinical picture but anger and irritability later replace it, and finally a state of severe panic ensues (Carlson & Goodwin, 1973, pp. 221–228). If this is indeed a general regularity, I wonder whether the progressive decrease in the *intensity* of the active instinctive activity and the concomitant increase in the sensitivity to incoming social mes-sages—first of all the realization of the inconveniences and harmful after-effects of the manic episode (concerning social relationships, status at the workplace, financial affairs, relationships within the family, etc.)—may facilitate or even bring about these transitions from euphoria to anger, irritability, and anxiety.

In the next chapter, a similar chain of events is referred to, this time as a result of antipsychotic treatment. I propose that the main thera-peutic effect of antipsychotics drugs consists of reducing the intensity of the overall instinctively based activity. Consequently, a manic individual treated with these drugs, as a result of decreasing instinctive intensities, will become first irritable and later anxious.

The next example shows quite similar emotional transformations but induced by a different kind of pathology. Consider the following description of the stages of recovery from acute reversible brain damage: "The period of maximal brain damage is associated with equanimity or euphoria. In the euphoric stage the patient's speech may resemble that of mild hypomanic reaction. . . . In intermediate stages of brain damage, the flow of associations tends to show a paranoid trend. . . . With fur-ther improvement in brain function, the entire capacity for denial of the illness may dissolve. At that point a depressive reaction, not entirely inappropriate to the clinical realities, may emerge. The emergence of a depressive reaction in a patient with reversible brain disease is evidence of improvement in brain functioning" (Linn, 1980, p. 1003).

In these last three examples, the stages of an untreated manic episode according to Carlson and Goodwin, the antipsychotic effect in mania, and the stages of recovery from reversible brain damage involve a more complex phenomenon than the simple transitions between the forms of active instinctive frustration. That more complex mental mechanism involves *the changing power relationship between acting mostly or exclusively according to active intrapsychic promptings versus increasing sensitivity and reactivity to external influences.* This process, the progressive moving from an uninhibited expression of internal motives (at the full-blown phase of a manic attack and the "period of maximal brain damage") toward a more balanced and integrated realization of the complex individuum-social environment interrelationship (based on a more equilibrated active/reactive/conforming behavior or anxiety interrelationship), seems to influence the transitions between the frustrated forms of active drives—from vacuum behavior and displacement to aggression to dysphoria (and/or anxiety). While the underlying mechanism is not entirely clear to me, I suspect that a kind of lawful interrelationship may exist between the transitions of the frustrated forms of active instincts, on one hand, and the degree of openness and reactivity to environmental stimuli on the other. The next example may also support this presupposition.

Alcohol intoxication (discussed in more detail in the next chapter) reduces the person's awareness and reactivity to constraining, disturbing, or potentially dangerous external influences or circumstances (experienced subjectively as reduced intensities of fear or anxiety), thus upsetting the balance between acting on internal promptings and heeding at the same time potentially troublesome external circumstances. This disinhibiting effect, as can be witnessed frequently at places of entertainment, may lead to euphoric, aggressive, or dysphoric behavior (crying, pounding one's head, etc.), or to all three of them in succession, while becoming sober is accompanied by dejection, worry, and possibly anxiety.

Amphetamines or similar types of psychostimulants can also upset the balance between internal individual motives and external influences on behavior in predisposed individuals by disproportionally strengthening the internal motivational aspect. The exaggerated or protracted use of these substances may lead to manic, paranoid, or dysphoric depressive states (including suicidality) of clinical proportions. This subject is detailed in the next chapter.

5.2.3.2. The Accentuation of a Single Form of Active Instinctive Frustration at the Expense of Other Forms in Normative Individuals

While some persons are predisposed to react with *all three forms of frustration of active instincts* with more or less equal frequency, other individuals *respond predominantly with only one of them*. This situation, in the framework of the present theory, must also be attributed to the relaxation of natural selective pressures and the resultant diversification of the instinctive constellation in the protective human environment. In contrast, natural circumstances, as we have seen, tightly control the form, strength, and duration of, and the flexible, reversible transitions between, these instinctive events in animals in natural conditions in accordance with their adaptive or intrapsychic tension-reducing value.

In the following, I cite some examples of human diversity regarding the predominance of a single form of active instinctive transformation under frustrating conditions:

We have seen in chapter 1 that, as early as the fourth century BC, the Hippocratic writings classified the "emotional orientation" or temperament of individuals into sanguine, choleric, melancholic, and phlegmatic (Kaplan et al., 1980, p. 23). While contemporary causal explanations consisted of naïve, materialistic speculations about the disequilibrium of certain body fluids, the *descriptive aspect* of this categorization has proven to be quite enduring. It reemerged in the eighteenth century in the writings of Johann Kasper Lavater (idem, pp. 51–52) and in fact has survived to the present day in our informal characterization of an individual's temperament. More recently, psychologist Hans Eysenk drew parallels between his own personality dimensions and the four classic Hippocratic temperamental types (Healy, 1999, pp. 170–173).

Three of these four classic temperamental types can also be assimilated into our theoretical framework. The *sanguine* temperament, preserving the instinctive motivations mostly in their original form among frustrating conditions, leads to efficiency in work, social relations, and goal-directed behavior in general, and induces a stable, pleasurable, optimistic emotional disposition. The *choleric* temperament has a predilection to respond promptly with the aggressive transformation of the original instinctive motive to annoying or disturbing stimuli, accompanied by the subjective experience of anger. These individuals, however, quickly and flexibly return to their original, unfrustrated emotional motives, and consequently their interpersonal relations, professional careers, etc., are not affected significantly as a rule, by this personality trait.

The *melancholic temperament* can be interpreted in our theoretical framework in two different ways. It may mean either a permanent low level of instinctive activity, as argued earlier in section 5.1.3, or it can be seen as the inclination to overemphasize the dysphoric form of frustration at the expense of the other forms (that is, these persons show a predilection to preferentially see the undesirable side of the surrounding realities, complain easily and lengthily, cry easily, etc.).

5.2.4. Pathological Mental States Relevant to the Transformations of the Active Instinctive Expression

While as a rule pathological mental states are not specific to or dominated exclusively by one single form of transformation of the active instinctive activity—that is, the diagnosis does not depend on it—they may powerfully influence the clinical picture. A manic attack, which is conceptualized here as a period of more than normatively intense instinctive activity, may be dominated by a euphoric, dysphoric, or aggressive mood, depending partly on the temperament of the individual and partly on the amount and nature of the frustrating circumstances. While the official psychiatric classification (DSM-5) admits only the variants with "expansive or irritable mood," as we have seen, current clinical practice also recognizes a mixed type with clear dysphoric elements.

Moreover, I encountered (albeit rarely) in my clinical practice manic patients with a predominantly angry mood and querulous or frankly aggressive behavior, accompanied by fleeting delusions of the persecutory type. These states may be diagnosed as acute paranoid psychosis. The diagnosis, in my opinion, depends on what received priority in the categorization of these states—whether it was the high energetic load of the behavior, the aggressive form of transformation of the active instinctive motives, or the *presence* or *absence of psychotic symptoms*. What seems more important, however, is the realization that these three elements of the clinical picture may *vary independently* from each other and represent different psychic mechanisms. (The psychotic mechanism is detailed in section 7.2.6.4.)

In our theoretical scheme, as we have seen, recurrent major depressive episodes involving grave psychomotor agitation, the expression of severe mental pain, intense guilt feelings, and suicidal preoccupations are more related to the manic state than they are to the undiscriminating Major Depression category. In addition, there exists another form of dysphoric depression that occurs mainly in women at the postmenopausal period and has a different evolutionary mechanism (detailed in section 7.2.1).

The quality of the underlying mood in schizophrenic patients, especially those with "flat" or "inappropriate" affect, is more difficult to grasp through observing their behavior, their emotion-expressing body language, or verbalizations than it is in normative persons or in the case of milder mental disorders. Nevertheless, we can try to deduce it from the content of the individual's predominant delusional or hallucinatory preoccupations. They may have predominantly grandiose content, hinting at a euphoric mood or a predominantly persecutory content hinting at a perceived inimical relationship between the patient and his human environment. Sometimes these delusional preoccupations may have a self-deprecating, self-accusing, somatic, or apocalyptic content, hinting at a painful, dysphoric underlying mood.

Paranoid conditions and the abovementioned depression of the postmenopausal period are the only mental disease categories I am aware of in which the peculiarities of *one single* frustrated form of the active instinctive activity determines by itself the clinical diagnosis. In these disorders, as we have seen, the genetic predisposition consists of the unusual intensity and tenacity of the respective form of instinctive frustration. These disorders are detailed in sections 7.2.1 and 7.2.6.

Finally, I want to mention here a mental disorder in which *the peculiar nature of the transitions* between the frustrated forms of the active instincts seems to be the basic psychopathological mechanism. I refer to Borderline Personality Disorder, in which these transitions are extremely brusque, complete, and unattenuated by a more balanced cognitive appraisal of interpersonal relations. This condition is detailed in section 7.2.4.3.

I want to also note in passing the existence of a fourth kind of transformation of active instinctive expression in our theoretical scheme—the

*modification of the active instinctive expression resulting from the impera-
tives of social coexistence and cooperation.* A good illustration of this kind
of active instinctive transformation in the animal kingdom would be
the reproductive strategy of a closely knit wolf or wild dog pack. In this
instance, as a rule, only the dominant pair mate, while other members
of the group contribute to feeding and caring for the dominant pair's
offspring, renouncing their own reproductive ambitions (cooperative
breeding) (Summers & Crespi, 2013, p. 66). Owing to its complexity and
lasting nature (contrary to the simple transformations discussed in this
section), this type of transformation of the active instinctive expression
is dealt with in section 5.4.

In a somewhat similar fashion, human individuals—sometimes after
a period of rebellion around adolescence—*submit or conform to, or even
"internalize," the prescribed ways of behavior in the social group they belong
to, including the group's prohibiting, restrictive, or modifying prescriptions
concerning active (as well as self-protective reactive) instinctive expression.*
While the individual thus renounces, modifies, or restricts some of her
active ambitions, this loss as a rule is more than compensated for by the
resultant gain in fitness from social coexistence and cooperation.

5.3. The Differentiated-Diffuse Spectrum of Human Instinctive Drives

5.3.1. General Considerations

As we have seen in previous sections of this work (introduction, instinct-learning interaction, relaxation of natural selection pressures,
primatology), a wide range of variations exist in the animal kingdom
in the ability of instinct to directly induce mature, adaptively effective,
overt behavior without modification by acquired knowledge. This range
of variability can be seen as an evolutionary trend from rigidly preprogrammed, differentiated instinctive behavior (in simple animals) toward
more malleable behavior in which the inherited predispositions are more
and more open to modification by acquired knowledge (see Ernst Mayr
on "closed" versus "open" genetic programs, section 2.2). At the strongly
"closed" or differentiated end of this spectrum, we find clear-cut, genetically fully preprogrammed instructions translated in an unmodified

or almost unmodified form into overt, observable behavior (FMPs). This genetic program is linked tightly to another well-differentiated genetic program, the innate releasing mechanism (IRM), which guides instinctive behavior toward specific, genetically rigidly preprogrammed environmental agents designed to supply critically important resources for sustaining or reproducing the organism's specific form of life (e.g., a restricted range of food source; a heterosexual, mature conspecific in its sexually receptive phrase; etc.). Feedback from practice, experimentation, or learning from other conspecifics has no role, or only a very restricted one, in perfecting this innate mechanism.

This kind of behavior is characteristic of *very simple or very immature organisms*. A precondition for it to work is the reliable availability of the needed environmental resources (releasers) in the organism's immediate surroundings without the need for elaborate searching, competing, cooperating, etc., during the appetitive phase of the active instinctive behavior.

One trend of evolution consists of the progression from simple to complex forms of organismic life, and, accordingly, from simple forms of genetically preprogrammed behavior toward complex, flexible, modifiable ones. In these complex, flexible forms of behavior, acquired knowledge of the environment and, at least in humans, acquired knowledge about oneself is capable of modifying innate behavioral predispositions in order to widen and perfect the organisms' adaptive abilities. The evolutionary rationale for this trend has to be the acquisition of increased abilities for flexible, effective exploitation of a more varied assortment of potential environmental resources in more complex, quickly changing environmental niches, as well as utilization of more sophisticated protective strategies (reactive behaviors) against varied harmful environmental influences. For this purpose, the rigid behavioral programs of simple organisms evolved into more malleable "propensities," "tendencies," "inclinations" to behave. As a result, these genetically less rigidly defined propensities became accessible during ontogeny to modification to an ever-increasing extent by such influences as experimentation, practice, classical and operant conditioning, imitation of other conspecifics' behavior, simple learning, and, in humans, complex, symbolic communication.

At the same time, the assortment of environmental releasers became larger and more varied. Potentially fitness-enhancing resources that are more difficult to recognize and secure (certain kinds of food, for example) may also become available with the aid of more sophisticated behavioral strategies. Remember, for example, Konrad Lorenz's account of the complex exploratory behavior of ravens and the possibilities it opened for the species to spread into very different kinds of environmental niches (section 2.2.1), or of the learned tool-using behavior that some chimpanzee populations, but not others, engage in for termite-fishing.[2]

This evolutionary development, from genetically rigidly preprogrammed behavior toward more malleable genetic predispositions that are open to modification by learning and directed toward a more variable and richer assortment of releasers, accelerated when highly evolved mammals began to live in groups with increasingly complex social structures. This development further improved their ability to exploit an ever-increasing variety of environmental resources for purposes of survival, defense, and reproduction, adding to their behavioral repertoire such novel strategies as concerted group efforts or transgenerational transmission of acquired forms of behavior (like termite fishing in chimpanzees). In addition, living in social groups entailed new forms of adaptation and a higher level of behavioral complexity as a result of the need to adapt to the behavior of other group members. Interactions between conspecifics living together in a tightly organized group (unlike most interactions with the physical and biological environment) are quickly changing, fluid, complex, and varied. To adapt behavior to these circumstances requires an even more accentuated reduction of rigid instinctive differentiation or preprogramming since rigid genetic programs can become increasingly maladaptive in these circumstances. For example, during this process the appetitive phase of active behavior has lost the *intermediary fixed motor patterns* indispensable for less evolved animals that do not live in complex social groups. Examples are ritualized courting behavior of birds; nest building; and ritualized sexual fighting between males in many vertebrates from fish to mammals, but not in primates. At the same time, an explosion took place in the extent to which instinctive predispositions could be modified by or channeled into behavior patterns strongly influenced by knowledge acquired during ontogeny. A relevant example is the communication between members of a group with the aid of body language and vocalizations, which became richer and more expressive and precise. (For example, vervet monkeys can emit three different alarm calls depending on whether a vulture hovers above the tree, a snake is detected between the branches, or a leopard approaches from below. Consider that the evasive reaction has to be different in each case [Gomez, 2004, pp. 175–180].) The primatologist Susan Perry described an even richer assortment of alarm calls in the capuchin monkeys of Costa Rica (2011, pp. 47–48, 83, 93–94, 181).

At the same time, the need arose to recognize and remember other group members personally by their physical and mental assets and liabilities and their veiled intentions, reflected in their behavior (and perceived through rudimentary "mind reading" or "theory of mind" abilities). This kind of knowledge strongly influenced the relationships between group members, enabling them to tell who might be a potential competitor, who is a reliable ally, and so on.

It seems quite clear that *the position of each animal species on this instinctive differentiated versus diffuse spectrum (in other words, the degree to which instinctive motivations can be modified by acquired knowledge during ontogeny) has to be quite rigidly predetermined by phylogeny.* The instinctive makeup of fish and reptiles seem to be more differentiated than that of birds (especially more intelligent birds like the raven), and the instinctive makeup of birds is more differentiated than that of the more evolved mammals; inside the class of mammals, the social ones, like wolves and dogs, are more modifiable by learning and training than solitary ones, like the wild and domestic cat; among the apes, the "lesser ape"—the gibbon—is less intelligent and educable then the "great apes." In the sphere of sexuality, bonobos' sexual instincts are much more diffuse and variegated than those of chimpanzees, while chimpanzee social organization is more complex than that of the bonobos. However, some exceptions exist to the assumption that complex social life invariably leads to more educability, and vice versa. The solitary Orangutan is "almost as responsive to education as Chimpanzees" (*The New Larousse Encyclopedia of Animal Life*, 1981, p. 513). (Nevertheless, there are suggestions in the literature that orangutans may have been more social during their evolutionary past. See Summers & Crespi, 2013, p. 293.)

Humans undoubtedly have by far the most diffuse instinctive predispositions in the animal kingdom, and in consequence the greatest behavioral flexibility and educability. It has been argued by social theorists and behavioral psychologists that in humans instincts became so atrophied (tentative) that their impact on overt behavior became negligible or nonexistent, or at least unprovable by accepted scientific methods. Instead, human behavior is guided predominantly or exclusively by acquired knowledge.

This standpoint can be understandable but not acceptable, at least in an evolutionary framework of reasoning. It is true that the differentiated-diffuse spectrum of instinctive predispositions is a highly concealed aspect of the behavior, even more concealed the closer we come to its diffuse extreme. It may even be said that, in humans, this behavioral incentive is *doubly concealed.* It directly affects *specific* instinctive inclinations (sexuality, reaction to danger, etc.), which, by themselves, may be more or less concealed incentives of overt behavior. (Recall, for example, how Freud's discovery that sexual drives influence human behavior much more extensively than was admitted by nineteenth-century Western culture came as a revelation.)

Now, if specific instinctive motives are concealed factors in overt human behavior, the comprehensive mechanisms—the main topic of this work—must be even more concealed, since they do not affect overt behavior directly but rather through modulating discrete instinctive motives. It is therefore understandable that various approaches to human behavior, like those of the social sciences, or psychological trends that accept only observationally verifiable phenomena as valid scientific data (behaviorist methodology, for example), find difficult to accept the idea that innate instinctive predispositions have a powerful impact on human behavior. Those *external* influences that can modify behavior are frequently observable to a much greater extent than *internal*, instinctive influences and thus are more suitable for scientific experimentation. However, in my opinion, the methodology of enquiry has to be adjusted to the nature of the phenomena studied, and not vice versa—that is, it is not sufficient to study only phenomena suitable for the methodology of exact sciences (see Hayek, 1955, pp. 13–14). The intention to consider as much relevant information as possible must gain preference over rigid methodological constraints; otherwise, critically important information may be lost. Therefore, in spite of these difficulties, it has to be accepted in an evolutionary framework of thinking that human behavior is not an entirely new entity that differs qualitatively from the behavior of other animals in which the instinctive motives are more apparent. Otherwise, no way exists to account for human populations' extraordinary success in biological survival and reproduction, especially in the last 10,000 years.

Regarding human behavioral evolution, it has to be hypothesized that, initially, the progress of our nonhuman social primate ancestors from more differentiated instinctive predispositions toward optimally diffuse ones in early human populations was induced by two parallel evolutionary developments: the gradual relaxation of natural selection pressures extrinsic to the group resulting from increasing ecological dominance (leading to diversification of the instinctive predispositions) and a simultaneous increase in directional intragroup selection pressures enhancing social coexistence and cooperation, or, more precisely, their instinctive foundations (chapter 4, sections 4.6.1. and 4.6.5.).

Some members of the group must have possessed *more than optimally differentiated*, egocentric instinctive predispositions—that is, instinctive predispositions whose appetitive phase showed clearly the shortest and most economical way toward achieving individual biological goals by promptly seizing opportunities offered by group life, without feeling the need to internalize the constraints of social coexistence. These individuals sooner or later found themselves at a disadvantage for survival and reproduction. That is because the strongly selective, narrowly goal-oriented behavior that these more than optimally differentiated instinctive predispositions facilitate tends to disregard agreed-upon social norms and compromises, the power relationships between individuals and clans inside the group, and the priorities of the group as a whole. It tends to exploit current opportunities without considering long-term consequences in the context of group life. While this kind of behavior may confer some initial short-term fitness advantages, sooner or later it incurs social (and individual) opposition with serious repercussions for long-term survival and reproduction.

Group members with *more than optimally diffuse* instinctive predispositions were also at a disadvantage regarding fitness, but for very different reasons. In their case, environmental stimuli, natural and especially social, were absorbed in an indiscriminate, less than optimally selective way (a phenomenon well known in schizophrenics)—without sensing with enough precision with the aid of an innate, inherited guiding mechanism (a "teaching mechanism" in Konrad Lorenz's term, 1965) their relevance to basic biological interests. As a result, these individuals were unable to achieve the right balance between expressing individual motives and at the same time heeding group norms and others' interests in varied, sometimes quickly changing, social situations. This lack of balance (combined with less than optimal differentiatedness of the innate strivings) rendered their behavior inconsequent, hesitant, unpredictable, or confused and their social relations problematic, unlike the social relations of those possessing flexible, normatively differentiated drives (their mental capacities being equal).

While behavior influenced by instinctive predispositions that are too diffuse attract as a rule less social opposition than behavior influenced by instincts that are too differentiated, the individual is unable to optimally exploit the opportunities for securing his interests in the ways group life allows it, or to secure for himself a safe place in the group's social network.

However, with the increasing ecological dominance of human populations, especially after the introduction of agriculture some 10,000 years ago, it can be hypothesized—as in the case of the other comprehensive instinctive mechanisms discussed in this chapter—that these intragroup selection pressures began to relax, and this relaxation has increased progressively with technological and cultural advances up to the present day. As a consequence of this development, the differentiated-diffuse continuum of instinctive predispositions (the D/D scale) in humans widened at both ends away from optimum, facilitating at the extremes a predisposition for increasingly dysfunctional forms of behavior.

I am keenly aware of the highly unusual and abstract nature of this line of reasoning, as well as the inability to explore its implications directly with scientific methods. Great difficulties exist in sensing by the mechanism of empathy (detailed in section 2.3) the position of the instinctive makeup of surrounding persons on this differentiated-diffuse spectrum in comparison with that of the other three comprehensive instinctive mechanisms discussed in this chapter. The fluctuations in the overall instinctive intensities and the three forms of transformation active instinctive drives undergo under frustrating conditions are experienced subjectively from time to time by everyone; hence the subjective experience of others in these instances can be approximated quite well with the mechanism of empathy. The same is true with regard to the active/reactive behavior dichotomy and the interrelationship of the active and reactive phases, discussed in the fourth section of this chapter. Comprehending the differentiated-diffuse spectrum of human instinctive drives, on the other hand, and pinpointing the specific place that one's own instinctive configuration occupies on this scale relative to that of known others seems to be a much more problematic mental exercise. People are not aware, as a rule, of this instinctive dimension, either in themselves or in others. Nevertheless, I am strongly convinced that this dimension in one's instinctive configuration (namely, the acuity with which one senses one's own instinctive interests and consequently that of others through empathy, as well as the degree of openness of the behavior to modification by social and cultural influences, even those remote from specific biological goals) powerfully influences the path of life of every person. Moreover, I believe that this concept may constitute a promising avenue toward understanding in evolutionary terms the nature of the inherited predisposition of two major categories in psychopathology—the antisocial personality disorder and schizophrenia, including the latter's genetically related spectrum disorders.

The differentiated-diffuse scale of instinctive predispositions in humans, so far as I am aware, is the most unfamiliar, most unexplored idea discussed in this work. To my knowledge, no other theories on human behavior have entertained it; at any rate I have not encountered this idea in the professional literature. However, once I began to contemplate this subject, I found a passage in a book suggesting that its author may have had, unwittingly, and without fully realizing its importance, a similar intuition. This passage came from an unexpected source, American anthropologist and social theorist Ernest Becker's Pulitzer Prize–winning book, *The Denial of Death* (1975). In accordance with the traditional standpoint of the social sciences, Becker repeatedly negates the existence of instincts in humans: "An animal [the human animal] who has no instincts has no programmed fears" (p. 18); "Man has no innate instincts of sexuality and aggression" (p. 96); "Life is an overwhelming problem for an animal free of instinct" (p. 177). Later in the book, however, in the context of anxieties resulting from that supposed comprehensive lack of preprogrammed behavior patterns and genetically predetermined relatedness toward specific aspects of the environment, Becker writes: "*Schizophrenia takes the risk of evolution to its furthest point in man. . . .* We cannot imagine an animal completely open to experience and to his own anxieties, an animal utterly without programmed neurological [instinctive] reactivity to segments of the world. Man alone achieves this terrifying condition that we see *in all its purity at the extremes of schizophrenic psychosis*. In this state each object in the environment presents a massive problem because one has no [instinctive] response within his body that he can marshal to dependably respond to that object [releaser]" (p. 219, emphasis added).

If schizophrenia is the purest, the most extreme form of "nonpossession" of (differentiated) instincts in humans, if it "takes the risk of evolution to its furthest point," it follows logically that *this nonpossession of instincts has degrees, of which schizophrenia is the extreme,* something that is quite in keeping with what is proposed in this theory. To be clear, however, I do not think that the extreme diffuseness of instinctive predispositions can be equated with the clinical picture of schizophrenic psychosis. It is argued in this section, as well as in section 7.2.7, that *the extreme diffuseness of instinctive intentions is the underlying inherited predisposing factor common to schizophrenia (inclusively its subcategories) and to its genetically related spectrum disorders* (schizoaffective disorder, schizoid personality, and so on). The final clinical picture of each one of these conditions is the outcome of a long chain of interactions between this predisposition, other unrelated inherited predispositions to behave, and environmental influences during ontogeny, both physical (like perinatal insult), as well as social, interpersonal ones. (These considerations will be detailed in section 7.2.8.9.)

Taking in account the wide gap between the position of the social sciences and those sciences influenced by evolutionary thinking concerning the place of instincts in human behavior, I feel that the concept of the differentiated-diffuse scale of instinctive intensions may serve as a kind of bridge between these two opposing standpoints, partially explaining the observed data in both kinds of disciplines.

Since the idea of the differentiated-diffuse spectrum (D/D spectrum) of instinctual drives may seem highly unusual to the reader, and has not to my knowledge been dealt with earlier by previous authors, I will illustrate it with some commonplace examples. However, before describing these examples, I would like to summarize succinctly the direct behavioral repercussions induced by the extreme ends of the D/D spectrum as contrasted with those induced by the normative middle section of this graded continuum. The reader should take into account that this presentation is highly simplified, since the overt behavior is influenced—in addition to the D/D scale—by additional unrelated inherited predispositions, including cognitive ones, as well as ontogenetic factors.

- At the *differentiated end* of the scale, the individual, life-sustaining, re-productive, or self-protective instinctive strivings are reflected in overt behavior in a more conspicuous, more transparent way, frequently accentuating narrow, biologically relevant intentions through body language, intonation, the use of emotionally charged words, or the propensity to promote instinctive goals through physical display or confrontation. In this respect, the overt behavior may resemble to a certain extent that of highly evolved social animals expressing similar instinctive motives (de Waal, 1982/2007). These persons are acutely aware of their own personal, ego-syntonic intentions, which are perceived as clearly distinct from what they are compelled to do as a result of social constraints and pressures. Their social learning is highly selective, sensitive primarily to information that is relevant to finding the straightest route toward the fulfillment of their instinctive intentions—their biologically relevant releasers—and the best tactics to avoid the undesirable or punitive social repercussions resulting from their deeds. On the other hand, their social learning is relatively insensitive to the complexities of longstanding social coexistence; rules of conduct; reciprocation; cultural, moral, and esthetic values; the importance of heeding others' interests; and so on. It seems that, at its extreme, this instinctive configuration may constitute one of the predisposing factors of the antisocial personality disorder discussed in chapter 7.

- The *normatively differentiated or normatively diffuse* human instinctive constitution ensures a desirable balance between successfully pursuing individual instinctive interests, on one hand, while being adequately malleable to accept—without having to suppress a strong internal resistance—or even to embrace the constraints of social coexistence, successfully exploiting at the same time the great assets of that very coexistence. This involves the inevitability of making compromises, reciprocating, making friends and allies, being part of a larger group of people with common interests, identifying with this group's culture, way of life, shared beliefs, and so on; perhaps also being receptive to knowledge or attitudes not strictly of a utilitarian nature, such as the artistic achievements, fashions, or mythology of the ethnic group of which the individual is a part. This instinctive configuration accepts more easily the postponement or renunciation of instant emotional gratification in exchange for better prospects of attaining more remote fitness-enhancing goals.

With further accentuation of the instinctive strivings' diffuseness (but still in the normative range) appeared an increasing ability to channel instinctive intentions and concomitant energies and emotions into activities progressively more remote from common biological goals. With the increasingly complex division of labor, this became one of the pillars on which specifically human cultural and technological advance rests. One of the earliest manifestations of this trend may be the prehistoric cave paintings of Paleolithic artists, whose purpose presumably was to set the stage for hunting, magic, or similar rituals. Energies of instinctive origin (in reality, no other kind of mental energies exist) in predisposed persons may be channeled into pursuits so distant from common biological needs that any relationship to them becomes unrecognizable. An extreme example is life of the solitary monks or hermits of monastic communities, beginning with the early medieval times, who "turning their back upon mundane things," that is, more biologically oriented pursuits, sought "to live close to God . . . devoting themselves to prayers and spiritual contemplation" (Stournaras, 1990, p. 8).

■ *The extreme diffuse polarity* of the D/D scale means that the instinctive intentions (at their appetitive phase) are so blurred, so undifferentiated, that they are no longer able to guide the person effectively through the intricacies of social life to achieve individual goals (biologic or social). This is due to such resulting inabilities as being unable to grasp others' concealed emotions, intentions, or expectations with the mechanism of empathizing, or being unable to choose between irreconcilable options or forms of advice that cannot be evaluated on purely rational grounds. This kind of "ambivalence" is a well-known clinical sign in schizophrenia, in which the emotional relatedness to other persons or groups of persons becomes uncertain, confused, or nonexistent.

In diametrical opposition to the behavior of individuals with prominently differentiated drives, persons with excessively diffuse drives are exposed, quite defenselessly, to social influences. Their social learning is not selective and goal-oriented enough to effectively serve the attainment of instinctively fueled individual goals, and they are overwhelmed with information that is frequently more confusing than helpful. The overt behavior—far from tellingly expressing emotions, attitudes, and intentions in body language, intonation, or the use of emotionally charged words or expressions—is characterized by an effaced nature suggestive of the chronic schizophrenic's blunted affect or emotional withdrawal. These individuals may be excessively self-conscious, remote, contrived, confused, or unpredictable.

Persons with prominently diffuse drives, lacking clear *internal* guidance, may be oversensitive and overresponsive to *external* guidance, in particular guidance from parents as well as from education institutions (leading to the "model child" phenomenon frequently encountered in pre-schizophrenic children). However, this external guidance is rarely effective, given the inability to integrate ontogenetically acquired contents with instinctive activity (more precisely its energetic, intentional, and emotional aspects).

In the following, I illustrate the D/D scale with concrete examples, one pertaining to active behavior, the other to reactive behavior.

5.3.2. Illustration of the Differentiated/Diffuse Scale of Active Behavior

The illustration from the domain of *active* instinctive activity concerns the arousal of sexuality in adolescent boys. I specify the sex in this instance since the instinctive reproductive strategies differ between the sexes (see Buss, 2008, pp. 105–195). In addition, I chose adolescence, since at this age the possible interference from subsequent experience and learning in this domain is relatively small.

It would be very tempting to contrast a direct, physical approach with a narrow intent to obtain sexual access, an example of the differentiated pole of the D/D scale, with falling in love—an all-embracing, devastating infatuation, an intense, diffuse, emotional attraction toward one specific representative of the other sex. The first alternative is a narrow, clearly defined behavioral intention toward a narrowly defined releaser, while the other is the prototype of the behavioral expression of prominently diffuse drives. In this latter case, one craves the presence of the adored person (whose exterior and intrapsychic qualities are grossly idealized by the predominant emotion), yet in her presence, the infatuated boy doesn't know what to do next. His behavior becomes awkward, inhibited, confused, losing the characteristics of an aim-directed, purposeful activity. Even the thought of a direct, sexual approach becomes repulsive, and the intense longing for her presence is not ameliorated by masturbation or casual sex. Possibly as a result of the inability to express the emotion effectively in overt behavior, this active diffuse instinctive arousal increases in intensity, suppressing other everyday activities, giving the impression that the infatuated boy is detached from his surroundings, living in a kind of constant daydream.

On a superficial glance, the direct sexual approach—at the differentiated end of the spectrum—seems the characteristic one for the differentiated animal reproductive instinct, while the ability to fall in intense love—at the diffuse end of the spectrum—seems to be an exclusively human attribute.

However, trying to put these two kinds of "mating" approaches at the ends of a scale that changes only *quantitatively* but is qualitatively identical for each approach has serious flaws. Indications show these alternatives are qualitatively different, too, both in humans, as well as in animals. The intention toward exclusively sexual access implies that attributes of the respective girl other than sexual appeal are relatively unimportant, the relationship is intended to be short and mainly or exclusively of a sexual nature, and further commitment to the relationship or to the possible offspring is out of question. In infatuation, on the other hand, the *uniqueness* of the loved person both in her physical and mental aspects is all important, other similar beings are excluded, the dedication to her is intended to last forever, and the sexual relationship is subordinated to a much more comprehensive kind of emotional bond.

In fact, these two different kinds of reproductive relationships can be clearly discernible in the mating behavior of many nonhuman animals. Most bird species (Barkow, Cosmides, & Tooby, 1995, p. 292) and in rare cases mammals (such as the lesser ape, the gibbon) are instinctively predisposed to form longstanding monogamous bonds. In the case of birds, for example, the male courts the female with ritualized movements, attempts to impress her with conspicuous adornments (like the peacock's tail), brings "gifts" to her (delicious pieces of food), and in some cases attracts the female with a well-built and sometimes ornamented nest. The female can choose to accept the male's advances or to move away. If she consents, a lifelong bond may be formed between the two. The disappearance or death of the mate, after the bond is formed, has serious negative consequences for the behavior and self-confidence of the widowed male, resembling human dysphoric response or mourning (as was illustrated by Konrad Lorenz in the case of the goose in section 5.2.3). During the mating season, the female is jealously guarded by her mate, who attempts to drive away strange males, and the nestlings are fed and defended by a common effort of both parents.

It seems to me reasonable to presume that intense love or infatuation in humans is related to this instinctive mechanism, constituting its subjective experience, and *represents the extreme diffuse end* of the respective D/D spectrum. The *normative* section of this spectrum, on the other hand, can, I think, be deduced from the description of pair formation in birds, while making the needed adjustments between entirely instinct-induced FMPs in birds and behavior that is enriched and altered by acquired knowledge in the case of humans. The human male courts the chosen female—not with innate, ritualized motor patterns, of course, but with culturally determined, tradition-sanctioned rituals. He offers her gifts and tries to impress her—not with a well-built ornamented nest or innate physical adornments, but usually with a carefully prepared appearance, conforming with the latest fashions of the culture he lives in, with his social status, his knowledge and wits, his capacity to secure the material means needed for a future family, his willingness to care for her and invest in the offspring, and so on (Buss, 2008, pp. 109–128).

In many cultures, the woman has the right to consent to or refuse the man's advances during courtship and during the marriage rituals. In spite of the incomparably richer behavioral repertoire of humans, the nature of the underlying instinctive predisposition in humans and the instinctive activity in birds closely parallel each other in this respect. Well-known metaphors in everyday parlance or literature depicting love between young people or a longstanding happy relationship of a mature couple are frequently taken from bird life (for example, "lovebirds," "making one's nest," and so on).

The *differentiated pole* of this instinctive mechanism, coupled with other innate or culturally determined influences, induces a more rigid, manipulative enactment of the respective instinctive intentions, especially evident in times of interpersonal conflict, or in the presence of possible rivals. I refer to such behavioral expressions as excessive possessiveness toward the mate, exaggerated jealousy, tight guarding (including seclusion), violent reaction to the mate's intention to break off the relationship or when infidelity is suspected, and so on. The excessively clear-cut instinctive intentions, especially when frustrated, may lead to well-differentiated, resolute action even against prevailing cultural norms or against the law. (For an impressive account of the parallelisms between birds and humans in this respect, see Daly and Wilson's *The Man Who Mistook His Wife for a Chattel* [1995, pp. 289–314].)

The second kind of reproductive strategy in animals, characteristic of most mammals, is different in many respects from that of most birds. According to this instinctive variant, males fight intensely with each other in the reproductive season for the right to impregnate the females of the group, a practice that leads to sexual dimorphism (that is, the males are considerably larger and stronger than the females and are equipped with body appendages—like large horns—that assist them in sexual fighting). The victorious males impregnate the majority of the females, while the losers will be mostly deprived of chances to reproduce in the mating season. The females have no options, or very few, to choose between males and instead simply accept the victorious ones. While the males impregnate as many females as possible, they do not as a rule invest in feeding, rearing, or guarding the young they sired, which remains an exclusively female task. Naturally, no longstanding bond is formed between the mating partners, and after insemination they part company.

The example given at the beginning of this section to illustrate the differentiated pole of the D/D scale of the reproductive instinct in humans—that is, the narrow, almost exclusive interest in sexual access alone—belongs, in my opinion, to this kind of reproductive relationship. For the sake of brevity, it will be referred to in the following as the characteristic "mammalian type" of reproductive strategy, as distinct from the characteristic "bird type." In the evolutionary psychological literature, these two kinds of reproductive strategies are called "long-term mating strategies" as distinct from "short-term sexual strategies" (Buss, 2008).

The boys at the *more differentiated end* of the D/D scale of the mammalian-type reproductive strategy are intensely interested from an early age in the sexual differences between boys and girls, as well as in the sexual act itself. They are predisposed to learn selectively on this topic and can sense by the mechanism of empathy which of the girls in their surroundings have similar predispositions and which do not. Later they became sensitive to the subtle (or not very subtle) signs certain girls emit in their body language or verbalizations regarding their willingness to engage easily in sexual activities. The narrow intention for sexual access will characterize these boys' relationship with the other sex to the detriment or exclusion of other, more nuanced, sensual, romantic, aesthetic, or nonsexual aspects of the relationship. No intimate bonds are formed as a rule, and the quantity of sexual partners and variety of sexual practices gain preference over the depth and exclusivity of the relationship. These individuals are probably the Don Juan type of seducers in either their more sophisticated or simpleminded versions.

At the diffuse end of the D/D scale, in the mammalian-type of reproductive strategy, we find those adolescents who clearly appreciate and desire sexual activity with the opposite sex, but the diffuseness of the *appetitive phase* of this instinctive intention—that is, its inadequate potential for guiding the boy through the complexities of the relationship between the sexes in a social network—precludes the attainment of its ultimate, consummatory goal. The choice of the appropriate sexual partner may be mistaken, the communication of the sexual intention, both through body language and verbally, may be misguided, awkward, or badly timed (as a result of the insufficiency of the empathizing mechanism), lowering significantly the chances for success. It can be presumed that the sexual life of these young persons will frequently become derailed toward displacement or vacuum activities. In other words, the appetitive phase of this instinctively fueled behavior is frequently short-circuited by masturbating, using pornographic material, or seeking the services of prostitutes. (I regularly encountered such details in the anamnesis of my male schizophrenic patients.)

It has to be stressed that these two different reproductive strategies, the bird and mammalian, are by no means exclusive of each other, either in animals or in humans, in spite of the very different kinds of relationship they induce toward the partner or offspring. In fact, they have to be quite closely related to each other regarding their underlying genetic machinery. The two strategies, for example, can be observed in different strains of the same animal species: "In voles, mouselike rodents, we find both a monogamous and a polygamous population. The monogamous kind maintains a long-term relationship with one mate, and both partners share in raising the young. In the polygamous strain, the male copulates with a number of females, each in a different territory, and does not participate in infant nurturing." The most striking difference between these two strains concerns the sexual and maternal hormone levels during the mating relationship (Lampert, 1997, pp. 92–93).

It should be noted that *the specific selection pressure in nature that determines whether the mammal or the bird type reproductive strategy will evolve or predominate in an animal population is thought to do with the differential survival of the offspring with or without both parents' cooperation in their upbringing* (Maestripieri, 2012, pp. 156–158).

In birds, living in monogamous long-term relationships, the female often consents to copulate with a strange male in the absence of her permanent mate, hence the need for male "jealousy." And vice versa: "Once incubation begins . . . mate guarding virtually ceases, and mated males reallocate their efforts toward the pursuit of whichever neighboring females are still fertile" (Barkow, Cosmides, & Tooby, 1995, pp. 293–295).

In chimpanzee populations, no fewer than five different kinds of sexual relationships were observed. The dominant males copulate most of the time with the receptive females at the right time of the sexual cycle for maximizing the chances of fertilization while frequently interfering with copulatory attempts of lower-ranking males. Another kind of mating strategy is the promiscuous one. While a female in heat copulates with a male, one can observe an "orderly, polite line of males waiting their turn. They jump her one after the other with no quarrels or animosity." Another type of heterosexual relationship is "exclusive consortship" for short periods, which has two different forms. The first begins with a female initiative toward a subordinate male; in this kind of relationship, the male pampers the female in the ways chimpanzee life permits; he grooms her and brings her food. The second is initiated by a male, and as a rule, involves coercion and aggressiveness. Another kind of relationship between the sexes involves exchange of food for sexual access (Lampert, 1997, pp. 85–86; for a more detailed description of chimpanzees' varied and complex sexual practices, see Goodall, 2000, especially section 9).

The multiplication of reproductive strategies and lack of rigid innate motor patterns during the appetitive phase (like the ritualized courting movements observable in birds, for example) indicate, in my opinion, more diffuseness of the sexual drive in chimpanzees and a resultant diversification of reproductive strategies, enhanced by both a wider assortment of temperamental types and the contributions of ontogenetically acquired knowledge. It has to be mentioned that chimpanzee sexuality differs from that of the bonobo in the different *kinds* of relationships between the sexual partners (enumerated above), rather than the great variety of partners and positions; the frequency of the sexual behavior; and the multitude of triggering circumstances characteristic of the bonobo (section 3.1). All the chimpanzee sexual strategies (as well as additional ones) can be found in humans, who in my opinion reflect an even greater diversification of the sexual drive's appetitive phase than chimpanzees, augmented by highly evolved capabilities for behavioral modification through learning.

Perhaps it is superfluous to mention that both the predisposition for the bird type longstanding pair formation as well as the mammalian type of uncommitted sex may be found together frequently in the same human individual. (As we have seen, the same may be said of the birds.) It is not an uncommon phenomenon that a committed husband or wife who has a good, stable, and providing relationship with a long-term partner and children indulges every now and then in an extramarital affair.

Furthermore, it can be presumed that when both types of reproductive strategies are present, the respective degrees of differentiation or diffuseness will be similar. Moreover, the place of one's instinctive configuration on the D/D scale influences at the same time a *wide variety of specific instinctive mechanisms* from different behavioral domains. Therefore, it will be expected that a person with an excessively diffuse instinctive predisposition, for example, will have difficulties with the "bird type" relationship," the "mammalian type" sexual strategy, and other domains of instinctively underpinned interpersonal relationships (friendship, occupational functioning, and so on).

5.3.3. The Differentiated/Diffuse Scale of Instinctive Predispositions in the Domain of Reactive Behavior

5.3.3.1. On Reactive Behavior in General

Although strictly speaking this topic pertains to the next section, in order to be able to discuss the differentiated/diffuse scale of reactive behavior, it is necessary to first make some general remarks about reactive behavior.

The reader may remember from the introductory section that reactive behavior represents the organism's attempt to deal with (escape from, neutralize, etc.) environmental influences that are deleterious or dangerous to its life processes. Therefore, it is distinct from active behavior, whose role is to exploit environmental resources needed for life-sustaining processes (survival, growth, reproduction, etc.). The neutralization of disturbing environmental stimuli by reactive behavioral tactics is needed when they are unpredictable, quickly changing, or complex. Excluded here, of course, are the deleterious environmental conditions that induce the behavioral strategies discussed in sections 5.1 and 5.2 of this chapter.

The reactive behavior of more evolved animals has three basic forms: *freezing, avoidance or flight, and counteraggression.* When a prey animal, for example, perceives an approaching predator, it suspends all activities it was engaged in before and becomes *motionless but highly alert,* directing all its telereceptors (visual, auditory, olfactory) toward the source of the potential danger. The reactive tactic of *freezing* has adaptive importance in several ways. It offers an opportunity to grasp more precisely the nature of the danger (whether the predator is alone or is part of a group, whether it intends to hunt or is simply passing by, etc.). Freezing makes the animal less conspicuous in its environment, which increases its chances of remaining undetected, and may save energy that would be expended unnecessarily on further reactive tactics if they are employed in an untimely way.

If the predator continues to close the distance, the prey animal resorts to the next reactive tactic, which as a rule is *flight,* with the obvious intent to put an obstacle (distance, the height and dense foliage of a tree, the narrow entrance of a burrow, etc.) between itself and the source of danger.

The third universal reactive tactic is *counterattacking.* Sometimes the prey animal skips the flight option and counterattacks at the beginning. This may happen when the difference in the physical power between predator and prey is small (a wildebeest against a lioness, for example). In other cases, the prey animal counterattacks only when its flight is obstructed by the conditions of the terrain or the presence of a second predator if the two are hunting together (Grandin & Dessing, 1998, pp. 73–74).

In animals, these three reactive tactics possess highly effective innate motor patterns for coping with a disturbing or dangerous situation that occurs frequently in natural conditions. Besides the behavior pattern itself, the precise *timing* of the discharge of the respective reaction, its precise *intensity and duration,* as well as *the flexible, swift transitions* between the reactive tactics, according to the quickly changing circumstances of a dangerous encounter, are equally important. If the timing of the freezing reaction is flawed—for example, too short or too long, which means that the switching to flight or counterattack comes too early or late—it may have negative consequences for survival. The *flexible switching* from one reactive tactic to another (and back to the former one when needed), sometimes in fractions of a second, can also mean the difference between escaping a lethal attack or being killed. Obviously, the animal under attack has no time for wavering, experimenting, or considering the benefits and drawbacks of one tactic over another. Maneuvers have to be executed instantaneously, automatically, in fractions of a second, whether the behavior is entirely innate or partly honed by practice and acquired knowledge.

I include this description as a starting point for comparison with human reactive behavior. Obviously, it illustrates the polarity of exclusively physical reactive behavior that is optimally differentiated to certain types of danger frequent in natural conditions. These kinds of behavior patterns can be observed in humans, too, in circumstances requiring prompt physical reaction to evade danger or tackle a physical challenge. For example, in the case of soldiers participating in combat involving shelling, shooting, and physical contact with the enemy, all the basic reactive tactics may be carried out. These behavioral reactions evidently have an innate, inherited foundation made more effective and more adapted to specific combat situations by training (including simulation), learning, or previous experience in similar circumstances. In some kinds of sports involving a direct physical contest between two opponents (boxing, wrestling, tennis, etc.), we can see a rapid alternation between trying to predict the opponent's next move, fending it off, and counterattacking. Here, we are talking about a more complex behavior pattern than a simple reaction to an attack, one that involves not only a chain of reactive techniques but *a rapid alternation between active and reactive* behavior of a physical kind. After all, the competitors' final goal is not only to "survive" but to win the contest.

All of these examples of human reactive behavior represent well-differentiated physical reactions based on innate predispositions perfected through training, experience, etc. In the following, I will try to show that emotionally charged verbal reactions in challenging situations may be built on the same phylogenetically evolved instinctive foundations as physical ones. An example is a verbal contest between, say, two parliamentarians with loyalties to opposing parties. In a typical case, one of them *listens* tensely to the other party's arguments (the equivalent of freezing) and carefully *evades* those topics for which he has no impressive or effective counterarguments, while *attacking* forcefully (trying to invalidate or ridicule) those points for which he has impressive replies. Such verbal contests are accompanied frequently by small, unconscious body movements toward or away from the opponent, as well as facial expressions and intonations appropriate to instinctive intentions to physically retreat or attack.

In trying to build a tentative hypothesis on how *human reactive behavior* evolved from characteristic nonhuman reactive behavior as a result of technological and cultural advances, increased ecological dominance, and a progressively more complex social coexistence, we could postulate the following scenario:

It is well recognized in the literature on the evolution of human behavior that selection pressures for effective physical reactions against natural dangers and harmful influences (predation, bad weather, etc.) relaxed progressively over the course of human evolution. In our theoretical framework (as outlined in chapter 4), this led initially to more accentuated quantitative diversification of the related mental and physical traits. Some of the relevant traits became atrophied to a certain extent, leading for example to more slender build, longer reaction times to disturbing external stimuli, and rudimentary body movements instead of the full-fledged behavior patterns representing instinctive intentions to escape or attack.

What interests us more, however, is the process of adapting nonhuman reactive behavior to the greatly changed conditions in the human-built artificial environment and complex social coexistence. As we have seen before (sections 4.4.4, 4.6.3, 4.6.4, and 4.6.5), adaptation to changed selective pressures in nonhuman animals involves relaxation or disappearance of the former natural environment's selective pressures, diversification of the respective behavior patterns, and ultimately the evolution of new or altered forms of instinctive behavior suitable to the new natural environment's different adaptive requirements. During this process, natural selection may make use of some elements of the former diversified adaptations.

However, in the case of humans, *the change in the nature of selective pressures consisted of a transition from those coming predominantly from a natural environment toward those from a predominantly social environment.* As a result, something similar to the evolutionary trajectory of nonhuman animals in changing environmental conditions became impossible for humans, largely owing to the great complexity, quickly changing nature, and relative unpredictability of the predominantly social environment. In consequence, instead of an evolution of new, differentiated motor patterns (as argued in this text), human instinctive behavior progressively lost most of its strictly preprogrammed motor patterns and innate releasing mechanisms, preserving only a more diffuse *propensity* to act in a certain manner toward a *more loosely defined* category of environmental agents. This propensity, however, *retained the basic adaptive motives of the respective instinctive domain*—in the present case, the propensity for *reacting with freezing, escaping, or counteraggression* in potentially harmful circumstances. On the foundations of these innate motives, a vastly richer, much more flexible, and adaptively adequate behavioral repertoire than that of the nonhuman animals can be built with the aid of accumulated knowledge and experience.

With further relaxation of natural and intragroup selection pressures (largely in the last 10,000 years, dependent on much-improved ecological dominance, the improvement of living conditions, and the rapid multiplication of social roles and "social niches"), the *originally optimally differentiated and interrelated reactive instinctive predispositions also became more diversified.*

5.3.3.2. Diversification concerning the Preferred Reactive Tactic

Although in nature an animal has to resort to all three reactive tactics at its disposal in accordance with changing circumstances, in the relaxed conditions of the human environment, in a significant proportion of the population, one of the three is used preferentially or overwhelmingly while the others are employed only tentatively, inadequately, or not at all. If the predominant reactive tactic is *flight*, it leads to exaggerated, appeasing, avoidant, evasive, or anxious behavior predisposing to such mental conditions as Avoidant Personality Disorder or certain Anxiety Disorders. On the other hand, the accentuation of the *counteraggressive* tactic (at the expense of the avoidant one) may predispose to choleric, impulsive, or defiant personality traits, or to the aggressive form of anti-social personality disorder.

The *freezing reaction,* too, may be present in differing amounts and intensities in different people. Its adaptive role is described in the language of neurobehavioral science in the following way: The "behavioral inhibition system" (freezing) both in the framework of the "reward-related behavioral activation system" (active behavior), as well as in the "fight-flight system" (reactive behavior), is hypothesized to have "an important role in directing attention and facilitating the processing of important environmental information" (Derryberry & Rothbart, 2001, p. 975). If the freezing reaction is too short and shallow, the collecting and processing of the relevant environmental information before deciding on the desirable course of action will be insufficient. Consequently, the ensuing behavioral reaction will be rushed, hasty, impetuous, or thoughtless. On the other hand, when the freezing reaction lasts too long—for example, before responding to a challenging remark during a flowing conversation—the eventual reply may come too late, thus losing much of its effect. In complex social situations, particularly when the respective reaction may have important consequences, the depth and length of the freezing phase may be critical.

The overaccentuation of one reactive tactic over another will naturally hamper the quick, adaptive transitions between them in response to changing circumstances, an ability factor that is so important in nature as well as in flowing social interactions.

5.3.3.3. Differentiated/Diffuse (D/D) Scale of Reactive Instincts in Human Populations

Our next topic (the one that led to this short discussion of reactive behavior) is the wide differentiated/diffuse (D/D) scale of reactive instincts in human populations.

The *differentiated segment* of this D/D scale was illustrated by the behavior of well-functioning soldiers' behavior in combat and by certain kinds of sport contests. However, most of the dangers, inconveniences, and derangements of everyday life are of a less physical nature, requiring less developed bodily fitness. We encounter dangers originating in our advanced technology (cars, electricity, home appliances, etc.) against which we possess no innate reactive tactics. The same applies to most inconveniences or harmful influences inherent in social relations (gossip, ridicule, misunderstandings, unfair exploitation of interpersonal relations, cheating, etc.). Therefore, in most instances, the average person can get along in a social environment with quite modest abilities to react physically to these everyday inconveniencies.

The subjective experience of *normatively diffuse* reactive behavior at its freezing stage consists of mental arousal and attentiveness directed toward the source of suspected harm or nuisance while trying to get more information about the nature of the disturbance, as well as cautious restraint so as not to react precociously. *The stage of the flight reaction in humans typically takes the form of verbal escape, avoidance, or retreat,* as when one acknowledges a mistake and apologizes. One may feel uneasy or fearful yet at the same time experience a kind of self-assurance that this maneuver may solve the conflictual situation, at least temporarily. *When the reactive tactic consists of counterattack (generally in the form of a counteropinion or verbal threat, implied or explicit), the subjective experience is, as a rule, confidence in one's ability to master the situation, feelings of assertiveness, or anger.*

The *diffuse pole of the D/D spectrum of reactive behavior* can be described as follows: The person is clearly aware of a real, possible, or imagined danger or inconvenience either of an imminent nature or expected in the future, which arouses the mental apparatus and activates the related physiological processes in the body in order to prepare it physically for carrying out reactive tactics in a way appropriate to a natural environment. The blood pressure increases, voluntary muscles become tense, etc. However, no clear, adequately differentiated innate guidance exists in the form of flexible instinctive predispositions such as the three tactics of freezing, fleeing, or counterattacking. This is true even when the ability to learn is normative or higher than average. Since the diffuse, directionless mental and physical arousal cannot be channeled into organized, effective activity, it tends to accumulate, leading either to a paralysis of overt behavior, or the accentuation of the reaction's physiological accompaniments (culminating possibly in a panic attack), or to diffuse, chaotic, adaptively completely useless, generalized agitation. The subjective experience of this kind of instinctive arousal is not accentuated focused attention, caution, fear, or anger, as in the more differentiated forms of reactive behavior, but *anxiety*. This interpretation of the concept of anxiety is quite in agreement with Ernst Becker's more extreme formulation of the same theme, quoted earlier in section 5.3.1; he refers to the human predicament of being "an animal completely open to experience and to his own anxieties, an animal utterly without programmed reactivity to segments of the world." For the remainder of the present text, the term anxiety will be used in this sense.

5.3.3.4. The Relevance to Psychopathology of the Diffuse/Differentiated Scale of Reactive Behavior

Discrete psychiatric disorders relevant to the D/D scale of reactive behavior are in fact scattered along the whole span of this gradated continuum. They are mentioned here only briefly and detailed in the seventh chapter.

Specific (simple) phobias consist of circumscribed fear reactions to potentially dangerous objects or situations in the natural environment (physical and biotic). It was hypothesized in evolutionary psychology and psychiatry that simple phobias represent specific adaptations to the environment in which early human evolution took place, the Pleistocene epoch. These causative factors or situations were common enough and relevant enough to fitness to enable natural selection pressures to build discrete defensive mechanisms against them. Fear of small, possibly venomous animals (snakes, spiders, insects, etc.); fear of open places (agoraphobia), which reduce the possibility to hide or the chance of remaining undetected; fear of enclosed spaces (claustrophobia), which might thwart an attempt to flee; and fear of high places (acrophobia), with the danger of falling down; fear from storm, deep water, and so on, are such examples.

Fears of certain situations brought about by modern technology's achievements may, in my opinion, be traced back to the reactions to features in the environment of our evolutionary past. Fear of flying, for example, or of elevators may be rooted in an innate fear of heights and enclosed spaces, respectively. On the D/D scale of reactive behavior, these fears have to be placed on the differentiated end and have been preserved unchanged from when they evolved in a natural environment—that is, they did not become generalized or diversified. Their motor pattern consists of a quite differentiated physical avoidance reaction to a specific, circumscribed natural trigger. They usually have a childhood onset, a familiar aggregation, and moderate heritability. They did not disappear entirely from the contemporary populations in spite of the relative rarity of the respective causative factors in the modern environment (snakes, venomous spiders) or the disappearance of the inherent danger that characterized them in the evolutionary past (closed or open spaces, for example). In fact, in subclinical intensity, they are "exceedingly common" (Pine & McClure, 2005, p. 1773). These reactions most of the time remain a circumscribed maladaptive phenomenon, and if the causative agent can be avoided they do not affect the respective person's daily routine.

Social phobias are induced by much more diverse, less circumscribed situations coming from the social instead of natural environment. Their possible deleterious effects cause less of a direct threat to survival and reproduction than the natural triggers of simple phobias. As a rule, they appear in situations that may incur potential humiliation, derision, disapproval, contempt, verbal threats, etc. The actual triggering activities range from public speaking (a stressful situation for many normative persons, too) to initiating or maintaining a conversation—in fact almost anything one may do in public, including eating, drinking (Sadock & Sadock, 2005, p. 1721), or urinating in a public lavatory (DSM-5, p. 203).

Social phobia has a more circumscribed form, when it is triggered by a single or restricted number of situations, as well as a more generalized form, which includes many or most of the social situations one encounters during one's daily routine (DSM-IV-TR, 2000, p. 216). The behavioral response in social phobia is not a circumscribed avoidance or escape reaction of a physical nature (flight), as in special phobias, but a more generalized alteration of social behavior, including embarrassment, behavioral inhibition, decline in occupational and social functioning in adults, and, in children, as crying, tantrums, and freezing (idem, p. 215). The complexity, unpredictability, and fluid nature of the social environment prevents natural selection from building well-defined, innate defensive behavioral patterns in spite of the instinctive roots of these reactions. Therefore, it seems evident that social phobia lies further on the D/D scale's *more diffuse end than* simple phobias, both regarding its motor pattern as well as its releasing circumstances. In fact, it occupies a more extensive section of this scale from more differentiated fears in the case of more circumscribed social phobias to more diffuse reactions triggered by more varied social circumstances in the more generalized form of the disorder.

Generalized Anxiety Disorder, in which neither a specific environmental causative factor nor a more differentiated behavioral reaction is discernible, has to be placed at the *extreme end of the diffuse pole* on the D/D scale of reactive behavior.

5.4. Active/Reactive Behavior Dichotomy and Interrelationship

5.4.0. Introduction

Before proceeding, for the sake of clarity, I will summarize the active/reactive behavior dichotomy and the interrelationship of active and reactive behavior.

Active behavior is induced by discrete life-sustaining or reproductive instinctive intentions. It has an appetitive phase (that is, searching for the appropriate resource or agent in the environment that is designed to satisfy the need—the releaser) and a consummatory phase (after the appropriate releaser is contacted). Active behavior is therefore initiated and sustained by an autonomous mechanism originating *inside* the organism and directly serving the implementation of its life processes. It is true that active behavior can be triggered by the perception of an appropriate releaser that is external to the body and part of the environment (delicious food or an attractive sexual object, for example), but even in this case the internal need, the "action specific excitation" (as Konrad Lorenz termed it), has to be present in various intensities, and the strength of the response depends, at least partly, on the level of intensity of this internal excitation at the given moment. Indeed, in ethology, a method known as *dual quantification* measures the relationship between the intensity of the internal readiness to discharge an active intent and the degree of "attractiveness" of a given releaser (Lorenz, 1982, pp. 115–117).

If some environmental agent or circumstance interferes with an active behavior in progress, it can transform into one of its frustrated forms: displacement, vacuum behavior, aggression, or dysphoria—as detailed in the second section of this chapter. In spite of the fact that these transformations are brought about by a *reaction* to a specific environmental agent (like a competitor for food resources or sexual access), or by the lack of a specific resource (like food or an adequate sexual partner), these transformations are nevertheless categorized in our scheme as an organic part of the domain of active behavior.

However, other deleterious environmental causes or circumstances can harm an organism *without* interfering directly with one specific active instinctive striving, or can affect the organism when no active instinctive behavior is in progress. Body integrity may be harmed or an animal may be killed by a falling rock in a mountainous region or by a predator, irrespective of the state of its active behavioral intentions at that time. At the end of the previous section, I detailed the three basic instinctive mechanisms that evolved as defenses against environmental dangers, derangements, or nuisances—freezing, flight, and fighting back. (Other, more specialized behavioral reactions to deal with these circumstances, like driving away annoying insects by whisking the tail, or feigning death when captured by a predator, do not concern us here.)

The active/reactive behavior dichotomy is valid in animals at all levels of complexity, from the simplest organisms that may be considered to possess behavior up to the evolved social animals and humans. A flagellate unicellular organism, for example, "accelerates its locomotion when traversing unfavorable conditions and slows down on entering more favorable ones and, by these simple reactions, achieves the goal of spending most of its time within the latter." And: "An amoeba can creep, with varying intensities, toward or away from a source of stimulation" (Lorenz, 1982, p. 154).

This dichotomy was first formulated by the early ethologist Wallace Craig in his seminal 1918 paper, which also proposed the dichotomy of appetitive versus consummatory behavior (Lorenz, 1982, p. 5). He labeled active behavior (more precisely its appetitive phase) as "appetites" and reactive behavior as "aversions": "Appetite is a state of arousal which continues as long as a specific stimulus situation . . . is not reached. When this specific stimulus situation is reached . . . the consummatory action is set free. . . . Aversion, on the other hand, . . . [is] a state of agitation which continues as long as a certain specific stimulus, referred to as the disturbing stimulus, is present, but which ceases . . . when the stimulus has ceased to act on the animal's sense organs" (idem, p. 190).

A basic difference between these two behavior forms is that active behavior, being the expression of the internal physiological needs of the organism, frequently has a kind of *rhythmicity* or *cyclic nature*; that is, the intensity of the need *waxes and wanes* periodically in accordance with the fluctuations of physiological processes (hunger, thirst, the reproductive cycle, or the sleep-wakefulness cycle, for example). When an active instinctive need is fulfilled, frequently a *refractory period* follows, during which time the active behavior pattern is inhibited and cannot be triggered, even in the presence of an attractive releaser. Reactive behavior, on the other hand, possesses neither rhythmicity nor a refractory stage. A prey animal, for example, eats and drinks periodically in accordance with its fluctuating internal, physiological needs, but naturally, it has no periodically accumulating needs to perform an escape reaction "in vacuo," even when a predator is absent for extended periods. And, conversely, if a predator charges many times in a short interval, the prey animal will try to escape each time. In addition, no separate appetitive and consummatory phases can be discerned in reactive behavior.

In summary, reactive behavior is induced by deleterious stimuli external to the organism, and its timing, intensity, frequency, duration, etc., depend on the nature of this external stimulation, while active behavior is induced and regulated by the intrinsic physiological needs of the organism.

One may argue that migration, hibernation, and diurnal variations in the intensity of daily activities may also be considered reactive behavior in this sense, since their cyclical and periodic nature is ultimately environmentally induced. Even the timing of reproductive activities in wild animals has seasonal rhythmicity, and migrating birds may desert their recently hatched chicks in order to migrate on time. Moreover, we have seen, for example, that in ground squirrels kept in laboratory conditions, overeating in autumn and hibernation in winter continue even when the temperature and daylight hours are kept constant through the seasons (Foster & Kreitzman, 2009, p. 101). In other words, *an originally reactive behavioral response to life-threatening environmental conditions (which fluctuate predictably) took upon itself during phylogenesis the characteristics of active behavior—that is, it appears autonomously, without the appropriate external stimulus, according to an evolved internal clock.* When we refer to reactive behavior that evolved throughout a long phylogenetic process in order to deal with rhythmically, predictably changing environmental conditions, the distinction between active and reactive behavior may become blurred. Therefore, the term reactive behavior in this text is used to mean only *a behavioral response to randomly occurring disturbing environmental influences that do not exclusively frustrate a circumscribed active intent in progress.*

Another general difference between active and reactive behavior is the subjective experience induced by them. Active behavior is sensed as *originating from one's own internal motivations*; that is, it is felt, as a rule, as ego-syntonic. (It has to be mentioned, however, that exceptions to this rule exist in clinical psychiatry. An obsession that expresses repressed sexual content, for example, may surface in consciousness in an intrusive form, not expected or wanted by the person, and felt as alien from her conscious, identified-with values and intentions.)

As noted earlier, if an active behavioral pattern is discharged in an undisturbed interaction with an appropriate releaser, the subjective experience will be a *pleasurable* one. When the same behavior pattern is discharged in its original form but displaced toward an improper releaser or in vacuo, the subjective experience still remains pleasurable, as we have seen. When the same frustrated active instinctive motive cannot be discharged in a displaced or vacuum form, the pleasurable subjective experience switches into *anger or dysphoria.*

The subjective experience of reactive behavior, on the other hand, is always unpleasurable. It feels as if the discharge of the respective behavioral sequence is forced upon the organism from without—that it is ego-dystonic. And, conversely, when reactive behavior achieves its purpose and subsides, the subjective experience may be pleasurable; it brings a sense of relief from former fears, tensions, constraints, or externally originating promptings and may be accompanied by the expectation to return to more active behavioral preoccupations.

When comparing aspects of active and reactive behavior, we find that both frustrated active behavior and reactive behavior contain an aggressive form or phase, and, moreover, that the specific motor pattern may be similar in both instances. I don't see any contradiction here. In both cases an attempt is made to neutralize an unwanted, interfering, or deleterious environmental agent. However, when the aggressive behavior is fueled by active motives alone, or by a blend of active and reactive ones, instead of being exclusively reactive, its intensity increases considerably. It is common knowledge that fighting between males during the reproductive period is much more intense than outside these periods when they may peacefully coexist; that an animal inside its territory is more aggressive than outside it; that a mother animal is more aggressive toward animals perceived as a threat to her cubs than toward other female conspecifics during ordinary quarrels.

In fact, we need not presume that the environmental stimuli that interfere with active drives versus those that lead to reactive tactics have to be sharply different in nature. Sometimes the difference may be more of degree than of kind. A simple example may illuminate this point. If, for example, one member of a couple involved in an affair (usually the man) pushes for more sexual access, while the other (usually the woman) refuses to consent, it can lead either to more accentuated pressure for sexual access on the man's part or to the transformation of the original sexual arousal into one of its forms of frustration: vacuum behavior (masturbation in the bathroom); displacement (looking for another, more accessible sexual object); aggressive coercion with the sexual intent preserved or even intensified; or dysphoria (disengagement while feeling offended or hurt). However, if the woman's opposition to the sexual advance reaches a high intensity, if she opposes it violently and physically or threatens to call the police, the situation may completely suppress the man's sexual desire and instead elicit reactive tactics, like retreat from the sexual intent, flight, or aggression—perhaps including physical violence without the sexual intent.

When considering the *interrelationship* between active and reactive behavior, what strikes us first is the clear priority of the reactive behavior over the active one. This subject was briefly discussed in the section on displacement behavior (5.2.1.1.3) and is enlarged upon here. When an animal perceives danger, it instantly suspends any possible ongoing active behavior and becomes frozen—think of a grazing herbivorous ungulate perceiving an approaching predator. It will not resume eating or any other active preoccupation until the danger is over. If the predator charges, the prey animal invests *all its energies* into escaping: "Only a few instances are known in which the activation of one behavior system excludes absolutely the activation of any other. Flight from a predator is an example of such activation having priority" (Lorenz, 1982, p. 245), and: "Very few instinctive motor patterns can be repeated right up to the limits set by the basic functions such as breathing, heartbeat, or general muscular fatigue. Yet the readiness to flee, the overpowering urge to get away, is still at work even when the peripheral organs fail to obey" (idem, p. 123).

No mentally healthy man, to my knowledge, is capable of performing sexual intercourse in the face of a perceived imminent external danger or derangement. (The animal kingdom does contain some known rare exceptions where an active instinct, like the sexual drive, has priority over escape—in other words, where evolutionary forces give priority to reproduction over survival [idem, p. 247].)

The priority of reactive over active behavior is understandable if we take into account that environmental danger may strike unexpectedly and instantaneously, while organismic needs, like hunger or the sexual drive, intensify relatively slowly over a period of time, and their gratification is usually postponable for some time without deleterious consequences. Therefore, it seems reasonable that evolution has built a mechanism that ensures the absolute priority of reactive over the active behavior (at least initially, as we will see below). The switch from active to reactive behavior therefore is *instantaneous* when danger is sensed, while the *return* to active behavior after the danger is over may take some time. For example: "[after the] provoking of a cat by confronting it with a large dog, so that it assumed the hunchbacked defense attitude, a half hour or more would be needed by the cat to quiet down enough to become interested in catching prey" (idem, p. 327). A similar regularity is described in the behavior sequences of some fish (idem, p. 217).

The subjective experience of active/reactive behavior interaction parallels their motor aspects. A sudden dangerous occurrence, while eliciting focused alertness, fear, or anxiety, instantaneously suppresses any active emotion present. When the danger is over, the reactive emotion slowly dissipates and the active one may gradually return.

However, the priority of reactive over active behavior, when the danger is not overwhelming or imminent, *has only relative, not absolute, validity.* If the active instincts are suppressed for longer periods of time, they will intensify and the animal (nonhuman or human) *will risk increasingly serious external danger in order to appease them.* As an example, I will describe *National Geographic* footage I saw of a waterhole in an African savanna infested with predators (lions, crocodiles). Thirsty prey animals (impalas, zebras) stood at a distance from the water fearing the predators. However, becoming increasingly thirsty, they began to risk greater proximity to the predators. While watching this footage, I wondered how precisely natural selection has to coordinate active and reactive behavior, a coordination that is perfected even more by experience, to enable these animals to avoid the predators and still manage to drink. Those with less precise innate coordination of active and reactive behavior, or less experience (the more immature prey, for example), will be the most probable victims of predation, or alternatively will remain thirsty, risking dehydration.

The same active/reactive behavior interrelationship can be observed in some wild animals that, in winter months, risk entering human settlements in search for food, something they don't do so long as food is plentiful in nature; or in humans who, in serious need of material means, commit theft or robbery. Another example of the tight coordination of active/reactive behavior interaction in nature would be the moment when one of two male animals who are sexually fighting decides to quit and move away. If this decision comes too early, reproductive opportunities will be lost, and if it comes too late, serious injury can be incurred.

Since the tight coordination of active/reactive behavior is of utmost importance for fitness in wild animals, we have to hypothesize a brain mechanism responsible for this coordination whose optimal functioning must be under rigorous control by natural selection. In fact, Konrad Lorenz argued on behalf of such a high level coordinative center, not only in the special case of active/reactive behavior coordination, but in the case of a much wider gamut of behavior forms: "A superior command locus obviously becomes necessary at the stage of evolution when an organism has developed more than one system of behavior patterns, of which only one can function at a given time" (Lorenz, 1982, p. 206).

The function of this center becomes more and more complex as we ascend the evolutionary ladder, approximating Freud's concept of ego: "In higher vertebrates, the 'superior command locus' has a much more complicated function . . . as it has to act as a connecting link between phylogenetically programmed and individually acquired behavior" (idem, p. 209).

I think it is best to imagine the function of this "superior command locus" or coordinative center as hierarchical in nature, because it manifests itself at different levels of behavioral complexity. For example, it has to decide about priorities of possibly conflicting individual active strivings at a given time (hunger, sleep, sexual arousal, parental instincts). In the case of reactive behavior, as we have seen in section 5.3.3.1, this center has to tightly coordinate the discharge of and transitions between the three basic forms of reaction to disturbance or danger: freezing, flight, or counterattacking. On a higher level of complexity, it has to decide on the priorities and interactions between the more comprehensive categories of active versus reactive behavior. On a still higher level, especially in humans, it has to decide about issues in which instinctive intentions and cognitive (learned) considerations are conflicting, or when individual instinctive strivings conflict with social situations in which submission or conformity is required. At this stage, instinctive and cognitive elements in behavior are inseparably intertwined. An example of this level of coordination would be the decision on how much of my saved money I wish to spend now to make my life more comfortable and pleasurable versus how much to put aside for needs that will most probably arise with aging (incapacitating diseases), or to help adult children.

We have to ask again what happens when the tight coordination of active/reactive behavior by natural selection, characteristic of wild animals in nature, began to relax during human evolution as a result of increasing ecological dominance and decreasing intensities of intragroup selection pressures for optimal social functioning. In addition, we have to keep in mind that, in a complex social environment, a great variety of "social niches" become available to accommodate instinctive configurations that exhibit increasing degrees of imbalance in the power relationship between active and reactive behavior.

This imbalance is manifest in human populations in either direction:

1. *When active behavior tends to overpower reactive behavior, the individual will strive to attain instinctively fueled individual goals without paying enough attention to the signs of dissatisfaction or opposition his inconsiderate behavior may induce in others.* Critical remarks, warning, threats, etc., will suppress active behavior to an inadequate degree and only for a short duration, if at all. These persons may be predisposed to develop certain kinds of personality disorders (see chapter 7).

2. In the reverse situation, w*hen reactive behavior tends to overpower active behavior to a more than desirable extent and length of time, these individuals will be oversensitive and overreactive to environmental disturbances that, in a social milieu, are caused mostly by other persons' behavior.* If the preferred form of reactive behavior is predominantly an *avoidant or flight reaction* (accompanied by the subjective experience of fear or anxiety), these persons may employ several kinds

of "avoidant or defensive tactics" in social life. They may attempt to behave very kindly with everyone to an extent unwarranted by the circumstances (in order to prevent dissent); they may observe prescribed etiquettes to an unnecessarily exaggerated degree; they may try to recruit the help of nearby persons much earlier than usual in a confrontation; they may avoid problematic social situations, or withdraw from social encounters altogether. Moreover, the active drives, being suppressed to a more than optimal degree and for longer periods than normative, will accumulate and manifest themselves ultimately in inadequate ways, intensities, or circumstances (as for example, obsessions with sexual content). This instinctive configuration may predispose to anxiety disorders (detailed in section 7.2.2).

When the preferred form of reactive behavior is predominantly a *counteraggressive* reaction, it can predispose to choleric personality traits, or to Oppositional Defiant Disorder.

It has to be stressed that, even in normative behavior, active and reactive motives are rarely kept in a constant balance of intensity for long time periods, with the possible exception of those rare individuals with a considerably inflexible mental constitution. As a rule, the active/reactive behavior power-relationship *oscillates* around an optimal balance dependent on circumstances. For example, we tend to behave more reactively toward our superiors (reactively here meaning submissively under the influence of caution or fear of authority), and more actively toward subordinate people; more reactively at our workplace than at home or among our friends; more reactively in unfamiliar places or toward unknown persons or toward those we expect to act undesirably than toward our family members; and so on. This does not imply that our active behavior will be always of a pleasurable kind. Active instinctive tensions suppressed and accumulated during long hours of work, particularly in a tense, authoritarian milieu, for example, may "explode" in a displaced manner at home in their aggressive or dysphoric form of frustration.

5.4.0.1. The Imbalance of Active/Reactive Behavior Relationship in Mental Disorders

In mental disorders, the imbalance in the active/reactive behavior power relationship may attain pathological intensities. The functioning of the "superior command center" posited by Konrad Lorenz, whose function is to balance optimally the active and reactive motives for maximal fitness gains, *may be diversified by the relaxation of natural selection pressures* in humans, leading to different grades of deviation from the optimum in both directions. Alternatively, *one participant in this behavioral dichotomy may become so intense or so weak* (as in mania, lethargic depression, or post-traumatic stress disorder) *that it surpasses an otherwise normative superior command center's ability to keep an optimal balance between them.* Some of the resulting mental conditions may represent a lifelong disability, as in avoidant, narcissistic, or antisocial personality disorder, while in others the imbalance may appear only episodically, as in mania, lethargic depression, or social phobia. These conditions are detailed in chapter 7.

The examples of mania and lethargic depression make clear that some of the comprehensive instinctive mechanisms may play a part in bringing about this imbalance. In the case of mania and lethargic depression, naturally I mean the fluctuations in the overall intensity of the instinctive activity that originally evolved as an adaptation to seasonal periodicity. Instinctive drives that are too diffuse or too differentiated also affect the optimal balance of active and reactive behavior (and, even more, their integration), as detailed in section 7.2.7.

Before leaving the topic of the active/reactive behavior dichotomy, we must make a further distinction between *individualistic reactive behavior*, which protects the egocentric interests of the organism in both natural and social environments, and the *submission to or conformity with the imperatives of social coexistence.* We must also to try to analyze how the resultant three comprehensive, genetically based motives in human behavior interact—namely, individualistic active motives, individualistic reactive motives, and the predisposition to comply with the imperatives of social coexistence, even to the extent of internalizing the interests of the group to which one belongs (ethnic, political, national, religious, etc.) at the expense of self-centered active and reactive strivings. This subject is briefly considered in the next two sections.

5.4.1. The Effects of the Social Environment on Active and Reactive Behavior in Nonhuman Animals

Our relatively simple scheme of the active/reactive behavior dichotomy, while certainly relevant to the behavior of both solitary and social animals as well as humans (although only with respect to less complex human behavioral patterns), must be complemented by additional forms of innate behavior or predispositions resulting from natural selective forces stemming from the organized group environment. These selective forces are "natural" in the sense that their effect conforms to the rationale of natural selection (see section 4.6.6).

In certain simple animals, like social insects, group living can induce, beyond behavioral alterations, far-reaching changes in body morphology and physiology leading to strongly reduced functioning of organs serving some basic life-sustaining aims, while exaggerating other organismic functions and structures. In an ant or bee colony, for example, evolution inside cooperative groups created members whose tasks solely concern reproduction: the queen and the drones. These insects are incapable of other important life-sustaining functions like foraging, building the nest, defense, or tending the eggs. Those functions are fulfilled by nonreproducing female workers. In some ant species, the members designed for these tasks may be further differentiated into workers and "soldiers"; and the workers themselves may be further divided into those whose task is tending for the nest, the eggs, and immature offspring and those that do the foraging. (It is presumed that ant and bee colonies originally evolved from solitary insects that were capable of fulfilling by themselves all the needed life—supporting and reproductive functions [see Maynard Smith & Szathmáry, 1995, pp. 264–265, 268–269; Tinbergen, 2003, p. 194].)

In more evolved social animals, the selective effects of group life may manifest themselves in peculiar, complex behavioral adaptations. I will illustrate this with a simple active/reactive/submissive behavior interaction in chimpanzee life: A lower-status adult male chimpanzee tries to copulate with a consenting female in heat, but the dominant male interferes repeatedly, thus preventing sexual intercourse. This instance in our theoretical scheme has to be considered frustration of active behavior in progress with the behavioral options of displacement or its discharge in vacuo, aggression, or dysphoria. In other instances, for reasons not related to this incident, the same dominant male displays violently, forcing the group members in his vicinity to flee repeatedly. This kind of behavior has to be considered reactive behavior (no active behavior is involved). Since counteraggression against the dominant male is not an option, our chimpanzee may choose to flee, keeping a considerable distance from the dominant male for longer periods (which precludes full participation in the group activities). This, however, seems to be a bad choice. Remaining an active member of the group has many advantages, such as the security in greater numbers of conspecifics, better chances of securing a territory with food resources, a relatively dominant position in the group with regard to lower-ranking members, the possibility of mating surreptitiously, and so on. Even more than this, the dominant male may intervene on his behalf if he is attacked by other members of the group (this example is further developed in section 5.4.2.2.3). Therefore, accepting the dominance-submission hierarchy of the group is almost always the better option. Even animals with a much simpler group life than chimpanzees prefer a dominance hierarchy over solitary life or continuous quarreling each time individual interests clash (as in the pecking order among chickens).

The extent of the sacrifices more evolved animals are required to make in exchange for the assets group life offers varies greatly. It ranges from some constraints in reproductive possibilities, as previously mentioned, to the complete renunciation of reproductive chances by most members of the group for an indefinite period. This latter phenomenon is encountered frequently in wolf packs, in which only the alpha pair reproduces, as a rule, while their infants are tended by all the members of the pack. This later adaptation to group life resembles somewhat that of the ant colonies, but there it is achieved exclusively through behavioral modification—that is, without changes in the somatic domain. However, the final result or adaptation is similar from an evolutionary point of view. While it is only the alpha pair that reproduces, the duties of the rest of the pack involve tending the offspring, hunting, and activities related to competition with other conspecific groups. The ant-type reproductive strategy is called *eusociality* in the professional literature, while the wolf-type strategy (which is also encountered in other mammals, like wild dogs and naked mole-rats) is termed *cooperative breeding*. (See more on this topic in Summers & Crespi, 2013, pp. 55–60 and 63–68.)

Edward Wilson succinctly summarizes the evolutionary rationale of these kinds of behavioral strategies, dubbing it "a primal rule of social life throughout the animal kingdom. It arises when *loss of personal advantage by submission to the needs of the group is more than offset by gain in personal advantage due to the resulting success of the group*" (Wilson, 1998, p. 245, emphasis added. See also, de Waal, 1996, p. 9).

———————

Before considering the effect of the human social environment on active and reactive behavior, we need to take a short digression to summarize some of the main points of the present chapter. Moreover, the active/reactive behavior dichotomy is not the only possible one when assessing the organism's relationship to, or interaction with, its environment. Niko Tinbergen, in his important book *The Study of Instinct* (written in the late 1940s but in print until 2003, presumably as a result of its lasting relevance), divides animal behavior into two aspects:

1. "Behavior as a Reaction to External Stimuli" (pp. 15–56 and 76–100).
2. "The Internal Factors Responsible for the 'Spontaneity' of Behavior" (pp. 57–75).

Tinbergen attempts a synthesis between these two aspects of animal behavior (pp. 101–127).

At the beginnings of the science of ethology, this (simpler) division was a fruitful way to obtain precise, observable data on animal behavior, uncomplicated by more complex theoretical considerations. However, seeking to clarify the innate underpinnings of mental disorders—and human behavior in general—the present work proposes a different, more complex, division of behavior, both nonhuman and human, a division in which each category comprises external stimuli, as well as genetically preprogrammed motor patterns or predispositions to behave in a certain way. In this classification, the external stimuli are divided into three different categories:

1. *Environmental stimuli desirable for the organism,* signaling the presence of environmental resources needed for the sustainment and reproduction of life (releasers of active motor patterns or strivings).
2. *Influences undesirable or deleterious to the life processes.* This category is divided further into three subcategories:
   a. Cyclic or very predictable environmental changes (like seasonal or diurnal fluctuations).
   b. Deleterious influences specifically frustrating a discrete active instinctive striving in progress (mostly in its appetitive phase).

    c. Deleterious or dangerous influences irrelevant or not exclusively restricted to a circumscribed active instinctive striving.

This model seems clearly more desirable for studying behavior of higher-level complexity, since, as we have seen, the innate behavioral tactics designed to deal with each of these categories are different. When the route toward contacting a desired life-supporting resource is open, the situation is dealt with through behavior guided by the relevant active instinctive mechanisms or predispositions—in animals, the respective motor pattern and releasing mechanism. When the route toward the desired resource is blocked, the active behavior tends to be discharged in one of its frustrated forms.

When the harmful environmental influence does not concern a distinct active instinctive striving but may produce harm of a different kind to the organism, it is dealt with through the reactive behavioral tactics of freeze, flee, or counterattack.

And when the harmful environmental influence is cyclic and very predictable, natural selection builds *specific* behavioral or somatic adaptations in order to counteract their effect (migration, dormancy, hibernation, etc.).

3. Environmental influences originating from the *imperatives of social coexistence*. These are more fully discussed in the next section.

5.4.2. The Effects of Human Social Environment on Self-Centered Active and Reactive Behavior

Active and reactive behavior, that is, behavior induced or guided by active instincts in either their original or frustrated forms and the three basic tactics used in reactive behavior, can be observed in both solitary and social nonhuman animals, as well as in humans. The same can be said about the interaction and interrelationship of active and reactive behavior. However, in organized groups, the unchecked, unmodified expression of both active and reactive individual instinctive motives may be highly disruptive to the social network. Therefore, additional, complementary, or hierarchically superordinate instinctive mechanisms evolved when needed in order to preserve a desired measure of group cohesion, enhance cooperation among members, or enable the settlement of ubiquitous conflicts of individual interests in an attenuated, rule-governed way that is acceptable for the majority of the group members—in a word, to ensure a kind of social coexistence whose assets surpass clearly that of the solitary existence.

The most conspicuous instinctive mechanism serving this purpose is *submission*. Its purely instinctive form can already be discerned in puppies who are under seven or eight months old and serves as an inhibitory mechanism against the aggressiveness of older dogs: "As soon as the approach of the adult [dog] seems at all dangerous, the puppy throws itself on its back, presenting its still naked baby belly and passing a few drops of urine" (Lorenz, 1974, p. 104). Simple submissive patterns (complemented by simple forms of learning) are also encountered in solitary mother/immature offspring pairs, ensuring the effectiveness of the mother's guarding, disciplining, or teaching activities.

In simple social networks, submissiveness ensures the acceptance of the dominance hierarchy; we have already noted the pecking order of chickens as an example. In more complex social organizations, like those of the chimpanzees, the acknowledgement of social hierarchy may be expressed by more elaborate behavior patterns, such as "submissive greeting" (de Waal 1982/2007, pp. 78–79).

In the vastly more complex human environment, in which learning and experience play a prominent role in behavioral modification, the simpler *submissive* behavior (meaning: "accept and yield to a superior force") may evolve (mature) into the more complex *conformity* ("compliance with the conventions, rules and laws"). Conformity in turn may further mature into *internalization* of the rules of social coexistence, which makes it "part of one's nature" (*Concise Oxford English Dictionary*, 2009). In the terminology of this text, this means that the submission (which must be considered an additional reactive behavioral tactic) may gradually take on some of the characteristics of active behavior. In other words, *behaving according to these internalized social rules and conventions is felt as an internal motive representing the interests of the individual* rather than the result of submitting to or complying with externally originating pressures. This kind of behavior becomes fueled by something analogous to Lorenz's "action specific excitation"—a characteristic of active behavior. Consequently, these social rules and conventions are transmitted willingly, along with other aspects of the culture, to the next generation as an internal need, even if this implies difficulty or external danger (as in the case of religions forbidden by the reigning social order), thus ensuring the continuity of cultural traditions. In this example, the attitude and resultant behavior is clearly *not* brought about by "conditioning" sensitive to the behaviorist concepts of prize and punishment, reinforcement, or extinction.[3]

In the following, we examine separately in some detail the social environment's effects on reactive and active behavior. In this discussion, the accent is placed on the *alteration of aspects of the instinctive activity* induced by the social environment rather than the *enrichment* of behavior as a result of accumulated knowledge transmitted from one generation to the next and the employment of sophisticated cognitive mechanisms.

5.4.2.1. Reactive Behavior

In solitary animals living in a natural environment, deleterious or dangerous influences are, as a rule, of a type that enables the animal to tackle them with the aid of the three basic instinctive reactions (this, of course, is why they evolved in the first place). Remaining motionless and inconspicuous, fleeing, or neutralizing harmful influences with a counteraggressive reaction, when successful, may resolve most such situations in nature. *However, many of the deleterious effects of the social milieu are not amenable to these tactics, especially in their crude, physical form.*

Moreover, two of the three "individual" reactive tactics, fleeing and physically counterattacking, may clearly be deleterious in many social circumstances, especially in humans. As mentioned, intragroup aggression may disrupt group cohesion and concerted cooperation, and fleeing each time conflicts arise in a group setting can mean living mostly a solitary life, thus losing the benefits conferred by the group.

While the *basic incentives* of these instinctive motives, felt subjectively as goal-directed arousal, fear, or anger as a response to undesirable, deleterious, or possibly harmful social influences, undoubtedly exist in humans, the final observable behavioral output must be drastically modified to fit into the framework of socially acceptable behavior. Think, for example, of the discipline required in an educational or military setting. Frequently, the basic reactive impulses (to flee, to refuse cooperation, to revolt against instructions or orders) have to be suppressed, or expressed verbally instead of physically, preferably in an attenuated or veiled form.

As an extreme example of the far-reaching modification of basic re-active instinctive motives by social imperatives, think of the behavior of soldiers in combat—that is, *intergroup* aggression. In this instance, the innate motor reaction, of fleeing from a very real danger to body integrity or life itself, may be completely reversed. The soldier is required to function in cooperation with his fellow soldiers; she frequently has to move *forward*, toward the enemy, which means *increasing rather than decreasing* the level of dangerousness of the situation. The instinctive underpinnings of the mental motives that *counteract* the basic reactive intention to flee, to save one's life, may be diverse: an equal or even greater fear of punishment or even execution as a result of insubordination (think of the practice of decimation in the ancient Roman army); (counter-)aggressive feelings directed toward the enemy; fear of social condemnation; conformity with the rules of socially expected or prized behavior; or internalization of the moral values of the military unit or of the bigger social entity (the nation, for example). In the latter case, the soldier may put the interests of the group ahead of her own life and may even be expected to try to induce similar feelings and behavior in less motivated soldiers.

We can witness the development of reactive behavior from its basic instinctive foundations toward more mature subordination to or conformity with the requirements of social coexistence at some of the "stepping-stones" of a maturing individual's integration into the community. Some children in their first days at a kindergarten, for example, may react to the new, still unknown, environment, as well as to the prospect of temporary separation from the mother, with motor patterns and vocalizations not unlike those of evolved social animals. They may anxiously cling to the mother, refuse to remain alone, try to flee when left alone, or have a temper tantrum. (By the way, young chimpanzees during weaning—at about four years of age—regularly show temper tantrums. A particularly difficult case of weaning of this kind was described by Jane Goodall, 1988, pp. 233–236.)

The well-known rebellion of many adolescents against the prevailing social order, especially against its constraining aspects, may also be fueled by aggressive reactive instincts (as well as frustrated active ones). Compulsory military service, especially in its initial stages, may elicit evasive reactions, passive noncooperation, or open defiance. However, most mature humans ultimately succeed in conforming to the requirements of social coexistence or even internalizing different aspects of those requirements; if and when the opportunity arises, they may make efforts to willingly encourage this attitude in others, particularly the younger generation.

In fact, most undesirable or potentially harmful influences of the social milieu are not of a clearly physical nature—that is, of the type that can be sensed directly and promptly with the sensory organs and neutralized with physical reactions. Examples of nonphysical threats include contamination by pathogens, being fired from a job with resulting unemployment, being cheated or exploited by other individuals or social organizations, degraded in social status, and so on. While in these instances the basic reactive instinctive predispositions (in the event the harmful influence is sensed or foreseen) are the same as in evolved animals, and felt subjectively as goal-directed arousal, fear, or anger, the overt behavior is almost completely the result of complex learning processes that are in turn selectively guided by the basic instinctive strivings.

However, this does not mean that effective reactive behavior in a social setting is always more involved and difficult to master. *Instances exist in which reactive behavior is made easier and much more effective in a social milieu by additional reactive mechanisms that are not encountered, as a rule, in nonhuman animals.* In a possibly dangerous interpersonal situation, the individual can appeal for help, for example, to bystanders (a simpler form of this occurs in the form of mobbing in animals living in groups). In addition, one may appeal for help to social institutions established for this purpose: law enforcement agencies, welfare institutions, self-help groups, etc. In more complex harmful circumstances, whose successful neutralization necessitates specialized knowledge and training, one can seek the advice and help of experts (like medical personnel). Developed technology can sometimes provide warning of danger or aggression in advance, and the media may supply further relevant information on how to behave in various kinds of impending or possible danger (dangerous weather conditions, burglary in the neighborhood, expected shelling of the civilian population in an armed conflict, etc.), enabling people to take preventive steps.

We can see the role of the basic individual reactive instincts in a social setting from another, very different, vantage point. Social pressures to conform and internalize (for example, to volunteer to perform socially prized activities) frequently come at the expense of individual interests. It seems logical to presume that the basic reactive instinctive motives—which evolved in order to safeguard the interests of the individual life, its survival and reproduction, and thus are frequently in conflict with social pressures to conform—may protect the individual from overinvolvement in socially desired activities at the expense of individual interests.

Ideally, the best thing seems to be to keep an optimal balance in this regard. However, this is easier said than done. This optimal balance may change according to circumstances (for example, during conflicts between groups, the involvement on behalf of one's respective group may prevail over individual interests). In addition, a wide scale of diversity of instinctive configurations is present in human populations. As a result, it can be presumed that the intensity of the pressure to conform or submit within a certain social niche is tailored to the strength of the instinctive individual predispositions to oppose conformity that characterize the average member of that niche. Consider the differences in this respect between a voluntary organization, a hospital, the army, and a typical prison environment.[4]

5.4.2.2. The Modification of Active Behavior by the Human Social Milieu

As we have seen, what distinguishes humans from other animals is the pronounced diffuseness of the active (as well as reactive) instinctive strivings, which makes possible their far-reaching modification through learning the constraints as well as the extended possibilities that social coexistence offers for active instinctive gratification. *This diffuseness refers both to the motor pattern of the instinctively fueled activity as well as to the releaser toward which it is directed.* As has been repeatedly stressed, preprogrammed fixed motor patterns disappeared almost completely from the human behavioral repertoire due to their rigidity. Instead, we can identify our instinctive strivings subjectively as more or less blurred intentions, dispositions, and aspirations. In accordance with this development, the *releaser* of these more diffuse intentions also became *more varied*, influenced by other considerations than the narrowly biological ones, such as social mores and cultural fads. Think, for example, of the culturally induced changes in what was considered beautiful and attractive in the female body in the time of ancient Greeks and Romans compared to the Renaissance ideal of beauty, or to what is considered attractive in present-day Western culture. The same is true with regard to the clothing that covers the female body.

The optimal diffuseness of active drives (with regard to motor patterns as well as their releasers) made possible the channeling of active intentions and their respective concomitant energies and emotions into the diverse preoccupations created by cultural evolution (for example, by the increasing division of labor). Formal education or other social influences may create sustained interests that may last a lifetime, directed toward some narrow technological, financial, political, or scientific field, sport, artistic endeavor, or other preoccupation in an appropriately endowed individual—in essence, releasers that cannot even be imagined in a natural environment. In fact, the overwhelming majority of human preoccupations and professions consist of more-or-less *narrow activities* created by cultural tradition, instead of by nature, and have no meaning outside the domain of social coexistence and cooperation.

Another development enabled by the considerable diffuseness of active instinctive intentions, in my opinion, is the widening of the scope and target of these intentions in certain instances. For example, the instinct of territoriality with respect to one's individual place (home) may be widened to comprise the territorial interests of the whole ethnic group, and individual instinctive motives to defend one's offspring or extended biological family may be widened to the aspiration to defend the whole ethnic group or nation. Such political rhetoric as the summons to defend "the Fatherland" or to protect the "mother country" or "homeland" probably aims at stimulating such widening of the scope of these instinctive intentions (Ehrlich, 2000, pp. 238–239). It should be mentioned, however, that a similar kind of widening in the scope of certain instinctive motives is already present in groups of other social mammals, in which adult males may collectively guard the territory of the group (in the case of chimpanzees) or protect the females and immature offspring (in the case of elephants and wildebeests).

Another peculiarity of the social influences that affect active instinctive expression is that these frustrating influences are more subtle and nuanced, do not as a rule completely block the instinctive intent and the way toward the appropriate releaser, but only *constrain its* discharge to some appropriate circumstance, place, time, form of expression of the innate motive, etc. For example, if we compare a sexual contest between male animals for access to fertile females in a natural environment versus sexual competition among human males in a rule-bound society, the difference becomes apparent. In many mammal species such as herbivores, walruses, gorillas, etc., the males fight intensely for access to fertile females. The winner gains the right to impregnate most or all the females of the group, while the loser, as we have noted, may lose his reproductive chances completely for that season (known as increased "reproductive variance" in the parlance of sociobiology).

In most human societies, on the other hand, particularly in modern ones, physical fighting is forbidden by law when competing for an attractive woman and even verbal abuse is undesirable. The institution of marriage encourages, or in certain cultures completely limits, sexual activities to the husband-wife couple and discourages or even completely forbids extramarital sexual affairs. The aggressive form of active instinctive frustration in traditional societies is discouraged against members of the same group but is frequently allowed toward competing groups or strangers.

In general, social conventions prescribe quite clearly which active instinctive motives are allowed to be expressed, in what form, at what level of intensity, and in what circumstances. Open expression of aggressiveness or sexual interest is forbidden in many circumstances but allowed in a more-or-less attenuated form in others. Loud expression of excitement, enjoyment, frustration, or disappointment—verbally and in body language—is allowed, even encouraged, when cheering one's favorite sports team, for example, while prohibited during solemn events and in most educational settings.

Regarding the discrete forms of frustration of active drives, the social milieu can lead to the following alterations in comparison with a natural environment (these examples are for illustration and not an exhaustive treatment of the subject).

5.4.2.2.1. Displacement

Frequently, the social milieu tolerates or even facilitates displacement activities when the interaction with the appropriate releaser is impracticable, undesirable, or impossible. For example, as mentioned before, modern societies offer various artificial means to gratify sexual drives in a displaced form, such as pornography, sensual literature, movies, fashion, and even sex dolls. At least part of the services prostitution provides also enter into this category. Another example of displacement is the "adoption" of pet animals as substitutes for biological offspring, a practice that is encouraged by animal welfare organizations.

5.4.2.2.2. Aggression

As mentioned, less-developed societies or traditional ethnic groups may encourage the redirection of aggressiveness as a result of frustrated active drives toward targets outside the social group (an overlap of aggressiveness and displacement). This is also true in the case of developed societies in times of international or ethnic conflict. Many kinds of sports, like those between two contestants or competing groups who engage in direct physical contact (from boxing to rugby) can serve as a rule-bound outlet for aggressive drives. Fans of these contests may express aggressiveness, too, while cheering their favorite athletes or teams and encouraging them to be more assertive, and even more aggressive, when disparaging the adversary group or athlete.

5.4.2.2.3. Dysphoria

As mentioned before, with the relaxation of intragroup selection pressures for continuous cooperative functioning among humans, periods of loss of function or dysfunction became tolerated much better than, for example, we see in chimpanzee groups. Remember Jane Goodall's description concerning the exclusion of ill members of a chimpanzee group from common activities, and even aggression toward them during a polio epidemic (section 4.6.3, endnote). Moreover, in the case of significant loss, human communities may regularly give emotional or material support. During mourning, for example, providing emotional (and frequently material) support has become an ingrained social ritual. Similarly, during the initial period of temporary unemployment, modern states offer financial support.

It was argued in this chapter that the predisposition to respond with dysphoria became diversified in human populations. It can be hypothesized in this context that the innate inclination to empathize and sympathize with other beings' suffering also became diversified, and that the normative segment of the diversified scale became wider, especially toward the more intense sympathizing end (this trend is most evident in welfare states). This inclination can be seen in the ease and willingness with which certain individuals volunteer to help populations struck by disasters such as war, famine, natural calamities, etc., or to alleviate the suffering of animals, to ensure they are humanely treated and are not killed for their fur, or even used for food.

Modern societies allow for the "sublimation" of dysphoric feelings through socially accepted or prized activities. For example, persons who have lost a loved one during an armed conflict or road accident may enroll in organized activities against warfare or campaigns to reduce the frequency of road accidents. During these concerted activities, the *dysphoric* form of instinctive expression, caused by the personal misfortune, may take on an aggressive quality directed against those supposedly accountable for the respective social adversity.

Besides the three forms of frustration of active incentives, both social circumstances and advanced cognitive abilities may create diverse additional behavioral tactics to express and achieve active instinctive aims among frustrating conditions. One of these tactics is concealment and cunning. This example from chimpanzee life was mentioned earlier in a different context: A low-ranking male tries to copulate with a consenting receptive female but the dominant male of the group repeatedly interferes with his attempts. The low-ranking male decides to use concealed tactics. He presents his erect penis to the female (which in chimpanzee communication means, in this instance, an invitation to copulate) at a place not visible to the dominant male. The couple chooses for copulation a place that is also hidden from the sight of the dominant male, and the vocalizations accompanying the sexual act are suppressed (deWaal, 2007, pp. 36–37; Gomez, 2004, pp. 209–211).

Humans have highly sophisticated means to achieve active instinctive goals in frustrating conditions; however, a discussion of this topic is beyond the scope of this book.

5.4.3. The Relationship between Egocentric Active and Reactive Instinctive Intentions and Conformity with the Human Social Milieu's Instinct-Modifying Effects

The social milieu's modifying effects on self-centered active and reactive predispositions vary widely. They may entirely suppress an instinctively fueled intention, partially suppress it, postpone it, or modify the releaser toward which it is intended to be discharged. They may modify its motor pattern, or (depending on its degree of diffuseness or malleability) rechannel its intentional, energetic, and subjective experiential aspects into culture-created activities that have no counterpart in the natural world. Of course, *in addition to the social environment's modifying effect, the final behavioral outcome depends greatly on the diversified instinctive configuration of the individual.* The interrelation between the intention to express self-centered drives in their original form as much as possible and the inclination toward conformity or internalization, depending partly on the degree of diffuseness of those drives, greatly influences the roles or tasks a person can successfully fulfill in the social milieu.

In fact, we can construct a continuous graded scale regarding the degree of priority that one component of this dichotomy gains over the other. At one end, we find persons who give almost absolute priority to their own individual interests (based mainly on biological motives) and submit (if at all) to contradictory requirements of the cultural milieu only under significant and continuous social pressure or the threat of severe punishment (for example, the discipline enforced in a corrective institution or prison). This inclination may be hypothesized to consti-tute one innate aspect of the antisocial personality.

At the middle section of this scale, we can place those normative per-sons who, perhaps after a period of youthful rebellion, achieve a more-or-less optimal balance or compromise or partial integration between individual strivings and conformity with the constraining requirements of social coexistence. Contrary to the previous category, in these persons the degree of conformity with social expectations or rules is not tightly linked to punishment or reward. Rather, compliance with those rules and expectations is an internally accepted, even embraced, way of life.

Here, I think, our theoretical scheme comes closest to Freud's tripar-tite structural model of the psyche. The Freudian concept of *id* parallels the intention to express in socially unchecked ways individualistic active and self-defending reactive instinctive motives; the *superego* represents that social environment's modifying, regulating, prohibiting effects on those same strivings (internalized to varying degrees in different per-sons); while the *ego* tries to attain the best possible compromise when conflicts arise between these two incentives to behavior.

At the other end of the scale, we find people inclined to give prefer-ence to the interests of the social group over their own personal inter-ests. The behavior of this kind of persons is least in agreement with an exclusively behaviorist view of socialization (and for that matter with the Freudian tripartite model of mental functioning as well). Extreme examples that may illustrate this point best are those individuals who, in a self-imposed manner, of their own "free will," in situations in which other, less self-damaging alternatives exist, are prepared to sacrifice the most basic interests of their individual life (survival or reproduction) for the needs or ideals of the group. I have in mind such examples (already partly mentioned) as early Christians who accepted torture and death rather than relinquish their faith; the Japanese kamikaze pilots of the Second World War; the Renaissance philosopher Giordano Bruno's refusal to retract his unorthodox theological and scientific beliefs in spite of long imprisonment and a death sentence by the Inquisition; Mother Teresa's lifelong dedication to helping the destitute, disaster-stricken, seriously ill, dying, and so on. Such feats cannot result exclusively from social learning. The intensity and perseverance of the intention to embrace a certain social ideal or extremely altruistic social program (necessitating the mental power to suppress or redirect basic biological interests), the resolution and energetic resources to persevere in spite of possible or already materialized harmful consequences, have to have an instinctive underpinning. As we have seen repeatedly, the intentionality, mental energies, and subjective emotional experience cannot be acquired through learning.

Besides being schematic and oversimplified, this discussion may strike the reader as irrelevant to clinical mental disorders. However, we will need the conclusions of this brief digression in upcoming chapters when we try to understand such aspects of psychopathology as the alternation of acute psychotic episodes with remission; the negative effects of pathologically intense instinctive strivings on conformity; the ability of antipsychotic drugs to suppress instinctive intensities and, as a consequence, to restore former levels of submission or conformity; and the innate sources of antisocial personality disorder.

6

The Effects of Psychotropic Drugs in the Context of the Present Theory

6.0. Introduction

Let me explain first why I treat the subject of psychotropic drugs before, and not after, the discussion of clinical mental disorder categories. The central idea of this section is that the *primary or direct* effect of the main psychotropic drug categories on behavior and mental functioning are the result of their effect not on circumscribed clinical disease entities, but rather on some of the comprehensive instinctive mechanisms discussed in the previous chapter, mechanisms that evolved in the distant evolutionary past of the animal kingdom. This is why it is possible to localize the biochemical action of these drugs in circumscribed brain structures and functions and why researchers are able to use animals in their experiments to develop new drugs. The neurotransmitters dopamine, noradrenaline, and serotonin—the target of major psychotropic drug categories—are already present in very simple organisms, even unicellular ones (Healy, 2009, p. 150). As a result, this primary effect of psychotropic drugs is apparent in the behavior of laboratory animals, in the behavior and subjective experience of normative humans, as well as in psychiatric patients, irrespective of what specific mental disorder they suffer from.

These primary effects in a complex organ like the central nervous system, in which many functions are interconnected and hierarchically organized, have *secondary consequences or repercussions* with regard to other mental functions. However, since the probability of the appearance and the nature of these secondary repercussions depends on a host of other factors besides the primary effect of the drug (augmented by the pronounced diversity of human mental functioning), *these indirect effects are much more variable and much less reliably predictable than the primary effect.*

A good analogy for this state of affairs is gene pleiotropy (discussed in section 4.1.1.1). As you may recall, gene pleiotropy refers to a gene having multiple effects on various traits of the organism. Examples are genetic disorders caused by one single gene with multiple somatic and possibly behavioral derangements, which calls into question the validity of Mendelian genetics. As already quoted in section 4.1.1.1, geneticist Theodosius Dobzhansky explains: "A dogma in molecular genetics is that each structural gene specifies one and only one polypeptide chain in a protein." In consequence, "As the traits are further and further removed from the primary gene action, the possibilities of epistatic interactions of different genes, as well as modifications due to environmental influence, increase" (Dobzhansky, 1970, p. 64). Put simply, the primary, direct action of the gene is to specify a single polypeptide chain. The diverse observable somatic and behavioral traits of a monogenic genetic disorder, on the other hand, reflect mostly *secondary, tertiary, etc., repercussions* of the primary gene action as a consequence of multiple interactions with other genetic and environmental factors.

Clinical trials of new psychotropic drugs are founded on principles that are entirely different from those just outlined. The standard procedure is the "randomized clinical trial" (RCT). In this kind of research, homogenous groups of patients are assembled (homogenous from the point of view of clinical diagnosis) according to more stringent criteria than those used in everyday clinical practice (Research Diagnostic Criteria). The patients' mental and behavioral state is checked with the aid of "standardized rating scales," which contain a list of the characteristic signs and symptoms of the disease that the drug in question is expected to improve. These symptoms may have degrees of severity by which the expected improvement during the trial can be quantitatively evaluated. After the trial, the results are statistically averaged and compared to a placebo effect or to the effect of another drug with already established efficacy (that is, whose efficacy was established in the same way). No symptoms have priority over others—that is, no symptoms reflect more directly than others the fundamental genetic predisposition underlying the disorder. Nor do these rating scales tell us whether the drug in question specifically targets these fundamental symptoms or, alternatively, has a more generalized, more nonspecific effect on mental functioning and behavior.

David Healy, an authority in the field of psychopharmacology, has severely criticized the widespread use of RCTs in psychiatry: "This type of trial works best where there is a homogenous patient population. This fact favors fitting patients into categorical disease entities. . . . *A response of these seemingly discrete disease entities to a drug then creates an illusion of specificity.* . . . RCTs also require efforts to reduce inter-rater variability. This leads to the use of operational criteria and rating scales, which in turn add to the illusion that the disorder being treated is responding in the same way that cultures of bacilli on a Petri dish shrink when exposed to an antibiotic" (Healy, 2002, p. 284, emphasis added).

A moment of reflection, I think, would convince anyone of how untenable the claim is that an antipsychotic drug could directly influence the whole range of symptoms of a psychotic disorder. The symptomatology of these disorders regarding brain functioning may contain widely different symptoms and signs—for example, disorders of alertness or attention, sensation and perception (hallucinations), emotion, or thought form and content (delusions). Each of these categories is dealt with by very different brain regions and neurophysiological and neurochemical mechanisms. It seems highly improbable that an antipsychotic drug with a relatively simple biochemical effect that modulates the level of activity of one or a small number of neurotransmitters can *directly and differentially* influence all these mental functions in a desirable direction—that it can remove hallucinations while leaving normal perception unaffected, abolish delusions while leaving normative thinking unchanged or even improved, etc. How on earth can an antipsychotic drug distinguish between delusions and normative thoughts? Consider that normative thinking frequently contains views about unrealistic or unprovable issues, such as prejudice or religious and mystical beliefs.

Robert H. Belmaker, an Israeli psychiatrist with a strong biological orientation, puts his disappointment with RCTs in the following words: "In almost every area of psychiatry controlled clinical trials that yielded statistically significant and promising results . . . [have] been found to be non-replicable. . . . It is hard for the clinician to decide whether one can base treatment on research evidence at all" (2015, p. 4).

Yet it has to be admitted that clinical practitioners regularly observe the disappearance of diverse mental disease symptoms in individuals under psychotropic drug treatment. Agitation, delusions, hallucinations, withdrawal from social interactions, bizarre behavior, etc., may disappear under antipsychotic treatment in the course of several weeks. However, in other patients with a seemingly identical pathology, under the same drug treatment, the psychotic symptoms will not disappear; only their underlying emotional charge and their impact on overt behavior will be reduced (the "deemotioning effect"). To my mind, only one reasonable explanation exists for this riddle. Antipsychotics do not act *directly* on the hallucinations, delusions, and other characteristic psychotic symptoms. Their alteration by the drug represents *secondary consequences* depending on other variables of the disorder, such as the degree of chronicity of the psychotic symptoms or the successful removal of a somatic disturbance in the case of organic psychosis. (These factors are detailed in section 7.2.6.4, on the psychotic mechanism.)

It has to be mentioned that no controlled drug trials were used to differentially study symptoms that may seem to be a more direct target of these drugs' biochemical effect, like the improvement of agitation: "What the antipsychotics most clearly do is reduce tension and agitation, regardless of diagnosis. But they are never put into a clinical trial for this indication—trials are always for a disease indication such as schizophrenia" (Healy, 1999, p. 103).

Another consideration that does not support the hypothesis of the specificity of the effects of psychotropic drugs is that, as is commonly known, the same psychotropic drugs are effective in the case of diverse, clinically distinct, disease categories. Antipsychotics have been used successfully apart from schizophrenia in other psychotic disorders, including those of an organic and toxic nature, nonpsychotic mania, agitation in mental retardation, dementia and autism, Tourette syndrome, etc. Along with depressive disorders, antidepressants have been found useful for obsessive-compulsive disorder, panic disorder, posttraumatic disorders, the "negative syndrome" in schizophrenia, pathological laughing and crying, and eating disorders. It seems highly improbable that psychiatric drugs are able to differentially target each of the various disorders that, in a modest proportion of cases, they are able to improve.

Some of the drugs used to treat the acute phase of a mental disorder are also frequently employed for prevention (antipsychotics, antidepressants), another practice that is quite rare in other fields of medicine. It seems reasonable to presume that in these cases the drugs do not target specific predisposing or triggering factors but act promptly when the clinical condition begins to deteriorate. They shorten the duration and severity of the illness and reduce mortality (Kaplan et al., 1980, p. 2867). And: "Given that secondary prevention essentially became a synonym for effective treatment, the 'secondary prevention efforts' of clinicians became indistinguishable from their regular clinical efforts." In other words, these drugs do not prevent the appearance of the disease but "rather minimize morbidity" (Mrazek & Mrazek, 2005, p. 3513).

It seems apparent that the principles of drug-employment in more advanced medical specialties are very different from those we encounter in psychiatry. In cardiology, for example, no such thing as an "anti-myocardial infarction drug" category exists. Myocardial infarction is not treated in a "holistic" manner; instead, specific functional and structural aspects of the clinical entity are targeted by specific drugs or procedures: Chronotropic or inotropic drugs may be used to improve cardiac efficiency, and specific invasive procedures may relieve an obstruction or constriction in the affected myocardial blood vessel. While dangerous arrhythmias may be prevented using the same drugs that are used for their treatment, other drugs used for prevention (anticoagulants, lipid-lowering drugs) are specific agents against discrete predisposing factors or possible complications (atherosclerosis, embolization), the mechanisms of which are well understood.

By now it must be clear that these great differences between cardiology and psychiatry in the rationale and efficacy of treatment procedures are the result of different levels of understanding regarding the chain of events that lead to the respective illnesses they are treating. In the case of cardiology, predisposing factors and their long-term consequences are understood; the interrelations between the structure and function of the heart are clear; the interrelations between the heart and the whole organism, as well as the role of the heart in the organism's adaptation to the physical environment, are well known. The interrelationships between successive levels of complexity (molecular, cellular, parts of the heart, and the whole heart) are also better understood.

However, this lag on the part of psychiatry is understandable. The brain is by far the most complex organ of those organisms that possess one, being responsible, among other functions, for the organism/environment interaction at its highest possible level of sophistication. Moreover, *in psychiatry (as in other behavioral sciences) the central nervous system is both the target of inquiry as well as the instrument that conceives the method of inquiry and tries to make sense of its results.* This is something we cannot sustain in the case of cardiology or of any other medical specialty. As already quoted (section 1.4.3), economist and philosopher von Hayek has argued that "'any apparatus . . . must possess a structure of a higher degree of complexity than is possessed by the objects' which it is trying to explain."

As noted earlier, philosopher of science Thomas Kuhn argued that psychiatry is in the "pre-paradigmatic" stage of its development. We have no universally accepted comprehensive theory of psychiatric disorders whose implications can be clarified by "normal science." In its more than one hundred years of existence, psychiatry systematized the phenomena in its domain into clinical disease categories and identified many of their empirical regularities or laws (average age of onset, expected course, the characteristic clustering of signs and symptoms into circumscribed syndromes and clinical entities, the extent of their prevalence in blood relatives, possible response to different treatment modalities, and so on). It seems, however, that this established scheme of reference came under increasing strain by findings discordant with it, like the existence of clinical pictures that are intermediary between two well-established disease categories, for instance schizoaffective disorder. Genetic studies only partially validated the reigning disease classification (genetically related schizophrenia spectrum disorders contain psychotic as well as personality disorders, as we have seen). Psychotropic drug effects seem increasingly nonspecific for discrete clinical entities. This is the present state of affairs in psychiatry, to my understanding. One has to impartially consider its assets and drawbacks before a more drastic change to its present ("pre-paradigmatic") state is proposed.

An additional reason must be mentioned here for treating whole disease entities rather than circumscribed symptoms with discrete psychotropic drug categories: the commercial interests of drug manufacturing and marketing companies. I familiarized myself with this subject by reading the work of English psychiatrist David Healy. Healy has a special interest in psychopharmacology, its history, and the complex interaction between commercial interests and therapeutic considerations during the process of development and marketing of psychotropic drugs. His ideas, and the facts he described in his books on the subject of psychopharmacology, are referred to throughout this chapter.

Healy argues that clinicians' erroneous belief that the specificity of clinical disease entities goes beyond a descriptive level, and that drugs improve such disease entities specifically and differentially, aligns with the commercial interests of drug companies. These companies encourage the widening of the boundaries of clinical disorders to include mild or subclinical cases, endorse preventive medication on questionable grounds, bias the results of their research by publishing favorable results only, and so on (2009, pp. 301–310). While Healy's arguments and the data in their favor seem convincing, I also think that such a marketing strategy can work in a medical field only when it is enabled by hazy conditions—that is by a lack of clear-cut therapeutic targets and unequivocally demonstrable therapeutic effects.

6.1. Antipsychotic and Antidepressant Drugs' Primary and Indirect Effects

The expectation of the clinician with regard to antipsychotic and antidepressant drugs can be easily inferred from their names: The primary effect of antipsychotic drugs should be the amelioration or relief of characteristic psychotic symptoms, and the primary therapeutic effect of antidepressants should be the conversion of a depressive mood into a euthymic one (or, in the case of overtreatment, to euphoria). A lot of clinical and research data on drug effects (as well as my own experience of more than thirty-five years with these drugs) suggests that neither of these effects is attainable as a rule. The alternative hypothesis, proposed in this text, is that *the primary effect of these drug categories consists of the decrease or increase respectively of the overall intensity of the energetic charge underlying behavior.* Since it cannot be learned or otherwise acquired during ontogeny, this energetic charge has to be considered an innate, instinctive aspect of behavior.

It can be inferred also that these drug effects are achieved by modulating the activity of that brain center that determines the constant overall intensity of behavior according to selective constraints with regard to the organism-environment interaction (discussed in section 5.1.3). Furthermore, clear indications exist that this modulating effect of psychotropic drugs on the constant overall instinctive intensities is achieved through the brain mechanism that evolved to deal with the fluctuations in life-sustaining resources in accordance with seasonal changes (hibernation and its milder versions). Indeed, some antipsychotics have a poikilothermic effect (explained in section 6.2.1), and the effect of both drug categories unfolds slowly, during a protracted period. Further evidence comes from the fact that artificially lengthening the daily hours of light (light therapy) has an effect similar to that of antidepressant drugs, as detailed in the following.

I do not mean to suggest that these drugs do not induce the clinically expected effect in a proportion of individuals—that is, amelioration or complete disappearance of the psychotic symptoms, or the conversion of a depressive mood into euthymia or euphoria.[1] However, it is proposed here that these effects are *secondary*, and consequently more unpredictable than the primary effect, since they depend also on parameters unrelated to the primary effect of the respective drug.

6.2. Antipsychotic Drugs

6.2.1. The Direct or Primary Effect of Antipsychotic Drugs

Antipsychotic drugs are hypothesized to reduce, that is, to weaken in a comprehensive way, the overall intensity of instinctive activity. In other words, this effect is not related to the reduction in intensity of a single instinctive domain ("action specific excitation" in K. Lorenz's terminology) but to the energetic level of instinct-fueled activity in its entirety. This manifests itself in the reduction in the intensity of multiple instinctive activities—for example, sexual activities, parenting, individualistic reactive behavior such as flight or reacting aggressively, etc. This does not mean that we will find no variations or oscillations in the intensity of these individual instinctive domains, particularly in the presence of an adequate releaser, but the *amplitude* of these oscillations will become considerably smaller. In the case of active instincts, this effect will be especially evident at their appetitive phase, that is, in the intensity of the striving to sexual access, to secure food, to guard a personal territory, etc.

In hospitalized, acutely ill psychotic patients, the reduction of the individualistic reactive behavior's intensity (but not a reduction in conformity or submissiveness) as a result of antipsychotic treatment manifests itself, for example, in the progressive diminution of aggressive responses to compulsory confinement or other undesirable stimuli of a crowded, closed psychiatric ward, as well as in the diminishing number of attempts to escape from the hospital.

Besides their effects described above, antipsychotic drugs may possess a sedative effect, too, that affects reactive behavior more strongly. This effect, whose intensity varies among antipsychotic drugs, will be detailed in the section on sedative drug category.

The overall reduction of the energetic charge fueling behavior will manifest itself in the slowing down of activity in a human or nonhuman animal; in humans, both physical and mental, activity will be reduced, reflected in subjective experience as a reduction in the *intensity* of emotions, intentionality, and willpower. This subjective state is labeled in clinical terminology (and partly in everyday parlance) as indifference, emotional blunting, "deactualizing," a "who cares" feeling, lack of interest, apathy, "a deemotioning effect," affective flattening, demotivation, and so on (Healy, 2009, pp. 9, 11–13, 31, 33–43; Healy, 2002, pp. 232–233). Please note that this indifference or emotional blunting does not target a particular instinctive domain, like sex or dominance-submissiveness or territoriality, but is a pervasive reduction of desires or aspirations toward *any* emotionally important goal. This effect is clearly different from the calming effect of sedatives, which, by reducing fears or anxieties, can even augment the expression of suppressed active emotions (as well as counteraggressive intentions), as will be discussed later in this chapter.

The effect of antipsychotic drugs to reduce the intensity of instinct expression has been demonstrated in mental patients, normative individuals, and experimental animals. With regard to experimental animals, it was found, for example, that "rats trained to climb a rope to a platform for their food, when given chlorpromazine, appeared to have little or no interest in climbing. As far as could be made out, this was not because they were sedated, nor was it because their coordination had been interfered with. They had what seemed to be a loss of interest" (Healy, 1999, p. 45).

In general, it was found that chlorpromazine, as well as other antipsychotics, block conditioned responses in laboratory animals. It is common knowledge that conditioned responses are built on unconditioned instinctive motives, both active (offering or withdrawing a releaser an animal appreciates, like food), as well as reactive (applying or withdrawing discomfort, fear, or pain-inducing stimuli, like electric shock or immersion into cold water). Therefore, it may be hypothesized that weakening the unconditioned instinctive intensity (that is, the innate drive to seek food, for example, or to escape disturbing or dangerous stimuli) will also reduce or extinguish altogether the intensity of the conditioned response. In fact, the conditioned (learned) response seems to receive its energetic charge from the unconditioned instinctive mechanism, instead of possessing its own, as argued repeatedly in this text (see section 5.2.1).

In laboratory animals, antipsychotics were found to inhibit pleasurable self-stimulation through intracerebral electrodes or intravenous self-administration of cocaine. These findings concerning the antipsychotic effect: "are typically interpreted as an induction of an anhedonic state" (Tasman & Lieberman, 1997, pp. 292–293). Antipsychotic drugs can also lead to *catalepsy* in experimental animals, which is a kind of paralysis of motor behavior—more precisely, the inability to correct an externally induced body posture while the animal seems to retain its consciousness. It may represent an extremely severe suppression of the capacity to act according to intrapsychic incentives. However, the subject of catalepsy and its relationship to catatonia, extrapyramidal side effects, hibernation, and grave lethargic depression is far from clear. This subject is detailed in the next chapter (section 7.2.8.8).

In what concerns the primary or direct antipsychotic effect on normal individuals, as well as, on the mentally ill, it is useful to look at examples from the early stages of development of these drugs, when their effect was observed in an unbiased way—that is, unbiased by preconceptions about their specific effects on discrete clinical entities and unaffected by the commercial interests of drug companies. In this respect, David Healy's books detailing the circumstances of the discovery of these drugs and the early period of their use on humans were of great help. On the effect of antipsychotics, Healy writes: "For the past two decades, under the influence of the notion that antipsychotics are antischizophrenic, interest in these drugs has focused almost exclusively on the fact that their use seems to get people out of hospital. There has been little interest in the changes the drugs bring about to get people out of hospital and as a consequence, despite sixty years of use, it is difficult to be precise about the beneficial effects of antipsychotics." And: "Everybody who takes an antipsychotic is affected by it whether they have a mental illness of any sort or not" (2009, pp. 12–13).

Shortly after antipsychotic use in humans for psychiatric (and anesthesiologic) purposes began, phenothiazines' ability to induce indifference, both in psychiatric patients and normative persons, became evident. This effect was blamed, for example, in cases of taxi drivers under the effect of chlorpromazine who drove through red lights (Healy, 2002, p. 341).

Another example of chlorpromazine's effect in normative individuals concerns soldiers participating in combat. The drug was included in the medical kits of U.S. soldiers participating in the Korean War, to be used in the case of a stress reaction. It was found, however, that soldiers under chlorpromazine became so indifferent that they did not react quickly enough in rescue situations (Healy, 2002, p. 82). The next example concerning the use of the atypical antipsychotic clozapine, shortly after its discovery, is even more impressive: "Although it does not immobilize people in the same way as neuroleptics do, clozapine has marked 'serenic' effects. It makes people *docile*. This characteristic of the drug led to its widespread use in Europe in aggressive populations long before its resurrection in the 1990s . . . by the 1990s up to 40 percent of hospital cases involved personality problems rather than the old-style schizophrenic or manic-depressive disorders . . . personality disorders, although not cured by neuroleptics, could be contained by them" (Healy, 2002, pp. 345–346, emphasis added).

The following concisely (as well as embarrassingly, for those believing in antipsychotics' specific "anti-schizophrenic" effect) summarizes their impact on experimental animals and normative humans: "Antipsychotic drugs, which are dopamine antagonists, produce behaviors suggestive of negative symptoms of schizophrenia in animals and humans free of mental illness" (Buchanan & Carpenter, 2005, p. 1338).

Antipsychotics were discovered during trials conducted on the use of some anti-histaminic compounds for surgical purposes. Surgical procedures constitute a considerable stress to the body, and the stress response itself can further decompensate the patient (Healy, 2002, pp. 78–79). Consequently, "Laborit had a hunch that stabilization could be produced by putting the body into a state of *artificial hibernation.*[2] When the body is made hypothermic, some components of the stress reaction can be prevented. But cooling the body is also stressful. The new antihistamines blocked compensatory responses to cooling. Their effectiveness in making both animal and human bodies poikilothermic led to a vogue for hibernotherapy" (idem, p. 80, emphasis added). (*Poikilothermia* means the inability to maintain constant body temperature during fluctuations in the ambient temperature—that is, to respond to undesirable cold or heat with shivering, sweating, etc.) Without neutralization of the compensatory physiological and behavioral responses to low temperatures, hibernation and related adaptations seem unimaginable in warm-blooded animals. Laborit experimented first with promethazine (a phenothiazine antihistaminic with a mild antipsychotic effect) but found that "chlorpromazine marked a significant advance on promethazine, particularly in causing indifference" (idem, pp. 81–82).

Taking in account the close relationship between hibernation, comprehensive suppression of instinctive intensities, and the compensatory physiological responses to low ambient temperature, the discovery of the "antipsychotic" effect of chlorpromazine (that is, its antipsychotic effect as interpreted in this work) was perhaps not as accidental as it appears at first sight. Hibernation, in its mild form, exemplified by winter depression; reduced instinctive activity, both behaviorally and in subjective experience; and the "antipsychotic effect" are all interconnected in the kind of evolutionary reasoning proposed in this text.

The wide-scale use of antipsychotics led to the recognition that they may induce, beyond an undesirable subjective experience, a whole psychiatric syndrome called "negative state" or "passivity syndrome" that was previously considered an integral part of the schizophrenic process. While the positive symptoms (agitation, aggressiveness, delusions, and hallucinations of recent onset, etc.) improve under antipsychotic treatment, a different but equally worrisome behavioral and mental state takes their place. It consists, in addition to the subjective phenomena already mentioned, of underactivity, slowness, lack of spontaneity, poverty of thought content, and withdrawal (Healy, 2002, pp. 268–270;

Healy, 2009, p. 14). Reducing the dose of traditional antipsychotic drugs, or substituting milder ("second generation" or atypical) antipsychotics, improves the negative syndrome as a rule.

From a neuroanatomical point of view, the dopamine-activity-reducing effect of antipsychotics (their indisputable primary effect) takes place in lower brain structures not concerned directly with thought processes or sensory activity. Herman van Praag, a Dutch psychiatrist best known for his contribution to biological psychiatry, found that the first symptoms that respond to antipsychotics are aggressiveness, agitation, and anxiety, and these effects parallel the changes these drugs induce in dopamine activity. No such correlation was found between the reduction of dopamine activity and the improvement of cognitive symptoms in psychosis. He concludes: "An increased dopamine turnover can be present in psychotic disorders but this phenomenon seems unrelated to the etiology of psychosis or the presence of 'genuine' psychotic symptoms such as delusion and hallucination" (van Praag, 1975, pp. 87–90). In other words, *the improvement of the characteristic psychotic symptoms cannot be seen as a direct effect of antipsychotics.*

All this having been said, we have to admit that we still are left with a great mystery regarding antipsychotic drug effect. While the suppressing effect of these drugs on psychomotor agitation seems to be an unquestionable clinical reality, a seemingly reverse effect of psychotropic drugs is also regularly observed. Withdrawn, autistic schizophrenics, possibly frozen in some bizarre catatonic body posture and stereotypically uttering some senseless or delusional verbalizations, can "wake up" under antipsychotic treatment and adopt a much more organized, adaptive, communicative behavior. They become more open to interpersonal transactions and to persuasion to conform to the daily routine of a psychiatric ward. David Healy describes some very impressive examples of individuals "waking up" from a long-lasting inaccessible psychotic state shortly after chlorpromazine administration (2002, 91–92). Less deteriorated patients "report beneficial effects on their ability to focus or concentrate on things. Subjects may find themselves more *alert* mentally, more able to *focus on tasks that need doing,* less in a daydream, less distracted by internal dialogues, strange thoughts, or intrusive imagery. The voices, thoughts, or obsessions may be described as being still present but having receded from center stage" (2009, p. 12, emphasis added).

How do we explain this puzzling paradox, which has for so long baffled efforts to translate the relatively simple impact of antipsychotic drugs on one or several neurotransmitters into a comparably simple effect on behavior and subjective experience, as Occam's parsimony principle demands? I will offer here my own attempt to solve this puzzle.[3] It seems that intense instinctive activity, experienced as strong arousal, emotion, or intention, *may either activate or alternatively inhibit* other mental processes and their behavioral or verbal expression. Since the ability of strong emotions to stimulate or activate behavior is taken for granted, I will present examples of their opposite effect. In normative persons, strong emotions of a pleasurable or painful nature may inhibit fluent speech, the usual level of intellectual output, and fine visuomotor coordination (think of certain sports like target shooting or pool), and may even completely paralyze motor activity or speech for short periods ("I remained speechless").

In neurotic disorders, like some conversion disorders, dissociative states, or panic attacks, this inhibition is much more pronounced. In general, the more normative, integrated, and equilibrated a mental apparatus, the more it can support intense internal tensions without considerable upset in its functioning. For example, a person with adequately differentiated reactive instincts may respond to a dangerous situation with a well-differentiated fear response. Even if it is not entirely appropriate for coping with the respective danger, the behavior will make sense, at least in an evolutionary context, and will be understandable through empathy by other normative people. On the other hand, intense anxiety may lead either to adaptively ineffective agitation or, alternatively, to paralysis of the overt behavior. An anxiolytic drug may suppress *both* of these states, at least partially and temporarily, and at the same time restore, as a *secondary* consequence of this primary effect, some better integrated and more socially adaptive behavior. Catharsis, in the form of an emotionally charged recount of the anxiety-producing event to a sympathizing person (a therapist, a close friend, or family member), may have a similar effect. A strong dysphoric reaction to some unexpected, tragic news can also induce either disorganized agitation or a reticent silent breakdown (the latter being more frequent in men, in my experience).

It would be instructive to compare in this regard a characteristic manic or hypomanic state with that of a characteristic acute schizophrenic psychosis. In a typical, good-prognosis, manic or hypomanic episode in a previously well-functioning person, the instinctive predispositions are normatively diffuse and the mental apparatus is well integrated. When such a mental apparatus finds itself under considerably increased instinctive tensions, the previously normative behavior at first becomes more accelerated (altered only quantitatively), and, later, with further intensification of the instinctive tensions, the behavior expressing these instinctive motives (sexual, self-assertive, or aggressive) will gain preference over its adaptive, conforming, or submissive aspects, leading to gradually increasing conflicts with the social environment. At the peak of the manic attack, the behavior may become disorganized and psychotic symptoms (predominantly mood-congruent) may appear. No withdrawal or serious inhibition of the overt behavior (motor or verbal), so familiar in schizophrenia, is characteristic in this condition.

In schizophrenia, however, the situation is very different. As hypothesized throughout this text (and further detailed in the next chapter), the innate predisposition in the schizophrenia spectrum of disorders consists of a more than normatively diffuse instinctive configuration. Some repercussions of this deficiency are: inability to select the relevant social information; inability to figure out through the mechanism of empathy others' emotionally charged intentions; and, in consequence, inability to interact normatively with surrounding persons on this emotional level. *Diffuse drives bring about a diffuse arousal of the mental apparatus, which leads to confusion about how one intends—as well as how one is expected—to behave, especially in emotionally or intentionally charged interpersonal encounters.* This mechanism, in my opinion, is responsible for the shy, awkward, confused, sometimes bizarre quality of pre-schizophrenics, which can be detected long before the appearance of the characteristic psychotic symptoms. When these diffuse drives become overwhelmingly intense for whatever reason, such as hormonal changes at puberty or the increased expectations placed upon a maturing person to function in diverse social roles, they can lead, as in intense anxiety, either to undifferentiated behavioral activation or to undifferentiated behavioral inhibition.

I came upon a psycho-physiological finding that may give support to the hypothesis that arousal in schizophrenics may be associated with behavioral withdrawal. P. H. Venables found that schizophrenic patients without coherent delusions showed strong correlation between psycho-physiologic arousal and behavioral withdrawal, the most aroused being the most withdrawn (Kaplan et al., 1980, p. 1135). It must be noted, however, that Venables's conclusions were criticized on methodological grounds (idem, pp. 347–348).

6.2.2. Indirect Effects of Antipsychotic Drugs

To be sure, the truly direct effect of antipsychotic drugs is their modulating effect on the activity of some neurotransmitters—primarily dopamine—at the nerve synapses. These neurotransmitters may also affect, besides the behavioral and emotional domain, neurologic and other somatic mechanisms that will not be considered in this text. However, since we are concerned here with behavioral and mental states rather than with the neurochemical events induced by the drug, the term *direct or primary effect* refers to those behavioral and mental changes that can be correlated in the most direct, straightforward way with the neurochemical events. As argued in the previous section, this direct effect, in the context of our discussion, is the reduction of the energetic intensity of a wide range of active and reactive instinctive activity and the reflection of this effect in the subjective experience. This effect is exerted on a brain center or mechanism hierarchically at a higher level than those regulating discrete instinctive activities. As we have seen, this center's function is to adjust the overall behavioral intensity of an animal to its environmental circumstances, and, when needed, this behavioral intensity (including the intensity of the body's concomitant physiological processes) is modulated according to predictable rhythmic variations in the abundance of the environment's life-sustaining resources.

This direct behavioral effect of the antipsychotic drugs has *secondary consequences or repercussions*. While the identification of the primary effect is naturally very important, the secondary repercussions are equally relevant to the concerns of this text (which deals with human behavioral diversity and its relevance to psychopathology). They are also more variable and unpredictable than the primary effect, as discussed in section 6.1 of this chapter.

Some of the secondary effects have already been noted. For example, Herman M. van Praag found (section 6.2.1) that the clinical symptoms whose improvement correlated directly—that is, appeared simultaneously with changes in the dopamine activity—were the reduction of agitation, aggressiveness, and anxiety; in our context, this reduction has to be seen as the direct consequence of these drugs' neurochemical effects. On the other hand, characteristic psychotic symptoms (delusions, hallucinations, thought disturbances), the improvement of which were more unpredictable, frequently only partial, and not correlated temporally with the drug-induced changes in the dopamine activity, have to be regarded as *secondary repercussions* of the direct effect. The possible mental mechanism that may be responsible for the reality-distorting effect of strong emotions in psychosis will be dealt with in section 7.2.6.4.

The next and probably most impressive example in this respect is the seemingly paradoxical effect of antipsychotics on the retarded versus agitated schizophrenic. As proposed, this paradoxical effect is caused by the ability of intense, diffuse instinctive tensions to either augment or alternatively inhibit overt behavior. In other words, the direct antipsychotic effect is the same in both alternatives, that is, the reduction of instinctive intensity. However, the *final outcome expressed in overt behavior, the observable effect, depends on whether the increased emotional intensity leads to activation or inhibition.* These effects, however, must depend on additional mental (brain) mechanisms that are not directly affected by the antipsychotics.

As mentioned, the characteristic psychotic symptoms—delusions, hallucinations, disordered thinking—do not respond uniformly to antipsychotic treatment. In some cases, they may disappear completely after several weeks of treatment; in other cases, we can see only a "deemotioning" effect. The complete disappearance of psychotic symptoms is a likely outcome in acute cases such as mania or a brief psychotic episode. In mania, the close interconnection between the strong emotional arousal and the psychotic symptoms is quite apparent, since the psychotic symptoms are mostly *congruent* with the prevailing mood (megalomanic delusions, for example). Therefore the disappearance of these psychotic symptoms when the intensity of the instinctive activity is reduced seems to be a close repercussion of the drug's primary effect.

In schizophrenia, on the other hand, the relationship between emotions (reflecting instinctive activity) and psychotic symptoms is less direct and more complex. As argued in the next chapter, in these individuals, as a result of pronounced instinctive diffuseness, a *partial dissociation*, rather than the normative integration, takes place between the mental and behavioral strivings induced by these diffuse instinctive predispositions, on one hand, and behavior induced by the requirements of social coexistence to adapt, conform, or submit, on the other.

Therefore, in schizophrenics, additional mental mechanisms are intercalated between the drug-induced changes in the instinctive intensities and overt, observable changes in psychopathology. Consequently, the antipsychotic drug effect on the characteristic psychotic symptoms in this case is a more remote secondary consequence than in mania, leading to different outcomes. This subject will be detailed in the section on schizophrenia in chapter 7.

The last secondary effect of antipsychotic drugs I would like to mention here is their ability to promote the discharge of patients from hospitals as we saw earlier in this section. This topic will be touched upon here only from the point of view of the secondary effects of antipsychotic drugs, omitting the ideological or humanitarian aspects. Hospitalization (especially compulsory hospitalization) of psychotic individuals is justified, as a rule, by the disruptive effect of the patient's behavior on the social coexistence in his narrower community, often due to gross transgressions of the community's moral code and desirable ways of behavior; by the dangers the patient's behavior may pose to himself or surrounding people; by the great economic or psychological burden on caretakers; and so on. Most families and communities will not oppose the patient's discharge, provided that the behavioral disturbances have been considerably ameliorated during hospitalization, without expecting a complete recovery. (In the context of the present work, this attitude is a further instance of relaxation of intragroup selection pressures.)

It is not difficult to conclude that some of the problems schizophrenic behavior poses on normative social coexistence do *ameliorate* as a result of antipsychotic treatment. Even if the dissociation between the idiosyncratic, unrealistic, diffuse, instinct-induced pathological behavior and the shallow, socially induced, submissive behavior persists, the power balance between them will change in favor of the latter one, thus considerably decreasing interpersonal tensions. The individual becomes less agitated, less aggressive, less negativistic. In general, under the "indifference inducing effect" of antipsychotic drugs, the patient may stop or considerably reduce the uninhibited expression of dissociated internal motives (most importantly, the self-destructive and dangerous ones). The spontaneous, unsolicited expression of the bizarre thought content will become less intense and less frequent. Bizarre cognitive contents will surface mostly during therapeutic sessions, when the individual is asked directly about their presence. She will become more open to persuasion to conform to some basic daily routine, rules, and prohibitions, which will enable her to live outside the hospital in sheltered conditions. While this situation is a much less impressive therapeutic achievement than those possible in some other medical specialties, for many patients, it is a definite improvement in living conditions over the alternative that existed before the introduction of antipsychotic medication—long-term hospitalization in a chronic psychiatric ward.

It is interesting to note how closely chlorpromazine action on rat behavior parallels the changes in schizophrenic behavior under antipsychotic treatment. The experimenters tested both a lower and a higher dose: "At both levels there is a slight reduction in *mating* behavior and *aggression* is even more reduced by both doses. *Submission* is very little interfered with but the larger doses increase very markedly the *escape* activity and there is a simultaneous further decrease in *attack* and increase in *flight*, but hardly amounting to a disruption of the social elements" (Chance & Silverman, 1964, p. 77, emphasis in the original).

Summarizing how the *four comprehensive instinctive mechanisms* discussed in the previous chapter are affected by the antipsychotic drugs—which ones are directly affected, which secondarily, and which not at all—we can conclude the following:

1. *The direct effect of antipsychotic drugs on behavior and related subjective experience consists of a reduction of the overall energetic intensity fueling active and individual reactive instinctive predispositions.* The hypothesized target of this effect is the brain mechanism (or mechanisms) that adjust this overall energetic level in order to maximize fitness in a specific environmental niche, as well as when appropriate, to the fluctuations due to seasonal variations in the amounts of life-sustaining resources.

2. *The three forms of frustration of active instincts and their reflection in subjective experience are not affected qualitatively by these drugs, but their intensity is reduced* in accordance with the drugs' primary effect. The euphoria, and resultant megalomanic distortions of reality, for example, do not change into euthymia, aggression, or dysphoria under antipsychotic treatment, only the intensity of the related behavior, of the subjective experience, and of the associated mental

events decreases. David Healy makes a similar observation: "All emotions may be blunted [by antipsychotic drugs], rather than just certain emotions that have been troubling. Many takers complain that all feelings, from joy to anger, are dulled" (2009, p. 34). A confusing factor here may be the following: It is difficult to distinguish the emotional tone induced by the primary effect of these drugs on the mental state from the reaction to the various side effects (mental, neurological, or somatic), as well as from the reaction toward the whole setting in which the treatment takes place (the psychiatric ward, the relationships with the personnel and with other patients, the compulsory or voluntary nature of the hospitalization, etc.). However, I think that, in general, the following can be said: Patients (with previous experience concerning antipsychotic effect) with a predominantly *euphoric* mood, as a rule, strongly oppose taking antipsychotic drugs (while they may at the same time consent or even ask for other kinds of medication, such as sedative/hypnotics or somatic medication), presumably as a result of not wishing to lose an intense, highly pleasurable subjective experience. Patients with an *aggressive* mood, like paranoid psychotics, also oppose antipsychotic treatment on a regular basis. I presume the reason for this opposition is the concern that the reduction of aggressive instinctive intensities will leave them powerless in the face of environmental pressures and dangers (imagined or real). Sometimes, in my experience, this effect was so strong that a transition from an aggressive attitude into an anxious one was clearly observable. With regard to predominantly *dysphoric* patients, as far as could be judged, they tended to welcome the antipsychotic treatment's ability to exchange indifference for intense mental pain. In any case, I have seen in dysphoric patients less resistance to antipsychotic treatment than in the other two categories.

3. *The degree of diffuseness versus differentiation of instinctive predispositions is clearly unaffected by antipsychotics* or for that matter by any other therapeutic means known to me.

4. *The effect of the antipsychotic drug on the active/reactive/conforming behavior interrelationship consists, as we have seen, in altering the balance in favor of conformity with external, social pressures* (as a secondary consequence of suppressing active, as well as egocentric reactive motives, particularly counteraggression).

6.3. Antidepressant Drug Effect

6.3.0. Introduction

Before entering the discussion of the antidepressant effect, the reader should be reminded of the distinction made in the previous chapter regarding the existence of *two different types* of depression, which were named the lethargic and the dysphoric ones. The first type represents considerably low intensities of the overall instinctive activity for a more protracted period, while the second type is induced by active instinctive motives in their dysphoric form of frustration. When this condition is intense and longstanding, it leads to the clinical picture of *agitated* depression. The nonrecognition by clinical psychiatry of the different nature of these two kinds of depression induced, in my opinion, by widely different evolutionary mechanisms, causes great confusion in the evaluation of antidepressant drug effects. For example, while antidepressants that energize behavior may improve the lethargic type of depression, they may aggravate the dysphoric type. (In the same vein, antipsychotics, which may worsen the lethargic form of depression, may exchange a blunted affect for an intense dysphoric mood, an outcome that may be welcomed by the individual, as mentioned earlier.)

Therefore, in order to avoid confusion, the antidepressant effect will refer in this text exclusively to an effect that *increases* the energetic level of the overall instinctive activity. This effect, in a clear, unambiguous form, can be claimed only in the case of the noradrenergic antidepressants (that is, antidepressants whose action involves the norepinephrine neurotransmitter at the neuronal synapses).

If antidepressants cause an *increase* in the energetic aspects of both active and (egocentric, not submissive) reactive behavior, it seems obvious their effect is directly opposite to that of antipsychotic drugs. Numerous clinical data, as well as neurochemical findings, support the presupposition that antipsychotics and noradrenergic antidepressants have opposite effects on catecholamine neurotransmitters as well as on behavior. The former group decrease behavioral intensity, the latter increase it. Antipsychotics indirectly suppress psychotic symptoms; the noradrenergic antidepressants augment them. Antipsychotics improve acute manic episodes; antidepressants aggravate them, or even induce a switch from lethargic depression into mania.

The activating effect of noradrenergic antidepressant drugs has been demonstrated in animals, in normative humans, and in mental patients irrespective of the clinical diagnosis, making it more convincing that their primary or direct effect is on a basic phylogenetic mechanism with a circumscribed brain substrate as their somatic base.

The *serotonergic* antidepressant effect on behavior and subjective experience seems multifarious, possessing even contradictory or irreconcilable aspects, thus making its interpretation along the lines of reasoning used in this work problematic. Clear indications exist that the serotonergic effect can activate behavior ("disinhibition is probably what underpins many cases of apparent manic reactions on SSRIs [selective serotonin reuptake inhibitors], which clear up on discontinuation of treatment"—Healy, 2009, p. 76). However, it seems that these drugs also possess the opposite effect, and can cause emotional blunting not dissimilar to that of the antipsychotics' effect. For example, some individuals complain of an inability to cry when the circumstances would justify crying (idem, pp. 76 and 163). A more comprehensive statement with a similar message is the following: "All motivated and active emotional behaviors, including feeding, drinking, sex, aggression, play, and practically every other activity (except sleep), appear to be *reduced as serotonergic activity increases*" (Panksepp, 2005, p. 111, emphasis added). It is obvious that this statement is exactly the reverse of what is proposed in this text as a direct antidepressant effect.

Another puzzling effect of serotonin is that serotonin activity is considerably higher in dominant than in nondominant primates. Moreover, when such an animal loses its dominant position, serotonin activity decreases and vice versa (Kramer, 1993, pp. 301–302).

It is claimed that SSRIs possess an inherent anxiolytic effect (Buspiron was marketed especially for this indication). Other serotonergic drugs are licensed for treating various anxiety disorders (social phobia, generalized anxiety disorder, obsessive-compulsive disorder, panic disorder). However, it is common knowledge that, paradoxically, serotonergic drugs may cause mental tension and sleeplessness (Healy, 2009, p. 70), in which case their combination with sedatives or hypnotics is indicated.

Serotonergic drugs (similarly to the noradrenergic ones) can increase suicidal ideation or behavior and rarely lead to violent acts (Kramer, 1993, pp. 301–313). The presupposition in this work, as mentioned, is that, in these cases, *rather than the apathy characteristic of lethargic depression, an already intense dysphoric mood or a combination of dysphoric and aggressive moods was further intensified by the antidepressant drug.*

Summarizing the serotonergic effect, it seems that it has both excitatory as well as inhibitory effects on both active as well as reactive behavior; it increases active instinctive intensity but may lead also to blunted emotions; it possesses sedative effects but can cause nervous tension and sleeplessness, too. The professional literature attempts to explain these inconsistencies in the serotonergic action by the existence of different types of serotonergic receptors, each with different effects on behavior. Since I cannot integrate the serotonergic effect in its entirety in the conceptual framework of this text, in the following, the antidepressant effect refers to the increase of active and self-centered reactive instinctive intensities, which is seen in its pure form in the case of noradrenergic antidepressants and partially in serotonergic (and some other antidepressant) drugs.[4]

It should be mentioned that dopamine, another catecholamine neurotransmitter, albeit one not used as an antidepressant in clinical practice, has similar energizing (and psychosis-inducing or aggravating) effects to noradrenaline. Dopamine closely resembles noradrenaline in its chemical composition. Noradrenaline is synthesized from dopamine by a one-step chemical reaction (oxidization) at synaptic storage vesicles (Sadock & Sadock, 2005, p. 53). English neurologist and clinical writer Oliver Sacks movingly described in his book *Awakenings* (1973) how the then newly discovered, dopaminergic antiparkinsonian drug L-DOPA in high doses succeeded in "awakening" patients "frozen" for decades by encephalitis letargica (sleeping sickness).

The energizing effect of catecholamines is also recognized when seen from a wider neurobiological point of view: "One of the normal functions of this system [the mesolimbic dopamine circuit] is to energize appetitive behavior" (Panksepp, 2005, p. 118). It is worth mentioning here that the antidepressant Bupropion (Wellbutrin) inhibits noradrenaline and dopamine reuptake (that is, increases their effect at the respective synapses) and was found to be effective in preventing major seasonal depressive episodes (Sadock, Sadock & Sussman, 2006, p. 83).

6.3.1. The Primary Effect of Antidepressant Drugs

The ability of catecholaminergic drugs to increase the intensity of active and reactive instinct- fueled behavior was well recognized from the early days of the discovery of these drugs. Roland Kuhn was one of the first psychiatrists to test imipramine. He replaced chlorpromazine treatment with imipramine in schizophrenic patients, expecting a similar antipsychotic effect in view of the close resemblance of the chemical composition of these drugs. His trial led to the following results: "Many of the patients previously on chlorpromazine began to deteriorate, becoming increasingly agitated. Some appeared to be hypomanic. When one patient escaped from the hospital and rode into town on a bike in his nightshirt, singing at the top of his voice, the company decided to discontinue the study. . . . Patients [who responded to the treatment] typically had a general retardation in thinking and acting . . . had a general loss of interest . . . and were preoccupied by fixed ideas of hopelessness, guilt, and despair" (Healy, 1999, pp. 51–53).

A more complete list of symptoms that, in Healy's opinion, respond to antidepressant treatment is the following:

"Loss of energy

Loss of interest

Feeling physically run down or ill

Poor concentration

Altered appetite

A slowing of physical and mental functions" (2009, p. 53).

It has to be mentioned here that in one of Healy's books I encountered an idea that seems akin to my attempt here to divide psychotropic drug action into primary and secondary effects. He calls the primary effect *antecedent*: "There is a considerable amount of evidence that suggests that the antidepressants may act on something that is an antecedent for mood and other neurotic disorders" (1999, p. 34). And: "For the most part, improvements in the sadness, hopelessness, guilt, and suicidal thoughts that may go with depression seem to occur as a *reaction* to changes in things such as sleep, energy, and interest" (2009, p. 60, emphasis added).

6.3.2. Antidepressant Effect on Laboratory Animals

Researchers primarily use rodents for testing antidepressant effects. Some of the test procedures exhaust the instinctive energies, not in a comprehensive way as in lethargic depression, but by overtaxing the escape response of reactive instinctive behavior. Both the "Forced Swimming Test," as well as the "Tail Suspension Test," prevent animals from escaping from a stressful or dangerous situation. These procedures are continued until the animal relinquishes further attempts to escape and becomes motionless, evidently as a result of the exhaustion of the respective instinctive behavior's energetic resources. Antidepressants administered *for longer periods* (but not for a short period) *prolong the duration* of the escape attempts in the treated animals.

In another category of tests designed to evaluate the antidepressant effects, "loss of interest, fatigue and loss of energy" are induced by submitting the animals to chronic mild stress. The resultant lethargic state is indicated by decreased self-stimulation through intracerebral electrodes. This state is interpreted by the researchers as *anhedonia*. Again, a protracted treatment (but not a short one) with antidepressants reverses the situation. The same results were achieved whether the chronic stress was biological or social in nature. (A possible explanation for the long interval needed for the antidepressant and antipsychotic effect to occur is discussed in section 6.4.)

The similarity, as hypothesized in this text, of the brain mechanism that induces the exhaustion of instinctive energies in rats versus the mechanism responsible for lethargic depression in humans is questionable; however, the behavior-energizing effect of antidepressants in these animal models seems quite clear.

6.3.3. Antidepressant Effect in Healthy Volunteers

The emotional reaction of normative persons to antidepressant drugs was found to be variable. Some appreciated their effect, some not. When serotonergic drugs were administered and they "suited" the individual, they made him more "serene or mellow." Others felt that the serotonergic effect was "unhelpful." When a noradrenergic effect was appreciated as helpful by normative volunteers, it was because of what the takers described as a "useful increase in energy and drive." This effect was detectable within forty-eight hours (that is, within a relatively short time of taking the drug, compared to the longer detection time in mental patients and laboratory animals) (Healy, 2009, p. 61).

6.3.4. Primary and Secondary Effects of Antidepressants in Mental Patients

I think that dividing the antidepressant effect into primary and secondary may facilitate the understanding of these drugs' therapeutic potential (as well as their therapeutic pitfalls) even more than is the case with antipsychotic drugs.

It was argued in the previous chapter that the lethargic form of depression does not cause mental discomfort *per se*. It is a period of low overall instinctive energy resembling that of rest after sustained intense mental or physical effort, which has to be seen as an adaptation enabling the recovery of the brain or body's energetic resources.

However, lethargic depression may have *secondary consequences*, which are not an integral part of it and which may be utterly disturbing or painful. Imagine, for example, the possible emotional state of a hibernating animal that, instead of being well hidden and protected by an optimally chosen or prepared den, is exposed to disturbing, demanding, or dangerous environmental disturbances. While these situations call for immediate self-protecting reactions, the animal evidently lacks the necessary instinctive energies to carry out such behavior.

While this scenario is presumably rare among animals, a similar situation is more the rule than the exception among humans suffering from lethargic depression. The technological and cultural (human-made) environment more than compensates for fluctuating natural resources such as food, temperature, and light and expects year-round, stable functioning from its members. Long periods of inactivity or low-level activity frequently incur unwanted intrapsychic or social consequences. Low instinctive energies interfere with carrying out one's daily duties, like working, looking after young children, fulfilling social obligations, housekeeping, and so on. As a consequence, the affected person may suffer (secondarily) from anxiety or feelings of inadequacy or guilt, especially those who have high expectations of themselves (in Freudian terms, possessing a rigid, punishing superego). It can be expected that in lethargic depression the *intensity* of these emotions, and especially the ability to act them out self-destructively, is lowered by the same biological mechanism. On the other hand, the intrapsychic sensitivities may be increased as a result of weakness or a lack of the outwardly directed, goal-oriented, selective intentionality characteristic of normatively intense instinctive activity.

The primary antidepressant effect will intensify whatever form of active instinctive motive is present (either in the unfrustrated or one of the frustrated forms). *The effect on subjective experience is reflected as an increase in the intensity of that predominant emotion or intention.* Euthymic lethargic apathy may change gradually into an increasing internal prompting to be active and increased confidence in one's abilities to face the daily challenges of one's life, which in turn will alleviate anxiety and guilt feelings when present. Alternatively, if the respective mental apparatus has a strong predilection to accentuate one of the forms of active instinctive frustration (which is a variable *independent* of the mechanism of lethargic depression *per se*, as well as of the primary antidepressant effect), the overt behavioral consequence of antidepressant treatment will be very different. In the case of an inclination to accentuate dysphoria or aggression, for example, an individual recovering from lethargic depression, instead of making a good recovery, will show progressively intensifying signs of dysphoric or aggressive mood and behavior.

When antidepressants are administered to a patient who already suffers from the dysphoric, agitated (instead of lethargic) form of depression, the outcome may be even more undesirable. As argued in the previous chapter, a strong, clinically diagnosable dysphoric depression occurs when two preconditions are present:

1. *A frustrating agent or situation blocks the discharge of an important active instinctive motive (in its original, unfrustrated form) toward the biologically adaptive releaser.* The example given was the inability to continue parenting behavior toward a beloved child who has died.
2. *The person has an innate predisposition to accentuate the dysphoric form of instinctive frustration at the expense of other forms* (like displacement toward a substitute for the biological child, or aggression). Also implied here is the considerably reduced ability or inability to make flexible transitions between the different forms of frustration, as well as between the frustrated and unfrustrated forms of the same instinctive motive. Having no easy outlet, the energetic charge of

the respective drive is intense in agitated depression even before the initiation of antidepressant therapy. *The antidepressant drug in this case, in contrast to what its name implies, will accentuate, instead of reduce, the intensity of the mental suffering!* The related cognitive and motor aspects of behavior will be secondarily (or tertiarily) affected, too, leading to an increase in agitation, to thoughts or delusions of guilt, or to suicidality. This further intensification of the intrapsychic and behavioral processes, when the respective patient possesses appropriate executive abilities, may lead to crossing the boundary between expressing helplessly intense subjective suffering and taking more effective, drastic actions in order to put an end to this suffering. In my opinion, this is the mechanism underlying the widely recognized increase in suicidality around the ten- to fourteen-day mark of the antidepressant treatment, both noradrenergic and serotonergic (Healy, 2009, p. 78).[5] In those fortunately rare cases in which intense dysphoria is coupled with strong aggressive impulses (or the intrapsychic state represents a transition between them), and self-control and good executive abilities are present, preplanned homicide coupled with suicide may occur (Kramer, 1993, pp. 301–313).

The reader may appreciate the amount of confusion created when, in the rating scales of randomized clinical trials on antidepressant drug effects, the symptoms of lethargic versus dysphoric depression (as well as the primary and secondary drug effects) are not sorted out. When the results are statistically averaged, the scores of the two different forms of depression—which respond in contradictory ways to antidepressant therapy—may counterbalance or nullify each other. This situation may account for a puzzling paradox regarding antidepressant effects. In such clinical trials, the antidepressant therapeutic effect on clinical depression, as compared with the placebo effect, was found to be very *weak* (roughly a 50 percent positive response rate to antidepressant treatment compared with a 40 percent improvement for the placebo [Healy, 2009, pp. 55–56].) However, from sources other than randomized clinical trials, it can be inferred that antidepressants have a *strong* effect on behavior. They are able to energize the behavior of normative persons as well as that of animals; they can switch lethargic depression into mania; they can increase suicidal tendencies in predisposed persons; and they can aggravate psychotic states.

It can be argued that the same discrepancy also applies to the antipsychotic drug effect. While their expected therapeutic effect compared to placebo, at least in schizophrenics, is modest, they show powerful effects on behavior when appreciated independently of clinical considerations. They induce catalepsy in animals; reduce the level of activity in animals, mental patients, and normative volunteers; and in humans, at least, may produce profound indifference.

I can see only one possible explanation for this puzzle: The clinical trials do not measure differentially what these drugs do directly, primarily, to behavior. Instead, they measure mostly therapeutically desirable effects on the whole, descriptively defined dysfunctional behavioral complex (which represent mostly secondary, tertiary, etc., repercussions of the primary pathological predisposition).

6.3.5. Antidepressant Effect on Other Major Psychiatric Disorders

In *bipolar disorder*, as we have seen, antidepressants can induce mania in a lethargic, depressed patient or intensify an already present manic episode. In my experience, the form of frustration of the active drives by and large remains the same. A predominantly elevated mood will become more intense euphoria, an aggressive manic patient will become more aggressive, and a mixed-dysphoric state will also intensify in its original form. However, the intensified instinctive energies can switch from a relatively harmless verbal *expression* of the underlying emotions and intentions into a more harmful *acting* in accordance with these mental motives.

In the same vein, *paranoid patients* under antidepressant treatment can switch from a defensive attitude in the face of imagined or exaggerated external threats to an offensive attitude, possibly involving violent actions.

In schizophrenics in remission or deterioration, the "deemotionalized" delusions (that is, deemotionalized either by the antipsychotic treatment or as a result of the natural course of the disease) tend to be "reemotionalized" when antipsychotic treatment is replaced by or supplemented with antidepressants (as an attempt to ameliorate negative symptoms), and, in consequence, acted out with therapeutically undesirable consequences. See Roland Kuhn's experiment with imipramine, described in section 6.3.1.

In schizophrenics (as in bipolar disorder) the antidepressant does not change the quality of the underlying mood. Predominantly euphoric schizophrenics with grandiose delusions will remain euphoric. A withdrawn paranoid schizophrenic may become agitated, accusatory, or violent under antidepressants, while one with a predominantly dysphoric mood may act out frustrated instinctive motives as unforeseeable self-injury or suicidal attempts.

6.3.6. Antidepressant Effect on Other Comprehensive Instinctive Mechanisms

As mentioned, no psychotropic drugs I am aware of, including antidepressants, can change the position of an individual's instincts *on the differentiated/diffuse scale*. However, when *normatively differentiated* instinctive predispositions, well integrated with ontogenetically acquired knowledge, are augmented by antidepressants, this may lead to intensified, purposeful, socially or individually useful forms of behavior (as can be seen in many depressed outpatients of the lethargic kind). If, on the other hand, *diffuse* instinctive predispositions are intensified by antidepressants, the result will be more intensely disorganized, maladaptive behavior, more disrupted from the surrounding social realities (that is, the reverse of the antipsychotic effect).

With regard to the *balance between active and egocentric reactive instincts*, on the one hand, *and socially expected conforming or submissive behavior*, on the other, the antidepressant effect may lead to an imbalance by accentuating individualistic active and reactive strivings at the expense of socially adaptive, conforming, or submissive elements (again the reverse of the antipsychotic effect). It is needless to say, of course, that these "antidepressant" effects are secondary repercussions of their direct, instinctive intensity-enhancing action.

6.4. The Delay in the Antipsychotic and Antidepressant Effect

The monistic belief in a strong causal interrelationship between mind, behavior, and brain, which characterizes behavioral sciences, presumes that when a biological treatment modality alters some relevant brain function or structure, within a short time behavioral and subjective experiential changes have to follow. In the case of psychoactive drugs, this relevant brain substrate is the neurotransmitter system. When its activity is altered by the drug, an instant change in behavior or subjective experience is expected to occur.

In the case of certain psychoactive drugs, like sedatives, hypnotics, narcotics, psychostimulants, and aphrodisiacs, this is indeed the case. However, the two drug categories discussed previously in this chapter, antipsychotics and antidepressants, are exceptions. In their case, the therapeutic effect on a mental syndrome or disorder is delayed by several weeks. This phenomenon is even more puzzling than it seems at first glance. Some of the effects of these drug categories on the mental apparatus (other than those defined as their primary or direct effects) may take place shortly after the drugs pass the blood-brain barrier. For example, the sedative "side effect" possessed by some representatives of both drug categories, or the anti-orgasm and ejaculation-delaying effect of serotonergic drugs and clomipramine (the latter also being a central, not peripheral effect; see Healy, 1999, pp. 208–209, 218), are instances of such immediate effects. Even more puzzling is the finding that the change in the subjective experience induced by antipsychotics and antidepressants in normal volunteers—a subjective experience reflecting their characteristic primary effect, that is, increased indifference or energy, respectively—appears within forty-eight hours of ingesting the drug and sometimes even earlier (Healy, 2009, pp. 9 and 61).

On the other hand, the alleged "anxiolytic" effect of some serotonergic antidepressants (Buspiron, for example, is marketed for this indication) takes two to four weeks to manifest itself (idem, p. 162), as opposed to benzodiazepines, barbiturates, alcohol, and narcotics, in which the anxiolytic effect is instantaneous once the drug contacts its respective target in the brain (benzodiazepine receptors, opiate receptors, etc.). In addition, the reader may remember that the previously discussed energy-enhancing antidepressant effect in laboratory animals appears only after protracted administration of the drug, not instantaneously (section 6.3.2). How can one make sense of these irreconcilable facts?

No doubt exists that the effects of these drugs (the antipsychotics and antidepressants) on the neurotransmitter activity are instantaneous after they contact the respective nerve synapses: "Reuptake-inhibiting antidepressants blocked reuptake mechanisms within an hour or so of the drug's being consumed, yet recovery took several weeks to appear" (Healy, 1999, p. 161). Previous attempts to explain the delayed effect of these drugs, so far as I know, argue for some kind of time-consuming "changes in the presynaptic and postsynaptic receptors" as a precondition for the therapeutic effect to take place (Nelson, 2005, p. 2957; Healy, 1999, pp. 161–169). However, it seems puzzling that sedatives acting on benzodiazepine receptors or narcotics acting on opioid receptors do not need those receptor changes in order to express their effect on behavior and subjective experience. Antiparkinsonian dopaminergics acting on the D2 dopamine receptors, like bromocriptine or pergolide, also lack the delay in their action (Victor & Ropper, 2001, p. 1134), while antipsychotics with antidopaminergic activity on similar D2 receptors (with affinity to the mesocorticolimbic dopamine system instead of the nigrostriatal one) do possess delayed action (Sadock & Sadock, 2005, p. 2823). Stimulant drugs like methylphenidate and pemoline, which act through release of dopamine from granular pools and blockage of reuptake, also have an instantaneous effect (idem, p. 2939).

In fact, the whole hypothesis of receptor change as a cause of delayed action of these drugs has been questioned, since it was found that such changes take place more quickly than was previously thought: "It is now apparent that such views of receptors were quite mythical, in that receptors themselves are in a state of constant flux and drugs affect their number and configuration within half an hour to an hour of their being administered—that is, as quickly as they act on reuptake mechanisms" (Healy, 1999, p. 161).

How can we make sense of the conflicting and confusing data concerning the time needed for some psychoactive drugs to exert their effect on behavior and subjective mental experience? While contemplating this puzzle, I began to realize that a parallel exists between the length of time a psychotropic medication needs in order to exert its effect on behavior and mental state, on one hand, and the speed with which the respective behavioral system or mechanism that is the target of the drug effect has to respond in order to secure adaptive gains, on the other. As will be argued in the following about sedative/hypnotic drugs, their effect in therapeutic doses concerns mainly reactive behavior. They attenuate and slow down the reaction to undesirable or dangerous stimuli. The reactive mental apparatus, in order to be adaptively effective, has to react instantaneously to danger-indicating environmental stimuli. Similar, though less extreme, considerations apply to behavior expressing specific active instinctive motives, like sexual or feeding ones. They also have to be available within the relatively short time interval during which the appropriate releaser is contacted in order to be effective. Such opportunities may be fleeting, or potential competitors for the same resource may be present. Therefore, it seems to me reasonable to presume that psychotropic drugs effects that bring about changes in one specific active behavioral inclination, like aphrodisiacs (sexual desire stimulants) or the ejaculation-delaying effect of serotonergic antidepressants, will be quickly apparent in behavior or subjective experience.

However, the pace of change brought about in behavior and physiology by the mental mechanism that, according to the hypothesis of this text, constitutes the target of antipsychotic and antidepressant effects, is an entirely different story. The transition from one season to the next is slow and may be irregular, depending on the weather. The changes reflecting a reduction or increase in life-sustaining resources (light, temperature, abundance of plants, or prey animals, etc.) happen slowly over a period of weeks or months. Animals also need time to prepare for winter conditions. They have to find or prepare a den and must gradually reduce the overall intensity of their behavior and physiology. With the arrival of the warm season, the transition to intense activity will happen at a similar pace. I read with understandable delight a similar line of thought in the following excerpt from David Healy concerning the antidepressant effect: "Clinically for the first two weeks of treatment, the standard view is that antidepressants do very little except cause side effects. The change that then occurs seems to creep up on people rather than to sweep in on them. . . . What usually happens is that there is a *slow* increase in energy, a *slow* return of interest, a *mild* increase in appetite, and an improvement in sleep. These occur *gradually* rather than clearly and they may be patchy, for example, with one good night's sleep followed by a poor one the night after. Rather *like a slow change of season*" (Healy, 2009, p. 60, emphasis added). By now it has to be clear to the reader that, to my mind, that last sentence is more than a metaphor.

In other words, in my opinion, *antipsychotic and antidepressant drugs begin to exert their effect, shortly after the treatment begins, on the mental mechanism that is able to change the overall intensity of instinctively fueled behavior according to seasonal variations. However, as with its reaction to seasonal changes, this mechanism reacts to the drug's effect slowly.* Cooperative volunteers with a normative behavior may sense even a small effect on the subjective experience (increased energy or indifference, respectively) from the first days of taking these drugs. However, for an individual in a state of lethargic depression of clinical proportions or in an acute manic or other acute psychotic state, the *quantitative change* needed in the overall intensity of behavior, in order to be considered clinically significant, has to be much more pronounced than the slight change in instinctive intensity and subjective experience that a cooperative volunteer is able to detect. Since the brain apparatus responsible for inducing this change works slowly, the expected clinical improvement, or the switch into the opposite extreme, from depression to mania, or from "deemotionalized" partial remission of a schizophrenic psychosis into a "reemotionalized" acute psychosis under antidepressants, takes time. The same is true in the case of laboratory animals. To document a clear, statistically significant, quantitative difference in the respective behavior's intensity with drug treatment, as compared to its natural baseline, requires continuous administration of the drug for a more prolonged period.

Note that the average delay in the therapeutic effect *is similar* for these two drug categories (antidepressants and antipsychotics). Both need a period of two to six weeks for their effect to take place, while in the case of the atypical antipsychotic clozapine, even more time is needed. This fact supports the presumption that *both drug categories act on the same brain mechanism but in opposite directions.* The exact time it takes for a specific drug to influence the mental state of a particular patient probably depends on multiple variables, including the extent of the deviation from normative instinctive intensities at the beginning of the treatment, or the speed with which the mental apparatus responsible for the fluctuations of instinctive intensities reacts to relevant influences. Remember that the sensitivity of this apparatus to environmental clues (and perhaps also to drug effects) is diversified in humans. It merits mention in this context that light therapy—effective in winter depression, and to a lesser extent in non-seasonal affective disorders—also needs several weeks to alleviate the depressive symptomatology. It seems unquestionable that the amount of daily light modulates (entrains) the mechanism of the brain responsible for adjusting the behavioral intensity to the abundance of seasonally fluctuating amounts of life-supporting resources.

As to the claim that the serotonergic drug Buspiron has an anxiolytic effect (but without sedation) that unfolds slowly over a period of two to four weeks, this seems to me to be a marketing gimmick. As described earlier, serotonergic drugs have a confusing mix of sedative, energizing, and emotion-blunting effects. Therefore, it seems that the claimed anti-anxiety effect of Buspiron, which takes several weeks to appear, is in fact an emotion-blunting one resembling that of anti-psychotics: "Clinically, it is very clear that when SSRIs work they make the taker more mellow, more docile and more serene or sanguine. In some cases, they do this to a greater extent than is desired, leading to complaints of emotional blunting or numbness" (Healy, 2009, p. 163; see also p. 76).

The anti-orgasm and ejaculation-delaying effects of serotonergic drugs have to target an entirely different behavioral (and brain) system than their antidepressant and emotion-blunting effects. Instead of a delay of several weeks, these effects take place shortly after the drug is absorbed into the bloodstream, and their duration is short-lived. In this case, however, the target is one single active instinctive mechanism (the sexual one) instead of the overall energetic level underlying a wide range of instinctive activities. As previously argued, discrete active instinctive systems have to respond promptly in order to be efficient.

In summary, I think that the timescale of the psychotropic drug effect depends at least as much on the functional variables of the evolved behavioral mechanism targeted as it does on the respective neurotransmitter system through which this effect is achieved. It seems that the respective neurotransmitter by itself indicates or defines only very approximately the brain area involved. However, when, in addition to the kind of neurotransmitter targeted, the respective drug's affinity to specific brain structures is also known (for example, antiparkinsonic drugs' affinity to the nigrostriatal dopamine system and antipsychotic drugs' affinity to the corticomesolimbic dopamine system), this combination of knowledge may improve the accuracy of our guess regarding which evolved behavioral mechanism the respective drug is influencing.

6.5. Anxiolytics, Sedatives, and Hypnotics

Drugs with anxiolytic, sedative, and hypnotic effects (in short, "sedatives") are less central in psychiatric practice than the antipsychotic (AP) and antidepressant (AD) drugs, since they are not believed to target a circumscribed clinical disease entity or psychiatric syndrome. They are widely used in any disease in which mental tension, agitation, or sleeplessness may aggravate the clinical picture, as well as for relieving tension and promoting sleep by the normative population. However, the interpretation of their effects on the mental state and behavior is important in the context of this work. Since the primary sedative effect is entirely different from the direct antipsychotic and antidepressant effects as understood here, contrasting the actions of these three drug categories may shed additional light on their unique effects, as well as on some of the already discussed phylogenetic behavioral mechanisms.

The psychoactive substances to which I refer here are the barbiturates, the benzodiazepines, and alcohol. Alcohol is not considered a medication, but its effect on behavior, subjective experience, and brain chemistry resembles that of the barbiturates and benzodiazepines enough that it can be seen in the context of this text as belonging to the same category of psychoactive substances.

The most comprehensive formulation of the direct or primary effect of these substances on the functioning of the central nervous system, to the best of my knowledge, is the following: "Barbiturates depress the activity of all excitable tissues. The CNS is extremely sensitive to these effects; peripheral tissues are not. In the CNS, barbiturates produce varying degrees of depression, from mild sedation to general anesthesia. . . . Barbiturates depress the sensory cortex, decrease motor activity, alter cerebellar function, and produce drowsiness, sedation, and hypnosis" (Nemeroff & Putnam, 2005, p. 2777).

The neurochemical target of the action of these sedative substances is the gamma-aminobutyric acid (GABA) receptor. GABA and glycine are the major *inhibitory* neurotransmitters in the CNS. While GABA is comparatively scarce in the brain stem and spinal cord, and abundant in the cortex, the reverse is true for glycine (idem, pp. 62–63). Benzodiazepines influence the GABA receptor through the "benzodiazepine receptor" (GABA—Benzodiazepine receptor complex) (idem, p. 2782). Barbiturates also increase the inhibitory effect of GABA receptors on the CNS, as well as inhibit the glutamate effect. (The neurotransmitter glutamate has an excitatory effect on neuronal activity [idem, p. 2776].) Alcohol at clinically relevant concentrations acts similarly to barbiturates by "enhancing the inhibitory action of GABA-ergic neurotransmitters . . . and reducing the excitatory actions of glutamatergic neurotransmitters" (idem, p. 1159; see also Panksepp, 2005, p. 217).

The degree of depression of CNS activity by sedative substances depends on the dose, the method of administration (parenteral or oral), as well as on the functional state of the particular mental apparatus at the time of the drug's administration. The progressively increasing degrees of depression brought about by these drugs on the behavioral and mental functions can be categorized in the following way:

1. Anxiolytic effect
2. Hypnotic effect
3. Sedation leading to drowsiness
4. Depression of consciousness of varying degrees from sopor to coma

Other consequences of these substances' inhibitory effect on the nervous tissue include anticonvulsant and musculorelaxant effects, amnesia, and reduction of motor coordinative abilities, which are not considered here. Our main topic is the primary as well as indirect anti-anxiety effect of sedative substances.

6.5.1. Anxiolytic Effect

The main idea here is that the sedative substances in anxiolytic doses reduce the intensity of the reactive behavior alone (more exactly its freezing and fleeing away aspects). As a secondary consequence, they may even *enhance or disinhibit* behavior fueled by active instincts when the two kinds of behavior are conflicting and the active instinctive expressions were formerly suppressed by reactive behavioral arousal. Considering, without clinical preconceptions, the effect of sedative substances on behavior, this statement seems obvious. Think of the *disinhibitory* effect of alcohol intoxication, or of some of the so-called "paradoxical" effects of benzodiazepines and barbiturates: agitation, aggression, violence.

From a theoretical point of view, however, this argument is in need of explanation. After all, both active and reactive behavior are the outcome of an interaction between internal promptings and some specific environmental agent or situation; both need the ability to recognize quickly the relevant environmental stimuli and the prompt employment of fitness-enhancing behavioral strategies. If this is so, a question arises: Taking into account that sedatives depress the CNS's activity in a comprehensive way, as we have seen, why is it that, at an anxiolytic level of sedation, they preferentially inhibit reactive behavior only—more precisely, the kind of reactive behavior motivated by underlying fears or anxieties, while leaving active behavior unaffected? Why don't they suppress active, as well as reactive, behavior to the same extent? My own attempt to answer this question follows:

1. *It could be that the moderate inhibitory effect of sedative substances at anxiolytic doses targets the most active neuronal activity present*, which, when anxiolytic drugs are used (and to a lesser extent alcohol), is an overactive, overtaxed reactive behavior. After all, we need and look for the anxiolytic effect when strong anxiety, fear, or at least a tenacious worried arousal suppresses or inhibits resting or other important or more pleasurable activities. The reader may remember the argument on active behavior's suppression under these circumstances discussed in the section on active/reactive behavior interaction (section 5.4.0).

    (I am not aware of studies concerning the anxiolytic effect on *counteraggressive reactive tactics*. However, it seems to me that, since the anxiolytic effect reduces the subjective evaluation of the extent of the environmental danger or inconvenience, the employment of counteraggressive tactics may be secondarily emboldened. Think of the violent behavior associated with drunkenness or the side effect of sedatives just mentioned.)

2. A second line of reasoning concerns the great difference between the kinds of organism-environment interrelationships characteristic of active versus reactive behavior. Reactive behavior, as was discussed before, is absolutely dependent on environmental stimuli in order to be released. In animals, these stimuli have to be of immediate or recent nature, and the reaction to them is prompt. In humans, who remember past events and experiences, and are able to foresee possible future harmful scenarios, fears and anxieties can be, and frequently are, long-lasting. Dangers that activate reactive behavior can strike unexpectedly; therefore, constant awareness and screening of the surroundings to detect them in time is an organic aspect of

reactive behavior, even when the danger is not imminent or concrete but is a more or less plausible possibility. As a result, *it can be presumed that a slight inhibition of this constant awareness of the environment will affect reactive much more than active behavior.*

Active behavior originates from the autonomous processes of the individual life, and in order to fulfill its fitness-enhancing role, active appetitive-behavior patterns have to find and contact the appropriate releaser. However, as we have seen previously, active behavior is often postponable for some time, can be discharged in the absence of the appropriate releaser (displacement), or even when no releaser of any kind is present (vacuum behavior)—a "luxury" that reactive behavior cannot afford. After an active instinctive motive is fulfilled or discharged, as a rule, a refractory period follows, during which time the organism is disinterested in the particular releaser (another phenomenon absent in reactive behavior). Therefore, it may be reasonable to assume that active behavior patterns are more resistant to the mild CNS depressing effect of sedatives at anxiolytic intensities than the reactive ones. A slight reduction in the level of arousal, of awareness to the environment, of the speed of sensory/perceptual processes, and of the motor response may affect the proper discharge of active behavior patterns much less than those that fulfill avoidant reactive tasks.

The behavioral and mental effect of sedatives, in contradiction to antipsychotics and antidepressants, is *instantaneous* when they contact the appropriate brain neurotransmitter. Diazepam administered intravenously begins to reduce the subjective and behavioral signs of anxiety, as well as the physiological signs of overarousal, in about ten minutes, attaining its maximal effect in twenty to thirty minutes (Kaplan et al, 1980, p. 2322). Compare this with the effects of antipsychotics (AP) and antidepressants (AD), which, as we have seen, may lead to therapeutic improvement only after a period of several weeks. Since APs and ADs modulate the activity of the brain mechanism that originally evolved to adjust overall instinctive intensities to seasonal changes, as hypothesized in the preceding sections, while sedative drugs at an anxiolytic level of sedation act preferentially on reactive behavior, this difference in the timescale of their effect makes evolutionary sense.

Actually, sedative drugs act on a wider range of neurobiological, neuropsychological, and mental activities, including arousal mechanisms, sensory-perceptual processes, encoding of memories (may cause amnesia), motor response (reduces coordination), and so on than AP and AD drugs. In consequence, their action on mental state and behavior is more difficult to categorize into clearly separable primary and secondary effects. However, in the case of the *anxiolytic* effect, we may try to discern the difference between the effects more directly related to the slight inhibitory effect on CNS functioning, on the one hand, and its more remote secondary consequences, on the other. The primary or direct effect of these substances, therefore, has to be this comprehensive inhibitory effect on CNS activity. Considering the increased sensitivity to, and dependability on, the environmental stimuli of reactive behavior, the anxiolytic effect must be closely and directly related to that comprehensive inhibitory effect.

A more distant secondary consequence would be the disinhibition of previously suppressed active behavioral motives. In the following, I try to illustrate with some simple examples the increasingly variable repercussions of anxiolysis by describing its more remote behavioral consequences.

After a day of satisfying work and perhaps some pleasurable time spent with the family, let's say that the active instinctive motives were adequately gratified for that day. However, tenacious worries about some unresolved or unresolvable difficulties do not permit us to relax or sleep properly. Drinking a small quantity of alcohol or taking a sedative in these circumstances will induce a pleasurable relaxation and perhaps sleep. However, if in the same evening we are looking forward to a meeting with some good friends, the same amount of alcohol (or most probably of a sedative drug) will have as a rule a disinhibiting effect, augmenting the excitement, talkativeness, frequent laughing, etc., usually accompanying such events.

A less dramatic but probably more desirable secondarily induced change in an active-reactive-conforming behavior relationship brought about by anxiolysis is the enhancement of cognitive functions and social performance in general in the appropriate circumstances: "The patient may perform better with than without the drug in cognitive functions related to his work and social life because the anti-anxiety effects of the drug free him from the disruptive influence of anxiety" (Davis, 1980, p. 2328).

However, suppressed active instincts released by anxiolytics may emerge not only as increased well-being and better functioning but in any one of their three forms of frustration, or sometimes these forms of frustration may follow each another during the same event. (This is another illustration of the more accentuated *unpredictability of the secondary effects* consequent to the direct action of psychoactive drugs.) That sedatives and especially alcohol can lead to or augment aggression, physical violence, reckless driving, and other antisocial acts is well recognized. These effects are categorized (erroneously according to our theoretical framework) as "paradoxical effects" (Davis, 1980, p. 2322; Arana & Rosenbaum, 2000, p. 205; Sadock, Sadock, & Sussman, 2006, p. 77).[6]

In fact, in our theoretical framework, aggressive behavior related to anxiolysis may have *two different* underlying mechanisms:

1. *Suppression of anxiety, fear, or even cautious arousal changes the adaptively desirable power-balance between active and reactive behavior.* It inhibits reactive behavior (its freezing and avoidant aspects) and may enhance secondarily active behavioral expression. If the active motive, emerging as a result of sedative-induced disinhibition, is for some reason in its *aggressive form of frustration*, it will manifest itself in overt behavior as such.

2. *Aggression may appear or be intensified as a result of anxiolysis altering the formerly existing relationship between the three basic reactive tactics (freezing, fleeing, counteragression).* The reduction of the intensity of fear or anxiety, as mentioned previously, will reduce the perceived level of dangerousness of an external stimulus or situation. In consequence, the freezing reaction will shorten and the individual will be predisposed to more often choose the aggressive reactive tactic over the more cautious danger-avoiding one than she usually would in a similar situation without anxiolysis.

To decide which of these alternatives is present in a particular case is not always easy. However, when no external danger or derangement is apparent and the individual gives the impression of *actively seeking trouble*, actively producing a pretext for an aggressive confrontation (as can be frequently observed in drunkenness), the aggression probably represents disinhibition of frustrated active behavior. A tertiary consequence of anxiolysis in this case is the possible counteragression of the unjustifiably challenged person, which thus *creates* the environmental danger for the reactive apparatus to deal with (very probably with the aid of aggressive tactics as well).

The emergence of active instincts in their dysphoric form of frustration as a result of disinhibition by anxiolysis can also be observed frequently in alcohol intoxication as crying, pounding one's head, and loud self-accusations. This secondary effect is more rarely noticed in the case of sedative drug treatment. However, I found a related clinical observation: "Diazepam use has been observed to be associated with an increase in suicidal ideation in a small proportion of patients treated" (Davis, 1980, p. 2328).

With regard to the differentiated/diffuse scale of reactive instincts, I find no reason to suspect that sedative drugs can change the position of the reactive instinctive predispositions on this scale, or that it targets only a certain segment of this scale. A differentiated and, in principle, adaptive fear reaction is probably suppressed by sedative substances to the same extent as diffuse anxiety. Nevertheless, the suppression of a differentiated fear reaction as opposed to diffuse anxiety has to make a great difference in terms of therapeutic usefulness. Suppressing a *differentiated avoidant reaction* with sedative drugs, especially when it is useful in tackling a real environmental disturbance or danger, may be useless or even harmful. It may suppress the instinctive foundations underlying circumspection or choosing the appropriate adaptive tactic. (Think of a dangerous occupation, like that of a construction worker who is working on a tall building.) In addition, by being expressed in behavior, these normatively differentiated—or normatively diffuse—reactive drives will decrease in their intensity and ultimately disappear, especially when they are successful in removing or avoiding the danger that elicited them in the first place. The suppression of *diffuse anxiety* by anxiolysis, on the other hand, clearly has a sound therapeutic rationale. Anxiety may interfere with useful mental and motor functions (cognitive abilities, coordination, etc.), and by being unable to fuel effective reactive behavioral strategies, it tends to perpetuate itself or accumulate. As mentioned, the reduction of the intensity of anxiety by anxiolysis to a level more or less similar to normative arousal and awareness to one's surroundings may enhance the emergence of more useful active and reactive behavioral patterns.

6.5.2. Hypnotic Effect

In spite of the argument maintaining that "benzodiazepine hypnotics act on [benzodiazepine] receptors in the reticular activating system" (Dubovsky, 2005, p. 2785)—that is, that they supposedly possess a differential affinity to a brain area controlling arousal—the general opinion seems to be that the hypnotic effect of sedative drugs does not differ qualitatively from their anxiety-reducing or sedative effects: "The benzodiazepine hypnotics are essentially the same as the benzodiazepine anxiolytics" (Healy, 2009, p. 183). Naturally, it is expected that a sedative drug used as a hypnotic will possess an effect with a quick onset and a duration that coincides with the desired length of sleep. I am not even convinced that hypnotics possess a considerably stronger sedative effect than anxiolytics, and surely in the usual doses they do not induce an irresistible somnolence. The sleeping pill user can wake up several times during the night (particularly one who is at a more advanced age) and use the toilet, for example, without being particularly drowsy, disoriented, or suffering from incoordination (any more than when awakening from sleep not facilitated by a hypnotic).

The sleep-inducing effect of hypnotics probably depends strongly on their working in synergy with other physiological and mental sleep-promoting mechanisms. The sleeping pill is taken so that its effect occurs when the onset of natural sleep is expected, which in turn depends on the diurnal cycle of the physical environment, the rhythm of melatonin secretion, ingrained psychological habits, and so on. Consider, for example, that hypnotics have only a modest ability to alleviate sleep disturbances caused by jet lag or shift work (Sadock & Sadock, 2005, pp. 2028–2029).

The interrelatedness of the sedative effect and preexistent physiological and mental states is valid the other way around, too. The more a person is overaroused, troubled, or anxious, the greater the dose of the sedative or hypnotic he needs to obtain the desired effect. While the average sedative dose of diazepam, when used repeatedly, is about 5–10 mg, "[t]he same individual going to have a tooth extraction, to an interview, or engaging in some anxiety-provoking procedure may be able to take 30–40 mg immediately before their ordeal without significant sedative effects" (Healy, 2009, p. 154).

6.5.3. Drowsiness, Somnolence, Sopor, Coma

With the progressive increase of the dose of sedative drugs, and the resulting progressive increase of their inhibitory effect on the central nervous system's functioning, the differential suppression of the reactive behavior alone (with the active behavior remaining relatively unaffected) gradually disappears. Both active as well as reactive behavior will become increasingly interfered with, disorganized, and suppressed until the individual experiences a complete inability to perceive environmental stimuli, process them, and produce an organized (or any other kind of) motor output.

6.5.4. Animal Models of the Sedative Effect

Testing laboratory animals for the sedative (anxiolytic) effect of benzodiazepines confirms the hypothesis that this effect reduces reactive behavioral intensity differentially, and secondarily releases suppressed active behavioral tendencies. This was observed both in the case of more basic, biologically oriented behavior as well as in the case of more complex social behavior.

The first representative of benzodiazepines, chlordiazepoxide, was found to be very effective in taming wild animals at a zoo (Panksepp, 2005, p. 219). This finding, in the context of this work, can be interpreted as a reduction of the intensity of flight or aggressive reactions in wild animals as a result of being confined to an entirely new environment involving close proximity with other wild animals and humans.

The rationale for developing new anti-anxiety drugs by using animal models in the early trials was that: "Anti-anxiety effects are generally *assumed* to exist when previously punished behaviors are released from inhibition" (idem, p. 209, emphasis in the original). It seems clear that punishment suppresses a certain active behavioral pattern by inducing a reactive behavioral response (avoidance) by exposure to a disturbing or painful stimulus. Or, alternatively, punishment suppresses an egocentric reactive motive (flight or counteraggression) by replacing it with submission.

In later, more effective, animal studies, the animals were put in circumstances in which naturally occurring adaptively effective flight reactions are prevented by the experimenter, which in turn increases the fear and reduces active behavior. For example, rats "naturally prefer to enter dark holes, *yielding the latency to enter dark hole task*. When forced to remain under bright light, rats also exhibit reduced social activity, yielding the *diminished social-interaction test*, an anxiety that is effectively counteracted by BZs [benzodiazepines]" (idem, p. 211, emphasis in the original). Another test measures the length of time animals "freeze" in response to "contextual cues" that have been paired with electric shock, a response that can also be reduced by benzodiazepines (idem, p. 211).

The most technologically advanced models use direct electrical stimulation to those subcortical areas that are known to respond to such stimulation with powerful flight behavior. The interesting finding, from our point of view, was that low currents induced a freezing reaction while more intense currents *at the same location* led to a flight reaction (idem, p. 211). This finding strengthens the presupposition that these two reactive behavior patterns are indeed strongly interconnected (sharing by and large the same somatic substrate), and may be two successive phases of the same phylogenetically evolved response to signs of possible environmental danger. Whether the freezing response escalates into the flight response or not depends, it seems, solely on the perceived *intensity* of the danger-signaling stimulus, at least in lower animals.

6.6. Psychostimulants (Amphetamines and Methylphenidate)

Since I treated mainly older, chronic psychiatric patients throughout my professional career, I have limited clinical experience with this drug category, as well as with attention-deficit/hyperactivity disorder (ADHD), the main therapeutic target of psychostimulants. Consequently, the following arguments are based mainly on data from the literature. However, the inclusion of this category seems to me important in the context of the present work. An attempt to decipher the primary behavioral mechanism targeted by psychostimulants, as well as this mechanism's evolved adaptive role, even in a tentative form, may be instructive. Furthermore, this discussion may show how uncertain our knowledge is of the relationship between a discrete neurotransmitter activity and its behavioral consequences.

Following is a concise enumeration of the basic psychostimulant effects in therapeutic doses: "Amphetamine . . . has excitatory effects on cortical function. Depending on personality and contextual factors amphetamine in adults can increase wakefulness, energy, alertness, initiative, self-confidence, and physical and mental performance; lessen fatigue; and produce euphoria. These effects occur shortly after dosing" (Fawcett, 2005, p. 2940). Some of these effects, like the increase in wakefulness and alertness, seem to be the reverse of those of sedatives. Additional stimulant effects, categorized as side effects, like nervousness, anxiety, and interference with sleep (idem, pp. 2940–2941), pertain to the same category. Other effects of the stimulants, however, like increase in energy, initiative, and physical and mental performance, as well as self-confidence, resemble those of noradrenergic antidepressants. One of these, increased energy, was categorized in our theoretical framework as the antidepressants' *primary effect*, while others like improved initiative, increased mental and physical performance, reduced fatigue, euphoria, and self-confidence, are more or less remote *secondary repercussions* "depending on personality and contextual factors." However, in spite of their beneficial effects, *psychostimulants were found to be ineffective in the treatment of clinical depression* (of the lethargic type, in our theoretical scheme), probably as a result of their short-lived action, as well as the serious side effects associated with their long-term use: anorexia, dysphoria, insomnia, paranoid psychosis, etc. (Fawcett, 2005, pp. 2940–2941).

The resemblance between the action of noradrenergic antidepressants and psychostimulants is expected considering the close similarity in the chemical structure between amphetamines and catecholamines (noradrenaline and dopamine), as well as the stimulants' neurochemical effect, which is an increase in noradrenergic and dopaminergic activity at the brain synapses (Fawcett, pp. 2938–2939). However, an important difference exists between the stimulant and antidepressant effect. Psychostimulant action, as is also the case with anxiolytic drugs, begins shortly after their ingestion and is short-lived. The antidepressant effect, on the other hand, takes several weeks to manifest itself in behavior or to subside after the cessation of the treatment. Why the behavioral and experiential repercussions of the same dopaminergic and noradrenergic effects are expressed slowly in the case of antidepressants (and antipsychotics), and promptly and for short periods of time in the case of stimulants, is to the best of my knowledge unknown.

To summarize, the stimulant effect is prompt, short-lived, and, in addition to the typical direct and indirect effects of antidepressants, affects other mental functions as well, like alertness, wakefulness, and sleep, all being targets—with a reversed effect—of the sedative drugs. Besides, as can be inferred from their effect on children with attention deficit and hyperactivity, *psychostimulants enhance higher cognitive functions like attention and learning, as well as inhibit behavioral expressions at lower levels of integration* (detailed in the following). To what extent these effects are primary or secondary is not clear to me.

This profile of the stimulant effect in the context of evolutionary reasoning suggests that stimulant drugs target an evolved behavioral pattern or mental function that needs to be promptly employed and requires a coordinated, intensive interplay between a wide range of mental and physical faculties ranging from differentiated alertness to incoming stimuli, improved and accelerated cognitive processing and integration with intrapsychic contents at the afferent side, to intense and precise physical or mental output at the efferent one. Furthermore, it can be presumed that the need to discharge this kind of behavior is relatively short-lived and is characteristic of animals with highly developed cognitive abilities living in a complex, frequently unpredictable, and quickly changing social milieu. It would therefore be expected, from a neurochemical point of view, that this stimulant effect, in contrast to the sedative one, should *inhibit* GABA-ergic neurotransmission (that is, GABA's inhibitory effect on cortical functions), or *enhance* the activity of the excitatory neurotransmitter glutamate—but the stimulant drugs in question do not possess such effects.

The suggestion that stimulants target the brain mechanism responsible for relatively short-lived, intense mental or physical activity, with possible adaptive significance in highly evolved, complex social circumstances, may receive support from the way normative adults typically use—or misuse—these stimulants. These instances arise when intense, high-level, precise mental or physical performance is needed for relatively short periods, as in the case of stage performers, movie actors, singers, some professional sportsmen, or students before and during examinations. In some cases, the acute need for high-level performance may clash with the usual ebb and flow of the level of awareness and of physical and mental effort, as in the diurnal rhythm or other kinds of regular oscillation between activity and rest. Stimulant drugs are also used in combat situations, when increased alertness and high-level functioning are needed in exhausting duties, as in the case of sleep-deprived combat helicopter pilots (Fawcett, 2005, p. 2938).

Equipped with these preliminary considerations, we will try to cope with the great puzzle of the stimulant effect in attention-deficit/hyperactivity disorder (ADHD). In this disorder, psychostimulants, rather than further energizing an already hyperactive and impulsive behavior, calm it down.

In fact, the main therapeutic gain of the stimulant effect, the improvement of the attention deficit aspect of the disorder, naturally follows from some of its effects on mental functioning. It is quite easy to see why "excitatory effects on cortical function" (quoted at the beginning of this section), especially the increase in wakefulness and alertness, may lead to improved attention, concentration, and learning performance. But how can a stimulant be the source of a *calming* effect on hyperactivity? It seems to me that the only logical possibility here is that *overactivity and impulsivity in these children has to be the consequence (and not the cause) of problems that are summarized by the term attention deficiency.* Attention is defined as the "[c]onscious and willful focusing of mental energy on one object or one component of a complex experience and at the same time *excluding other emotional or thought content*" (Campbell, 2004, p. 60, emphasis added). That is, the exclusion, inhibition, or suppression of irrelevant emotional, intentional, or cognitive contents is an organic part of attention. And, conversely, lack of focus and disorganized behavior, fidgeting, restlessness, excessive talking about irrelevant issues, answering before a question is completed (that is, a very shallow or nonexistent freezing stage of reactive behavior in our theoretical scheme), which are so typical of ADHD (Sadock & Sadock, 2005, pp. 3186–3187), may be *consequences* of the attention deficit. Thus, it seems that the psychostimulant effect, while improving higher level, integrated mental functioning, may suppress at the same time the uninhibited expression of these more elementary, more immature emotional, intentional, motor, or cognitive expressions.

It has to be stressed that in most cases ADHD represents a *maturational lag* with an inherited component (Sadock & Sadock, 2005, pp. 3184–3185). It appears early in the course of life (before age seven) and, in a considerable proportion of cases, it improves with age. Maturation in this respect is a universal phenomenon in humans that consists of a gradual transition from the more spontaneous, physically active, more uninhibitedly emotion-expressing behavior of the child, who has a shorter attention span, to the more emotionally restrained, more thoroughly thought-over and frequently future-oriented behavior of the typical mature adult. (In a considerable proportion of clinically clear-cut cases of ADHD in children, hyperactivity may improve or even disappear completely around adolescence, which is also true to a lesser extent for the attention deficit [Kaplan et al., 1980, pp. 2542–2543].) Why this maturational process stopped or became hampered in ADHD children is not clear. Genetic causation is well substantiated, while organic brain insult during pregnancy and delivery and chronic stressful environmental conditions were implicated, too (Sadock & Sadock, 2005, pp. 3183–3186).

From an evolutionary point of view, the above maturational process seems indispensable for achieving successful adaptation to a complex social environment; therefore, its genetic predisposition must have been subject to intense natural selective pressures during the prehistory of the species. In fact, a similar maturational trajectory can also be observed in more evolved animals. Compare, for example, the behavior of puppies or kittens with the behavior of mature dogs and cats. In humans, however, the relaxation of intragroup natural selective pressures, especially after the agricultural revolution, probably affected this aspect of mental functioning, too, as it did the comprehensive instinctive mechanisms discussed in this text. Consequently, it may be hypothesized that the ensuing diversification of the respective trait (more exactly one pole of its diversified continuum) constitutes the innate predisposition to ADHD.

Why stimulant drugs, by enhancing noradrenergic and dopaminergic activity, activate a wide range of mental functions (including awareness, wakefulness, and learning ability) promptly and for short periods, while antidepressants with a similar direct action on noradrenergic activity act slowly and on a more restricted range of mental phenomena, is not known. (It should be mentioned that antidepressants are also used, as a less desirable alternative, in the treatment of ADHD when stimulant use is impracticable.)

It seems that the precise dose of a stimulant drug an individual mental apparatus needs in order to improve its functioning is narrowly determined and varies according to the unique characteristics of the respective mental apparatus, as well as the state of the same mental apparatus under various circumstances. Stimulants can improve mental and physical functioning in normative individuals as already mentioned, while in other instances, their use may have the reverse effect and induce a behavioral change reminiscent of ADHD. In these instances, the effect is categorized, as we have seen, according to a typical preconception of clinical thinking, as a paradoxical side effect. Healy describes it as follows: "One [of these side effects] is an increasing hyperactivity, leaving an individual with too much motor energy and drive. The energy is relatively unfocused so that, rather than getting a lot of useful things done, the person is left pacing up and down restlessly" (2009, p. 126).

In several respects, the clinical picture of a *manic episode* resembles ADHD, including the latter disorder's main symptoms: hyperactivity, attention deficit, and impulsivity. (Attention deficit in mania manifests itself as distractibility and clang associations.) However, in the case of mania, these dysfunctions are the result of the effect of unnaturally intense instinctive energies on a normatively mature mental apparatus. In ADHD, on the other hand, they are the outcome of the effect of normal intensities of instinctive energies on an immature mental apparatus that—at normal levels of arousal—is unable to integrate and adequately synchronize emotional, intentional, and cognitive intrapsychic contents with more complex, socially provided stimuli. Accordingly, in mania, similar overactivity and attention deficit improve under treatment with antipsychotics, which *reduce* the intensity of instinctively based tensions and intentions (whereas stimulants and antidepressants *aggravate* it). ADHD, on the other hand, *improves* under drugs that *enhance and stimulate* those mental functions that enable a more effective, more mature integration of affective, volitional, and cognitive contents in a circumscribed domain, namely psychostimulants (and to a lesser extent antidepressants).

We can summarize the main points proposed in this chapter as follows:

- *Antipsychotics and noradrenergic antidepressants directly modulate the intensity of active and egocentric reactive instinctive activity in a comprehensive manner in opposite directions,* through their effect on the catecholaminergic (noradrenergic, dopaminergic) neurotransmitter systems, which are located in subcortical brain areas. Since their effect is attained through modulating the brain mechanism responsible for the synchronization of instinctive intensities with the seasonally changing abundance of life-supporting resources, this effect unfolds slowly.

- *Sedative-hypnotics act mainly through the GABA-ergic neurotransmitter system, which is located predominantly in the cortex.* In anxiolytic doses, they reduce the intensity of the egocentric reactive instinctive activity, more precisely freezing and flight. Since this effect targets reactive behavior, it manifests itself almost instantaneously.
- *Psychostimulants target a wider assortment of mental functions, including alertness, attention, speed of processing incoming stimuli and their integration with intrapsychic contents, and motor coordination; at the same time, they suppress mental output and overt behavior at lower levels of integration.* It seems that they exert their effects through the same, or a similar, catecholaminergic neurotransmitter system as the antipsychotic and noradrenergic antidepressant drugs. However, their effect manifests itself promptly and is short-lived. The necessary conclusion regarding this puzzle has to be that *neurotransmitters specify only very partially and imprecisely the brain substrate underlying discrete behavioral mechanisms.*

The implications of this discussion of the manner in which psychoactive drugs should be employed for therapeutic purposes may be summarized as follows: Psychiatric drugs should be used first and foremost for their *main effect* on behavior and subjective experience, *irrespective* of the diagnosed clinical entity. The therapeutic goal has to be to reduce the maladaptive behavior's deleterious influence on fitness and social adaptation as much as possible, as well as to alleviate subjective suffering. The expected therapeutic gains have to be balanced carefully against possible harmful effects on the targeted behavioral and mental dysfunctions (as well as on unrelated mental or somatic functions).

Discrete Clinical Disorders in Evolutionary Perspective

7.1. General Considerations

It goes without saying that innate predispositions alone, except in the case of very primitive animals, cannot determine most aspects of the overt, observable behavior of mature individuals. During maturation (and even earlier, during intrauterine growth), the genetic blueprint interacts with environmental influences, both desirable and deleterious, and biological, physical, or social in nature. The outcomes of these interactions will determine the regularly reappearing behavior patterns or behavioral predispositions characteristic of an animal species or an individual animal. (Section 2.2 detailed the arguments on behalf of the necessity for *fusion* of inherited and acquired aspects of behavior, contrary to traditional psychoanalytic thinking, which stresses the *antagonism* between the inherited and socially induced elements of behavior.) These mature behavior patterns or predispositions will interact with the influences coming from the animal's or human's *current* environment. Only at this stage will the characteristic dysfunctional behavior patterns general psychiatry deals with emerge.

It has to be stressed that the environmental conditions both during maturation and at the outbreak of the clinical disorder (with the partial exception of posttraumatic disorders) are as a rule *normative*, in the sense that most humans in similar conditions do not develop a mental disorder of any kind. Even after a highly traumatic event, not all the individuals involved will develop posttraumatic symptoms. Therefore, *it is the inherited predisposition that decides whether a mental disorder will develop in certain conditions*, and of what type. This does not mean that ontogenetic influences do not play an important role in triggering a mental disorder, in influencing its course or level of gravity, in bringing about complications, or in determining the verbal expression of the pathological mental contents (words are learned), but these effects are always dependent on the nature of the innate predispositions.

With this general scheme in mind, we examine the following common psychiatric disorders: affective disorders, personality and anxiety disorders, paranoia, and schizophrenia.

7.1.1. The Intertwining of the Four Comprehensive Instinctive Mechanisms in Concrete Clinical Disorders

The four comprehensive instinctive mechanisms discussed in chapter 5 having critical adaptive value are present in their entirety in the instinctive configuration of every person and consequently participate in shaping the clinical aspects of every mental disorder.[1] As a rule, either the comprehensive instinctive mechanism that is the most conspicuous, that lies at the most extreme end of its normative/pathological continuum, that leads to the gravest form of dysfunctional behavior, or is the most deleterious to social coexistence will determine the main clinical diagnosis in a particular case. For example, if the alternations of pathologically high and low instinctive intensities are the most conspicuous characteristic of the disease, while other general instinctive attributes, like the diffuse-differentiated scale, the ways active instincts transform into their frustrated forms, and the active/reactive behavior interactions, are all in the normative range in their respective scales, the diagnosis will be a characteristic, uncomplicated bipolar affective disorder. When very diffuse instinctive predispositions dominate the clinical picture leading to serious disturbances in interpersonal relationships and social functioning, the main diagnosis will be schizophrenia or one of its spectrum disorders. In this case, if *additional symptoms exist* in the clinical picture induced by another comprehensive instinctive mechanism (besides those induced by the mechanism determining the diagnosis), it may lead to the addition of specific *subcategories* to the main diagnosis, or to the specification of an additional syndrome not characteristically present in the respective clinical disorder—for example: mania, predominantly euphoric, aggressive, or of a mixed type; or schizophrenia, paranoid type or accompanied by an obsessive-compulsive syndrome.

One of the remaining comprehensive instinctive mechanisms (not relevant to the main diagnosis, or to a subcategory of it, or to an additional syndrome) may cause other alterations in the normative functioning, which, although a direct consequence of the malfunctioning of the main instinctive mechanism in the respective disease, is not specified in the diagnosis. For example, manic episodes disrupt the normative balance of active and reactive behavior, as well as the desired balance between pursuing individual (active as well as reactive) instinctive interests and modifying, suppressing, or postponing them according to the requirements of social coexistence. Active instinctive aims will be pursued more intensely, with less consideration for transgressing social norms or others' interests; reactive behavior will move toward predominantly counteraggressive tactics instead of freezing or engaging in avoidant reactions; alternatively, the individual may completely disregard any environmental factors that oppose the unchecked instinctive gratification. This latter change means a reduction in the sensitivity to disturbing or opposing environmental (mostly social) stimuli, which consequently reduces the conforming and submissive aspects of behavior.

The position of the instinctive predispositions on the diffuse/differentiated continuum and the resultant premorbid level of social functioning will determine whether the manic or hypomanic episode will be a benign one—that is, one showing symptoms that seem like exaggerations of normative aspirations or indulgences (spending money thoughtlessly, striving for multiple sexual relations, etc.)—or, conversely, in the case of more diffuse drivers and more problematic premorbid functioning, whether the manic behavior will become more bizarre, and contain mood-incongruent psychotic elements.

It seems to me quite probable that this complex situation, in which different comprehensive instinctive mechanisms participate in different proportions in the generation of the same clinical disorder, is responsible for the peculiar situation existing in psychiatry regarding genetic inheritance. I have in mind such peculiarities of inheritance as the following: Schizophrenia as a comprehensive nosological entity has a strong genetic component, but its *subtypes* (paranoid, catatonic, etc.) have not (Plomin et al., 2008, p. 203); on the other hand, schizophrenia as a psychotic disorder has a genetic predisposition in common with some personality disorders (schizoid, schizotypal) that belong to an entirely different comprehensive nosological category. In addition, type II schizophrenia has a more pronounced heritability than type I, and clinical pictures exist in between schizophrenia and bipolar disorder (schizoaffective disorder). When the clinical picture is more characteristic of bipolar disorder, the inheritance for bipolar disorder is the stronger and vice versa. In my opinion this situation strongly supports the presupposition that *what are inherited in mental disorders are the quantitative variations in the four (and maybe more) general instinctive mechanisms, and not discrete clinical pictures.* This subject is elaborated further in the next section.

7.1.2. From Genetic Predisposition to the Observable Clinical Disorder

Another undertaking that has to be attempted is to trace the succession of events from the hypothesized innate predisposition to the final clinical picture of a disorder. In some mental disorders, for example in good-prognosis mania, in typical, uncomplicated lethargic depression, and in many cases of posttraumatic stress disorder, the succession of related events from predisposition up to the full-blown clinical picture seems short and simple. In these instances, the predisposition does not significantly affect the maturation of behavior in normative social conditions, and the individual attains, as a rule, good social adaptation and functioning. The innate predisposition for the disorder is activated later in life and alters an apparently normative premorbid behavior. In posttraumatic disorder, the triggering factor seems obvious, and the innate predisposition seems to consist of an inability to revert an intense stress reaction to the previous (normative) balance between active and reactive behavior after the traumatic event is over. In seasonal affective disorder (like winter depression), whose course is synchronized with the succession of seasons, the trigger has to be the same as that in hibernating animals, chiefly the shortening of the daily hours of light. The advent of the spring or light therapy improves or terminates the affective disorder and the individual usually reverts to the premorbid, normative behavior (we disregard here the long-term intrapsychic and social consequences of protracted periods of dysfunctional behavior).

In other mental disorders, however, of which schizophrenia is the paradigm, the dysfunctional instinctive predisposition will powerfully affect the maturation of the behavior long before the characteristic clinical disorder becomes apparent. It will be argued, for example, that in the case of schizophrenia and its genetically related disorders, the more than normatively diffuse instinctive drives interfere considerably in the fusion of inherited behavioral incentives with acquired, mainly socially induced contents long before the appearance of the characteristic clinical disease.

The less than adequately differentiated instinctive predispositions are unable to induce body language or verbalizations that can convey with the needed precision the child's emotional or intentional states, thus preventing the parental figures from correctly grasping the child's needs and responding efficiently using the mechanism of empathy. The same more than normatively diffuse drives will prevent the schizophrenia-predisposed individual herself from figuring out others' emotional states or veiled intentions by the same mechanism of empathizing, leading to social incompetency (that is, serious difficulties in interpersonal communication concerning topics that are not strictly observational or logical in nature).

In the following, a hypothetical chain of events can be built leading to the appearance of psychotic clinical disorder and beyond it to the chronic deterioration and impoverishment of mental life, especially in the more heritable type II schizophrenia (Plomin et al., 2008, p. 204). This topic is detailed in the section on schizophrenia. At any rate, it seems evident that, in the case of schizophrenia, the chain of dysfunctional events from the inherited predisposition to the full-blown clinical disorder begins early in life, and it is more complex and protracted than in the case of affective or posttraumatic disorders.

The innate predisposition to personality disorders may affect maturation, too, albeit in a less dramatic way than in schizophrenia. The first signs of these conditions may appear in childhood and crystallize during adolescence, and they may powerfully affect social adaptation or discrete interpersonal relations long before a clinically diagnosable personality disorder emerges.

In the following, individual clinical disorders are discussed from the perspective of the present theory.

7.2. Discrete Clinical Disorders

7.2.1. Affective Disorders: Bipolar Disorder and Dysphoric Depression

Bipolar disorders, as mentioned, can be explained without the involvement of a long chain of interactional events between the innate predisposition and ontogenetic influences. Most of what I want to say on the affective disorders has already been discussed in the first section of chapter 5 concerning the effect of seasonal fluctuations on behavior.

Since detailing a long chain of maturational events is not necessary, I want to enlarge here on the role of additional comprehensive instinctive mechanisms in bipolar disorders, especially the forms of frustration of active drives, in shaping the overall clinical picture. It was stressed, for example, that a manic episode can be predominately euphoric, aggressive, or dysphoric in nature. The high intensity of the instinctive activity accentuates the expression of the respective emotional state in overt behavior.

Lethargic depression, that is, the low-intensity phase of overall instinctive activity, can be influenced by the predominant form of active instinctive frustration (as well as the reactive instinctive configuration), too. Compare, for example, the following personal accounts of lethargic depression: "It was the beginning of autumn and I became sad, gloomy, and susceptible. By degrees I neglected my business and deserted my house to avoid uneasiness. I felt feeble. . . . I became irritable. . . . I suffered also from insomnia and inappetence. . . . At length I fell into profound apathy, incapable of everything, except drinking and grieving" (quoted in Foster & Kreitzman, 2009, p. 201). The sad mood and irritability despite the reduction of social burdens (by leaving home and abandoning his business), and the use of alcohol to relieve his anxieties, seems evident from this description. Besides, it may be hypothesized that the well-known clinical finding that some lethargic patients suffer from inappetence and insomnia rather than overeating and excess of sleep is the result of an additional genetic predisposition, namely to respond excessively with anxiety and/or dysphoria to the undesirable consequences of the low level of instinctive activity (like failing to live up to social and/or intrapsychic expectations).

The next account of lethargic depression has an entirely different emotional tone: "So what if we slow down as winter comes, and are less excited by going out to see people? Is it bad if, for a period of the year, our high-powered jobs thrill us less, or if we become more contemplative and wonder at our place in the grand scheme of things? Rather than denying this process, might there not be some merit in embracing it, in seeing it as the natural cycle of seasonal changes" (quoted in Foster & Kreitzman, 2009, p. 207). To me it seems that in this case of lethargic "depression" the underlying mood remained euthymic.

A more problematic clinical dilemma arises when the overall intensity of the active drives is high and these drives manifest themselves predominantly in their *dysphoric* form of frustration. This happens in the agitated, dysphoric type of depression characterized by a constant enactment of being in severe mental pain and expressing guilt feelings sometimes of psychotic proportions as well as suicidal preoccupations. Do we have here a bipolar manic state of the dysphoric subtype or a separate category of depressive disorder?

Emil Kraepelin, the founder of modern psychiatric nosology, had exactly the same dilemma. He initially separated this condition (which he saw in older patients) from bipolar disorder and called it "involutional melancholia." Later, he reunited this condition with the bipolar group of disorders, seeing it as a "special form of mixed state," when his pupil demonstrated during follow-up "unmistakable excited phases" (Akiskal, 2005a, p. 1564).

In my opinion, it is questionable whether enough essential differences exist between agitated, dysphoric depression and mania of the dysphoric type to warrant a separate classification solely on clinical grounds. However, the *evolved brain mechanism* predisposing to agitated depression, in particular when it has no clear episodic (and, even less, a seasonal) course, and has its first appearance at a more advanced age, must be different from that which predisposes to dysphoric mania. We can no longer suspect that the increased intensity of active instinctive activity is related to the phylogenetic adaptation to seasonal changes. It seems more probable that specific negative life events characteristic of involution and old age (retirement, deterioration of the physical appearance, the "empty-nest syndrome," serious medical disorders, death of a beloved spouse or of close friends, and so on) frustrate and obstruct active instinctive expression in their original (unfrustrated) form in a personality predisposed to express these drives when frustrated, predominantly in their dysphoric—and perhaps also aggressive—form.

7.2.2. Anxiety (Neurotic) Disorders

Although in personality disorders we see a predilection to express a dysfunctional instinctive constitution overtly, including in circumstances of strong social disapproval or opposition, anxiety disorders are characterized by considerable *oversensitivity* to specific undesirable aspects of the environment—social, biological, or physical. The oversensitivity to these environmental insults in neurotic disorders manifests as a strong fear reaction, exaggerated worries, or a more generalized short-lived or longstanding anxiety.

As we saw in section 5.3.3.4, the nature of the external agent that triggers the dysfunctional reaction may vary and determines the name given to the respective anxiety disorder. *Specific phobia*, with its many subcategories, is released by discrete agents of the physical and biological environment (for example, closed or high places, or small, unexpectedly occurring, possibly venomous animals). *Separation anxiety* is triggered by the separation or prospect of separation from an attachment figure. *Social phobias* are aroused by various social situations in which the individual may be scrutinized and possibly humiliated by other persons or a group of persons. In *illness anxiety disorder* (hypochondriasis) and *somatic symptom disorder* (somatization), the source of worries is related to the body's health and integrity, while in *posttraumatic stress disorder* the pathological response is triggered by a sudden, highly traumatic experience. Furthermore, the nature of the releaser may determine the extent to which natural selection can narrowly specify the behavioral reaction. This reaction may be highly specific (differentiated) in the case of specific phobias or in separation anxiety disorder but much less differentiated when the source of disturbance is of a social nature.

A secondary consequence of the fear or anxiety reaction, in the context of this work, is the suppression of the active instinctive expressions (as discussed in connection with active/reactive behavior interaction, section 5.4.0). In principle, the active behavioral predispositions of a neurotic individual (regarding their place on the diffuse-differentiated scale, or concerning their transformations in frustrating conditions) should lie inside the normative range. However, their longstanding or frequent suppression, especially when the source of fear is of a social nature, may prevent the normative maturation of the active instinctive predispositions during ontogeny. I mean here the adaptive channeling (through negotiation, or trial and error, etc.) of the active instinctive intentions into socially desired, approved, or tolerated ways of expression. Consequently, these suppressed active instinctive motivations, when their intensity exceeds a certain threshold, may "surface" in behavior or conscious mental experience in a crude and intrusive form (felt as alien and ego-dystonic), of which obsessions and some compulsions are probably the best example.

Now, I hope, we are in a position to see more clearly the basic differences between personality and anxiety disorders. *While personality disorders uninhibitedly express a dysfunctional self-centered instinctive configuration in spite of social disapproval or resistance to which the individual is relatively insensitive, in anxiety disorders, active instinctive motivations are suppressed to a greater extent and for a longer duration than in normative persons as a result of increased sensitivities to various disturbing aspects of the natural and especially the social environment.*

Besides categorizing anxiety disorders according to the nature of the releaser, we can categorize them according to the extent of diffuseness or differentiation of the underlying fear or anxiety reaction. In this respect, in anxiety disorders, we encounter a wide range on the diffuse/differentiated continuum. This subject was briefly mentioned in section 5.3.3.4. The next sections examine it in somewhat more detail.

7.2.2.1. Specific Phobias

Specific phobias are interpreted by evolutionary psychology/psychiatry as differentiated fear reactions in response to circumscribed dangers of a physical or biologic nature that were present in our evolutionary past for a long enough time to allow natural selection pressures to build specific adaptive behavioral reactions against them.[2] The overt behavioral reaction induced by these circumscribed releasers is well differentiated. It consists as a rule of a prompt escape or avoidance reaction limited strictly to the specific situation or triggering agent (frequently accompanied by a loud shriek whose evolved function is probably the recruitment of help from other conspecifics). Having effectively fulfilled its adaptive function, the reaction subsides quickly (including the physiological arousal and accompanying subjective experience), enabling the reemergence of other life-sustaining activities.

If the person can avoid contact with the specific releaser in the future, he can lead, in this respect, a trouble-free life. It is worth mentioning that specific phobia is the only anxiety disorder in which no specific *cognitive* response accompanies the fear reaction. "A specific cognitive ideation is not featured in this disorder, as it is in other anxiety disorders. The fear, anxiety, or avoidance is almost always immediately induced by the phobic situation" (DSM-5, pp. 189–190). This observation, in my opinion, reinforces the argument that we are dealing here with an ancient differentiated instinctive reaction whose phylogenetic origins predate the evolution of elaborate human cognitive abilities. Some of these circumscribed fears, like that of snakes, exist in other primates, too. The childhood onset of these fears may be another hint pointing in the same direction.

In my opinion, some specific phobias triggered by some aspects of our modern environment represent the *displacement of these circumscribed ancient fears* into new situations brought about by our complex technology. Fears of being in an elevator or flying in an airplane represent, to my mind, the combination of two ancient fears, that of heights and that of enclosed places from which escape in times of perceived danger is impossible. In a plane full of passengers, even the way from your seat to the door is obstructed. When such situations involve the close proximity of other people, they may also trigger *social phobia* (the fear that the people around us will unfavorably judge our behavior as unworthy of a grown-up person).

Specific phobias connected to receiving injections, to seeing blood during laboratory testing, or to the prospect of undergoing an invasive medical procedure also represent, in my opinion, ancient fears of bodily injury displaced into the conditions of modern life.

However, the amount of danger these specific releasers (ancient or modern) represent are negligible compared to much more widespread and dangerous ones inherent in the modern, technologically advanced environment. I have in mind such potentially dangerous causes as electricity, radiation (from medical procedures, nuclear power, electronic gadgets, and other consumer products), cars, harmful chemicals in widely consumed food items, epidemic infectious diseases brought about by the high density of human populations and insufficient sanitation, and so on—dangers of whose nature we can have no "innate knowledge" and no differentiated inherited behavioral response (that is, for which we possess, respectively, no innate releasing mechanism and no fixed motor pattern).

To cope with these kinds of dangers, we have to rely on a more complex mechanism that welds together innate predispositions and acquired knowledge. Over the course of human evolution, the innate fears of physical, biological, and social dangers from our environment became more diffuse, providing only the energetic and subjective experiential aspect of the behavior and a vague direction to react—for example, by freezing, avoidance, or aggression, or, frequently, only a very generalized, directionless arousal or worrying that must be integrated with extensive acquired knowledge about the exact nature of the dangers and the effective behavioral strategies for coping with them.

7.2.2.2. Separation Anxiety Disorder

While specific phobias concern evolved responses to simple physical and biological triggers, separation anxiety disorder has its inherited roots in a more complex evolved behavioral mechanism concerning the relationship between two whole organisms. Separation anxiety can be detected by the mechanism of empathy only in highly evolved mammals (like apes), although offspring dependence on parenting activities exists in birds and some reptile and fish species, too.

The involved instincts are still of a quite differentiated nature. The parenting animal(s) has differentiated instincts or at least differentiated instinctive predispositions to provide nutrients to the immature offspring, to protect it from predators or from the elements, and so on. The immature offspring also has innate differentiated instinctive behavioral patterns (like the "distress call") that reliably trigger the parenting activities. Here we have to clarify a moot point in our theoretical scheme concerning the active/reactive behavior dichotomy. While some parenting behaviors, like feeding, as well as the offsprings' tactics to elicit it, have to be categorized as *active* behavior, the offsprings' fear of losing parental support and becoming exposed to an unproviding and possibly inimical environment has to be categorized as *reactive* behavior (clinging behavior, for example). The same is true in the case of the risk of losing an offspring (for example, to predation), which elicits in the parent(s) defensive or aggressive reactive tactics.

These bidirectional, complementary instinctive mechanisms in the parent-offspring dyad ensure a close, dependent tie between the immature offspring and the parenting figure. In animals, such instinctive behaviors are tightly checked by natural selection in order to optimize their adaptive value. If these instincts are not strong or differentiated enough, the immature offspring can lose this indispensable tie with the parenting figure. On the other hand, if this tie is too strong or long-standing it may hamper the independent exploration of increasingly extended ranges of the environment by the offspring during the appropriate stages of maturation, leading to the persistence of immature behavioral patterns. Frequently, the parenting animal refuses to nurture the young mature offspring any longer and may even drive them away. Young who are not prepared to give up the dependent relationship can show intense opposition, a situation that occurs frequently in adolescent chimpanzees (Goodall, 1988, pp. 232–235).

It should be mentioned in this context that in highly evolved social animals, like chimpanzees, the parental instinct is more diffuse than in less evolved animals, requiring learning in order to be perfected. As already mentioned, young female chimpanzees are very interested in newborn babies, as well as in the mother's parenting behavior toward them. They handle and caress the baby when permitted by the mother and imitate her behavior. Cases are known in which an orphaned infant was adopted by an older sister. It goes without saying that the human mothering instincts have to be even more diffuse in order to allow the assimilation of a vast amount of acquired knowledge about the mothering practices and traditions of a mother's respective human community.

The great diversity in humans' instinctive parenting predispositions (as a result of RfNSPs) is expressed in the mother in the extent of variations of their strength and duration, as well as the wide span of their positions on the differentiated/diffuse spectrum. In the child, the excessive strength and tenaciousness of the dependent tendency (augmented possibly by complementary traits in the mother) represents to my mind the innate predisposition to separation anxiety disorder (as well as dependent personality disorder), as will be detailed later in this chapter.

7.2.2.3. Social Anxiety Disorder (Social Phobia)

The reactive instinctive predispositions underlying social anxiety disorder, or social phobia, are more diffuse than they are in the case of specific phobias or separation anxiety disorder. The precise evaluation of the nature and seriousness of possible disturbance or harm coming from close proximity to an unfamiliar individual or great number of unfamiliar individuals, in crowded conditions, for example, or from a more homogenous group of people, such as an audience during a performance, is next to impossible. Even extensive knowledge on and experience with similar situations in the past may allow only a very approximate evaluation. In contrast, in the case of specific phobias and separation anxiety disorder, as we have seen, a person is endowed by heredity with quite specific innate "knowledge" for recognizing the respective danger, and with effective behavioral strategies—albeit strategies aimed partly at prevailing conditions existent in our evolutionary past. Social inconveniences, disturbances, and dangers that appeared later in human evolution than those involved in the simple phobias and separation anxiety are more complex and less predictable. To grasp them requires, often, the mechanism of empathy or theory of mind—that is, the capacity to read others' intentions, emotions, or covert thoughts beyond observation alone, or logical reasoning. This capacity is acquired by accumulating considerable experience and knowledge of the human community in which one lives. Usually, an individual has no better option than to hope that surrounding persons will act according to the accepted norms of peaceful (or even cooperative) social coexistence. In these circumstances, no possibility exists for natural selection to build specific innate knowledge (innate releasing mechanisms) or fixed adaptive behavioral strategies. One has to rely on a vast amount of acquired knowledge and experience, as we have seen, which is energized and vaguely guided by normatively diffuse reactive instinctive predispositions.

Social anxiety disorder is the most clear-cut opposite of personality disorder in the sense that it consists of an abnormally strong sensitivity to the social environment's evaluation of one's behavior (in contrast to considerable *in*sensitivity to the social evaluation that characterizes personality disorders). The individual has an excessive fear of being negatively evaluated, and in consequence being rejected, ridiculed, humiliated, etc. He may make exaggerated efforts to meet others' expectations and try to please at the expense of behaving according to active, as well as more differentiated, self-centered reactive, instinctive motives. These unnaturally intense fears of negative evaluation are frequently accompanied by physiological signs of overarousal—blushing, for example

(DSM-5, pp. 202–203), which by themselves constitute additional causes for embarrassment.

The genetic predisposition for this condition has been found to be quite strong (DSM-5, pp. 204–205). This innate predisposition, using our theoretical frame, seems to consist of an imbalance of the mental function that, in normative persons, strives to attain the optimal compromise between following egocentric active and reactive motives—against social pressures, if that is required, on one hand, and their alteration, postponement, concealment, etc., according to prevailing social norms, on the other hand. The aim of this mechanism (which loosely parallels the Freudian concept of ego) is to balance the gains and losses caused by uncompromised expression of individual instinctive motives with the gains and losses obtained by altering or forsaking individual aspirations, either entirely or partially, in order to attain better social acceptance. The optimal balance of these two contradictory strivings may change quickly in the continually changing flux of social life.

Owing to its high fitness-enhancing value in social animals with evolved mental capabilities, this mechanism has to have genetic roots. The relaxation from normalizing selection pressures in human populations created a wide spectrum of diversity in this mechanism's efficiency, both of whose extremes may be highly maladaptive. Any imbalance favoring the unchecked expression of individual interests in diverse social circumstances may incur strong social opposition and the withdrawal of various benefits resulting from social coexistence. On the other hand, an exaggerated imbalance toward pleasing others, observing to a disproportionate degree others' or the whole community's interests at the expense of suppressing to a more than necessary extent individual aspirations, is equally maladaptive. The dysfunctional and subjectively distressing consequences of this latter kind of imbalance are reflected in its extreme form in the pathology of social phobia.

It has to be stressed that the degree of complexity and scope of the particular social situation(s) that elicits the social anxiety reaction has a wide range of variability. Some persons show pathological levels of social anxiety only in a restricted range of complex social situations: for example, *performance anxiety*, which occurs during performances or lectures before an audience. These situations need, as a rule, adequate levels of self-confidence, self-control, and social competence (at least in their respective domains) and may be stressful to many normative individuals as well. At the other extreme of the continuum are persons who show intense anxiety when facing simple, commonplace social situations, like initiating a conversation, eating in public places, or urinating in a public lavatory—situations in which anxiety in normative persons is much less common. Social anxiety is further detailed in the next section.

7.2.2.4. Generalized Anxiety Disorder

At the diffuse pole of the diffuse/differentiated continuum of the reactive instinctive predispositions (meaning here the fear reaction, not counteraggression) lies generalized anxiety disorder. In this disorder, no specific natural or social triggers can be identified to which the individual differentially, preferentially, and consistently reacts with anxiety, and in consequence no specific reactive (avoidant) behavioral strategies can be built by either natural selection or acquired knowledge. The patient seems excessively anxious or worried concerning a wide range of natural/social circumstances or causes, from health and financial problems to facing job responsibilities or fulfilling household chores: "The focus of worry may shift from one concern to another" (DSM-5, p. 222). The clinical symptoms of this disorder (fatigue, problems with concentrating, irritability, muscle tension, sleep disturbance) are the outcome of this diffuse and continuous reactive arousal.

7.2.3. Posttraumatic Stress Disorder

POSTTRAUMATIC STRESS DISORDER

Diagnostic Criteria 309.81 (F43.10)

A. Exposure to actual or threatened death, serious injury, or sexual violence in one (or more) of the following ways:
   1. Directly experiencing the traumatic event(s).
   2. Witnessing, in person, the event(s) as it occurred to others.
   3. Learning that the traumatic event(s) occurred to a close family member or close friend. In cases of actual or threatened death of a family member or friend, the event(s) must have been violent or accidental.
   4. Experiencing repeated or extreme exposure to aversive details of the traumatic event(s) (e.g., first responders collecting

human remains; police officers repeatedly exposed to details of child abuse).

Note: Criterion A4 does not apply to exposure through electronic media, television, movies, or pictures, unless this exposure is work related.

B. Presence of one (or more) of the following intrusion symptoms associated with the traumatic event(s), beginning after the traumatic event(s) occurred:

1. Recurrent, involuntary, and intrusive distressing memories of the traumatic event(s).

Note: In children older than 6 years, repetitive play may occur in which themes or aspects of the traumatic event(s) are expressed.

2. Recurrent distressing dreams in which the content and/or affect of the dream are related to the traumatic event(s).

Note: In children, there may be frightening dreams without recognizable content.

3. Dissociative reactions (e.g., flashbacks) in which the individual feels or acts as if the traumatic event(s) were recurring. (Such reactions may occur on a continuum, with the most extreme expression being a complete loss of awareness of present surroundings.)

Note: In children, trauma-specific reenactment may occur in play.

4. Intense or prolonged psychological distress at exposure to internal or external cues that symbolize or resemble an aspect of the traumatic event(s).
5. Marked physiological reactions to internal or external cues that symbolize or resemble an aspect of the traumatic event(s).

C. Persistent avoidance of stimuli associated with the traumatic event(s), beginning after the traumatic event(s) occurred, as evidence by one or both of the following:

1. Avoidance of or efforts to avoid distressing memories, thoughts, or feelings about or closely associated with the traumatic event(s).
2. Avoidance of or efforts to avoid external reminders (people, places, conversations, activities, objects, situations) that arouse

distressing memories, thoughts, or feelings about or closely associated with the traumatic event(s).

D. Negative alterations in cognitions and mood associated with the traumatic event(s), beginning or worsening after the traumatic event(s) occurred, as evidenced by two (or more) of the following:

1. Inability to remember an important aspect of the traumatic event(s) (typically due to dissociative amnesia and not to other factors such as head injury, alcohol, or drugs).

2. Persistent and exaggerated negative beliefs or expectations about oneself, others, or the world (e.g., "I am bad," "No one can be trusted," "The world is completely dangerous," "My whole nervous system is permanently ruined").

3. Persistent, distorted cognitions about the cause or consequences of the traumatic event(s) that lead the individual to blame himself/herself or others.

4. Persistent negative emotional state (e.g., fear, horror, anger, guilt, or shame).

5. Markedly diminished interest or participation in significant activities.

6. Feelings of detachment or estrangement from others.

7. Persistent inability to experience positive emotions (e.g., inability to experience happiness, satisfaction, or loving feelings).

E. Marked alterations in arousal and reactivity associated with the traumatic event(s) beginning or worsening after the traumatic event(s) occurred, as evidenced by two (or more) of the following:

1. Irritable behavior and angry outburst (with little or no provocation) typically expressed as verbal or physical aggression toward people or objects.

2. Reckless or self-destructive behavior.

3. Hypervigilance.

4. Exaggerated startle response.

5. Problems with concentration.

6. Sleep disturbance (e.g., difficulty falling or staying asleep or restless sleep).

F. Duration of the disturbance (Criteria B, C, D, and E) is more than 1 month.

G The disturbance causes clinically significant distress or impairment in social, occupational, or other important areas of functioning.

H. The disturbance is not attributable to the physiological effects of a substance (e.g., medication, alcohol) or another medical condition.

Specify whether:

With dissociative symptoms: The individual's symptoms meet the criteria for post-traumatic stress disorder, and in addition, in response to the stressor, the individual experiences persistent or recurrent symptoms of either of the following:

1. **Depersonalization:** Persistent or recurrent experiences of feeling detached from, and as if one were an outside observer of, one's mental processes or body (e.g., feeling as though one were in a dream; feeling a sense of unreality of self or body or of time moving slowly).

2. **Derealization:** Persistent or recurrent experiences of unreality of surroundings (e.g., the world around the individual is experienced as unreal, dreamlike, distant, or distorted).

Note: To use this subtype, the dissociative symptoms must not be attributable to the physiological effects of a substance (e.g., blackouts, behavior during alcohol intoxication) or another medical condition (e.g., complex partial seizures).

Specify if:

With delayed expression: If the full diagnostic criteria are not met until at least 6 months after the event (although the onset and expression of some symptoms may be immediate).

Posttraumatic disorder (PTSD) represents a lasting maladaptive or dysfunctional response to a highly traumatic event. The diagnostic criteria and clinical features of the disorder are based on DSM-5, (see above) from which the quotations in this section are taken, if not otherwise specified.

The main idea of the present evolutionary interpretation of PTSD concerns a central aspect of active/reactive behavior interaction. As detailed in section 5.4.0, environmental stimuli of a dangerous or harmful nature promptly activate reactive behavior, which in turn suppresses any active behavior present. After the dangerous situation is over, the strong reactive behavior slowly subsides and the previously existent balance and flexible transitions between active and reactive behavior are reinstated. Since the disastrous consequences of the malfunctioning of this behavior-regulating mechanism can be easily imagined, it seems clear that in wild animals this mechanism's optimal functioning is tightly controlled by natural selection. However, as hypothesized throughout this text, RfNSPs in the protective human environment may relax and diversify a significant number of behavioral mechanisms. The symptomatology of PTSD strongly suggests that this is what happens in this mental condition. The subsidence of strong reactive behavior after a traumatic event is over, and the reemergence of active preoccupations or of the preexistent balance of active/reactive behavior, is considerably hampered or blocked altogether for a long period or indefinitely in this mental disorder.

Using this explanatory context, in the following sections, I discuss the symptom clusters that characterize this disorder.

7.2.3.1. The Nature of the Traumatic Event

The traumatic event has to be "severe enough to be outside the range of human experience usually considered as normal" (Andreasen, 1980, p. 1519). It is, as a rule, of a sudden, catastrophic, or recurrent nature. The person *has to experience it directly in a sensory way*.

Human tragedies of a much larger scale than those needed to elicit posttraumatic disorders in predisposed persons have no such effect if the information is conveyed in written language, verbally, or through the media (for example statistics of killed and wounded in a bloody war or in a large-scale natural calamity), or even lifelike broadcast reports (including photographs or motion pictures of such events).

This means that our highly evolved imagination and symbolic communicative abilities play no part in the causation of this disorder. Instead, it depends on an evolutionarily more ancient, direct, unmediated transaction with the physical/social world. The traumatic event has to target either the person herself, in which case physical trauma aggravates the disorder (car accident, natural disaster, rape), or affects other persons whose grave injuries or desperate or painful behavioral responses the patient directly witnesses. In this case, in my opinion, the mechanism of empathy (and sympathy) comes into play. The empathizing person "feels" acutely the directly involved persons' plight. That the empathizing mechanism may play a role in this case is reflected, too, by the only exception to this rule—that is, when PTSD can develop *without* the person being exposed directly to the traumatic situation. It may occur when one is informed that "a close family member or a close friend" was involved in a violent or accidental traumatic event. (As emphasized in the section on empathy, emotional closeness strengthens empathizing while emotional distance weakens it.) In all other instances, the person has to get the shocking information through some direct sensory-perceptual channel in order to be deeply influenced emotionally. In other words, our symbolic communication is immaterial here. For example, the verbal information that one suffers from a "life-threatening illness or debilitating medical condition is not necessarily considered a traumatic event" in this sense (DSM-5, 274).

7.2.3.2. The Oversensitivity to Cues That Remind One of the Traumatic Event, as Well as Its Vivid Reexperiencing

Reexperiencing aspects of the traumatic event and oversensitivity "at exposure to internal or external cues that symbolize or resemble an aspect of the traumatic event(s)" (DSM-5, p. 271) include the following: intrusive, distressing memories of the event, dreams with similar content, dissociative reactions (flashbacks—that is, brief, intrusive visual or other sensory experiences), overreaction to sudden unexpected noise if it was an aspect of the trauma, and so on. These cues, reminding the individual of aspects of the traumatic situation, evoke intense psychological distress and in consequence are avoided when possible. Such symptoms seem to be exclusive characteristics of posttraumatic disorders in humans. However, in a simpler form, expressed in behavior, as well as probably in simple emotionally charged memories, it may be present in more evolved nonhuman animals as well (although conditioned avoidant responses are already present in very simple organisms; see Lorenz, 1982, p. 279).

In cognitively more evolved animals, learned sensitization to clues of an experienced traumatic situation is well known. K. Lorenz described tellingly such an instance: "One of my dogs once got caught in a revolving door. Because of this single experience she not only meticulously avoided revolving doors wherever they happened to be, but she also fought shy of the whole vicinity of the building in which the mishap had occurred" (Lorenz, 1982, p. 279). He comments on the phenomenon in general: "The conditioning is irreversible—probably the effect of the shattering, unforgettable, terror-eliciting nature of the experience" (idem, p. 278). As was stressed in the section on the mechanism of empathy, no one has direct knowledge of the subjective experience of another person, and even less of a nonhuman animal. However, the similarity of the behavior induced by the traumatic event (that is, similar behavior in similar circumstances) suggests strongly that we are dealing with similar instinctive mechanisms (and probably similar emotional accompaniments in the subjective experience). It has to be mentioned here that one of the symptom clusters of human PTSD concerns the "[p]ersistent avoidance of stimuli associated with the traumatic event(s)" (DSM-5, p. 271).

I think that the longstanding or permanent sensitization to subtle cues of a dangerous situation, after an animal luckily escapes its worst consequences, has to be an adaptive behavioral phenomenon. It may aid the same animal in detecting earlier the possible recurrence of a similar situation, and thus improve its chances of escaping it (for example, sensitization to the smell, sound, and subtle visual clues of a surreptitiously approaching predator). Witnessing the fate of another conspecific who has fallen victim to the same dangerous event (predation, for example), or seeing the escape behavior of more mature, more experienced fellow conspecifics and empathizing with this behavior (remember the discussion on the "mirror neurons" in section 2.3), may also contribute to the early realization of the seriousness of the respective danger.

However, as Lorenz's example shows, *in animals, longstanding sensitization is narrowly restricted to the traumatic circumstances*, and, even more importantly, it does not lead to a protracted, comprehensive reactive arousal and, as a result, to a comprehensive suppression of active behavioral manifestations. In other words, the sensitization does not hamper the recapturing of the formerly existent balanced active/reactive behavior interrelationship outside the domain of the traumatic situation.

7.2.3.3. Marked Alterations in Arousal and Reactivity

The next symptom cluster associated with posttraumatic states is *"marked alterations in arousal and reactivity."* In posttraumatic disorders this means, in the context of this text, a protracted increase in the reactive arousal only (but not of active arousal, which happens as a rule when sensing the presence of an appropriate releaser). Both the continuous (reactive) arousal and the reactivity to stimuli that remind the individual of the traumatic event may attain high levels of intensity, and, even more importantly for us, may last for long periods after the traumatic event comes to an end. The diagnostic criteria in this symptom cluster comprises "irritable behavior . . . [h]ypervigilance . . . [e]xaggerated startle response . . . [s]leep disturbance . . . [m]arked physiological reactions to internal or external cues that symbolize or resemble an aspect of the traumatic event(s)" (DSM-5, diagnostic criteria, see above). The impression is that these individuals continue to live (partially) in the traumatic situation and are unable to return to their former (pretraumatic) balance of active/reactive behavior interaction.

In my opinion, this symptom cluster holds the clue to the understanding of the evolutionary mechanism that brought about in humans the possibility of developing posttraumatic disorders. In order to make this clear, let me remind the reader of the relevant points regarding active/reactive behavior interaction in general.

When a disturbing or dangerous stimulus is sensed, the animal instantly suspends whatever active behavior it is engaged in and does not resume it until the danger was dealt with successfully. Instead, in reaction to the danger, if needed, the organism mobilizes all its energetic resources even to the point of their complete exhaustion (K. Lorenz in section 5.4.0). After the danger is over, the reactive arousal (physiological, behavioral, and mental) gradually subsides, giving way to rest if practicable, to the redressing of possible somatic damages, and to recuperation of the loss in energetic or other resources caused by the traumatic event. (Reactive behavior, unlike active behavior, is not built by evolution to last for long periods or to wax and wane periodically. Apart from continuous low-level alertness for signs of danger, the actual reaction, as a rule, is relatively short and its intensity is tailored to the level of dangerousness of the precipitating cause.)

Active behavioral preoccupations (feeding, parenting, and so on) are also resumed. While the switchover from active to reactive behavior, if needed, may happen in fractions of a second, reassuming active behavior after the danger is over takes some time. Even in animals, the return to former active behavior (or active/reactive behavior balance) may take much more time than the transition from active to reactive behavior. Remember Lorenz's example of cats frightened by dogs needing about half an hour to be interested in catching prey again (section 5.4.0). Since environmental danger may strike unexpectedly and may endanger physical integrity or survival itself (while active needs are, as a rule, postponable for shorter or longer intervals), this timescale in the interaction between active and reactive behavior makes adaptive sense.

However, unnecessary prolongation of the reactive arousal after the traumatic event is over may be costly to the organism in energetic terms, as well as suppress unnecessarily badly needed rest and the resumption of active preoccupations, which is clearly maladaptive. Consequently, it is presumed that natural selection in natural conditions would promptly eliminate any considerable unwarranted delay in the subsiding of strong reactive arousal.

In the human environment, however, especially in a technologically advanced one, these selective pressures became considerably relaxed. No one in normal conditions needs to continuously scan her environment for signs of unexpected danger to which she is compelled to react physically in fractions of a second. The same is true concerning the unnecessary prolongation of the reactive arousal after the dangerous event is over. The fitness-reducing effect of this prolongation of the reactive arousal, in a biological sense—that is, with regard to survival and reproduction—is greatly reduced in humans. Therefore, the relationship between active behavior, reactive behavior, and rest, while controlled tightly by natural selection in the wild, is far less so in human populations. It seems reasonable to assume that the ease with which a strong reactive arousal can be elicited in a human individual; the intensity, efficacy, and speed of its somatic and behavioral elements; and, even more importantly for our present subject, the rate of its progressive subsidence after the traumatic event is over and the rate of recapturing former active intentions, show a wide scale of quantitative diversification.

Some persons are aroused easily and strongly by dangerous situations or by witnessing the effect of a traumatic event in others, and the return to the pretraumatic mental and physical functioning is prolonged. (It may be that this prolongation of the physiological aspects of this reactive arousal contributes to some psychosomatic disorders.) The extreme end of the diversified scale of the reactive instinctive configuration in this respect probably constitutes the innate predisposition for grave posttraumatic disorders. Whether the greatly reduced pace of the fear reaction's subsidence is a continuous, graded phenomenon in these individuals (subsiding more slowly the greater the fear reaction's intensity), or alternatively constitutes a kind of threshold phenomenon (that is, if the fear reaction's intensity surpasses a certain level, its disappearance will be greatly hampered), I cannot judge. Nor am I aware of any theoretical formulations of this evolutionary mechanism and related research findings.

At the other extreme of the same scale are situated individuals who can function exceptionally well under dangerous conditions and may even enjoy their ability to do so. Naturally, they react efficiently in these situations and return quickly to their former balance of active/reactive behavior interrelationship. Examples include those who are employed to provide first aid at the places where traumatic events occur; those who enjoy extreme sports; or soldiers who, after healing from a serious physical injury, return willingly to the same combat activity that caused the injury in the first place. It is common knowledge that in cases of PTSD "the stressor is a necessary but not sufficient cause, because even the most severe stressors do not produce a posttraumatic stress disorder in all persons experiencing that stressor" (Andreasen, 1980, p. 1519).

Moreover, many individuals with acute stress reactions recover within several days or months. Even among those who suffer from a more protracted posttraumatic reaction, a "large minority of patients develop complete remissions, whereas another large group exhibit only mild symptoms" (Pine & McClure, 2005, p. 1777). All these data speak on behalf of the dimensional and not categorical nature of the disorder (that is, on behalf of the existence of continuous gradations between exceptional resistance to traumatic situations at one end and unrecoverable damage to normative functioning at the other).

Of course, not all the variables that predispose to PTSD are of the nature just described. The relevant literature stresses the importance of the availability of social support, the age of the subject (very young and old persons being at increased risk), former traumatization, the predisposition to anxiety and depression, physical injury during the trauma besides the mental stress, and the prospect of monetary compensation. Nevertheless, *the genetic element specifically targeting the innate reaction to traumatic situations seems to me to be the decisive consideration.* An extensive twin study concerning veterans of the Vietnam War found heritability for fifteen PTSD symptoms to be around 40 percent, and no evidence of shared environmental influence (Plomin et al., 2008, p. 218).

7.2.3.4. The Effect of the Long-Lasting Suppression of Active Instinctive Motives

The next symptom cluster concerns *the effect of long-lasting suppression of active instinctive motives by the strong and protracted reactive arousal.* When active instinctive motives (in their appetitive phase) are suppressed for longer periods of time, several things can happen (I refer here to active instinctive predispositions that were of a normative kind before the trauma):

1. They may intensify progressively and surface in consciousness in an intrusive manner or erupt in overt behavior in a poorly controllable fashion. This can happen in the least suitable social situations.
2. *Active instincts* suppressed for longer periods tend to transform into one or more of their forms of frustration: displacement, aggression, or dysphoria. When they surface in overt behavior, the frustrated form contributes further to the inappropriateness or undesirability of the behavior induced by them. In posttraumatic patients, active drives tend to transform mostly into their aggressive or dysphoric form of frustration or into a transitional state between the two.
3. Active instincts in their appetitive phase suppressed for long periods have no opportunities to contact and interact with suitable releasers and, in social situations, to "negotiate" their most appropriate expression. Therefore, when they ultimately erupt in overt behavior (I mean sexual, self-assertive, parenting, etc., motives), this lack of previous experience also contributes to their inappropriateness.

Some relevant diagnostic criteria of posttraumatic disorder in the DSM-5 classification illustrate the behavioral and subjective mental outcomes of these mechanisms: "Persistent inability to experience positive emotions (e.g., inability to experience happiness, satisfaction, or loving feelings)," "Markedly diminished interest or participation in significant activities," "Reckless or self-destructive behavior," "Angry outbursts (with little or no provocation)" (DSM-5, diagnostic criteria, see above).

7.2.3.5. Secondary or Tertiary Consequences

The last symptom cluster I wish to mention here refers to secondary or tertiary consequences the dysfunctional instinctive processes may have in the conscious awareness of the individual regarding his current environment and self-appreciation. It can be expected that both changes will be negative. The environment will be seen as less satisfying and more dangerous than it actually is, and the person will appreciate himself less than before the trauma. The relevant diagnostic criteria that describe these changes are: "Persistent and exaggerated negative beliefs or expectations about oneself, others, or the world (e.g. 'I am bad,' 'No one can be trusted,' 'The world is completely dangerous')" (DSM-5, see above).

7.2.4. Personality Disorders

How can we define from an evolutionary (rather than clinical) perspective the common characteristics of that comprehensive nosological category called personality disorder as distinct from anxiety (neurotic) disorder? What would be the common characteristics of such strongly dissimilar behavior patterns as those encountered in obsessive, antisocial, schizoid, or borderline personality disorder?

On a descriptive level, the basic guidelines for answering these questions are provided by clinical psychiatry. Personality disorders are pervasive, inflexible behavior patterns that crystallize in childhood or adolescence and continue, as a rule, to characterize an individual's demeanor throughout his life span. Most, perhaps all, behavioral characteristics that lead to the diagnosis of a personality disorder may also exist as personality traits in normative individuals at an intensity that does not cause serious problems in social coexistence, everyday functioning, or (less characteristically) in the intrapsychic domain. On the other hand, a dysfunctional, inflexible behavior pattern, at the intensity at which a diagnosis of personality disorder is warranted, does lead to serious interpersonal tensions, defective socialization, other difficulties in functioning in a social setting, and sometimes (as in borderline personality, for example) to internal distress.

No strong internal incentive exists in these disorders to please, conform, or adapt, in contrast to normative persons or those with certain anxiety disorders, like social phobia. When these inflexible behavior patterns are suppressed, as a rule for short periods of time, such a maneuver is always employed in the expectation of an immediate external reward that will ultimately gratify these inflexible instinctive motivations (for example, when an intelligent, nonviolent, antisocial psychopath flatters a naïve prospective victim, or when, in dependent personality disorder, there is exaggerated compliance to preserve a dependent relationship).

The realization, when present, by the patient of the maladaptive nature of the behavior is not a strong enough incentive for altering it toward a more socially adaptive one. The patient feels, as a rule, that the intention leading to the dysfunctional behavioral traits of the respective personality disorder originates from a deep "layer" of her own being (that is, it is "ego-syntonic"), while the pressures to change the dysfunctional behavior come from an outside source, from impersonal social rules or directly from surrounding people (rather than from internal emotional incentives like anxiety, or the need to please, to be accepted). In the context of the present theory, what originates from deep inside "layers" of one's being are first of all the innate instinctive predispositions, including their subjective emotional and intentional aspects. According to this viewpoint, personality disorders can be redefined as the *behavioral and intrapsychic manifestations of a dysfunctional instinctive configuration that cannot be adequately modified by, adapted to, or synchronized with social requirements to conform.* However, it has to be stressed that the behavior induced by this kind of imbalance remains realistic enough to be legally accountable.[3]

The reader may remember the discussion of the extent that social influences in a normative human population can give rise to the wide quantitative diversification of individual active and reactive motives (section 5.4.3). At one end of this spectrum are those individuals who are unwilling to suppress or modify these individual motives, if at all, without considerable social pressure, such as threats of withdrawal of the benefits of social coexistence, threats of ostracism, or actual punishment. At the other end of this scale we can identify those who willingly, by their own free choice, sacrifice individual interests for the needs or ideals of the group. It was proposed also that submission (from which more mature forms of socially desirable behavior, like conformity or internalization, can evolve) has a clear innate foundation (section 5.4.2).

The attainment of a desirable *equilibrium* between self-centered instinctive motives and the pressure for their modification in order to suit group interests, or, alternatively, the longstanding *failure* to attain such equilibrium, seems to depend on one or both of the following factors:

1. The first factor concerns that high-level brain center whose existence has to be hypothesized in order to account for the need to coordinate different, partially innate, behavior patterns. As mentioned before, Freud called this center the "ego," while Konrad Lorenz, intending a more encompassing concept, called it the "superior command center." It seems reasonable to assume that in the relaxed conditions of natural selection of the human environment the synchronizing/synthesizing capabilities of this center also became diversified, allowing for a less than optimal interrelationship between self-centered and socially prized forms of behavior.

2. The second factor seems more relevant to our subject matter. It concerns the possibility that one or more comprehensive instinctive mechanisms (of those discussed in this text) lie at one of the extremes of their diversified continuum, overstraining the capabilities of an otherwise normatively functioning "superior command center" to an extent that this brain center can no longer adequately compensate for the resultant imbalance.

I have in mind such extreme instinctive states or configurations as long-standing, excessively high or low intensities of active and self-centered reactive drives leading to hyperthymic or dysthymic personality traits or disorder; prominently diffuse or differentiated drives predisposing to schizoid or antisocial personality disorder respectively; the predilection of active and reactive drives to be expressed mostly in their aggressive form, leading to choleric or impulsive personality traits; or the tendency of these drives to be stuck for long periods in their aggressive form of frustration after the aggression-eliciting situation is over, leading to paranoid personality traits or paranoid personality disorder (detailed later).

In the following, I illustrate this reasoning with a short consideration of antisocial, schizoid, and borderline personality disorders.

7.2.4.1. Antisocial Personality Disorder

Perhaps the most telling illustration of the foregoing considerations can be found in antisocial personality disorder. It is hypothesized in this theory that underlying the predisposition for antisocial personality (in the clinical, not legal sense) are the following instinctive predispositions: The active instinctive drives lie at the extreme differentiated end of the differentiated-diffuse continuum. A person with this type of active drives "knows" clearly from an internal source what exactly his needs for active instinctive gratification are; what the appropriate behavioral tactics are that lead most straightforwardly to contact or achieve the desired releaser (in a social setting), as well as what kind of releasers best suit his differentiated, goal-oriented instinctive intentions (sexual object, material goods, a certain subculture, drugs that intensify the subjective experience of instinctive gratification, and so on). His mating strategy is predominantly or exclusively of the "mammalian type" (that is, focused on sexual access, instead of the "bird type"; see section 5.3.2), and manifests itself early during sexual maturation. This consideration may explain the great difficulties that an antisocial personality has in sustaining a long-term relationship with a mate and investing in bringing up offspring.

In consequence, his perception of the social milieu becomes excessively selective (the complete opposite of the predisposition to the schizophrenia spectrum of disorders), disregarding social stimuli, or consciously transgressing social requirements irrelevant to the narrow path he is taking toward his instinctive goals. Stimuli originating from the more nuanced, more complex, more tradition-and-culture-bound prescriptions and practices of the respective human community are bypassed. (As a more general statement, it seems that to become adequately susceptible to more sophisticated cultural influences, a certain measure of *uncertainty* about one's internal, more biologically oriented, incentives is needed.)

In the aggressive type of antisocial personality, the empathizing process is weak or almost nonexistent. In the nonaggressive type, the empathizing process may be well developed, but instead of facilitating adaptive relationships, the information obtained about other persons' mental constitution is exploited for manipulation or cheating (Brüne, 2008, p. 312). When obstructed in their intent to contact the desired releaser, the active instinctive intentions tend to transform into their aggressive form of frustration rather than into displacement, vacuum behavior, or dysphoria.

With regard to reactive behavior, the instinctive predispositions in this domain are also more than normatively differentiated. The person possesses clear instinctive guides to how to escape, counteract, or disregard unwanted or harmful social stimuli or circumstances, and feels no need to learn socially desirable practices in order to avoid or tackle these situations (such as avoiding physical violence when possible; calmly defusing vehement aggressive emotions in others; appealing to bystanders or law enforcement agencies for help; and so on). However, these overly differentiated reactive instinctive predispositions (mostly of a physical nature—that is, threatening or acting violently, or fleeing) of the antisocial personality are useful only in rather simple disturbing or dangerous social situations, resembling those that occur as a rule in natural conditions (such as physical confrontation). They are quite useless or even detrimental in the case of more complex interpersonal conflicts or contradictory personal interests. For solving these later conflictual situations, organized human communities always have approved-of rules and practices.

In most cases of the aggressive type of antisocial personality disorder, the preferred form of reaction is an aggressive one, although escape reactions of a physical kind (in which the individual does not apologize or explain that he did not mean it) may also occur. In the nonaggressive type of antisocial personality disorder, the avoidance reaction rather than confrontation seems to be the rule. In both cases, however, the differentiated reaction expresses the instinctive arousal effectively, leading to its quick subsiding and the prompt recapturing of former active instinctive motivations.

Since the dysphoric form of active instinctive frustration is weak, short, or entirely absent, these persons do not suffer from regret, guilt, or self-accusation. On the other hand, because the reactive instincts are more than normatively differentiated, they do not experience any diffuse anxiety either, neither during a confrontation nor in its anticipatory stage. (It is perhaps superfluous to mention that conformity to or internalization of social values is nonexistent or barely developed in these individuals, and submission is achieved, as a rule, only with the aid of severe social pressure or punishment.)

It seems that the instinctive constitution just described can account quite well for the clinical symptoms and signs of Antisocial Personality Disorder. (See DSM-5, Diagnostic Criteria, Diagnostic Features, and Associated Features Supporting Diagnosis, pp. 659–661.)

In spite of the grave social consequences that frequently incur stringent forms of punishment, this behavior pattern remains a lifelong disability (with partial improvement, as a rule, in the fifth decade of life when the intensity of instinctive strivings begins to diminish).

The rigid nature of this behavior (a characteristic of all personality disorder categories), which is not amenable to considerable modification by social learning and experience (by operant conditioning, for example), suggests the strong involvement of genetic predispositions in the causation of the disorder. Genetic studies have indeed demonstrated a clear innate component in antisocial personality disorder (DSM-5, pp. 661–662; Plomin et al., 2008, pp. 255–259).

7.2.4.2. Schizoid Personality Disorder

SCHIZOID PERSONALITY DISORDER

Diagnostic Criteria 301.20 (F60.1)

A. A pervasive pattern of detachment from social relationship and a restricted range of expression of emotions in interpersonal settings, beginning by early adulthood and present in a variety of contexts, as indicated by four (or more) of the following:
1. Neither desires nor enjoys close relationships, including being part of a family.
2. Almost always chooses solitary activities.
3. Has little, if any, interest in having sexual experiences with another person.
4. Takes pleasure in few, if any, activities.
5. Lacks close friends or confidants other than first-degree relatives.
6. Appears indifferent to the praise or criticism of others.
7. Shows emotional coldness, detachment, or flattened affectivity.

B. Does not occur exclusively during the course of schizophrenia, a bipolar disorder or depressive disorder with psychotic features, another psychotic disorder, or autism spectrum disorder and is not attributable to the physiological effects of another medical condition.

Note: If criteria are met prior to the onset of schizophrenia, add "premorbid," i.e., "schizoid personality disorder (premorbid)."

The next personality disorder reflecting the common characteristics of personality disorders is the schizoid personality disorder.[4] The clinical picture consists of a "pervasive pattern of detachment from social relationships and a restricted range of expression of emotions in inter-personal settings" (diagnostic criteria, see above).

As argued in this text, *the basic inherited predisposition for schizo-phrenia and its spectrum disorders (including schizoid personality) is a more than normative extent of diffuseness of the active and individual-istic reactive instinctive predispositions on their differentiated/diffuse continuum.* At this level of diffuseness, the instinctive predispositions lose much of their ability to become fused with socially originating mes-sages and in consequence to guide the person effectively toward fitness-enhancing avenues amid the intricate web of social life. The more than normatively diffuse nature of the instinctive predispositions has sec-ondary consequences that may interfere powerfully with good inter-personal communication and social adaptation. The diffuse nature of the appetitive phase of active drives hampers the selective perception of those stimuli that inform the individual about the social resources present for her gratification, the socially sanctioned routes to approach the desired releasers, the potentially opposing influences, and so on. This is true particularly when the goal is a distant one and the route toward it, among complex, changing social circumstances, is tortuous. In ad-dition, the diffuseness of the drives hampers the (mostly unconscious) nuanced expression in body language and verbalizations of diverse in-stinctive intentions and their accompanying emotional states, reducing the possibility of their adequate decoding by others with a normatively differentiated instinctive constitution through the mechanism of em-pathy. Conversely, patients with diffuse drives cannot grasp with the same empathizing mechanism others' emotional and intentional states, especially when they are not expressed explicitly and unequivocally by spoken language.

Attempting to cope with this situation, the affected person has in principle several alternatives, of which the following two are used most frequently:

1. The first alternative may consist of organizing one's life in a way that allows for or facilitates the expression of diffuse drives in behavior or mental activity while minimizing the undesirable consequences of social inaptitude. The best strategy in this sense seems to be *the reduction of social encounters* to an indispensable minimum, espe-cially those of an intimate or emotionally charged, intrusive nature,

and instead the search for ways to express the individual's peculiar instinctive constellation in interactions with less-complex, less-demanding objects than human conspecifics—that is, in a *solitary lifestyle*. This lifestyle may include interactions with nature, with pets, or with other domesticated animals; investing mental energies in artistic or scientific pursuits (when the person possesses the necessary mental endowment for it) that do not require close contact with other conspecifics; and experiencing passively the life and deeds of imaginary characters while reading fiction, seeing movies, etc. In this way, the schizoid person can attain a certain mental stability, in which active instinctive motives are expressed on a more-or-less permanent basis. This stability may be lifelong or fragile, in which case it may be interspersed by acute psychotic episodes.

2. The second main option for avoiding the expression of diffuse drives in a social setting is a peculiar kind of *dissociation*. The overt behavior in social settings will be guided by submission to or conformity with the parental and later with the wider community's clearly stated expectations, without channeling into it active or individual reactive instinctive motives. Since the level of intelligence (abstract, not social) and learning abilities of most schizophrenics is in the normative range, and since the reactive tactics of many pre-schizophrenics are fueled by diffuse anxiety, this avenue is a real possibility indeed. Several psychiatrists identifying with the psychodynamic or existential traditions have recognized and described it, as will be detailed in the section on schizophrenia.

The schizoid personality evidently "chooses" the first alternative, and, maybe as a consequence, escapes or at least reduces the probability of developing a psychotic disease with a chronic or deteriorating course. Of course, this is not a conscious choice but may be determined by some specific aspects of the instinctive configuration or the environmental influences or both. For example, the reactive instinctive predispositions, instead of accentuating diffuse anxiety, may tend to induce avoidance of undesirable social pressures; or the active instincts may be less extremely diffuse than in those predestined to take the second alternative, allowing a higher level of independent functioning; or the external (parental, social) pressures are less intense, permitting a wider latitude of personal choices.

The DSM-5 diagnostic criteria for schizoid personality disorder support, to my mind, this reasoning. A person with schizoid personality disorder: "Neither desires nor enjoys close relationships . . . [a]lmost always choses solitary activities . . . [l]acks close friends . . . [a]ppears indifferent to the praise and criticism . . . shows emotional coldness, detachment or flattened affectivity . . . [h]as little, if any, interest in having sexual experiences with another person" (DSM-5, diagnostic criteria, see above). In that last criterion, the accent is on the "with another person." Many schizophrenics enjoy sexual self-stimulation, since their problem pertains to the appetitive phase of the respective instinctive striving, not to the consummatory phase.

With regard to the quality of affect, described in DSM-5 as "restricted range of expression of emotions . . . indifference . . . coldness, detachment or flattened affectivity," I am inclined to think that the basic problem here, as mentioned, is not with the *intensity* of the affect, but rather with the inability or reduced ability to express vividly and precisely in body language or verbalizations (including tonality of the speech and the choice of emotionally laden expressions) distinct, nuanced emotional states recognizable as such through the mechanism of empathy. Diffuse instinctive predispositions lead to diffuse, poorly differentiated behavioral arousal, which, as mentioned in the discussion on antipsychotic drug effects (section 6.2.1), may induce either diffuse behavioral activation or alternatively inhibition (paralysis) of the overt behavior. Persons with a different instinctive configuration will find it difficult to interpret these emotional expressions through the mechanism of empathizing.

Moreover, the social withdrawal itself prevents the partial compensation for these innate disabilities by acquiring though learning, experience, and imitation the more nuanced emotional expressions that normative persons are equipped with through inheritance, or by perfecting these abilities when the inheritance is inadequate. (In spite of these efforts, these learned or perfected emotional expressions may always seem somewhat unnatural or clumsy. However, even in this imperfect form, they may still improve interpersonal communication.)

7.2.4.3. Borderline Personality Disorder

BORDERLINE PERSONALITY DISORDER

Diagnostic Criteria 301.83 (F60.3)

A pervasive pattern of instability of interpersonal relationships, self-image, and affects, and marked impulsivity, beginning by early adulthood and present in a variety of contexts, as indicated by five (or more) of the following:

1. Frantic efforts to avoid real or imagined abandonment (Note: Do not include suicidal or self-mutilating behavior covered in Criterion 5.)
2. A pattern of unstable and intense interpersonal relationships characterized by alternating between extremes of idealization and devaluation.
3. Identity disturbance: markedly and persistently unstable self-image or sense of self.
4. Impulsivity in at least two areas that are potentially self-damaging (e.g. spending, sex, substance abuse, reckless driving, binge eating). (Note: Do not include suicidal or self-mutilating behavior covered in Criterion 5.)
5. Recurrent suicidal behavior, gestures, or threats, or self-mutilating behavior.
6. Affective instability due to a marked reactivity of mood (e.g., intense episodic dysphoria, irritability, or anxiety usually lasting a few hours and only rarely more than a few days).
7. Chronic feelings of emptiness.
8. Inappropriate, intense anger or difficulty controlling anger (e.g., frequent displays of temper, constant anger, recurrent physical fights).
9. Transient, stress-related paranoid ideation or severe dissociative symptoms.

It would seem we must look for the basic genetic predisposition for borderline personality disorder in the way the three forms of frustration of active instinctive drives (displacement and vacuum behavior, aggression, and dysphoria) alternate and express themselves in interpersonal relationships and subjective experience. (This is in addition to the predisposition for personality disorders in general—the predominance of self-centered active and reactive drives over conformity with the requirements of social coexistence.) The predisposition for the peculiar pathology of borderline personality disorder consists of *brusque, unattenuated, complete transitions* between the three forms of frustration, while the respective frustrated forms are intense and may last for a considerable time.

In normative behavior, cognitive processes moderate and counteract to a certain extent the distorting influence of crude instinctive strivings, both in their original as well as their frustrated forms. The memories on how we experience our social environment, our close relationships, and ourselves when uninfluenced by strong emotions confer a certain measure of stability on the way we grasp the surrounding realities and on ourselves in spite of occasional emotional fluctuations.

In the borderline personality disorder, exactly the opposite happens. These brusquely changing, intense, active emotional motives, instead of being moderated in their impact on cognitive processes, overpower and distort the individual's cognitive assessment of the surrounding realities, giving to this mental condition a pronounced instability in regard to interpersonal relations and self-appraisal.

Let's see how these mental events are reflected in the diagnostic criteria of borderline personality. The following quotations are some of the diagnostic criteria of the disorder according to DSM-5 (see above): Borderline personality features a "pattern of unstable and intense interpersonal relationships characterized by *alternating between extremes of idealization and devaluation*" (emphasis added). I interpret this statement in the following way: These individuals have an instinctive predisposition to form strong loving or attachment bonds in which, under the influence of these instinctive motives, only the desirable aspects of the love object are seen, accompanied by the dominant emotion of pleasure. While the desired aspects of the loved person are grossly exaggerated (or even invented), the qualities that are discordant with the predominant emotional expectations are minimized or disregarded. This relationship toward an environmental agent, distorted so grossly in imagination, seems to me to resemble displacement or even, in a certain sense, vacuum behavior. The qualities of the adored person exist mostly in the mind of the individual, not in the reality. In other words, under the pressure of a strong active instinctive motive, he sees what he wants to see, not the real person.

In a situation like this, the surfacing of disappointments is only a question of time. When they occur, the exaggerated, unconditional love or attachment brusquely changes into devaluation, hatred, or aggressiveness toward the same person, or into dysphoria mixed with aggressiveness (that is, an intense mental pain frequently expressed in this condition as an intention to damage one's own body). In the DSM-5 diagnostic criteria, these changes are described in the following way: "Affective instability due to a marked reactivity of mood . . . Inappropriate, intense anger or difficulty controlling anger . . . Recurrent suicidal behavior, gestures, or threats, or self-mutilating behavior . . . Identity disturbance: markedly and persistently unstable self-image or sense of self" (see above).

Since the active instinctive energies are expressed uninhibitedly in their various forms of frustration in stormy behavior, no or only a reduced possibility exists of canalizing them into more stable, more balanced, more socially mature forms of relationship and activities. Therefore, after the uninhibited emotional discharge is over, the patient suffers from "[c]hronic feelings of emptiness" (DSM-5, diagnostic criteria, see above).

It is worth noting that in borderline personality disorder all three forms of frustration are expressed in approximately equal frequency and intensity, contrary to other mental disorders in which one of the three forms of frustration is preferentially accentuated, as in paranoid disorders (the theme of aggression), or dysphoric agitated depression (the dysphoric form of frustration).

Borderline personally disorder is highly heritable (DSM-5, p. 665), in spite of being a strongly dysfunctional behavioral pattern, exemplifying once again the effect of relaxation of natural selection pressures in humans on the transmission of genes underlying behavioral predispositions.

It seems strange that such a highly unstable behavior like that encountered in borderline personality can lead to the pathology of personality disorder at all. Personality disorder is defined as "an enduring pattern of inner experience and behavior . . . [which is] pervasive and *inflexible*" (DSM-5, p. 645, emphasis added). I suspect the explanation has to be looked for in the uninhibited nature of the expression (acting out) of the dysfunctional active instinctive constitution in the interpersonal sphere in spite of causing great damage, the characteristic of personality disorders in general. This is probably the reason why an episodic course (with or without genuine psychotic symptoms) is lacking in such a labile and strongly reality-distorting condition.

––––––––––

At this point of our discussion it is worthwhile to summarize succinctly the repercussions (both desirable and maladaptive) of the diversification (and partial disintegration) of the second comprehensive instinctive mechanism in our theoretical scheme, namely the three forms of transformation that active instinctive motives undergo in frustrating conditions.

1. The flexible *transitions* between the *unfrustrated and frustrated forms* of active instinctive motives, as well as the flexible transitions *between the three frustrated forms* (displacement and vacuum behavior, aggression, and dysphoria), may be adaptive, even desirable, in the right circumstances and to the right extent, especially when an individual has to function well in complex, challenging, and quickly changing social circumstances. All three of the frustrated forms prevent the accumulation of undesirable mental tensions; in addition, the aggressive transformation may remove obstacles in the way of active instinctive gratification; and the dysphoric form may mobilize help from surrounding people.

2. The possibility of displacing or redirecting diverse elements of the instinctive activity (intentionality, energetic charge, rudimentary innate motor patterns, choice of a releaser) from a naturally adaptive target (as in lower animals) toward a naturally irrelevant one (concerning fitness), in the case of humans makes feasible the channeling of these instinctive elements into culturally created avenues, avenues that may be highly desirable in an individual's human community. However, the same possibility, that is, displacing adaptive instinctive motives in frustrating conditions, or discharging them "in vacuo," in relaxed selective conditions, may divert the behavior (and subjective experience) toward sterile daydreaming or other unprofitable avenues.

3. If a person's original active instinctive motives are predisposed to transform readily into one of their frustrated forms (that is, at a low threshold of frustration), their fitness-enhancing potential may be prematurely lost. If these transformations are intense and

longstanding—that is, if the original, fitness-enhancing motive cannot be recaptured readily after a frustrating event is over—it may lead to psychopathology. As we have seen, this is the case with agitated, dysphoric depression. Later in this chapter, it is argued that a similar mechanism is present concerning the innate predisposition to paranoid disorders, this time involving the aggressive from of frustration.

4. As we have seen, the brusque, complete transitions between the three frustrated forms, unmitigated by the correcting or compensatory effect of cognitive mechanisms, underlie the pathology of borderline personality disorder.

7.2.5. Comparing and Contrasting Corresponding Pairs of Personality Disorders and Episodic Mental Conditions

In my early years of studying psychiatry, I was puzzled by the multitude of corresponding pairs of clinical disorders, one of which consisted of a lifelong, stable, dysfunctional behavior pattern, while the other appeared periodically, was interspaced by periods of remission, or had a waxing and waning or deteriorating course. These matching pairs of mental disorders have striking similarities with regard to characteristic symptoms and frequently got the same denomination followed by the qualification of personality versus anxiety, affective or psychotic disorder. Here is the list:

1. Hyperthymic personality disorder—Bipolar disorder, predominantly manic episodes
2. Dysthymic (depressive) personality disorder—Depressive episodes
3. Cyclothymic personality disorder—Bipolar disorder
4. Paranoid personality disorder—Delusional disorder (formerly delusional [paranoid] disorder, DSM-III R), persecutory and jealous subtypes
5. Schizoid or schizotypal personality disorder—schizophrenia or schizophreniform psychotic episodes
6. Avoidant personality disorder—Social anxiety disorder (social phobia)

7. Dependent personality disorder—Separation anxiety disorder
8. Obsessive-compulsive personality disorder—Obsessive-compulsive disorder

Sometimes both members of a pair may exist in the same individual. Obsessive-compulsive patients also have obsessive-compulsive personality disorder in about 23–32 percent of cases (DSM-5, p. 242), and a schizophrenic patient may also have a schizoid premorbid personality, but this is by no means the rule.[5]

The search for an explanation of this phenomenon may prove worthwhile, since it may hold the key to why the same or a similar underlying pathological mental mechanism may manifest in some individuals as a stable inflexible behavior pattern lasting for a lifetime, yet in others as an episodic, fluctuating, or deteriorating condition.

I believe that underlying each pair of the list (with the possible exception of the last) are the *same basic originally adaptive mechanisms* that evolved during our phylogeny. With regard to the first three pairs, the underlying evolutionary mechanism is the one that determines the optimal overall intensity of the instinctively fueled behavior in the long run, as well as its fluctuations according to the seasonally changing amounts of life-sustaining resources. This mechanism became dysfunctional as a result of the changed conditions of the specifically human environment (when contrasted with the natural one), leading to RfNSPs and consequent diversification (as detailed in sections 5.1 and 5.1.3).

In the fourth pair of personality versus episodic disorders, the underlying adaptive mechanism concerns the aggressive reactive tactic, as well as the aggressive form of frustrated active drives. One kind of diversification of this mechanism, resulting from the relaxed selective pressures of the human environment, created at one extreme of its graded continuum an instinctive configuration in which the aggressive transformation of these drives became unduly protracted. In other words, *these drives became unable to revert to their original form (in the case of active drives) or to dissipate (in the case of aggressive-reactive tactics) for a long time after the aggression-eliciting situation is over.* This line of reasoning is further elaborated in the section on paranoia (7.2.6.).

In the fifth pair, the more than normative diffuseness of the instinctive predispositions constitutes the common underlying innate trait.

In the next two pairs of personality versus anxiety disorders, the basic underlying instinctive motives are excessive fears concerning discrete areas of interpersonal relations. In the sixth pair these fears concern social acceptance, while in the seventh pair, the loss of the dependent relationship with a parental figure originally evolved to secure the provision of resources needed by an immature offspring.

The relationship between the two members of the eighth pair being more complex is discussed in detail in the following.

If this line of reasoning seems to be valid, our next question has to be: *Why does the same originally adaptive mechanism* (at its extreme on its diversified quantitative scale) *induce in one member of the pairs a stable, lifelong disability but in the other an episodic disorder?* At this stage of our discussion, the list of corresponding personality versus episodic disorders seems to become a less homogenous group.

The first three pairs represent dysfunctionally high or low intensities of the overall instinctive activity, longstanding or episodic. As discussed in sections 5.1 and 5.1.3, we have to presume that normative stable levels of instinctive energies of an animal species are determined by their unique ways of interaction with the environment and are closely checked by natural selection. When the abundance of life-sustaining resources oscillates on a seasonal basis, a second evolutionary mechanism modulates and synchronizes the behavioral intensity with the changing environmental conditions. It was also argued (in section 5.1) that both mechanisms became diversified as a result of the relaxed selective conditions of the human environment. It is likely that the complex, multifarious interactions between these two diversified evolutionary mechanisms (one of which induces *permanent* levels of behavioral intensity while the other *fluctuating* ones) are responsible for the existence of these three mental disorder pairs. However, since this interpretation to my knowledge is novel and no research has been done in this domain, I cannot be more specific in this respect.

The next two pairs of longstanding versus episodic dysfunctional behaviors pose a new problem, hitherto unmet in this text. The episodic member of these pairs represents a *psychotic* disorder. The mechanism of psychotic disorders from the perspective of the present work is discussed later in this chapter (section 7.2.6.4). We need this discussion before trying to disentangle the complexities of psychosis versus corresponding personality disorder dichotomy.

We are left with the last three mental disorder pairs. One constituent of the pair is a personality disorder, while the other one is a corresponding anxiety (neurotic) disorder. In the following, we discuss them in some detail. Concerning the first two pairs (avoidant personality disorder/social phobia, and dependent personality disorder/separation anxiety disorder), the main idea I want to propose in order to answer our question—namely, why the same underlying adaptive mechanism can lead either to a longstanding, rigid behavior pattern or an episodic one—is the following: The anxiety disorder counterpart consists mainly of the behavioral expression of the high levels of felt fear or anxiety, while the personality disorder counterpart represents mostly *active* attempts to manipulate the immediate environment in a way that enables better, more stable social functioning, or attenuates the triggers of these intense fears. That is, *the pairs represent different levels of power-balance between active and reactive behavioral intensities.* If the reactive behavioral expression predominates, the condition tends to be episodic, triggered by the relevant environmental influences. If, on the other hand, the active behavioral strivings gain precedence, the disorder tends to be more stable.

The obsessive-compulsive personality disorder/obsessive-compulsive disorder pair constitutes an exception to our theoretical scheme. These disorders seem to be the only mental disorder categories discussed in this text whose inherited aspects evolved *relatively recently* as a result of directional selective pressures of the specifically human environment. In addition, the relationship between them, in spite of the similar denominations, is less clear-cut than in the formerly discussed mental disorder pairs. They are discussed in the following section in some detail.

7.2.5.1. Avoidant Personality Disorder versus Social Anxiety Disorder

AVOIDANT PERSONALITY DISORDER

Diagnostic Criteria 301.82 (F60.6)

A pervasive pattern of social inhibition, feelings of inadequacy, and hypersensitivity to negative evaluation, beginning by early adulthood and present in a variety of contexts and indicated by four (or more) of the following:

1. Avoids occupational activities that involve significant inter-personal contact because of fears of criticism, disapproval, or rejection.
2. Is unwilling to get involved with people unless certain of being liked.
3. Shows restraint within intimate relationships because of the fear of being shamed or ridiculed.
4. Is preoccupied with being criticized or rejected in social situations.
5. Is inhibited in new interpersonal situations because of feelings of inadequacy.
6. Views self as socially inept, personally unappealing, or inferior to others.
7. Is unusually reluctant to take personal risks or to engage in any new activities because they may prove embarrassing.

SOCIAL ANXIETY DISORDER (SOCIAL PHOBIA)

Diagnostic Criteria 300.23 (F40.10)

A. Marked fear or anxiety about one or more social situations in which the individual is exposed to possible scrutiny by others. Examples include social interactions (e.g., having a conversation, meeting unfamiliar people), being observed (e.g., eating or drinking), and performing in front of others (e.g., giving a speech).

Note: In children, the anxiety must occur in peer settings and not just during interactions with adults.

B. The individual fears that he or she will act in a way or show anxiety symptoms that will be negatively evaluated (i.e., will be humiliating or embarrassing; will lead to rejection or offend others).
C. The social situations almost always provoke fear or anxiety.

Note: In children, the fear or anxiety may be expressed by crying, tantrums, freezing, clinging, shrinking, or failing to speak in social situations.

D. The social situations are avoided or endured with intense fear of anxiety.
E. The fear or anxiety is out of proportion to the actual threat posed by the social situation and to the sociocultural context.
F. The fear, anxiety, or avoidance is persistent, typically lasting for 6 months or more.

G. The fear, anxiety, or avoidance causes clinically significant distress or impairment in social, occupational, or other important areas of functioning.

H. The fear, anxiety, or avoidance is not attributable to the physiological effects of a substance (e.g., a drug of abuse, a medication) or another medical condition.

I. The fear, anxiety, or avoidance is not better explained by the symptoms of another mental disorder, such as panic disorder, body dysmorphic disorder, or autism spectrum disorder.

J. If another medical condition (e.g., Parkinson's disease, obesity, disfigurement from burns or injury) is present, the fear, anxiety, or avoidance is clearly unrelated or is excessive.

Specify if:

Performance only: If the fear is restricted to speaking or performing in public.

If we examine more closely the clinical description of this pair of mental disorders, we cannot escape the impression that they are very similar. Social anxiety disorder is characterized by "marked fear or anxiety about one or more social situations in which the individual is exposed to possible scrutiny by others" (the quotations in this section are taken from DSM-5, pp. 202–228, and 672–675). Avoidant personality disorder, on the other hand, is defined as "a pervasive pattern of social inhibition, feelings of inadequacy, and hypersensitivity to negative evaluation."

Both disorders are characterized by prominent symptoms of anxiety in certain social circumstances (expressed in subjective, physiological, and behavioral domains). One may wonder whether enough grounds exist to separate these two disorders nosologically. Even DSM-5 admits that "there appears to be a great deal of overlap between avoidant personality disorder and social anxiety disorder" (p. 674).

Avoidant personality disorder in addition seems to contradict the basic characteristic of personality disorders in general proposed here—that they consist of expressions of dysfunctional instinctive dispositions in spite of social resistance or disapproval as a result of *decreased sensitivity* to these social influences, a contradiction that seems to be partially true. However, if we try to understand why social anxiety disorder is seen as consisting of discrete anxiety reactions to specific social situations, while avoidant personality is conceptualized as a stable personality disorder, that is, "an enduring pattern of inner experience and behavior ... which is pervasive and inflexible" (DSM-5, p. 645), we can find some helpful indications in the descriptions of these two disorders in the enlarged edition of DSM-5, which supports this distinction.

In social anxiety disorder, the accent seems to be on the subjective experience of anxiety reaction and its (secondary) effects on the cognitive processes: "The individual is concerned that he or she will be judged as anxious, weak, crazy, stupid, boring, intimidating, dirty, or unlikeable" (p. 203). Little is said in the DSM-5 (besides mentioning avoidance in general) about the specific ways the individual attempts to overcome his problems in social situations in order to alleviate the anxiety and enable better expression of active drives.

In avoidant personality disorder, this last consideration seems somewhat different. We can see how the person tries *actively* to choose or restrict aspects of the social environment in order to allow active instinctive gratification in spite of her shortcomings. These patients: "avoid work activities that involve significant interpersonal contact. . . . Offers of job promotions may be declined because the new responsibilities might result in criticism. . . . These individuals avoid making friends unless they are certain they will be liked and accepted without criticism . . . will not join in group activities unless there are repeated and generous offers of support and nurturance . . . they are able to establish intimate relationships when there is assurance of uncritical acceptance" (DSM-5, p. 673).

Beyond the multitude of constricting stipulations these persons use in order to avoid social situations they aren't able to cope with, we nevertheless get a glimpse of *active efforts* on their part to find or create the social circumstances that will enable better functioning despite the dysfunctional instinctive configuration. In these special conditions (for example, during a warm intimate relationship or relaxed working relations), these individuals are able to express active emotions more or less continuously, and to achieve a certain degree of functional stability (reducing the frequency or gravity of conspicuous anxiety states or reactions). Might this be exactly the difference between unstable, episodic anxiety disorders and stable personality disorders? (By the way, the same also seems to be true in the case of schizoid personality, where the individual "chooses" a solitary way of life and restricts his interactions mainly to nonhuman objects in order to eschew those social situations he is unable to handle effectively, achieving thus a better behavioral and intrapsychic stability [section 7.2.4.2].)

7.2.5.2. Dependent Personality Disorder (DPD) versus Separation Anxiety Disorder (SAD)

The similarities and differences between DPD and SAD resemble those of the above discussed mental conditions pairs. Separation anxiety disorder means: "Developmentally inappropriate and excessive fear or anxiety concerning separation from those to whom the individual is attached," while dependent personality disorder consists of " [a] pervasive and excessive need to be taken care of that leads to submissive and clinging behavior and fears of separation." (All the quotations in this section are taken from DSM-5, pp. 190–195 and 675–678.) The differences between these two disorders concerning the diagnostic criteria and clinical description are the following: In separation anxiety disorder, the emphasis is on the subjective experience brought about by separation or the prospects of separation from an attachment figure (distress, worry, fear), and its secondary effects on the imagination, anticipation, and physiological states of the body. Although clinging behavior and reluctance to be away from home or from attachment figures is mentioned, the accent is on the fear of losing such relationships— manifesting, for example, in imagining situations that may lead to such loss (illnesses, disaster, accident, death, kidnapping, getting lost, etc.), nightmares with the same content, and related physiological symptoms of anxiety: headaches, abdominal complaints, cardiovascular symptoms, etc. Typically, the disorder is characterized by exacerbations and remissions presumably induced by the vicissitudes of the dependent relationship.

In dependent personality disorder, although the diagnostic criteria stress excessive fears of separation, the emphasis lies on the actively initiated behavioral measures taken by the affected person to preserve the dependent relationship and prevent separation from that attachment figure, or to find another as quickly as possible when the need arises. These behavioral measures are the following: submissive and clinging behavior over a long period of time (already mentioned in the main criterion for diagnosing the disorder); "volunteering to do things that are unpleasant"; "difficulty expressing disagreement"; and "agree[ing] with things that they feel are wrong." "They may make extraordinary self-sacrifices or tolerate verbal, physical, or sexual abuse," and when a close relationship ends, these persons "become quickly and indiscriminatingly attached to another individual." To me, it seems quite clear that the main difference between separation anxiety disorder and dependent personality disorder is the following: In the former, the person suffers *passively* (reactively) from signs of anxiety when separation threatens or actually occurs, while in the latter a person *actively* safeguards a dependent relationship at almost any price, and when separation seems inevitable, she urgently seeks another attachment figure.

The individual, like the avoidant personality, actively creates the social environment (in this case a two-person relationship) that enables her to be as anxiety free as possible. Securing the social environment that compensates best for the respective mental frailty, the individual may attain better functional stability.

It seems clear that the basic instinctive predisposition for this pair of mental disorders is the innate mechanism that secures the dependent relationship between an immature offspring and a parenting figure, a universal phenomenon in the class of mammals. That this dependence becomes more protracted and complex as we ascend the evolutionary ladder is common knowledge. Furthermore, it is known that in conditions of relaxed selection pressures, as in the case of domesticated animals or pets, infantile behavior patterns may persist into adulthood. (One of my cats, at an advanced age, still tries to stimulate my completely undeveloped mammary glands by pushing rhythmically on my belly with its forelegs).

Therefore, considering the great diversification of human instinctive predispositions in this domain as well, I think it can be hypothesized quite confidently that the innate predisposition for this pair of mental disorders has to be an uncommonly strong instinctive attachment to parental figures or their substitutes that survives into adulthood.

7.2.5.3. Obsessive-Compulsive Personality Disorder versus Obsessive-Compulsive Disorder

We are dealing here with a pair of mental conditions that are more complex than the previous two pairs. In spite of each of them having the same denomination, these two disorders differ greatly clinically and the relationship between them is less obvious than in the conditions just discussed. The phylogenetic origin of the underlying instinctive predisposition is much less evident than in the case of attachment disorders or social phobias.

I propose that these predispositions have a more recent evolutionary origin. Obsessions and compulsions are dependent on the evolution of specifically human cognitive abilities, like the ability to imagine yet unactualized dangers and project them into frightening future scenarios. The obsessive-compulsive personality disorder, on the other hand, seems to be the upshot of specific selective pressures originating in the agricultural way of life, the appearance of growing permanent settlements, social stratification and its hierarchical organization, and increasing division of labor—selective pressures that were nonexistent in the social milieu of prehistoric hunter-gatherers.

On the other hand, the fundamental difference between personality and anxiety disorders, as hypothesized in this theory, is more striking in the present pair of mental conditions than in the previous ones. Obsessive-compulsive personality disorder evidently expresses a behaviorally very stable dysfunctional instinctive predisposition expressed inside a social setting, even if it incurs social dissatisfaction, interpersonal tensions, or lowered efficiency at work. The normative range of this predisposition in the respective gradated continuum is highly adaptive. Obsessions and compulsions, on the other hand, relate mostly to subjectively felt intrapsychic manifestations leading to anxiety and the behavioral or mental measures taken by the individual to alleviate them.

7.2.5.3.1. Obsessive-Compulsive Personality Disorder (OCPD)

OBSESSIVE-COMPULSIVE PERSONALITY DISORDER

Diagnostic Criteria 301.4 (F60.5)

A pervasive pattern of preoccupation with orderliness, perfectionism, and mental and interpersonal control, at the expense of flexibility, openness, and efficiency, beginning by early adulthood and present in a variety of contexts, as indicated by four (or more) of the following:

1. Is preoccupied with details, rules, lists, order, organization, or schedules to the extent that the major point of the activity is lost.
2. Shows perfectionism that interferes with task completion (e.g., is unable to complete a project because his or her own overly strict standards are not met).

3. Is excessively devoted to work and productivity to the exclusion of leisure activities and friendships (not accounted for by obvious economic necessity).

4. Is overconscientious, scrupulous, and inflexible about matters of morality, ethics, or values (not accounted for by cultural or religious identification).

5. Is unable to discard worn-out or worthless objects even when they have no sentimental value.

6. Is reluctant to delegate tasks or to work with others unless they submit to exactly his or her way of doing things.

7. Adopts a miserly spending style toward both self and others; money is viewed as something to be hoarded for future catastrophes.

8. Show rigidity and stubbornness.

OCPD is defined in DSM-5 as a "pervasive pattern of preoccupation with orderliness, perfectionism and mental and interpersonal control, at the expense of flexibility, openness, and efficiency" (all the quotations in this section, if not otherwise specified, are taken from the comprehensive edition of DSM-5, pp. 678–680). The disorder is characterized also by preoccupation with small details to the detriment of grasping the "major point of the activity" and "perfectionism" in the execution of these small details on the expense of "task completion." Even more unusual for a personality disorder with an instinctive foundation is the excessive devotion to work to the exclusion of leisure activities and friendships, as well as overconscientiousness about matters of morality and values (which as a rule are mostly *socially induced, acquired* behavioral characteristics that tend to suppress instinctive strivings). The adherence to social rules of behavior and to the fulfillment of one's occupational and familiar duties is exaggerated at the expense of spontaneity, reliance on emotions and intuitions, and improvisation and adventurousness—all predispositions with a clear instinctive foundation. Uniformity in social life is preferred over pluralism: "When rules and established procedures do not dictate the correct answer, decision-making may become a time-consuming, often painful process." These persons are inclined to "maintain control of their physical and interpersonal environment . . . [and] . . . express affection in a highly controlled or stilted fashion."

The obsessive-compulsive personality "[a]dopts a miserly spending style toward both self and others; money is viewed as something to be hoarded for future catastrophes." This last point seems to hint on continuous underlying anxieties about the security of one's material resources in the present and future. OCPD is one of the most frequent personality disorders in the general population, with prevalence rates estimated between 2.1–7.9 percent (DSM-5, p. 681). In spite of the fact that the obsessive traits strongly suggest ontogenetically acquired behavior patterns, one small twin study found substantial genetic influence in OCPD, and another study found that obsessional personality traits aggregate in genetic relatives (15 percent versus 5 percent in controls) (Plomin et al., 2008, p. 255).

It is impossible to imagine how natural selection could positively select such personality traits during the long evolutionary period characterized by the hunter-gatherer lifestyle. However, with the advent of agriculture, permanent settlements, surplus goods, and hierarchical social organization, the personality traits that enhance fitness in these changed environmental circumstances became considerably different from those favored by natural selection in nomadic hunter-gatherer groups, and consequently these traits began to spread in human populations. Since anthropologists G. Cochran and H. Hapernding proposed these hypothesized changes in behavioral predispositions, to my mind, quite convincingly in their book, *The 10,000 Year Explosion* (2010), I will quote them here in some detail: "Foragers could walk away from trouble, but farms were too valuable (too important to the farmers' fitness) to abandon. So farmers had to submit to authority" (p. 111). A desirable "trait was the ability to defer gratification for long periods of time. This was a practical requirement for farmers, since they had to save a portion of their crop for seed and some of their domesticated animals for breeding stock. . . . It takes a certain type of personality—with traits including patience, self-control, and the ability to look to long-term benefits instead of short-term satisfaction—and natural selection must have gradually made such personalities more common among peoples that farmed for a long time" (p. 114).

In fact, OCPD seems to be the only mental disorder (besides OCD) this work deals with whose instinctive predispositions resulted from the effect of *directional* selective pressures of the more advanced human environment, and apparently is not rooted in previous evolutionary processes in the animal world or human prehistory. These hypothesized evolutionary changes in personality traits *at the extreme pole of a diversified continuum* may account for several diagnostic criteria encountered in OCPD. They include: excessive devotion to work and productivity at the expense of leisure activities; inflexibility concerning rules, morality, social hierarchy, and values; rigidity and stubbornness; restraint in spontaneous instinctive (emotional) gratification; miserliness; and perhaps reluctance to discard useless objects.

On the other hand, this reasoning cannot account for another central characteristic of the obsessive-compulsive personality—the exclusive attention given to details at the expense of grasping the whole situation. However, further cultural development, especially in the occupational domain, can point toward the fitness-enhancing value of this personality trait, too. With increasing social stratification and division of labor, with the appearance of more and more specialized "occupational niches" (at the expense of grasping better the "grand scheme of things"), and with increasingly complex forms of cooperation, attention to details in a restricted field began to pay off in terms of fitness. Good specialists or experts in a narrow field are highly appreciated, while ignorance of other fields in a cooperative social setting is more than compensated for by the possibility of consulting the proper expert or written sources of information when the need arises. Progress in the field of science, for example, is characterized by increasing specialization into smaller and smaller "parcels" of a scientific domain, with the result that relatively few experts possess a wide, comprehensive grasp of the whole field in question. Perfectionism in details at the expense of not seeing the whole picture pays off well in less prestigious occupations like clerking or working on an assembly line. Obsessive-compulsive personality traits "may be a great asset to their owners, and society owes much of its stability and its efficiency to its obsessional members" (Nemiah, 1980, p. 1513). Taking into account the effect of the RfNSPs in this domain as well, it makes sense to propose that the exaggerated preference for details in obsessive personality disorder represents the extreme end of a diversified continuum.

The last point we touch on in relation to OCPD has to do with the strain it poses on our relatively simple scheme of classifying instinctively fueled behavior into active, self-centered reactive, and submissive or conforming categories. This scheme, as we have seen, can work well in the case of animal behavior and in many aspects of human behavior and its pathology. However, as a result of complex interactions between the principles governing the evolution of fitness-enhancing behavior in the *natural* world on one hand, and on the other the peculiarities of gene transmission that enhance behavioral fitness in a complex, highly cooperative *social* world considerably shielded from natural selection pressures, the rigid division of these instinctive predispositions (into active, self-centered reactive, and conforming ones) begins to loosen.

At first glance, it seems that what characterizes the behavior of the obsessive personality is the suppression of evolved active instinctive predispositions as a result of an exaggerated submission to or conformity with socially imposed rules of behavior that are unconditionally accepted. As a result of this mechanism, the spontaneous expression of active motives and the accompanying emotional states (sensual or sexual motives, signs of other kinds of pleasure, impulsivity, or despair) are suppressed.

Moreover, with further examination of obsessive-compulsive behavior, it becomes evident that the basic tactics of individual reactive behavior (flight, avoidance, aggression) are prevented from free expression in this behavioral style, which relies excessively on rules, moral considerations, and social norms. Even submissive behavior is strictly restricted toward persons higher in the social hierarchy.

However, most amazing to my mind, the behavior of the obsessive-compulsive personality clearly possesses some central characteristics of *active* behavior. It is almost never a *reaction* to an actual, circumscribed external pressure or threat but it is, like personality disorders in general, "an enduring pattern of inner experience and behavior . . . which is pervasive and inflexible" (DSM-5, p. 645). It is not elicited or intensified by reward, it is not suppressed by punishment, and it does not become extinct when positive external reinforcers cease to act. In addition, it is not influenced negatively by its own undesirable or harmful consequences with regard to the individual's social adjustment, as would be expected according to the behavioristic mechanism of "operant conditioning" (which predicts alteration of behavior by its own positive or negative consequences). Instead, it is a behavioral pattern experienced by the individual as originating from his own deep intrapsychic intentions, that is, it is "internalized" or "ego-syntonic," and is resistant to change resulting from social disapproval or other social pressures. These characteristics are in turn typical of the behavior induced by active, life-sustaining instinctive motives. Therefore, *the only satisfactory explanation I can see for the traits of obsessive-compulsive behavior is that some kinds of natural selection pressures (in this case intragroup natural selection pressures) are able to channel active instinctive features into originally submissive or conforming reactive behavior patterns.* In fact, we encountered a similar phenomenon earlier in this text in the case of animals. In section 5.1 on seasonal fluctuations of life-sustaining resources, we saw the example of ground squirrels that evolved an *autonomous* biological rhythm ("circannual clock") of physiological and behavioral processes that, while originally induced by seasonal fluctuations, became independent of external circumstances. This clock continued to be functional in constant laboratory conditions, even in those ground squirrels born in captivity that had never experienced a natural circannual photoperiod. At the end of the same chapter, in the section on active/reactive behavior dichotomy (section 5.4.3), it was hypothesized that *humans with a certain instinctive constitution are able to channel active intentional, energetic, and subjective experiential motives into culture-created activities remote from or even antagonistic to individual, biological goals.*

The preconditions that enhanced the evolution of such an inherited mechanism are probably the considerable diffuseness and malleability of human instinctive predispositions; the long period of individual behavioral maturation; and the long-lasting prehistory and history of living in cooperative groups. However, it has to be mentioned that this instinctive channeling isn't always smooth, complete, or conflict free. Obsessive-compulsive personality frequently co-occurs with obsessions and compulsions, which is our next topic.

7.2.5.3.2. Obsessive-Compulsive Disorder

In spite of having almost the same denomination, Obsessive-Compulsive Disorder (OCD) and Obsessive-Compulsive Personality Disorder (OCPD), in contrast to the previous two pairs of anxiety/personality disorders, are clearly dissimilar on clinical grounds. OCPD or obsessive personality traits lack obsessions and compulsions, while in OCD the obsessions and compulsions are the central, in fact exclusive, signs of the disorder, whereas obsessive personality traits per definition are absent.

7.2.5.3.2.1. Obsessions

The main sources of the present section are: DSM-5, pp. 237–239; and Nemiah, 1980, pp. 1504–1505 and 1510–1513. The chief characteristic of obsessions is their *intrusive* nature. They consist of thoughts, urges, or images that appear suddenly and unexpectedly in consciousness, without the individual's intentions or conscious will being involved, and frequently contradict the same person's usual habits, beliefs, values, or consciously intended ways of thinking and behaving (that is, they are "ego-alien" or "ego dystonic"). Frequently, the individual herself recognizes their absurd or exaggerated nature.

It seems clear from the perspective of this text that two different kinds of obsessions exist:

1. The first type is induced by suppressed active instincts (including their frustrated forms) or individual aggressive reactive impulses whose direct unaltered expression in behavior is prohibited by cultural norms, at least in the respective situations in which they emerge intrusively in consciousness. Examples are obsessions with a sexual content that collide with social norms or with the individual's own moral values, like an intrusive image of incest (Nemiah, 1980, p. 1511), or an urge to shout obscenities at a religious or other solemn ceremony, a ceremony with whose solemn nature the individual consciously identifies.

   What could lead to the dissociation of some active or individual reactive instinctive motive, preventing their integration into mental life or behavior? These instinctive motives, from our point of view, seem normative; that is, they are normatively differentiated, they have no long-lasting fluctuations in their intensity, and the active motives do not gravitate excessively toward one of their forms of frustration. It seems that a disequilibrium exists in this condition in the power balance between individual active and self-centered reactive behavior, on one hand, and socially induced submission or conformity, on the other, in the favor of the latter. Consequently, these suppressed innate egocentric predispositions cannot mature (by experimentation, trial and error, imitation, and fusion with learned elements) into culturally expected normative ways of expression. When they surface in consciousness, they do it brusquely, in a crude form, and frequently in inappropriate circumstances. The possibility that they may escape conscious control and become expressed in overt behavior constitutes an additional source of anxiety. This interpretation of obsessions seems quite in agreement with the traditional psychoanalytic view.

2. *The second kind of obsession concerns the intrusive emergence in consciousness of fears or anxieties not warranted as a rule (at least in their intensity) by current environmental realities.* From an evolutionary viewpoint, the eliciting factors of these fears are diverse. Sometimes the source of the fears is real and their possible materialization may have serious consequences for the mental equilibrium of the individual; for example, a gravely ill mother to whom the individual is attached may die. (Most examples of obsessions and compulsions in this section are taken from Nemiah, 1980, pp. 1504–1513.) Sometimes these fears concern a real source of danger, but the probability of its materialization is grossly exaggerated. An illustration of this kind of fear is that of a mother who has had intrusive imageries of her daughter being hit by a car (that is, seeing her small, disfigured body lying on the road) each time she leaves home. Other such fears, very frequent in OCD, are fears of contamination, theft, or burglary.

In other cases, the source of fear is a possible inadvertent, harmful action by the individual himself (which never materializes in OCD), an action that could cause great tragedy or destruction of property: for example, fear that one accidentally had bumped into a passerby on the street or in a subway, thus causing serious injury or death, or forgot a burning cigarette at home that may set the house on fire. Still other fears concern transgressions regarding one's own belief system, like that caused by a religious person's intrusive doubts about the existence or sacred nature of God. These fears in fact may overlap with those mentioned in point 1, meaning that they may ultimately be caused by suppressed defiant or aggressive impulses.

In summary, from an evolutionary point of view, the source of fear in obsessions may concern the physical environment (inadvertently setting the house on fire), the biological environment (fear of contaminations), the possible loss of an attachment figure (the mother may die), fear of transgressing one's moral standards, fear of losing control and acting according to an unconquerable internal impulse, fear of damage caused by lawbreakers (burglary, for example). Therefore, it seems that *in OCD the exact nature of the danger leading to anxiety or fear and the specific innate motor predisposition to neutralize it (such as avoidance reactions in specific phobias or clinging behavior in a dependent personality) are not determinant factors.* The individual "chooses" one or a very restricted range of sources of danger from the multitude of possible or imaginable ones inherent in her physical, biological, or social environment, or among her intrapsychic impulses and reacts to it differentially and exaggeratedly, while disregarding all other possible dangers as sources of obsessive fears.

In summary, the essence of the obsessions mentioned under point 2 seems to be the following: irrational or exaggerated fears from one source of danger among the multitude of those a person with normal intelligence can identify in his environment, seemingly "chosen" quite at random, that is, not according to the actual extent of its harmful potential. The individual "sticks" to this specific fear for long periods of time during which the condition may have an episodic waxing and waning, or (rarely) a deteriorating course (DSM-5, p. 239; Nemiah, 1980, pp. 1513–1515). Explanations of the irrationality of these fears, or attempts to disregard the obsessions, are ineffective, but specific *compulsions* intended to alleviate specific obsessions are temporarily effective.

Therefore, it seems that in the case of obsessive-compulsive disorder, contrary to the anxiety disorders previously discussed, the exact nature of the fear-eliciting situation is irrelevant. Instead, the central issue has to be a peculiar intrapsychic mechanism employed to provide an "anchor point" for some essentially directionless fears or anxieties. The complementary aspect of this mechanism consists of employing an idiosyncratic ritual that, in spite of not being an effective antidote against the imagined or exaggerated fear-eliciting agent, nevertheless is capable of temporarily alleviating the fear. This line of reasoning will be further developed after the discussion of compulsions.

7.2.5.3.2.2. Compulsions

Compulsions are defined by DSM-5 as repetitive acts either manifest in observable behavior or "mental acts" (that is, praying, counting, repeating certain words silently in the head) that the individual feels compelled to perform either in response to an obsession—that is, to alleviate the anxiety brought about by the obsession (as a rule, of the *second* type discussed earlier)—or in accordance with a "belief" that the compulsion can prevent a dreaded event in the future. Sometimes the causal connection between the obsession and the compulsion is obvious, even logically tenable. For example, one individual obsessed with the thought that he inadvertently knocked someone off a subway platform felt compelled to keep his arms tight against his sides (in order to prevent such accidents in the future); in addition, he called the transport authority to reassure himself that no such accident happened (Nemiah, 1980, 1512).

In other cases, compulsions consist of the widely practiced, more-or-less effective, measures against the respective danger; these compulsions (namely washing, locking, and checking, which serve as measures against obsessive fears of contamination, theft, and lapse of memory, respectively) are the most frequent. In these cases, the dysfunctional nature of the behavior is revealed by its disproportionate intensity (a great number of repetitions) at the expense of more useful daily activities. The same can be argued about compulsions of ordering, cleaning, or keeping symmetry.

In other cases, the connection between the obsessive fear and the reassuring compulsion is not causal but symbolic and may be part of a traditional belief system. For example, in the earlier case of a mother obsessed with the imagery of her little daughter being involved in a car accident after having left home, the distressing obsessions calmed down after repeated prayers. Similarly, obsessions with blasphemous content are "counteracted" by compulsive reiteration of prayers (idem, p. 1511). In still other cases the connection between the obsessive fear and resultant compulsion is obscure: for example, when the compulsions consist of counting or uttering meaningless words.

It is believed by its practitioners that psychoanalytic investigation may in some of these cases uncover hidden symbolic connections between the obsessive fear and the reassuring compulsion. An example is the case of a boy who, fearing the impending death of his beloved, sick mother, felt compelled to perform a dangerous climb in order to touch the topmost branch of a high tree, believing that this feat could prevent his mother's death (despite being well aware of the absurdity of this belief). Perhaps this desperate compulsion symbolized self-sacrifice or a desire to feel closer to the divine being who can better "see" his suffering and maybe provide help (idem, pp. 1504–1505).

It is important to note again that the compulsions do temporarily reduce the subjective feeling of fear or anxiety in spite of being, as a rule, completely ineffective as countermeasures against the threat (in fact, against any threat). Indeed, this is well recognized by the majority of the affected individuals.

In summary, it seems that the central aspect of OCD is not a clearly recognizable, discrete evolutionary mechanism as in the previously discussed anxiety/personality disorder pairs, or differentiated fear reactions to simple, concrete sources of danger in the environment of our evolutionary past, as in simple phobias. Instead, obsessive fears comprise a wide variety of fears that seem to be randomly and unconsciously "selected" and have nothing to do with the hierarchy of seriousness or immediateness of different possible harmful influences inherent in one's environment. Nor are they the result of exposure to a traumatic situation. Similarly, the compulsions, which are meant to alleviate obsessive fears, are highly varied in nature, ranging from the exaggeration of conventional measures taken against simple, real sources of danger (washing, locking), to compulsions whose supposed effect on the source of the fear is obscure, symbolic, or nonexistent. In a word, *obsessive-compulsive behavior has no or only a very precarious relevance to the cause-effect relationship as we are realistically aware of it in the natural and social world.*

Again, as was the case with the OCPD, I cannot think of any natural selective mechanism relevant to the animal kingdom that may explain the positive selection of mental and behavioral characteristics underlying obsessive-compulsive behavior (which could have been subsequently diversified by RfNSPs).

However, the existence of a genetic predisposition for OCD is well established. The condition is twice as frequent in the first-degree relatives of adults with OCD, as it is in the first-degree relatives of those without the disorder. In the case of childhood onset OCD, the rate is increased tenfold (DSM-5, p. 240). OCD of clinical severity is rarer than OCPD (the prevalence of OCD is 1.1–1.8 percent as compared to 2.1–7.9 percent for OCPD, DSM-5, pp. 239 and 681). However, mild forms of obsessive-compulsive behavior are widespread: "It is probably a rare person who has not at some time in his life experienced a fleeting bit of obsessional thinking or compulsive behavior . . . a transient unreasonable compulsion to check a gas jet that one knows is closed or a momentary urge to shout out an obscenity during a solemn ceremony" (Nemiah, 1980, p. 1505).

If this is indeed the situation, namely, that OCD has a clear genetic foundation, that its gravity is diversified, and its mild form is widespread, the following question seems justified: What adaptive advantages does the obsessive-compulsive mechanism confer (or what advantages did they once confer in our evolutionary past) at adequate intensities?

As a very tentative hypothesis, I propose the following evolutionary scenario: As a result of the accelerated evolution of their cognitive abilities (brought about by intragroup selective pressures during the Pleistocene epoch), human beings became able to *realize or imagine an infinite number of possible environmental dangers or harmful influences* (in the present or projected into the future, coming brusquely or slowly, in overt or covert form, threatening one individual or the whole community), against which these early humans possessed no effective countermeasures. While the situation in this respect is better today, it was much graver during human prehistory and early history.

Besides threats from external sources (natural or social), the individual also has to cope, as we have seen, with "internal impulses" to behave dysfunctionally originating from innate diversified predispositions that may deleteriously affect social adaptation. The progressively improving *self-awareness* (due, at least partially, to social condemnation of inappropriate behavior) facilitated introspection, or, indeed, it made it possible in the first place (see N. Cameron's quotation on this topic in section 7.2.8.6.3).

The realization of both the multitude of possibly harmful external influences for which no effective countermeasures exist as well as the possibly harmful impulses coming from within may pose a serious threat to a balanced, effective everyday mental functioning. It may cause longstanding anxieties and grave uncertainties as to how to behave in different circumstances, what order of priorities to adopt, and so on. Human ingenuity, however, conceived of a way to tackle this situation in a way that, while being utterly absurd (at least from the viewpoint of pragmatic or scientific reasoning), worked amazingly well. This "solution" bypasses both the need to search for, identify, and understand the real sources of hardships and dangers concerning human well-being (or even individual or collective human existence), as well as the need for working out effective ways to tackle with them. I am referring here, of course, to institutional or other widely held belief systems, first of all religious ones, and the accompanying rituals that express them in behavior. It is postulated in these belief systems that external harmful influences do not come ultimately from natural or social sources but represent the will or anger of some superhuman entity, the God or gods, the ghosts of the ancestors, etc., and in order to alleviate these hardships, the believers have to placate those entities by performing the appropriate rituals.

I am not proposing, of course, that religious beliefs *supplanted* overwhelmingly or even partially realistic behavior when that behavior proved to be effective in solving or alleviating troublesome or harmful situations. However, in cases of experienced or expected serious hardships or dangers against which no effective realistic solutions exist, or even in order to ensure that effective realistic practices in the present continue to be effective in the future, and that abundant resources in the present (huntable animals, rain during the appropriate seasons in agricultural communities, etc.) will be continuously provided in the future, the appeal to imaginary superhuman entities through rituals remains a very common practice in almost all human communities on the planet. It seems unquestionable, moreover, that these religious practices do alleviate anxiety in healthy believers (Ehrlich, 2000, pp. 215–218).

The belief that superhuman entities may be the cause of these hardships or calamities, may also have the power to relieve them, and may be influenced by appropriate rituals to do so seems much more reassuring than utter helplessness, hopelessness, and despair. Obsessive-compulsive disorder, in my opinion, represents a similar "self-deceptive double bypass" of realistic problem-solving (that is, with regard to the identification of the true sources of individual anxieties and the effective remedies). Despite this, it fulfills its intrapsychic goal—namely, the alleviation of subjective suffering and improvement of the overall mental functioning, at least temporarily.

It has to be mentioned that we have already encountered in this text inherited behavioral mechanisms whose aim is not the securing of adaptive gains concerning the organism-environment interaction, but solely or mostly *the restoration of a more desirable equilibrium* between different aspects of the mental functioning. Two of the three basic forms of active instinctive frustration, displacement or vacuum behavior and dysphoria, were hypothesized to serve mainly the same purpose. The subject of institutional belief systems and their additional benefits is elaborated upon in section 7.2.6.4.

All this having been said, we have to admit that a centrally important question concerning OCD remains unanswered. Even if our hypothesis about an evolutionarily adaptive, innate inclination for the "self-deceptive double bypass" in humans seems feasible, we still remain with the following puzzle: Why do normative persons channel this innate predisposition (diffuse instinctive intentions and emotions) into institutionalized or other widely held belief systems and rituals whose cognitive and behavioral details are acquired from external, social sources, while the obsessive-compulsive individual "invents" her own private fears and rituals? Considering the multitude of other mental, and some neurologic, disorders that are co-morbid (co-occur) with OCD, the solution to this puzzle seems to be a difficult and complex task.[6] We have to be content, I think, with the notion that in our theoretical scheme *obsessions and compulsions are not considered discrete, circumscribed behavioral adaptations to certain environmental conditions, and even less a circumscribed mental disease, but nonspecific mental mechanisms whose role is the attenuation of intrapsychic tensions from widely different sources.* It merits mention in this context that, in DSM-5, contrary to its predecessors, OCD was removed from the inclusive group of anxiety disorders and transferred into a separate nosological category.

7.2.5.4. Afterthought

If the arguments given here concerning OCPD and OCD prove to be tenable, it may be hypothesized that obsessions and compulsions appeared *earlier* during human phylogeny than obsessive-compulsive personality disorder. While obsessive-compulsive personality traits in our theoretical framework appeared as a result of the agricultural revolution (beginning about 10,000 years ago), it is believed that magical or religious rituals already existed in the Upper Paleolithic period (circa 40,000–12,000 BC) as prehistoric cave paintings and other archeological findings from this era suggest.

7.2.6. Paranoia

7.2.6.0. Introduction and Definition

Paranoia is not a discrete mental disease entity but a certain quality of the behavior and subjective experience that may be the predominant theme in some mental disorders, a secondary or additional disturbance, or a trait in otherwise normative persons.

7.2.6.0.1. Interpreting the Phenomenon of Paranoia

How can we make sense of the concept of paranoia in a framework of evolutionary reasoning?

Paranoia is evidently closely related to the subject of *aggression* or some similar kind of harmful intention. However, it does not consist of excessive or easily arousable aggressive or counteraggressive behavior. This trait may characterize other mental disorders, as well as the behavior of normative persons. Examples are: oppositional defiant disorder, intermittent explosive disorder, conduct disorder (DSM-5), as well as the aggressive type of antisocial personality disorder and the (unclassified) choleric personality. In these disorders, after the conflictual situation is over, the person returns to pursue his former unfrustrated active instinctive preoccupations.

If we propose alternatively *that paranoia is induced by oversensitivity to possible harmful influences coming from the human environment*, we probably can make some additional progress in the right direction. Excessive suspiciousness concerning possible malicious intentions of others is indeed a frequent descriptive trait in paranoid conditions. In addition, some anxiety disorders (especially social anxiety disorder) are also characterized by oversensitivity to possible harmful environmental influences. However, in psychiatry, anxiety disorders and paranoia are sharply distinguished on clinical grounds. Moreover, dangerous, even life-threatening, situations—like those encountered in combat—are not recognized as predisposing or causative factors for paranoid disorders. Therefore, in order to make sense of paranoia, we have to hypothesize an additional innate propensity, namely that which *predisposes differentially to the aggressive transformation of active instinctive strivings in frustrating conditions* (instead of to displacement or dysphoria), *or to counteraggressive tactics* with regard to reactive behavior (instead of freezing, avoidance, or escape) as a preferred behavioral response to perceived harmful external influences. However, here we again come dangerously close to the unrelated group of disorders mentioned earlier—that is, the mental disorders and normative traits characterized by easily arousable aggressive behavior (the aggressive form of antisocial personality, choleric personality traits, etc.).

In consequence, we have to hypothesize a third innate predisposition, which to me seems the most important in defining paranoia—namely that the central innate predisposition in this disorder consists of a considerably *protracted duration* of the aggressive transformation of both active instincts when frustrated, as well as of the reactive counteraggressive tactics, lasting long beyond the resolution of the conflictual situation that elicited them in the first place. This maladaptive trait therefore precludes either the return to the original unfrustrated active motive, a switch to another form of its frustration, or the use of other reactive tactics, thus leading to the unfounded subjective experience of *a long-lasting bidirectional inimical relationship* between the paranoid person and his human environment.

That last presupposition may explain, at least partly, another peculiar characteristic of paranoid disorders. The paranoid—that is, inimical—relatedness to aspects of the social environment (with regard to the person's aggressive or counteraggressive intentions, as well as the expectation of similar intentions from surrounding people) is present for protracted periods: months, years, or even a lifetime. Paranoid thoughts induced by the paranoid emotional relatedness can be elicited easily during therapeutic sessions. However, actual aggressive or violent acts on the paranoid individual's part are relatively rare. When they do occur, they are, as a rule, out of proportion in terms of the triggering situation, and the individual may have untenable justifications for them. In addition, the discharge of the aggressive or violent act does not alleviate the paranoid disposition, in contrast to other forms of aggressive behavior, both normative and pathological, that do reduce the accumulated mental tensions.

In my opinion, the inability to confine the expression of aggression to proper circumstances, that is, only to situations in which an external challenge or other kind or disturbance is present, is the result of a *permanent mismatch* between the *internal mental experience,* which relentlessly signals the presence of harmful external influences and prepares the organism to act or react aggressively, and the *actual perceptual input* from the social environment that *does not confirm* as a rule the presence of such intentions. Consequently, paranoid individuals frequently adopt a defensive (rather than offensive) attitude in the face of these felt external harmful influences. They are as a rule lonely people who stay at home behind locked doors most of the time. This is not the *avoidance* of the anxious individual but represents *defensive* measures against expected aggression.

The innate predisposition to develop paranoid disorders will be attributed in our theoretical framework to the same RfNSPs that were proposed to account for most of the dysfunctional behavior patterns previously discussed. The reader may remember the discussion of the adaptive rationale of the aggressive transformation, both when an active instinctive intention is frustrated as well as in the case of the aggressive form of reactive behavior. Its goal is the removal, neutralization, etc., of an environmental disturbance. As soon as that goal is attained, the aggression subsides and gives way to the reappearance of active instinctive motives in their original unfrustrated form, or to relaxation, resting, and restitutive biological processes (sections 5.2.2 and 5.3.3.1).

I have also detailed how closely natural selection controls the innate aspects of aggressive behavior in animals living in nature in view of its great fitness-enhancing value when discharged in the right circumstances, at the right moment, and at the appropriate intensity and duration; and, conversely, its disastrous fitness-reducing potential when employed at the wrong time and with inappropriate intensity and duration. Also stressed was the great importance for survival of *flexible transitions* between the aggressive and other forms of reactive behavior according to the quickly changing circumstances of a dangerous encounter, as well as the flexible transition between active and reactive motives. The reader may remember Jane Goodall and deWaal's examples of how quickly aggressive behavior alternates with reconciliation and friendly relationships in chimpanzees, as well as de Waal's interpretation of this phenomenon as a highly adaptive one with regard to group life (section 5.2.2). The switchover between "affectionate attraction to anger to renewed friendliness" is already recognizable in infant behavior, which suggests the innate foundations of this mechanism in humans (Eibl-Eibesfeldt, 2010, pp. 218–219).

As proposed, we must look for the inherited foundations of the predisposition to develop a paranoid disorder in the relaxation of the constraints of natural selection and the consequent diversification of the aggressive behavioral mechanism. The main reason for RfNSPs, as proposed throughout this text, is the ecological dominance and protective nature of the social/cultural environment, especially that of the modern, technologically advanced democracies, which precludes a reproductive disadvantage in the presence of less optimal variants of the aggressive behavioral predispositions. The relevant variant for paranoid disorders is one that permits the continuation of the aggressive transformation of instinctive motives for long periods after the need for aggression is over. In other words, in paranoid humans, an instinctive behavioral mechanism "designed" by evolution to consist of a fleeting episode that ceases when the frustrating or dangerous factor is removed induces a protracted, even lifelong, form of dysfunctional relatedness to aspects of the social environment.

At the opposite pole of the continuum regarding the readiness to respond with aggression or the suspicion of malicious intents in others are situated those naïve, indulgent, lenient persons who cannot properly perceive the ubiquitous presence of conflicting interests (biological and social) between individuals and groups of people even after repeated abuses and disappointments. They are not predisposed to suspect harmful intents in others. They are seen as lacking healthy skepticism, as credulous, unsuspecting, gullible persons. Even when they recognize harmful acts done by others, they tend to minimize the importance of those acts or find excuses for them. This predisposition, albeit much less damaging to social relations than paranoia or excessive aggressiveness, is also maladaptive, as it may be exploited by unscrupulous individuals.

It goes without saying that, as a consequence of the predominance and protracted nature of the aggressive transformation, other forms of active and reactive motives—that is, the pleasurable unfrustrated expression of active instincts; dysphoria involving self-accusations; fear, anxiety, freezing (while evaluating more thoroughly the nature and seriousness of suspected harmful intentions)—are proportionally reduced or nonexistent in paranoid mental disposition and behavior.

It should be noted that in the present chapter we have already encountered two other categories of mental disorder with a similar underlying evolutionary mechanism—that is, an underlying mechanism consisting of the *unduly prolonged* innate behavioral and intrapsychic response after the adaptive goal is fulfilled. The first category is the agitated, dysphoric form of depression in which the dysphoric transformation of active instinctive motives becomes unduly protracted and intense (section 5.2.3.); the second is posttraumatic stress disorder, in which the reactive arousal (its avoidant or fleeing tactic) persists long after the traumatic event is over (section 7.2.3.).

While the role of the protective and diversified social environment (being able to contain and even exploit a wide range of instinctive predispositions) seems predominant in bringing about paranoid disorders, we have to consider another factor that *enhances or amplifies* the same mental dysfunction. In nature, the aggressive intentions of an animal are observable in its demeanor and in certain bodily signs, like piloerection (bristling of the hair) or change of skin color in certain fish, etc., for some time *before* the actual aggressive behavior is released. Such threatening displays of aggressive intentions may be adaptive in that they may prevent actual violent encounters with possible harmful consequences for both sides involved.

In humans, however, interpersonal interactions with intentional emotional underpinnings (in the present case, of an inimical kind) became much more complex during evolution. Humans lost almost all the uncontrollable bodily signs that signal aggressive intent, and those relevant to behavior (but not to physiological changes) are under conscious control. Thus, harmful intentions frequently remain covert in order to prevent, postpone, or divert possible counteraggression or some other kind of countermeasure, or in order not to violate openly social norms. Therefore, the aggressive intentions of other conspecifics frequently have to be guessed from faint hints, discrete behavioral signs, intonation, previous knowledge about the psychology of the respective person, or imagining how one would feel in the other person's situation. That is, *the uncertainties as to the other person's intentions may invite the use of the empathizing mechanism.*

As mentioned in section 2.3, the empathizing mechanism works best in a neutral emotional state, and any strong or longstanding underlying emotion will distort the cognitive conclusions drawn from it. A person whose underlying emotional state consists mostly of the expectation of an inimical attitude toward him, and as a result prepares himself to respond with aggression, will inevitably distort the conclusions of the empathizing process (especially in indistinct or equivocal situations) toward expecting harmful intentions (hence the increased suspiciousness in paranoia). Moreover, as mentioned in the section on empathy in chapter 2, aggressive intentions reduce the intensity of the empathizing process or may abolish it completely. This theme is detailed in the section on the paranoid personality disorder.

The clinical entities (as well as paranoid symptoms in normatively functioning individuals) are discussed in the order of gravity of the dysfunction they induce. The paranoid suspiciousness of normatively functioning persons are dealt with first, then the following mental disorder categories: first, paranoid personality disorder, then delusional disorder, or, more precisely, its persecutory and jealous subtypes. Since in the paranoid subtypes of the delusional disorder the paranoid predisposition is coupled with the phenomenon of psychosis, the basic mechanism of psychosis is also discussed in some detail. The final disorder in which paranoia is predominant should be paranoid schizophrenia. However, since in this disorder *three* causative factors are intertwined—the innate predisposition to paranoia in general, the predisposition to schizophrenia and to its genetically related spectrum disorders, and the mechanism leading to psychosis—this disorder is mentioned only briefly here and discussed in detail in the section on schizophrenic conditions.

7.2.6.1. Paranoid Fears in Normative Persons

It is well known in clinical psychiatry that paranoid personality traits, paranoid suspiciousness, or paranoid thinking may be present in otherwise normatively functioning people, in which case the diagnosis of a paranoid disorder should not be made. The subject I want to consider here, however, is an attempt in the professional literature on paranoia to widen the meaning of the concept to a degree that seems to me problematic.

In *Paranoia, the Twenty-first Century Fear* (2008), clinical psychologist Daniel Freeman and Jason Freeman define paranoia as "the unrealistic belief that other people want to harm us" (p. 23). In their interpretation, paranoia is intimately connected with fear or anxiety: "Paranoia is in essence a form of anxiety. Both are connected with the anticipation of danger" (idem, p. 158). With this interpretation of paranoia in mind, the authors conducted an interesting virtual reality experiment that simulated a four-minute ride on a London underground train. Volunteers were chosen randomly but mental patients with paranoid pathology were deliberately excluded.

The authors found that 45 percent of the participants in the experiment had at least one "paranoid" thought (idem, pp. 66–73). Consider that a ride in a crowded underground train involves being in close proximity to a lot of strange people in a relatively small, closed space without the possibility of escape when the train is between stations. This situation can arouse ancient fears (fear of closed spaces) as discussed in the section on specific phobias, as well as more recent fears connected with the social environment, as discussed in the material on social phobias and avoidant personality disorder. I can understand the advantages of a wider definition of paranoia, especially in studying normative individuals in stressful conditions of modern life, a definition that does not require an aggressive attitude in the face of perceived danger. After all, in the DSM-5 criteria, the presence of an aggressive attitude toward, or aroused by, the suspected malicious intents of others isn't explicitly specified as an indispensable requirement for the diagnosis of paranoid disorders. Nevertheless, without it, the clinical distinction between paranoia and anxiety disorders (especially social anxiety disorder) becomes very problematic. The typical behavioral and mental response to perceived danger of the paranoid individual will not be differentiable from the response characteristic of an anxiety disorder. This can be illustrated with another quotation from Freeman and Freeman: "The most common type of safety behavior [of the paranoid person] is *avoidance*. So someone worried about being attacked might avoid leaving the house after dark." Or: "We might try to placate the people we think are out to get us (this is called *appeasement*) . . . safety behaviors are a typical feature of paranoia" (p. 77, emphasis in the original).

While avoidance of open animosity may be part of the paranoid behavior, appeasing others is clearly uncharacteristic of paranoia, though it may occur frequently in anxious or normative persons. It has to be mentioned that the authors themselves recognized the problem of differentiating paranoia from anxiety disorders according to the foregoing conceptualization (see Freeman & Freeman, 2008, pp. 35 and 38–39).

In summary, what paranoia and some anxiety disorders do have in common is the excessive awareness, oversensitivity, and overreactivity (mostly intrapsychic) to perceived or suspected harmful or dangerous intentions of surrounding persons or groups of persons. What differentiates these two psychiatric disorder categories, in my opinion, is the way the active and reactive mental apparatus respond, respectively, to the stress of these situations. *While in anxiety disorders the characteristic response is avoidance or appeasement as an attempt to distance oneself from or mitigate the perceived hostile intention, in paranoia the characteristic behavioral response is defiance or counterattack.* This aspect of the paranoid disorders will be further discussed in the following two sections on paranoid personality disorder and delusional disorder, persecutory type.

7.2.6.2. Paranoid Personality Disorder

PARANOID PERSONALITY DISORDER

Diagnostic Criteria 301.0 (F60.0)

A. A pervasive distrust and suspiciousness of others such that their motives are interpreted as malevolent, beginning by early adulthood and present in a variety of contexts, as indicated by four (or more) of the following:
1. Suspects, without sufficient basis, that others are exploiting, harming, or deceiving him or her.
2. Is preoccupied with unjustified doubts about the loyalty or trustworthiness of friends or associates.
3. Is reluctant to confide in others because of unwarranted fear that the information will be used maliciously against him or her.
4. Reads hidden demeaning or threatening meanings into benign remarks or events.
5. Persistently bears grudges (i.e., is unforgiving of insults, injuries, or slights).
6. Perceives attacks on his or her character or reputation that are not apparent to others and is quick to react angrily or to counterattack.
7. Has recurrent suspicions, without justification, regarding fidelity of spouse or sexual partner.
B. Does not occur exclusively during the course of schizophrenia, a bipolar disorder or depressive disorder with psychotic features, or another psychotic disorder and is not attributable to the physiological effects of another medical condition.

Note: If criteria are met prior to the onset of schizophrenia, add "premorbid," i.e., "paranoid personality disorder (premorbid)."

Paranoid personality disorder, in my opinion, is the paranoid disorder that expresses in the most clear-cut, unaltered manner the specific inherited predisposition for paranoid disorders. This predisposition is acted out directly in the observable behavior in an interaction with surrounding normative people without being blurred or covered up by the presence of additional mental mechanisms, such as the mechanism causing psychotic disorders in the delusional forms of paranoia, or the innate genetic predisposition to schizophrenia in paranoid schizophrenia.

Indeed, the diagnostic criteria for paranoid personality disorder accentuates the propensity toward distrust and suspiciousness as well as the inclination to interpret others' motives as malicious or malevolent. The paranoid personality: "Suspects without sufficient basis that others are exploiting, harming, or deceiving him or her . . . Reads hidden demeaning or threatening meanings into benign remarks or events . . . Has recurrent suspicions, without justification, regarding fidelity of spouse or sexual partner" (DSM-5, diagnostic criteria, see above).

Other diagnostic criteria, however, hint at a protracted underlying propensity toward an inimical disposition or intention to retaliate: "Persistently bears grudges . . . Perceives attacks on his or her character or reputation that are not apparent to others and is quick to react angrily or counterattack" (idem). This last theme is developed further in the comprehensive edition of DSM-5: "Individuals with this disorder persistently bear grudges and are unwilling to forgive the insults, injuries, or slights that they think they have received. . . . *Minor slights arouse major hostility, and the hostile feelings persist for a long time* . . . they may be 'pathologically jealous' and . . . want to maintain complete control of intimate relationships to avoid being betrayed . . . they may express 'argumentativeness,' they have a . . . combative and suspicious nature . . . they may be litigious and frequently become involved in legal disputes" (DSM-5, pp. 650–651, emphasis added).

From these excerpts, it seems to me quite clear that the most characteristic feature of paranoia is not the excessive or unfounded expectation that other persons want to hurt us. As we have seen, this feature of the disorder is shared with other psychiatric disorders (predominantly anxiety disorders). The most typical element in paranoid behavior seems to me to be the propensity to respond to those expectations or suspicions in a hostile manner for unrealistically protracted periods of time.

In common with other personality disorders, this behavior pattern is discharged in social settings while being resistant to change in the face of undesirable or even punishing reactions from surrounding people. The paranoid individual tends to overvalue herself and feels profound indignation when, in accord with his interpretation, somebody dares to relate to her in an outrageous way.

It has to be stressed again that similar distortions or exaggerations regarding the harmfulness or maliciousness of surrounding humans can be encountered in other disorders besides paranoia—disorders characterized by intense, longstanding, or easily arousable anxiety or dysphoria. In these cases, however, instead of rejecting and opposing these felt negative evaluations, as the paranoid person does, these individuals tend to be resigned to their plight. In social anxiety or avoidant personality disorder, the individual feels helpless, as if the validity and consequences of the exaggerated or imagined negative verdicts or opinions are agreed with or at least are seen as irrevocable. In the case of dysphoric depression, the felt negative opinion of others is not only accepted but the individual believes that he deserves it.

The paranoid predisposition manifests itself early in life and powerfully influences interpersonal relationships, which tend to be tense or even inimical, thus negatively affecting socializing processes. These persons are frequently single or divorced, have no close friends, and show a predilection to develop delusional paranoid states (the theme of the next section) at an older age.

7.2.6.2.1. The Empathizing Mechanism in Paranoia

The unanimous observation that paranoid suspicions are directed toward persons, groups of persons, or institutions that possess deliberate intention, and never concern possible dangers actually originating in the physical or biological world, which possesses no such intentionality, invites the possibility of using the mechanism of empathizing in order to figure out the other's covert intentions.

It is frequently maintained, especially in the psychodynamic literature, that the suspiciousness of the paranoid individual, instead of representing an exaggeration of existent environmental disturbances, is in fact *projection* of intrapsychic contents. Even DSM-5, a text with a strong descriptive orientation, accepts this reasoning: "Individuals with this disorder seek to confirm their preconceived negative notions regarding people and situations they encounter, attributing malevolent motivations to others that are projections of their own fears" (DSM-5, p. 651).

(Instead of "fear" I would prefer to use the word "hostility.") In the following, I reformulate the psychodynamic concept of projection in terms of the empathizing mechanism applied to the paranoid phenomena.

As we have hypothesized, the predisposition to paranoid disorders consists of a predilection to respond with defiance or hostility when threat or malevolence is sensed from the human environment (with no regard to whether an active instinctive motive in progress is differentially involved or not). The inimical reaction, accompanied by the subjective experience of anger, grudge, indignation, etc., in paranoid individuals is more easily elicited than in normative persons and, even more characteristically, lasts much longer than justified by the circumstances—in other words, much longer than its biologically adaptive function (that is, the removal of an environmental impediment) warrants. In consequence, the recapturing of the original active instinctive motive, or relaxation and rest after the actual aggression-eliciting encounter is over, are considerably delayed or lost altogether. As an example of the biological role of the aggressive response, consider the "bird type" mating relationship in pigeons disturbed by a male intruder. The male will try to chase away the intruder (aggression), and if it is unable to do so, removes or distances the female from the intruder's vicinity—that is, it resorts to a kind of defensive action. In either case, the frustrating element in the way of the active motive, the "bird type" mating relationship, is neutralized. The aggressive or defensive responses of the male to the situation that frustrates this mating strategy (which closely resemble the behavioral pattern of jealousy in humans) subside, giving way to the presumably pleasurable mood of the mates being in each other's company, and they may resume some common survival-oriented or reproductive activity.

Now consider how different a paranoid individual's behavioral responses and mental states are amid the frustrating conditions of the social environment. In a human community, the interaction between individuals is much more complex than that occurring in nature. Well-socialized humans are as a rule acutely aware what responses (desirable or undesirable) their emotional or intentional displays may produce in others. As a result, emotions and intentions are frequently hidden, downplayed, exaggerated, or faked, depending on the impression one wishes to give. Even clearly stated intentions cannot be taken for granted, as the exaggerated or frankly misleading promises of politicians during elections (about what they intend to do in case they are elected) amply demonstrate. Therefore, in social circumstances, one frequently has to resort to more complex mental mechanisms than simple observation in trying to figure out, in time, others' intentions, such as remembering previous similar instances and their outcomes, considering relevant accumulated knowledge, and making recourse to the mechanism of empathy.

Empathy, however (as discussed in section 2.3), works best when the individual is in a neutral emotional and intentional state. Then she can observe neutrally, unselectively, and unbiasedly the target person's behavior and the circumstances that elicited it. At the second stage of the empathizing mechanism, emotional neutrality is even more important than in the first stage. Strong emotions or intentions that would suppress, supplant, or bias the surfacing in consciousness of the emotional, intentional state the impartial empathizer expects to experience must not be present.

In the case of a paranoid individual, however, we hypothesized that grudge, combativeness, aggression, or related emotional and intentional states are more easily elicited than other emotions/intentions and last much longer than necessary, gravely compromising the optimal functioning of the empathizing mechanism. The inevitable result is a misinterpretation or strong bias toward interpreting others' behavior as malevolent or aggressive.

To make the paranoid person's resultant social disabilities more conspicuous, I will mention some examples of interpersonal interactions expected from socially well-functioning persons in conflictual situations. In such cases, the flexible transitions between the frustrated forms of active behavior and the original unfrustrated instinctive motive, as well as the optimal balance of active/reactive behavior, can frequently be observed. For example, if a subordinate accepts a superior's criticism, it is often quickly followed by the superior mentioning some praiseworthy aspects of the criticized person's work or qualities, or by a short talk about some emotionally neutral matter, signaling that the contentious issue is closed, at least for the present, and that the relationship between the two continues as usual. (Recall the examples of Jane Goodall and de Waal with regard to how quickly aggression is followed by reconciliation or friendly behavior in chimpanzee communities.)

Between friends, mocking, teasing, insulting, and even playful physical sparring are commonplace and do not damage the relationship.

Actually, this kind of behavior signals, as a rule, an emotionally close relationship. (Evolutionary biologist Amotz Zahavi proposes that the rationale for this kind of behavior is to test the strength of the bond between the participants in Zahavi & Zahavi, 1997.)

When an incident has to be clarified between two neighbors, for example, the insulted party, if he possesses good socializing abilities, proceeds with caution. Before he meets the blamable party, he tries to gather more information about the incident and possibly to attain the cooperation of other neighbors who may suffer from the same inconvenience. During the meeting with the troublesome neighbor, the insulted party tries to suppress the expression of his own angry feelings and to talk with the other party in as matter-of-fact a tone as he can, exploring at the same time that neighbor's attitude toward the incident. The main goal of this approach, I think, is to achieve an optimal compromise between two antagonistic intentions. On one hand, the insulted neighbor wants to make clear that he is unwilling to drop the matter without an acknowledgment that the blame was appropriate, followed perhaps by a promise from his neighbor for change, redress, or restitution. On the other hand, he tries to avoid arousing unnecessary counteraggressive feelings that may "spill over" into irrelevant areas of the relationship.

Paranoid persons seem to have serious problems regarding this kind of interpersonal transaction. Being stuck in a permanent state of expecting as well as intending to express an inimical attitude, they create unnecessarily strong interpersonal tensions; lose possible friends, allies, and even mates or sexual partners; and frequently will withdraw as much as possible from social encounters. However, the paranoid individual's withdrawal from the social sphere is very different from, say, an agoraphobic's refusal to leave her home, or that of someone suffering from social phobia or avoidant personality disorder. In the paranoid's case, the withdrawal happens when he feels overwhelmed with social disapproval or animosity or fears losing control of his own aggressiveness. His withdrawal resembles a medieval population's retreat into the settlement's stronghold where it can defend itself better against the enemy that may overpower them in open territory. Many paranoid individuals, in my experience, secure the entrance door with multiple locks and frequently keep some object at hand that can be used as a weapon in case the need arises. An anxious person, on the other hand, uses her home more as a *hiding place*, in which she remains inconspicuous.

7.2.6.3. Delusional Disorder (Persecutory and Jealous Subtypes)

Delusional disorders (formerly referred to as paranoid disorders) consist of stable, non-bizarre, well-systematized delusional beliefs in a certain domain with a chronic course. They lack the grave deterioration in personality traits and functioning characteristic of schizophrenia (Fennig, Fochtmann, & Bromet, 2005, p. 1525).

Delusional disorders are categorized, according to the central theme or content of the delusions, into: erotomanic, grandiose, jealous, persecutory, somatic, mixed, and unspecified subtypes, with the most widespread being the persecutory one. Our theme in this section being paranoia, the only categories relevant to our concerns are the persecutory and jealous subtypes. The decisive psychic mechanism involved in delusional disorders, the one determining the diagnosis, is the delusional (that is, psychotic) nature of the way the central theme of the false belief is adhered to and expressed verbally or in behavior. Therefore, we have in this section two different subjects to deal with—the *content* of the clinical disorder, on one hand, and the *delusional nature* of this content, on the other. Each has a wider scope than that expressed in delusional disorders.

Regarding the central topic of the delusional disorders (paranoid, jealous, grandiose somatic, etc.), all of them have nonpsychotic counterparts in psychopathology and normative behavior. The nonpsychotic counterpart of the persecutory delusional subtype consists of the paranoid personality, paranoid traits, or normative suspiciousness; the nonpsychotic counterpart of the grandiose subtype is narcissistic behavior or nonpsychotic hypomanic or manic behavior; the nonpsychotic counterpart of erotomania is the infatuated, daydreaming person's behavior; the nonpsychotic counterpart of the jealous subtype of delusional disorders is excessive but nondelusional, as well as normative, jealousy; and the nonpsychotic counterpart of the somatic subtype is hypochondria or somatization disorder. In my opinion, this aspect of the delusional disorders is amenable to interpretation along the lines of evolutionary reasoning used in this text. That is, they are originally adaptive behavioral tactics, underpinned by instinctive predispositions whose intensity and tenacity in delusional disorders is excessive as a result of diversification due to the RfNSPs.

The second topic central to delusional disorders—that is, their *psychotic nature*—also has a much wider scope in psychopathology and normative behavior besides the role it plays in delusional disorders. Psychosis or psychoticlike states can be elicited in normative persons in conditions of sensory and social deprivation, they can be aspects of other "functional" mental disorders (like schizophrenia or mania), they may also appear in a host of organic conditions (like dementia and delirium), result from drug side effects, or be elicited by the use of some psychoactive substances (like hallucinogens). The central mental mechanism that brings about psychotic states in general, to my mind, cannot be approached by the kind of evolutionary reasoning adopted in this text, although a formulation of its basic mechanism—underlying all forms of psychosis—is attempted in the following.

The specific theme or content of the jealous subtype of the delusional disorder is related to the "bird type" mating strategy (discussed in section 5.3.2).[7] Jealousy is induced by this active instinct's aggressive form of frustration (aggression toward a competitor, forceful removal of the female from the vicinity of other males, or, more commonly in humans, aggression toward an unfaithful spouse). Compulsory segregation of the wife or young unmarried women is a feature of many traditional cultures. While in human behavior the expression of jealousy is more complex than it is in birds, the underlying intentional motives have to be the same. (See a more detailed discussion of this subject in the context of evolutionary psychology in Barkow, Cosmides, & Tooby, 1995, pp. 289–314.) When the belief in the infidelity of the spouse has no realistic basis, and cannot be changed by contradictory evidence, a diagnosis of delusional jealousy is appropriate.

In the *persecutory subtype* of the delusional disorder, the same reasoning used in the jealous subtype can be applied, although the kind of delusional belief, with regard to the harm it is imagined that surrounding people intend to inflict, has a wider scope than in the jealous subtype. It may, for example, concern one specific person, a neighbor, or a group of persons or institutions (the police, the CIA, etc.) whose task may be to follow or spy on people. As in the entire paranoid disorder category, the instinctive constellation accentuates the aggressive form of the active frustration or the aggressive tactics of the reactive behavior. The person believes without plausible reason that he is hindered or thwarted in the attainment of (long-term) goals or that he is the target of other kinds of harmful intentions (see DSM-5, pp. 90–93). The aggressive disposition in this case is as evident in overt behavior as in the paranoid personality disorder, although its expression may be more unpredictable, out of context, or more disproportional in its intensity: "The persecutory beliefs are often associated with querulousness, irritability, and anger, and the individual who acts out his or her anger may at times be assaultive or even homicidal. At other times such individuals may become preoccupied with formal litigation against their perceived persecutors." The behavior of these individuals is further characterized by: "determination to succeed against all the odds . . . endless crusading spirit to right a wrong; a driven quality . . . quarrelsome behavior; and the belief that defeat is unacceptable" (Fennig, Fochtmann, & Bromet, 2005, p. 1527).

In the *erotomanic subtype* of the delusional disorder, the underlying active instinctive motive is the same as in the jealous subtype, that is, the "bird type" mating strategy. However, while in the jealous subtype it is expressed in its aggressive form of frustration, in erotomania the underlying instinctive motive retains its original, unfrustrated form. The delusional belief consists of the unrealistic confidence that while the feelings of the person (the few cases I have encountered were all women) are reciprocated by the chosen man, external impediments prevent the materialization of their "reciprocal" attraction or love. That unrealistic confidence is due to "projection" in the psychodynamic parlance or, using the approach of the present text, to the strong bias of the empathizing mechanism.

The *grandiose subtype* of delusional disorder, in my opinion, has an underlying emotional base similar to the erotomanic but in a wider arena of interpersonal relatedness. The grandiose individual desires to be appreciated, esteemed, or admired by surrounding people or by a more anonymous public for qualities, abilities, or talents he does not possesses, or for some accomplishment he did not achieve. The delusional (psychotic) mechanism allows him to convert these desires into the belief that they have already materialized. The underlying mood is a strongly pleasurable one ranging from euphoria to exaltation.

7.2.6.4. The Concept of Psychosis (General Considerations)

In this section we deal with the second aspect of the delusional disorder, that is, the *psychotic* or *delusional nature* of the subtypes just discussed. The unrealistic quality of the cognition (delusions, disturbances of reasoning and judgment), and such other possible signs and symptoms as illusions, hallucinations, and unrealistic behavior signals that we are now dealing—for the first time in this text—with the psychotic phenomenon. Owing to its great importance to our concerns, it is discussed in the widest possible context.

The concept of psychosis is not adequately defined in psychiatry: "As a result of conflicting usage, there is no single acceptable definition of what psychosis is." Psychosis often means whole clinical disease categories or groups of disease categories, like psychotic disorders, which are contrasted with nonpsychotic disorders like anxiety and personality disorder. Sometimes the word *psychosis* indicates the "severity" of the disorder: "The psychoses are 'major' disorders that are more severe, intense, and disruptive. . . . In contrast, some schools in psychiatry, in particular psychoanalysis, argue that the difference between psychotic and nonpsychotic disorders is only one of degree not of kind" (Campbell, 2004, p. 541).

The causative agents of psychotic states, when known, vary significantly, with seemingly little similarity in how they bring about the psychotic phenomena (pathogenesis). *Somatic disorders* with widespread effects on brain biochemistry and neurophysiology, like infections, metabolic disorders, endocrine conditions, fluid and electrolyte imbalances, certain drug intoxications, etc., may induce "organic" psychotic states. *Neurological conditions* as varied as epilepsy, especially of the temporal lobe origin; dementia; localized or widespread cerebrovascular diseases; and electrical stimulation, for example, during brain surgery can induce different psychotic symptoms. In addition, psychosis (or at least psychosislike states) may be induced in healthy, normatively functioning persons by drastically reducing the level of information input from the physical/social environment, or by taking hallucinogenic drugs. *Sensory/social deprivation* may occur in various circumstances: during solitary confinement in prison, in the case of lonely explorers or shipwrecked sailors, hospitalized patients in respirators, or even during a long drive alone at night. Controlled sensory deprivation experiments amply demonstrate the emergence of psychotic symptoms in these circumstances: "It was intriguing to think that simply doing nothing and being cut off from the outside world could bring about a transient psychotic-like state" (Solomon & Kleeman, 1980, p. 601). How can such widely different disorders, circumstances, or conditions lead to an assortment of mental and behavioral dysfunctions that are similar enough to be united into a single nosological category?

It is questionable whether the subject of psychosis *per se* pertains to the main concern of this work, the inherited predisposition to develop mental disorders as a result of RfNSPs. I am not aware of psychotic states in animals, and even if they exist (I have in mind such cases as transient intoxications from plant toxins or venomous bites in more evolved animals), the possibility of obtaining meaningful information about them is probably nonexistent. Furthermore, it is highly questionable whether a unitary *innate predisposition* to develop psychosis, a predisposition that can diversify under relaxed selective conditions, exists. It can be presumed alternatively that the potential to develop psychosis is present in *every* human brain, and shows up when the appropriate conditions or constellation of conditions (intrapsychic, organic, environmental) are present. However, we cannot avoid discussing the topic of psychosis in a section on delusional disorders, and we also need the fruits of this discussion to unravel the complexities of schizophrenia, the topic of the next section.

In the context of this work, psychosis is defined as a *syndrome* (not as a mental illness category or some attribute of it such as its gravity), which implies some kind of distortion in grasping the surrounding realities (natural/social), a distortion that is clearly in excess of what is acceptable in normative persons in normative circumstances or in non-psychotic mental disorders. Moreover, this distortion is not correctable by contradictory evidence.

The complexity of these distortions in the context of consensually shared reality varies from very simple perceptual distortions (isolated illusions, simple hallucinations) to complex hallucinations (hearing a whole conversation), gross misinterpretation of others' intentions (like persecutory delusions), or serious misjudgments about the nature of the social network and the individual's real place in it. The psychotic misbelief may concern a single person (delusions of jealously), several persons, or a social institution (involving more comprehensive delusions of persecution), or it may have an all-encompassing nature, as in grave schizophrenic delusions that include "the whole universe" and the individual's special (delusional) place or role in it. The individual may or may not *act* in accordance with the psychotic experience, but even in cases when he does not act them out or relates to them verbally, his distorted grasp of the surrounding realities may be reflected in some form of inadequate behavior.

If we attempt to categorize the various causes of the psychotic syndrome according to the specific ways they disrupt or lead to an imbalance in the higher-level functioning of the mental apparatus—especially in the function of integrating or balancing external input with already present intrapsychic contents—we can divide them into three more or less homogenous groups.

1. *Factors that substantially and for longer periods reduce the input from the external (physical/social) environment.*

2. *Causes that disproportionally augment the effects of the internal, intrapsychic contents in the final, overall mental output at the expense of the externally induced contents.* To this group belong those causes that strongly intensify the instinctive activity encountered, for example, in grave mania or dysphoric depression with psychotic symptoms. Other causes pertaining to this group include *excitation* of the appropriate brain tissue by psychostimulants like amphetamines, hallucinogenic narcotics, organic brain disorders like temporal-lobe epilepsy, or intra-cerebral stimulation during brain surgery.

3. *The organic causes that remain after the exclusion of those belonging to the first two groups.*[8] They act as a rule in a widespread (not localized) manner on the brain, like those causative factors that may bring about delirium or dementia with psychotic symptoms or like certain kinds of drug intoxications with widespread mental symptoms, including psychotic ones. I argue in the following that the psychotic symptoms in this third group of causes are induced mainly by interfering with integrative functions of the brain (more precisely, with the integration of the external input and intrapsychic contents).

To again summarize, all the causative factors that lead to psychotic symptoms or syndromes cause an imbalance or strong bias in the final integrated output of the mental functioning by either considerably reducing the impact of the external stimuli (social or physical), or by augmenting disproportionally the impact of the internal, intrapsychic contents (which may possess a strong underlying emotional or intentional component), or both. While all three groups of causes produce ultimately the same final result regarding the kind of imbalance or bias they induce in integrative mental functioning, they achieve it via different routes. As a corollary effect of this imbalance, the mind can erroneously interpret internally originating mental contents as coming from the external natural or social environment (illusions, hallucinations, delusions).

In the following, I discuss these three groups of causative factors.

7.2.6.4.1 Factors That Excessively Reduce the Influence of the Environmental Input on the Final Mental Output

Normative persons (without a relevant psychiatric or somatic disorder) can suffer from psychosislike symptoms or a whole psychotic condition as a result of reduced levels of input from the physical/social environment alone, when this reduction is significant or longstanding enough. Some of the circumstances that may induce such states (the isolation of explorers, prisoners kept in isolation, and hospitalized patients in respirators) have already been mentioned. It is important to stress that isolation from the *social* environment may have the same psychosis-inducing effect as the reduction of the physical aspects of environmental stimuli. In fact, it is impossible to differentiate between these two kinds of environmental deprivation. Sensory deprivation (that is, a reduction in the visual, auditory, tactile, etc., external stimuli that reach the sensory organs) *inevitably induces social deprivation as well.* The ensuing discussion of the present subject relies on, and the quotations are drawn from, two sections in Kaplan, Freedman, & Sadock (eds.), *Comprehensive Textbook of Psychiatry* (1980), written by J. G. Miller (pp. 98–114), and P. Solomon and S. T. Kleeman (pp. 600–608).

The details of an individual's mental experiences during sensory deprivation vary according to the circumstances and the unique reaction to them of the respective mental apparatus. However, the characteristic succession of events, from a balanced processing and integration by the mind of externally induced and internally activated mental contents to a considerable bias resulting from a strong reduction in external stimuli that overaccentuates the internal contents, seems to have a predictable course. This succession of mental events "became recognized as characteristic of the sensory deprivation state: anxiety, tension, inability to concentrate, or organize one's thoughts, increased suggestibility, vivid sensory imagery—usually visual, sometimes reaching the proportions of hallucinations with delusionary quality—body illusions, somatic complaints, and intense subjective emotional accompaniment" (Solomon & Kleeman, 1980, p. 600). It is also important to note in a text that deals with human diversification that: "[t]here are individual differences in the optimal rate of information input for a person to maintain arousal, think well, act effectively, and feel good" (Miller, 1980, p. 111). In spite of these individual differences concerning the ease with which a person can develop a psychosislike state, it seems to me that no indications exist for a unitary inherited predisposition in the animal kingdom in this respect that might have undergone diversification as a result of RfNSPs.[9]

One conclusion drawn from the effect of sensory deprivation on the mental functioning is very close to that espoused in this text with regard to the interpretation of psychotic phenomena. This view stresses the role of "the organism as an information-processing machine whose purpose is optimal adaptation to the perceived environment. With insufficient information, the machine cannot form the cognitive map against which to match current experience, and disorganization and maladaptation ensue. Continuous feedback is necessary to monitor the organism's own behavior and to attain optimal responsiveness. *Without this feedback, the person virtually lives inside a Rorschach inkblot, forced to project outward individually determined themes having little relationship to the reality situation.* This is similar to what many psychotics do" (Solomon & Kleeman, pp. 603–604, emphasis added).

Sensory deprivation is not the only mechanism that leads to reduced sensory input from the external world. Reduced acuity of the sensory organs, especially hearing and sight, also belong to the present group of factors that predispose to psychotic states through the same mechanism. Visual hallucinations are well known in blind persons, and moderate to severe hearing loss has been found in 40 percent of elderly patients suffering from paranoid delusions (Kaplan et al., 1980, p. 1293). The same applies to damage along the neural pathways of the sensory organs or to damage to the cortical areas where the primary sensory processing takes place.

Circumstances associated with certain psychiatric diagnostic and therapeutic practices have also been implicated in inducing sensory/social deprivation (of a milder intensity) that may enhance the expression of intrapsychic contents that are unrelated to the present environmental realities. Examples include the impoverished daily routine of chronic psychiatric wards, projective psychological tests (like the Rorschach test), as well as the classic psychoanalytic setting. It may be mentioned in this context a clinical trial that found "that the percentage of socially atypical behaviors (e. g., peculiar postures, talking to oneself, repetitive behaviors) among persons hospitalized in a psychiatric inpatient facility [is] inversely correlated with the number of people in functionally defined areas" (e.g., dining room, recreation room) (McGuire & Troisi, 1998, p. 76). Dreams, too, of course, are "constructed" solely from internal mental contents, while the dreaming person (during dreaming) interprets them erroneously as interactions with a real external environment.

It has to be stressed that the brain begins to integrate the crude sensory information (visual, auditory, etc.) that it gets directly from the outside world according to innate mechanisms at a very early stage—that is, during the crude sensory data's "aggregation" into more comprehensive percepts. These percepts then become integrated in turn with different intrapsychic contents, including predispositions to select and categorize according to the adaptive significance of various percepts; innate predispositions to be interested in certain aspects of the environment and ignore others; memories of former similar perceptual experiences and their consequences; other relevant ontogenetically acquired knowledge; relevant emotional or intentional states; and so on.

Moreover, it often happens that the brain has to *reconstruct* an external object or an impending event from very faint or partial external cues. Face recognition from difficult viewpoints or the recognition of a predator approaching in a stealthy manner concealed by vegetation are such examples. In these circumstances, the complementation of sensory percepts with intrapsychic contents has to be even more massive than in cases when the environment provides most of the necessary sensory cues, thus facilitating the distortion of the picture that one forms of aspects of the environment, especially when strong emotional or intentional intrapsychic motives or expectations are also present.

An additional relevant consideration in this context is that the whole unified percept or gestalt evidently has to be *projected back* to the same place in the environment the separate sensory cues originated from, and not sensed by the person as assembled inside the brain where the putting together of the gestalt actually happened. Otherwise, it would of course completely lose its adaptive value.

This line of thought may justify to a certain extent the brain's predisposition to complement more and more massively with internally originating material the insufficient or missing external data and then to project it back into the external world. This process may constitute one of the ways illusions, hallucinations, and delusions may be formed.

7.2.6.4.2. Factors That Intensify Disproportionally the Participation of Intrapsychic Contents in the Final Mental Output on the Expense of Environmental Influences

The second group of causative factors leading to the psychotic phenomenon includes those cases in which a strong intrapsychic arousal, excitation, or stimulation, rather than the scarcity of external stimuli, throws the desirable relationship between the external stimuli and related internal mental contents out of balance. The bias created is in the same direction as in the first group of causative factors. As we have seen, organic factors that stimulate the appropriate brain areas, such as psychostimulants, hallucinogenic drugs, temporal lobe epilepsy, and so on, can also cause this kind of imbalance.[10]

Our main topic in this section is a similar kind of imbalance in mental functioning, but one caused by an underlying dysfunctional instinctive constellation rather than organic or drug-induced factors. The most conspicuous examples discussed up to now, besides paranoid psychosis, are severe manic episodes and grave agitated dysphoric depression, both of which may display psychotic symptoms. (Schizophrenia's relationship to psychosis is discussed in section 7.2.8.4.)

The following discussion of the *delusional phenomenon* refers only to these "functional (not organic) psychotic disorders" and especially to paranoid psychosis. In this respect, our first and foremost task is the clarification of the concept of *belief* as well as the distinction between normative and psychotic forms of belief, the latter being the central theme of delusional disorders.

The word *belief* is defined in the *Concise Oxford English Dictionary*, eleventh edition (2009) as "an acceptance that something exists or is true, *especially without proof*" (emphasis added). The things people believe as being true and the nature of the mental mechanisms that lead to belief are diverse. We will briefly discuss three widely different kinds of belief: those attained with the aid of the scientific method; unprovable beliefs of normative persons; and the delusional beliefs of mental patients.

The belief most remote from the delusional one is the belief in knowledge achieved by employing the *scientific method*. We saw in section 1.1 the two different paths to scientific knowledge. The first is the Baconian route, based strictly on observation and experimentation in a framework of cautiously formulated empirical regularities and theoretical laws. No effort is spared to neutralize or exclude possible bias caused by the subjective incentives of the scientist: expectations, emotions, financial interests, and so on. Some of the methods that considerably reduce the subjective element in scientific research are double-blind experimentation, the use of control groups, and the condition that other research groups different in their loyalties, cultural backgrounds, etc., be able to replicate the results of an experiment.

The second path leading to scientific knowledge, and as we saw in chapter 1, the method embraced by philosopher of science Karl Popper, is the reverse of the first. It starts from a hunch, an intuitive belief, a hypothesis, whose observable implications have to be submitted to rigorous testing. However, the belief in the correctness of such a hypothesis or theory, even when the results of its testing support it, will never be absolute or unconditional. The validity of a theory is always seen as being temporary. A formerly well-validated theory may be challenged or refuted by new experimental data, in which case the theory may be modified or entirely replaced by a new theory that better explains the existing data (Thomas Kuhn, *The Structure of Scientific Revolutions*).

Beliefs in everyday life are much less substantiated than those in science. People believe in (and rely on) the permanence of regularities observed in nature, in the opinion of an authority or specialist in a special field, in the teachings and promises of a charismatic leader, or in whatever the majority accepts as true.

The ubiquitous belief in *unprovable* religious teachings and rituals, superstitions, and political and social programs, or the less widespread belief in such highly improbable matters as extraterrestrials, ghosts, and the imminence of doomsday by countless normative persons, shows clearly that the difference between delusions and normative beliefs has little or nothing to do with how implausible, unfounded, or unprovable a delusion is in contrast to normative beliefs that are supposedly better rooted in the surrounding realities or provable by logical reasoning.

The extent of the rigidity with which one adheres to erroneous or unprovable beliefs in spite of logically convincing reasoning or observable counterevidence (another argument used in psychiatry to differentiate between delusions and misplaced or unprovable normative beliefs) isn't tenable either. The adherence to a religious belief or faith is frequently so strong that it resists not only rational reasoning but highly painful attempts by other social groups aimed at getting these "believers" to relinquish the respective dogma (think of the ineffectiveness in this respect of the early Christians' persecution by the Romans, or of the Sephardi Jews' exile from Spain and Portugal for not accepting Christianity).

Francis Galton (Darwin's cousin) demonstrated the ineffectiveness of prayer in the fulfillment of human wishes or in the attenuation of human suffering using a comparative statistical experiment. He compared statistically the realization of wishes of two groups—one religious, who prayed for their fulfillment, the other atheistic—and found no differences between them. No one but the insurance companies were interested in his findings; nor did they shake the faith in prayer of the devout (Medawar, 1969, pp. 3–6; Youngson, 1998, p. 181).

If the difference between deluded and normative persons' unprovable beliefs is *not* the higher probability that the latter is true or realistic, and if the strength of the adherence to the belief in the face of lack of evidence or contradictory evidence cannot differentiate between them either, what then is the significant difference between them?

Religious, political, and other unprovable beliefs are inculcated in the mind of the believers by an *external source*: family atmosphere, an authoritative or charismatic teacher or leader, and so on, and thus *are shared* by a smaller or larger group of people. The *common belief serves as a unifying medium* for the members of the group; frequently it offers a common worldview and an explanation of socially important matters, such as the (divine) origin of the respective group; it may contribute to the sanctification of that group's hierarchical organization; and ensures acting in concert to achieve common aims (like territorial interests). It can be mentioned here that "self-deception" is a central aspect of behavioral biologist Richard Alexander's theory on human social evolution: "Self-deception explicitly plays a role in fostering and maintaining group unity, and this role is intricate with the practice and prominence of familial, tribal, ethnic, racial, or regional myths, including organized religion" (Alexander, 2013, p. 283). Anthropologists Emile Durkheim and Bronislaw Malinowski regarded religious rituals as "integrative factors increasing social cohesion" (Durkheim) and "as an integrative force to the individual but also as an organizing force to the society" (Malinowski, quoted in Dobzhansky, 1962, p. 218). See also Maynard Smith & Szathmáry, 1995, pp. 272–273.

Furthermore, religious belief may alleviate common fears (like the fear of death, dealt with by all religions), offer consolation for various kinds of human suffering, and promise redress of injustices and other human miseries (as a rule in an afterlife), thus giving it a strong emotional appeal. It may also promise ample rewards for self-sacrificing deeds that the respective religious community sees as desirable. For many, the common belief system attains the status of a *social and cultural reality*, perceived as not less real than the physical and biological realities of the group's natural environment. After all, the brain (any brain) is created by evolution first of all to assist the organism in fulfilling the imperatives of the individual life through adaptation to a specific physical and social reality, not to understand this reality, as well as itself, more deeply than is needed for successful survival (Wilson, 1998, pp. 61, 96).

Delusion, on the other hand, is an *unlearned, idiosyncratic, individual* creation expressing the influence of a certain kind of inappropriate instinctive configuration underlying cognitive functioning. Indeed, if it is not externally induced, acquired knowledge, what else can it be? Instead of welding the individual to the community he lives in, delusion estranges him from it: "What is characteristic of the delusion is that it is *not* shared by others; rather it is an idiosyncratic and individual misconception or misinterpretation. Further, it is a thinking disorder of enough import to interfere with the subject's functioning, since in the area of their delusion they do not any longer share a consensually validated reality with other people" (Campbell, 2004, p. 174, emphasis in the original).

The strictest adherence to the clinical data does not completely justify even this conclusion. In some cases, the delusion of a mental patient can "infect" others; that is, it can be "shared" with surrounding persons just as shared, unprovable beliefs spread in a normative population. A rather rare mental disorder, *folie à deux* or shared psychotic disorder, consists of the "infection" of a normative but suggestible person with the delusions of a mental patient. In this instance, the person "infected" with the delusion is not considered a mental patient suffering from a delusional disorder. Her belief in the delusion is the equivalent of the normative person's belief in some unprovable shared dogma. Sometimes a whole family or members of a cult can be infected in this way: "Many psychiatrists consider group delusions to be present in some cults . . . but exactly where the cutoff points occur between delusions and other zealous beliefs held by larger, more traditional and well-organized religious, political, and other groups is arguable" (Yager & Gitlin, 2005, p. 979).

Since even the possibility of transmissibility does not differentiate between normative persons' beliefs and delusions, we remain with only one sure difference between them—that is, the *internal origin of the delusion*. At this point, then, we can formulate more exactly the difference between the mechanisms that lead to the unprovable, shared beliefs of normative persons as opposed to those that lead to delusions.

Normative persons' shared beliefs are inculcated by an external social or cultural source: for example, persons, institutions, books, or media. During this process—and this seems indispensable in genuine shared beliefs—the purely cognitively acquired material is *secondarily* "impregnated" or "saturated" (in different amounts, in different persons) with the energetic, emotional, and intentional charges (aspects of the instinctive activity), enabled by the considerable diffuseness and malleability of human instincts, as well as by the unifying, emotion-gratifying, or mental-pain-reducing effect of many shared belief systems. In this way, the shared belief acquires its tenacity and resistance to change, as we have seen.

During the emergence of *delusional beliefs*, on the other hand, the role of the acquired versus inherited aspects, as well as the sequence of events, is entirely different from that encountered in the case of shared beliefs. As we have seen, *delusional beliefs originate from the impact of certain dysfunctional instinctive configurations that are intense enough to be able to throw off-balance the normative equilibrated integration of external stimuli with intrapsychic contents.* In *mania*, for example, the individual instinctive energies and intentions possess disproportionally high intensities that overpower the intrapsychic impact of contradictory environmental realities. In consequence, they may be expressed in a way dissociated from external input (megalomanic delusions). In agitated, dysphoric depression and paranoid disorders, the underlying frustrated instinctive motive (the dysphoric or aggressive one) is expressed with disproportionate intensity and duration. Depending on the intensity of these instinctive promptings, as well as on the integrating and balancing capabilities of the respective mental apparatus, each of these conditions may be expressed as either a psychotic or a nonpsychotic disorder.

In the case of schizophrenia, as will be argued in the following, the extreme diffuseness of instinctive predispositions will, from an early age, gravely hamper their integration with external, socially induced, learned contents, leading to pronounced difficulties in social assimilation. This disassociation between the internally emerging and externally induced mental contents may lead either to a psychotic schizophrenic disorder or to a nonpsychotic spectrum disorder, as is discussed next in some detail.

It should be mentioned in this context that both the normative believer, as well as the psychotic patient of the same social group, have to express their thoughts in the same shared language. However, the sequence of events with regard to acquiring or finding the appropriate verbal expressions (or rituals) is again entirely different. In the case of members of a shared belief system, the verbal formulation and other forms of expression of the shared belief, including the verbal content and body language of the prescribed rituals, shared symbols, and written and other "proofs" of the "truth" of the belief, etc., are received ready-made from the respective social source.

The situation of the solitary mental patient developing delusions is very different. He, naturally, gets no assistance whatever from the social environment in finding the appropriate expressions in words or body language of that internally felt emotional, intentional "reality." He has to find or create them by himself. Regarding this endeavor, we have to stress two important considerations:

1. *First*: The individual has no alternative but to choose from the assortment of words, expressions (especially those with an emotional load), sequences of reasoning, etc., that he formerly learned from surrounding cultural influences. Schizophrenic neologisms (new "words" created by the patient) are an exception in this respect.

2. *Second*: Since he is in a psychotic state, at least concerning the psychotic topic, he will overemphasize the expression of the emotional, intentional contents, and at the same time downplay, disregard, and distort incongruous external influences like contradictory evidence. While being concerned to express his internal mental state as tellingly, as perceptibly, as he can, he is much less preoccupied with being correctly understood by others—for example, by constantly adjusting his own explanations through a feedback mechanism to the reactions he gets from the conversational partner. He may borrow details from

one or more shared belief systems, but use them in an unusual, id-iosyncratic way. For example, he may believe in telepathy and claim that it enables him to have direct communication with God.

Nevertheless, while the individual expressions of delusion may vary widely and depend on cultural influences, the underlying basic emotional, intentional themes are limited to only a few, and are independent of these influences (megalomanic, persecutory, erotomanic and somatic delusions, etc.).

7.2.6.4.3. Psychotic Symptoms in Generalized Organic Brain Disorders

Organic mental disorders have a known organic cause, and, as a rule, a temporal correlation exists between the effect of the organic factor and the ensuing mental syndrome or disorder. I include here those psychotic disturbances brought about by psychoactive drugs or by drugs targeting somatic disorders, such as antibiotics, steroids, anticholinergics, anti-neoplastic agents, etc. (Sadock & Sadock, 2005, p. 1060).

Earlier versions of the DSM (DSM-III and DSM-IV) specified the specific psychiatric syndromes induced by drugs or somatic causative factors in more detail than DSM-5. For example, in DSM-IVR, the amphetamine-induced psychiatric disorders were subclassified into psychotic disorders with delusions, psychotic disorders with hallucinations, mood disorders, anxiety disorders, etc., while in DSM-5 they are subclassified only according to their severity (mild, moderate, severe), as well as whether perceptual disturbances are present.

DSM-IIIR still retained the subcategories of organic delusional disorder, organic hallucinosis, organic mood disorder, and organic anxiety disorder (as subdivisions of the more inclusive category of organic mental disorders), which were abandoned in later DSM versions.

In any case, the purpose of the classifications was descriptive, in order to serve clinical or research aims. They contain important observational data to support the theoretical propositions advanced in this work, but do not indicate the specific brain mechanisms responsible for the psychotic symptoms.

In both the first and second group of causative factors of psychotic disorders, I included, besides environmental and intrapsychic factors, disorders with a known organic nature. For example, in the first group were those organic disorders that interfere with the transmission of externally induced stimuli from the sensory organs to the cortical areas responsible for their primary processing, or with the primary processing itself. Similarly, in the second group of causative factors were mentioned those organic causes—hallucinogens, amphetamines, temporal lobe epilepsy—that differentially stimulate or irritate brain areas concerned with those intrapsychic contents or mechanisms whose excessive intensity overturns the desirable balance between the effect of external stimuli and the relevant intrapsychic mental contents. What remains to be discussed in this section are those organic mental conditions that induce a more generalized, widespread effect on brain functioning, conditions that do not fit neatly in either of the first two groups. In DSM-5 these conditions are classified as neurocognitive disorders (formerly organic mental disorders), and comprise delirium, dementia, and amnestic syndromes of diverse causation. While the specific psychotic symptoms may cause serious problems with these individuals' management, they are considered to be of secondary importance in these disorders. Their presence is not obligatory for the diagnosis, and when they are present they frequently have a fleeting, unsystematized nature.

While the mental mechanism that leads to psychotic symptoms in this third group of causative factors, delirium and dementia (amnestic syndromes do not concern us here), is less clear-cut or apparent than in the first two, it can be hypothesized, in my opinion, that the final outcome of its effect is similar to that encountered in the former two groups. All lead ultimately to the reduction of the impact of the external input on the mental functioning or the increase of the impact of intrapsychic contents, or both.

7.2.6.4.3.1. Delirium

In delirium, the main disturbance in mental functioning concerns the alteration of awareness and attention—that is, a "reduced ability to direct, focus, sustain, and shift attention" (DSM-5, p. 599). It seems clear that these disturbances first of all reduce the ability to acquire precise, relevant, selected information from the external environment. The psychiatric symptoms of delirium include: emotional disturbances; perceptual ones like illusions and hallucinations (typically visual); and cognitive ones like delusions. These symptoms are "especially prevalent at night and under conditions in which stimulation and environmental cues are lacking" (DSM-5, p. 600). The resemblance of the resultant mental dysfunctions to the effect of sensory/social deprivation in normative individuals seems to me quite clear. I do not mean, however, that this is the *only* mechanism in delirium relevant to the appearance of psychotic symptoms. The disturbance in mental functions seems to be more generalized and severe than that of normative persons who are in conditions of sensory deprivation, obviously affecting *higher level integrative capacities* of the mind as well, which may lead to *disinhibition* of suppressed archaic behavioral predispositions and their accompanying emotional states.

7.2.6.4.3.2. Dementias

The central feature of *dementias* is cognitive decline, especially impairment of memory, orientation, learning, and verbalizing abilities. However, especially in moderately severe dementia (for example of the Alzheimer's type), behavioral disturbances, like "psychotic features, irritability, agitation, combativeness, and wandering are common" (DSM-5, p. 612).

It may be hypothesized that the mechanisms actually underlying the appearance of psychotic symptoms in dementia are similar to those already discussed, namely:

1. The acquisition, primary processing, and especially the retention and proper use of information of external origin are reduced and defective, mainly as a result of the main mental dysfunctions in this syndrome: impairment of memory, orientation, etc.
2. The higher order integrative and coordinating mental activities progressively weaken and ultimately disappear (particularly in neurodegenerative dementias targeting widespread neocortical areas).
3. As a result of points 1 and 2, more primitive inherited behavioral programs (crude instinctive predispositions), formerly inhibited from directly influencing behavior, may be released. That is, they may surface in overt behavior without being integrated or synchronized properly with externally induced or ontogenetically acquired mental contents.

Of course, dementia's symptomatology, including that of psychotic dementia, varies with its severity, with the causative factor, and with the localization of the brain damage it causes. Behavioral disinhibition is most prominent in Frontotemporal Neurocognitive Disorder (Pick's disease), while hallucinations (which "are well formed and detailed") and delusions (which may be "systematized") are characteristic of Dementia with Lewy Bodies (DSM-5, pp. 618–619).

7.2.6.4.4. Summary of the Psychotic Mechanism

It seems that psychotic phenomena ultimately represent a serious imbalance in the higher-level synthesizing functions of the mental apparatus—namely, those which correlate, coordinate, integrate, and synchronize the processing of incoming stimuli with contents already existing in the mental apparatus, either as innate predispositions or as knowledge accumulated during ontogeny (possessing, as a rule, an emotional or intentional charge). For interpreting and effectively using information coming from the natural/social environment, these integrating functions are naturally indispensable. Psychotic phenomena, as stressed repeatedly in this section, represent a serious disequilibrium in favor of expressing internal mental contents to the detriment of their proper integration with adequately processed external input. We have to hypothesize that always underlying the cognitive contents of *functional* psychotic disorders is a strong instinctive activity felt subjectively as emotional or intentional states. Thus, intense euphoria (expressing high intensities of active instincts most probably displaced toward irrelevant releasers or discharged in vacuo) may lead to unfounded grandiose beliefs; the longstanding aggressive form of frustration may give rise to paranoid distortions in cognition and behavior; the dysphoric form of frustration may lead to delusional self-accusations; and intense anxiety may trigger unrealistic fears of nonexistent or strongly exaggerated external harmful influences.

It has to be stressed that some common nonpsychotic mental disorders, like anxiety and personality disorders, can also be interpreted as a kind of mental disequilibrium. Some anxiety disorders, like social phobia, represent an excessive predilection to respond to external natural/social stimuli of a disturbing nature by grossly exaggerating their harmful potential, while at the same time repressing active instinctive motives. That is, they may lead to a disequilibrium that seems *the reverse* of that seen in psychosis. Most personality disorders, on the other hand, can be interpreted as the acting out of active or individualistic reactive instinctive motives with insufficient compromise or integration with social requirements or expectations.

However, it seems that, in psychotic disorders, this imbalance is especially severe, even overwhelming. The contradictory external input relevant to the psychotic domain is either disregarded completely or excessively distorted in order to fit the intrapsychic assumptions. Notwithstanding, since any brain inherently distorts current realities to a certain degree as a result of biases induced by inherited predispositions or previously acquired mental contents, and because the gravity of psychotic distortions varies from simple, circumscribed illusions or hallucinations in normative persons to all-encompassing delusional distortions in grave schizophrenia, we must see the psychosis as an essentially *quantitative*, not qualitative, phenomenon.

7.2.6.5. Schizophrenia, Paranoid Type

Paranoid schizophrenia is the last clinical entity we have to mention in a section on paranoid disorders.

If delusional disorders of the paranoid subtype are the outcome of two different dysfunctional mental mechanisms, one consisting of an inherited predisposition to paranoid disorders and the other underlying the psychotic distortion, then paranoid schizophrenia has to be the outcome of the interaction between three different dysfunctional mental mechanisms:

1. An innate predisposition to paranoid disorders in general.
2. The mental dysfunction characteristic of psychotic states.
3. The innate predisposition common to schizophrenia and to its genetically related spectrum disorders.

In fact, the generation of the overt clinical picture of paranoid (or any other form of) schizophrenia is even more complex. It includes not only the innate predisposition to develop paranoia and schizophrenia but the effect of these predispositions on mental maturation during ontogeny. These topics are dealt with in the following sections.

7.2.7. Schizophrenia in Evolutionary Perspective

7.2.7.0. Introduction

Schizophrenia is the mental disorder whose genetic foundation is probably best documented in psychiatric literature. While its lifetime risk in the general population is somewhat less than 1 percent, the risk for first-degree relatives of schizophrenics is about 9 percent and for second-degree relatives 4 percent. When one parent is schizophrenic, the risk that her children will develop the disease is 13 percent; when both parents are schizophrenic the risk increases to 46 percent. When one of the parents is schizophrenic, the offspring, when adopted by genetically unrelated people, has a risk of 11 percent of developing the disease—similar to that of the nonadopted offspring of biological parents, one of whom is schizophrenic. This suggests that a different, healthier, parenting environment does not reduce the prevalence of the disorder.

The concordance rate for schizophrenia in monozygotic (MZ) twins is about 48 percent while in dizygotic (DZ) twins 17 percent. While the great difference in the concordance rate between MZ and DZ twins strongly suggests the existence of a genetic element in the causation of schizophrenia, the concordance rate of "only" 48 percent in MZ twins (who share 100 percent of their genes) equally strongly suggests the role of nongenetic factors in the causation of the disorder. These figures and interpretations are taken from Plomin et al., 2008, pp. 195–201. However, reasons exist to question their validity. As mentioned repeatedly in this work—and in the work of researchers who, to appreciate heritability, look for "endophenotypes" instead of the full-blown mental disease (section 5.1)—the overt clinical picture of schizophrenia is not a direct reflection of the inherited predisposition to schizophrenia. This is especially true of schizophrenic psychosis (which in the psychiatric genetic research exclusively determines the target population studied). Nor does the clinical picture of the genetically related schizophrenia spectrum disorders directly reflect that innate predisposition. This topic is detailed in the following section. In my opinion, these considerations do not question the presence of a strong genetic element in the causation of schizophrenia, but rather suggest the importance of using different parameters in establishing the genetic predisposition to the schizophrenic spectrum of disorders than those derived exclusively from the existence of psychotic schizophrenic pathology.

7.2.7.1. The Puzzle of the Directional Selection of Schizophrenia

If a genetic predisposition is involved so clearly in generating schizophrenia and its spectrum disorders, then without doubt natural selection of some kind was active in transferring the relevant genes to successive generations. The prevalence of schizophrenia is remarkably similar in different human populations globally, averaging a little less than 1 percent. If we include the genetically related spectrum disorders, which seem to have the same basic genetic foundation, the prevalence rises to about 5 percent (Buchanan & Carpenter, 2005, p. 1330).

Since natural selection in medical genetics means as a rule directional selection, the question inevitably arises: How could it happen that genes contributing to a behavior so highly dysfunctional in practically all the parameters of fitness (both biological and social) are not selected out from the gene pool, but, on the contrary, are fixated in the human genome at a considerably high frequency? Schizophrenics are clearly unfit in social, occupational, interpersonal functioning; most studies show a serious decrease in their reproductive rate; they are as a rule solitary people, requiring almost always some kind of sheltered conditions to survive, being unable to sustain themselves, and when they have children, their parenting abilities are highly deficient. Evolutionary psychiatrists are well aware of this paradox concerning the genetics of schizophrenia and have offered various explanatory hypotheses. Many of these hypotheses (involving, in accordance with traditional evolutionary thinking, *directional* selection) try to find some covert fitness advantages in the predisposition to schizophrenia or its spectrum disorders that counterbalance its fitness-reducing effects, analogous to the genetic predisposition to sickle cell anemia, for example, a grave autosomal recessive disorder. As we have seen, heterozygotes in this illness are usually healthy people with an increased resistance to malaria (Stevens & Price, 2000, p. 146). The hypotheses of this type were mentioned in the section on evolutionary psychiatry (section 3.2.2).

Since some mental disorders, including bipolar disorder and schizophrenia spectrum disorders, were found to be more frequent in creative individuals than in the normative population, it was proposed that schizophrenia represents a trade-off regarding human language acquisition and creativity (Brüne, 2008, p. 193). The schizotypal personality's idiosyncratic ways of thinking were hypothesized to be fitness enhancing in early humans because it suggested that such persons may have some occult access to divine forces, empowering them as prophets or leaders of ancestral human groups (Stevens & Price, 2000, pp. 147–155).

I won't analyze these hypotheses, but it seems to me that they explain only very partially, if at all, the complex mechanisms, both innate and acquired, that lead to the schizophrenic clinical picture or to that of its spectrum disorders.

7.2.7.2. The Present Theory's Proposal on the Innate Predisposition for the Schizophrenia Spectrum of Disorders

The psychiatric literature includes a tentatively formulated idea that may be seen as a forerunner of the explanation proposed in this text. This view argues that schizophrenia and its spectrum disorders are not rigid categorical entities but represent a dimensional quality, a trait that in adequate proportions is adaptive and becomes dysfunctional only in its "extreme variations" (Brüne, 2008, p. 193): "It is possible that schizophrenia represents the extreme of a quantitative dimension that extends into normality" (Plomin, 2008, p. 205). Yet, the exact nature of this trait or dimension, to the best of my knowledge, has never been specified.

As mentioned repeatedly, this text proposes that the dimensional trait underlying the schizophrenia spectrum has to be the outcome of *two* different evolutionary processes, one that characterizes an evolutionary path of behavior characteristic of the whole animal kingdom, and the other concerning only human evolution. The first evolutionary trajectory refers to the progressive weakening of the ability of the instinctive drives to induce exclusively overt behavior toward a narrowly defined, innately determined fitness-enhancing agent in the environment (the releaser). At the same time, this process allowed the gradual increase of the impact of knowledge acquired by learning and of individual experience accumulated during ontogeny to modify and enrich, to an ever-increasing degree, the instinctively based behavioral patterns. This evolutionary trajectory attained excessive proportions during human evolution. Naturally, however, despite its greatly increased flexibility, the *fitness-enhancing value* of the instinct-based behavior wasn't lost. Quite the contrary, the increased flexibility of the behavior enabled more efficient exploitation of increasingly varied environmental resources and, above all, allowed the organization of evolved animals (first of all humans) into more and more complex and efficient cooperative groups. This same idea was formulated repeatedly by theorists of evolution.[11] In my opinion, *this is the much sought after innate dimensional behavioral trait* submitted to natural selection that, *in excessive degrees in humans, leads to the predisposition to the schizophrenia spectrum of disorders.* (Point two will address why the genes responsible for these excessive degrees of instinctive diffuseness are not selected out in human populations by normalizing selection of polygenic traits.) In any case, it is intriguing to contemplate the possibility that the same evolutionary trajectory that, more than anything else, made us human and opened the path for unprecedented prosperity as a biological species, at its extreme, brought on us the most devastating mental illness.

The intuition that schizophrenia is somehow connected to evolutionary mechanisms central to human behavioral evolution has been expressed by philosophers as well as psychiatrists. Freud believed that "mental disorders . . . are side effects of highly useful human traits." And again "some of the greatest thinkers in philosophy have argued that our vulnerability to mental disorders distinguishes humans from other animals, and, to account for this vulnerability, many of them embedded their claims in an explicitly Darwinian framework." Psychiatrist Jonathan Burns argues that schizophrenia "'this most human of maladies' may represent a costly downside to the emergence of embodied social consciousness" (Adriaens & De Block, 2011, p. 25). Anthropologist Ernst Becker's formulation of the same topic was quoted in section 5.3.1.

This leads us directly to the second evolutionary mechanism proposed in this text for the genetic predisposition underlying schizophrenia (already described in section 5.3.1). The considerably diffuse quality of the instinctive predispositions of early humans (Mayr's widely "open genetic programs")—with the progressively increasing ecological dominance and consequent relaxation of natural and intragroup selection pressures during subsequent human evolution—led to a wide diversification of this instinctive trait, a diversification that can be conceived as a graded continuum from relatively strongly differentiated instinctive predispositions up to dysfunctionally diffuse ones. Because of the human social environment's great diversity and the presence within it of many different "social niches" that are able of containing, and even exploiting, the widely diversified human instinctive makeup, most loci on this continuum lead to normative behavior.

In later stages of human evolution, law-governed constrictions on the resolution of conflicts began to protect persons who had dysfunctional instinctive predispositions in certain areas, especially if they possessed mental or physical assets in other domains—for example, inadequate socializing abilities but high abstract intelligence or talents in an artistic or other creative domain. The effective cooperation of members with different instinctive, intellectual, and physical genetic endowments enabled human group accomplishments that would be unimaginable without this evolutionary mechanism. Think, for example, of the chain of cooperative steps beginning with a creative theoretical scientist (possibly with reduced socializing abilities and practical sense) who succeeded in discovering some basic scientific laws in a certain domain, laws that were exploited by more practically oriented scientists to conceive the prototype of a new invention, such as a new electrical device, or the airplane or telephone; which then, through the efforts of astute businesspeople with high social intelligence, was transformed from a small-scale development into large-scale production and distribution, while securing employment for a large number of workers with average mental skills but good cooperative abilities, affecting thus the standard of living of increasingly greater segments of the population.

———————

Returning to our original subject, the genetic predisposition to schizophrenia, it is hypothesized in this work that *the progressive widening of that differentiated/diffuse scale of instinctive drives in humans at its most diffuse end attained so pronounced a degree of undifferentiation that the drives became unable to guide the individual effectively any longer through the complexities of social life toward the achievement of basic biological goals.* (I mean here the *appetitive,* not the consummatory, phase of instinctively underpinned activities.) In other words, at this degree of pronounced diffuseness, instinctive motives lost the ability to guide the selective processing of external information, as well as the ability to show a more-or-less approximate adaptive direction to internal intentional motives, an ability that, as mentioned, normatively diffuse drives still retain. It seems self-evident that viable human societies have to be built on the foundations of the normative spectrum of instinctive diffuseness, which allows for the acquisition of at least the basic skills for both biological existence and social coexistence. A functional human society consisting predominantly of members with a schizoid or schizophrenic genetic predisposition (as hypothesized in this text), or alternatively of members with overly differentiated instinctive drives predisposing them to antisocial behavior, cannot be imagined. However, it can be presumed that with increasing relaxation of intragroup selective pressures, and the evolution of more complex, humanitarian, and technologically advanced societies, the two extreme poles of the differentiated/diffuse spectrum became increasingly represented in the human populations.

A recurring epidemiological finding may support this idea—namely, that dysfunctional levels of instinctive diffuseness are perpetuated by the protecting effect of the social order. It was found repeatedly in epidemiological research that, while the prevalence of schizophrenia in developing and developed countries was by and large equal, the course of the disease in developing countries was more benign than in the developed ones (Tsuang & Tohen, 2002, p. 379). This finding may suggest a natural intragroup selective effect. In view of the serious fitness-reducing effect inherent in schizophrenia, it can be presumed that those with a heavier genetic load should have a graver survival and reproductive disadvantage. It seems clear that technologically advanced democracies can offer a wider assortment of social and occupational niches and therefore a more adequate niche for their disabled citizens, as well as better psychiatric and social welfare services, than developing countries: "Available figures for *severe* disorders in primitive societies may be lower than those reported elsewhere . . . because under difficult living conditions the severely ill are less likely to survive" (Redlich & Freedman, 1966, p. 5, emphasis in the original; see also the discussion of the same finding in Buchanan & Carpenter, 2005, p. 1342, as well as pp. 1348–1351 with partially different interpretations).

After having discussed the kind of innate impairment hypothesized in this theory as central to the schizophrenia spectrum of disorders, it is important to stress, too, which mental abilities with a proven inherited aspect *remain intact* in most schizophrenics as well as in those suffering from genetically related spectrum disorders:

1. Language acquisition abilities.
2. The ability to learn and to manipulate by logical reasoning factual information conveyed explicitly.

3. Intelligence. I mean here *abstract* intelligence and not the practical, social, or "emotional" intelligence (Goleman, 1995) that presumes an ability to read other persons' emotions and covert intentions with the aid of the empathizing mechanism. (The average IQ of schizophrenics is somewhat lower than that of the general population's but with a wide range of variability from the subnormal to the exceptionally high.)

Of course, also included here are the mental capacities that depend on the interaction of these three abilities. Many schizophrenia-predisposed children are good pupils and students, particularly those discussed in the context of the "model child" phenomenon (section 7.2.8.2).

The retention of learning and related cognitive abilities in schizophrenia is not surprising considering the decisive role of instinctive diffuseness in enabling behavioral modification through learning. In fact, it evolved precisely for this purpose in the first place. I believe that *the excessive diffuseness of the instinctive predispositions, combined with these retained mental abilities, reflects best the basic genetic configuration underlying schizophrenia and its spectrum disorders.*

The reader may wonder at this point of the discussion how these ideas, even if they are convincing in the framework of evolutionary logic, can be relevant to a clinical disorder containing, besides social malfunctioning, such symptoms as bizarre delusions and hallucinations, confused thinking, unintelligible speech and behavior, catatonic symptoms, and progressive deterioration. Therefore, in the following, I propose a hypothetical chain of events that I hope bridges at least partially the gap between the hypothesized genetic predisposition and the observable clinical disease.

7.2.8. From the Inherited Predisposition to the Schizophrenic Clinical Picture

7.2.8.1. The Direct Effects of the Pathologically Diffuse Instinctive Drives on Prepsychotic Behavior and the Bidirectional Failure of the Empathizing Mechanism

Unanimous agreement exists among clinicians that, in most cases, long before the outbreak of the schizophrenic psychosis, signs of mental and behavioral disturbances can be observed. Whether these disturbances also include the period of infancy is doubtful. I am inclined to regard the infant's behavior as mostly fixed motor patterns, while the schizophrenic disturbances relate mostly to the appetitive phase of the instinctively underpinned behavior. However, according to some observers, "The presence of social behavior disturbances has been picked up as early as infancy by workers who have noticed a lack of responsiveness and emotional expression in infants who later developed schizophrenia" (Buchanan & Carpenter, 2005, p. 1341). In any event, the finding has not been contested that about 25 to 50 percent of predisposed individuals exhibit clear abnormalities in behavior before the outbreak of the characteristic schizophrenic clinical picture. These abnormalities include: "diminished social drive; decreased emotional responsivity; withdrawn, introverted, suspicious, or impulsive behavior; idiosyncratic responses to ordinary events or circumstances; . . . short attention span" (idem, p. 1341). In the following, I try to derive these observable behavioral attributes in interpersonal relations from the possible consequences of the excessive diffuseness of drives underlying pre-schizophrenic behavior.

It can be inferred that if the drives are too diffuse, it will lead to *uncertainties* on the part of the individual with regard to his own intentions. The individual will feel uncertain about his own basic personal interests and whether these interests are similar to, complementary to, or different from the interests of surrounding persons. This assumption seems particularly true when the interpersonal transactions do not consist of the exchange of clearly formulated informational material—such as purely observational phenomena; knowledge arrived at exclusively by logical reasoning, as in mathematics; or clearly stated instructions about how one is expected to behave, as is customary in school or the army, for example. This lack of objectivity and clarity is frequently the case in social interactions, particularly when the subject or situation is emotionally charged or biased by personal interests. For example, a casual conversation between two acquaintances begins with a friendly, seemingly pointless overture like, "How do you do," or a remark on the weather or some other neutral theme. One of them may report on some joyful event that happened since their last meeting that is in tune with the pleasurable reaction (sincere or exaggerated) to the present meeting, or on some sorrowful event that automatically triggers expressions of compassion. In such situations, many messages are *implied* rather than precisely formulated because of the ease with which the majority of humans can understand each other well enough in this incomplete form of communication, and because it avoids riskier, more clear-cut statements that touch on possible conflicting interests, obligations, or painful issues, or may transgress cultural norms, and so on. Psycholinguist Steven Pinker calls this kind of communication "indirect speech" and describes it in the following words: "Ordinary conversation is like tête-à-tête diplomacy, in which the parties explore ways of saving face, offering an 'out,' and maintaining plausible deniability as they negotiate the mix of power, sex, intimacy, and fairness that makes up their relationship" (Pinker, 2007b, p. 23; see also pp. 373–425.) In these circumstances, the *accurate realization of one's own personal interests or intentions in social circumstances vis-à-vis the interests of another conspecific is a basic prerequisite for employing effective behavioral strategies.* The following quotation, to my mind, captures well the pre-schizophrenic as well the schizophrenic individual's difficulties in this respect. His behavior is characterized by "a loss of the usual common-sense orientation to reality . . . that normally enables a person to take for granted [accept intuitively or instinctively] so many of the elements and dimensions of the social and practical world" (Sass, 2001, p. 258). As mentioned earlier, schizophrenics, and to a lesser extent those suffering from schizophrenia spectrum disorders, are unsuccessful in almost any area of activity involving, in social circumstances, life-sustaining or reproductive instinctive incentives in their appetitive phase. These include securing a steady source of income, private territory, and mate or sexual partner; bringing up children; forming friendships and alliances; and striving to acquire a stable or ascending social status—the instinctive underpinnings of which are already well developed in chimpanzees (de Waal 1982/2007; Goodall, 2000, sections 5, 6, and 13). The inability to precisely and spontaneously express, with flexible transitions, emotional and intentional states in common interpersonal situations; the inability to make flexible, quick transitions between behavior motivated by active and reactive instincts, or to find optimal compromises between them; the inability to make flexible transitions or compromises between individually motivated versus group-interest-influenced behavior; and so on all give the well-recognized impression of shyness, awkwardness, inhibition, and confusion that normative persons have of (pre)schizophrenic behavior. I think these qualities are a *direct behavioral consequence of the more than normative diffuseness* of the instinctive predispositions. These diffuse drives cannot guide well the unconscious use of body language or the choice and intonation of emotionally charged words in order to express emotions or intentions in a nuanced way (or to hide or fake them if necessary). Even when the pre-schizophrenic identifies this disability in herself and is motivated to perfect or learn some of these abilities by imitation from persons better endowed genetically in this respect, the unnatural, contrived quality of the performance is easily detectable.

7.2.8.1.1. Insufficient Selectivity regarding Social Stimuli; Overinclusiveness

A further direct consequence of more than desirably diffuse drives is the difficulty in selecting from the multitude of social and natural stimuli those that are relevant to an instinctively fueled intention. Such an intention, if differentiated enough, confers a definite direction to one's strivings. It gives adequately precise guidance in the right appreciation of the environment's stimuli—which ones may interfere with the respective intention, which may facilitate its attainment, and which are irrelevant to it. It is well known that individuals with schizophrenia, and, to a lesser extent, those suffering from schizophrenic spectrum disorders, have difficulties in this respect. For example, it was found that: "schizophrenic persons tend to be affected or distracted by irrelevant aspects in the context of a stimulus. In consequence, their ability to focus on and attend to the relevant aspects of the stimulus that demands a response is impaired." And again: "Schizophrenic anxiety occurs because the person has been flooded by unassimilable percepts" (Weiner, 1980, pp. 1134–1135).

7.2.8.1.2. Bidirectional Failure in Figuring Out Emotional and Intentional States with the Aid of the Empathizing Mechanism

Since the person possessing excessively diffuse drives cannot express emotional and intentional motives clearly in overt behavior, *a normative person cannot "read" with the aid of the empathizing mechanism the (pre) schizophrenic's internal, emotional, intentional states.* The first two stages of the empathizing process thus become problematic or disrupted. Observation of the pre-schizophrenic behavior does not supply the necessary data for comparison with a normative person's own emotional behavior in a similar situation; therefore, the observing person cannot experience corresponding emotional or intentional states through the mechanism of empathizing.

A problem exists with the empathizing process the other way around, too. A person with overly diffuse drives cannot discern with precision the emotional and intentional motives of normative people using the mechanism of empathy, or else he will grasp it with only a very rough approximation. The reason seems simple. He cannot elicit in himself the appropriate emotional/intentional state because he himself does not possesses it in an adequately differentiated form, despite its being clearly expressed by the other, normative party in observable behavior. In the language of neuropsychology: "Individuals with schizophrenia demonstrate multiple impairments of social cognition and theory-of-mind ability. These include eye gaze and facial affect recognition deficits, general emotion-recognition deficits, impaired mentalization and 'mindreading' ability, and social perception and attributional errors" (Burns, 2011, p. 301).

In summary, drives that are more than normatively diffuse lead to a bidirectional interference with or inadequacy in communication when matters are not expressed verbally in a clear, informative way. A similar idea is expressed, albeit in the case of psychotic schizophrenics, by Norman Cameron, a psychiatrist with a psychodynamic orientation: "The disorganized or scattered schizophrenic . . . has become unable any longer to share genuinely in the attitudes and perspectives of those around him, to take their roles when mutual misunderstanding arises, and so to be able to assume their point of view, grasp their difficulties, and modify his own behavior to meet them. On the other hand, his own asocial development has brought him to a point where no one else seems able to take his role and share his perspective, either" (Cameron, 1944/1964, pp. 50–51). While Cameron here refers primarily to cognitive processes, I wish to remind the reader of the importance of the instinctive motives in the selection, creation, and cognitive interpretation of ontogenetically acquired knowledge, discussed in chapter 2 under the heading of instinct-learning interaction. (Remember Lorenz's example of the exploratory behavior of the raven—that is, the successive application of a range of instinctive behavior patterns in order to create the biologically relevant "meaning" of a hitherto unknown environmental stimulus in section 2.2.1.)

Consider that, during childhood and adolescence in many families and especially in peer groups, emotionally charged and informatively not-very-clear transactions are the rule rather than the exception. In such circumstances, the pre-schizophrenic behavioral disturbances—"diminished social drive; decreased emotional responsivity; withdrawn, introverted, suspicious or impulsive behavior; idiosyncratic responses to ordinary events and circumstances" (Buchanan & Carpenter, 2005, p.1341; quoted in section 7.2.8.1)—become more understandable. Friendship on the basis of an emotionally mutually gratifying relationship, partnership based on common interests, or even a competitive or antagonistic relationship based on clearly sensed conflicting interests is not an option for these individuals.

What then are the practicable options for a pre-schizophrenic child, adolescent, or young adult in this situation? The possible behavioral strategies may take several forms, depending on the severity of the innate deficiency (that is, how excessively diffuse the drives are), other inherited aspects of the behavior, acquired life experience, and the nature of the social environment in which the respective person lives. By the other inherited predispositions, I mean primarily the remaining comprehensive instinctive mechanisms discussed in chapter 5—such as *the quality of the active-reactive behavior interaction*.

If the active behavior (mainly in its unfrustrated form) predominates over the reactive one, and the reactive instinctive predispositions accentuate the avoidant behavioral tactics, the child or adolescent may choose to withdraw from social encounters and seek gratifying interactions with some less complex agents than human conspecifics: interaction with toys, pet animals, computers; fictional literature; or solitary walks in nature. Furthermore: "He is not interested in petting or other heterosexual activities but is often disturbed by masturbation. He avoids all competitive sports but he likes to go to movies, watch television, or listen to hi-fi music. He may be an avid reader of books on philosophy and psychology" (Lehmann, 1980a, p. 1180).

If the reactive behavioral predispositions accentuate diffuse anxiety and the active ones immature dependence, the overt behavior of the pre-schizophrenic child will tend to show "patterns of clinging to the mother, shar[ing] her bedroom until late adolescence, ha[ving] nightmares . . . enure[sis]; and becom[ing] fearful and panicky away from home" (Weiner, 1980, p. 1136). If the *tactic of defiance* (that is, a kind of unselective resistance to social pressures) is predominant in the reactive instinctive constitution, and especially when the social environment is intrusive, the pre-schizophrenic child's behavior tends to be "from an early age, asocial, shameless, and lacking propriety" (idem, p. 1136). And when the reactive motives accentuate *submission,* it leads to a pattern of behavior known in clinical psychiatry as the "model child" phenomenon. It is through this phenomenon, in my opinion, that the peculiar pathological dissociation between the mental contents induced by diffuse, self-centered instinctive motives, on the one hand, and, on the other, the behavior induced by social expectations to conform that is so characteristic of many pre-schizophrenics, can be best studied. This dissociation is so important to the concerns of this text—especially for its contribution to the understanding of the psychotic breakdown and the acute episode/remission alternations—that it is discussed next in some detail.

7.2.8.2. The Dissociation between the Genetically Induced Behavioral Predispositions and the Externally (Socially) Induced Behavior; the "Model Child" Phenomenon

In normative persons, the main categories of mental content—adequately differentiated individual instinctive strivings, on the one hand, and socially induced knowledge and experience as to the possibilities of and restraints on instinctive expression and gratification in a social context on the other—are well coordinated or integrated in behavior. They are also optimally balanced—that is, they are balanced with regard to the instincts' action-inducing tension versus the action-modifying, action-suppressing effect of the socially inculcated knowledge.

In bipolar disorder, in some anxiety disorders like social phobia, and in some personality disorders, like antisocial personality disorder, the instinctive motives and the socially induced knowledge of their acceptable forms of expression are adequately represented, but the balance, the optimal power relationship between the instinctive pressures and the suppressing, modifying environmental influences is compromised. In a good-prognosis manic or hypomanic episode, for example, the instinctive tensions intensify and overthrow the satisfactory balance with the opposing, socially induced mental contents. When the instinctive tensions drop to normative levels (as a result of spontaneous remission or medication), the former normatively balanced behavior returns. The disease has no inherent potential to induce mental and functional deterioration (although secondary consequences of recurrent manic or depressive attacks in a social environment may lead to worsening occupational, familial, etc., functioning).

It should be clarified that by "coordination" I do not mean arriving at a conflict-free state between the internal and external incentives to behavior. Conflicts and their more optimal or more dysfunctional resolution in overt behavioral output is a daily occurrence, both in normative persons as well as in mental conditions like anxiety and (to a lesser extent) personality disorders. Instead, by "coordination" I mean that the two basic incentives of behavior are simultaneously dealt with by the mental apparatus, and, when they are conflicting, a kind of (more-or-less balanced) solution is sought. Moreover, even when the chosen solution is strongly biased, the two basic components are still simultaneously present in the mind. This has to be the consideration behind legal accountability for psychopathic behavior, for example, and its punishment's (partial) efficiency.

In the case of schizophrenia and its spectrum disorders, however, it is hypothesized here that the frequently opposing forces (active and self-protective reactive instinctive motives versus the socially induced influences on behavior), instead of achieving a kind of coordination, synchronization, or partial synthesis, or being dealt with by flexible "to and fro" transitions as circumstances demand, become permanently *dissociated* to a pathological extent *before* the eruption of the overt psychosis. Being dissociated, the conflict between these two different incentives of behavior (instead of being accentuated, as in some of the abovementioned mental conditions) is significantly reduced or even nonexistent in schizophrenia and related disorders. This is a central assumption in this text, without which the behavior and mental experience of schizophrenic patients cannot be comprehended. Formulated succinctly, *the extreme diffuseness of the schizophrenia-predisposed individual's instinctive predispositions precludes coordination or partial integration with socially induced knowledge, a step that is critically needed for adaptive actualization within the respective social environment.* Moreover, it can be expected that, as a result of this dissociation, the psychotic state resulting from the intensification of internal mental contents (at the expense of processing and reacting to external stimuli) may be brought about more easily than in persons in whom the internal and externally induced mental contents are well integrated; that it will be more bizarre (less understandable by others than a psychotic state appearing against the background of previously better integrated behavior, as in a good prognosis mania or in delusional disorders);

and that, when the psychosis clears up, the behavior cannot return to a normative, coordinated state.[12]

7.2.8.2.1. The "Model Child" Phenomenon

The "model child" phenomenon seems to be the best illustration of the abstract line of reasoning just presented. In these cases, the child carries out, without reservation or opposition, what is required of him in a structured social environment where desirable and undesirable, valued and unwanted behavior is clearly, explicitly stated (as in school or in certain families): "The child is often reported as having been especially good because he was always obedient and never in any mischief" (Lehmann, 1980a, p. 1179).

R. D. Laing was a Scottish psychiatrist with an existential orientation, which emphasizes the individual's subjective experience of himself, his existence, and the surrounding world instead of relying exclusively on observable clinical data. The kind of dissociation proposed in this text parallels I think quite closely the dissociation of mental contents in schizophrenic and schizoid individuals proposed in Laing's famous book, *The Divided Self*. Laing's treatment of the subject (beside his elegant descriptions of its clinical manifestations) may help us to clarify the subjective experiential aspect of this type of mental dissociation.

Laing refers to the socially induced aspect of the model child's behavior and subjective experience, which the individual does not experience as originating from his own true being or self (and which in our theoretical scheme represents dissociated reactive behavior of the submissive or conforming type), as the "false self": "The false self of the schizoid person is compulsively compliant to the will of others, it is partially autonomous and out of control, it is felt as alien" (Laing, 1960, p. 102). That is, in our interpretation, it is *dissociated* from active and self-centered reactive motives, not *in conflict* with them, as Arieti believes (see note 12 in this chapter).

Laing gives a particularly vivid illustration of such "model child" behavior, related by the father of one of his patients, an eighteen-year-old boy: "He had always been a very good child, who did everything he was told and never caused any trouble. His mother had been devoted to him. He was inseparable from her. He had been 'very brave' when she died." What is peculiar in this particular case, and seems characteristic of the model child phenomenon, is that no pressure or coercion on the parents' part is evident in inducing the described behavior. "I had simply been what she wanted," explained the patient. Of his mother's death he said: "As far as I can remember I was rather pleased." (That is, excessively dependent personality traits cannot explain his behavior. Rather his answer points toward a complete lack of positive emotional ties with the mother.) When Laing met him, he "was playing" in a contrived, artificial manner the role of a young, eccentric male university student studying philosophy (which he indeed was): "His speech was made up largely of quotations." He explained that "he always felt shy, self-conscious and vulnerable. By always playing a part he found he could in some measure overcome his shyness, self-consciousness, and vulnerability" (idem, pp. 72–74). It seems clear that in this case the behavior consisting of excessive compliance with impressions, influences, or expectations coming from the immediate environment was not induced by excessive social pressures to conform. It seems, rather, that the main reason for the excessively conforming behavior was the presence of highly diffuse active drives that failed to guide the patient toward those particular avenues that represented his own deep, identified-with, individual interests. In consequence, this intelligent boy took on an attitude that was probably valued or expected in his university (as well as home) milieu.

Laing named the subjective experience of active instinctive motives that can be identified by the person as coming from his own true being the "real (or inner) self." In the schizoid and schizophrenic dissociation, according to Laing, the real self never manifests itself in actual interactions with the social environment. Instead, it manifests itself only in fantasy: "Once [it] commit[s] itself to any real project . . . it suffers agonies of humiliation . . . simply because it has to subject itself to necessity and contingency" (idem, p. 88). I interpret this excerpt in the following way: The excessively diffuse active drives cannot be coordinated or integrated with externally originating mental contents; they can only be supplanted by them. Consequently, active drives could be expressed only in a way considerably disrupted from external realities (in their different forms of frustration, mostly as vacuum behavior).

Laing argues that the above dissociation inevitably leads to mental impoverishment (a process that I elaborate on when detailing deterioration in schizophrenia). In order to illustrate this impoverished state, Laing describes a patient who "oscillated between moments when he felt as though he was bursting with power and moments when he felt he had nothing inside and was lifeless. However, even this 'manic' feeling was expressed by imagining himself as a container full of air under tremendous pressure, in fact, nothing but hot air" (idem, p. 96). I think this description (or delusion?) is a good metaphor for accumulating active instinctive tensions, or "action-specific excitation" in Konrad Lorenz's terminology (1982, pp. 110–112), but of a kind that possesses no "action specificity" at all, representing thus the extreme diffuse pole of active instinctive activity.

7.2.8.3. The Prepsychotic Period

The fragile equilibrium between the two dissociated aspects of the pre-psychotic patient's behavior—conformity, submission, or some other kind of reactive tactic like avoidance or defiance in interpersonal relations versus expressing diffuse active drives outside the interpersonal sphere—becomes less and less tenable with the advent of adolescence and early adulthood. This fragile equilibrium is threatened both by maturational events within the organism as well as changing social conditions characteristic of this life-period. The strengthening of sexual drives in adolescence requires the abandonment of the pre-schizophrenic's solitary lifestyle.[13]

At the same time, society's increasing expectation that a maturing or mature individual function adequately in its framework may force him out of the relative isolation and safety of the parental home into more intense, more inevitable proximity to peers or other adults. Examples are boarding school, compulsory military service, holding a job, marriage, and parenthood, etc. In a word, while diffuse active instincts in the sexual sphere (and possibly in other instinctive domains as well, like rebelliousness toward the establishment) seek expression more intensely, the social environment also tends to increase the pressure for progressively more appropriate and complex adaptive functioning (see also McGuire & Troisi, 1998, p. 211).

The great strain on the overaroused reactive mental apparatus, which was not created by evolution for sustained effort, becomes apparent in some cases in neurasthenialike symptoms before or concomitantly with the eruption of the overt psychotic episode: "During the trema phase the patient is anxious, irritable, and often depressed" (Lehmann, 1980a, p. 1157). And again: "In the early stages of the disease, patients often complain of a multitude of symptoms—headache, rheumatic pains weakness, and indigestion. They are sometimes treated for months for neurasthenia or are thought to be hypochondriacs or malingerers" (idem, p. 1164). At a certain stage, the conforming reactive behavior collapses and emerging diffuse active and self-defensive reactive motives begin to dominate the conscious subjective experience and overt behavior, inducing at first perplexity and disorganization: "The patient becomes aware that something ominous is happening to him. . . . He may make desperate attempts to regain control." This transitional period between the prepsychotic and florid psychotic behavior has been called the "trema" phase (idem, p. 1157).

After the eruption of the florid psychosis in most schizophrenics, an alternation between acute episodes and remissions (that is, periods of accentuated conformity or withdrawal coupled with relative scarcity of the florid psychotic symptoms) will characterize the course of the disorder. However, it has to be stressed that the conforming behavior of prepsychotic proportions, that is the typical model-child behavior, never returns.

7.2.8.4. The Relationship between the Genetic Predisposition Common to Schizophrenia and Its Spectrum Disorders and Schizophrenic Psychosis

Schizophrenia in clinical psychiatry is considered the *paradigm of the psychotic* mental disorder. The first and foremost requirement for a diagnosis of schizophrenia according to DSM-5 and its predecessors is the presence of at least two of the following five symptoms, of which four are psychotic ones: delusions, hallucinations, disorganized speech, grossly disorganized or catatonic behavior, and negative symptoms. (Negative symptoms are discussed in the section on simple schizophrenia, 7.2.8.4.2.) Only *after* the detection of some of these features may additional requirements for the clinical diagnosis, such as a decrease in diverse aspects of social functioning, the duration of the disturbance, and exclusion criteria, be considered.

In this situation, it is tempting to draw the conclusion that the genetic predisposition to schizophrenia has to consist of an inherited mental inclination to develop psychosis of the schizophrenic type. After all, most genetic and epidemiologic research on schizophrenia is based on first establishing the presence of the schizophrenic psychotic disorder in blood relatives (as against a control group). We have to note, however, that the problematic relationship between the genetic predisposition and the full-blown clinically observable disorder is well recognized. The reader will recall the search for an "endophenotype" (discussed in section 5.1), which attempted to identify an inherited somatic or behavioral trait that is more intimately related to the inherited predisposition to schizophrenia than the ultimate observable clinical disease. However, hitherto, to my knowledge, this approach has not produced concrete results.

Nevertheless, to my mind, quite convincing evidence exists that *what is inherited in schizophrenia is not a predisposition to develop a certain kind of psychosis* but a more obscure innate trait that is much less evident in overt behavior, a predisposition that influences behavioral maturation in a deleterious way. After many years of interaction with social stimuli, the resultant intrapsychic and behavioral disturbance may or may not lead to a psychotic disorder.

This controversy over the place of the psychotic syndrome in the schizophrenic disorder between a more descriptively oriented and a more intuitive, more psychodynamic approach is not new. In fact, it has existed from the very beginnings of identifying schizophrenia as a unitary nosological entity: "Virtually the entire clinical picture that Kraepelin had described as typical of schizophrenic pathology was seen by Bleuler as consisting of *secondary* symptoms—for example, hallucinations, delusions, negativism, and stupor" (Lehmann, 1980b, 1107, emphasis added). Eugen Bleuler (who coined the term schizophrenia) suspected a dissociative process in basic mental functions as more relevant to schizophrenia, which manifests in disturbances of affect, associations, and volition (ambivalence), and in autistic thinking (which means "emphasis on subjectivity, rather than objectivity, and without regard to reality" [Gelfand, R. 1980, p. 3312].)

The psychiatrist Jonathan Burns's interpretation of Bleuler's primary symptoms is even closer to the standpoint adopted in this text: "He [Bleuler] coined the term 'schizophrenia' to describe what he considered *a fundamental split or dissociation between inner thoughts and cognitions, and the emotional contact or engagement with the world*" (Burns, 2011, p. 300, emphasis added). This dissociation (despite its downplaying of the complex, bidirectional relationship between emotional and cognitive contents) seems to me to represent a *more direct* consequence of the excessively diffuse instinctive drives' effect than psychosis *per se*, which, as we have seen in the section on delusional disorders (7.2.6.4), can potentially be induced by widely different causative factors and pathologic mechanisms in nearly all persons, including normative ones. It is worth mentioning here that Bleuler also suggested that the schizophrenic pathology is of a dimensional nature and lies on a continuum with normative behavior (Kirkpatrick & Tek, 2005, p. 1418).

Bleuler's four primary symptoms, however, are very difficult to translate into the routine of clinical and research practice. In other words: *"It is difficult to express those disturbances in operational terms and to recognize them reliably as clinical symptoms"* (Lehmann, 1980a, p. 1153, emphasis added).

The controversy between the requirement of exact sciences to deal only with strictly observable, "objective," or "inter-subjectively testable" phenomena (Popper, 1968, pp. 46–47), on one hand, and the need to make use of our subjective experience in order to gain knowledge of another person's (the patient's) subjective mental states, on the other (the central aim of the empathizing process, as we have seen), cannot in my opinion be resolved by taking a resolute position on behalf of one of the alternatives. The requirement for objectivity in scientific work is self-evident. However, the second alternative is indispensable for gaining less precise but vitally important information on another person's mental states. Clinical psychiatry, as well as research, is unimaginable without such terms as *anxiety, depression, euphoria, suspiciousness, inappropriate affect,* etc., terms that refer to essentially unobservable subjective mental states. In the same way, the subjective experience of the activity of excessively diffuse drives cannot be even grossly approximated without the empathizing process. A theoretic approach aiming to promote a more precise understanding of the phenomenon in question cannot be obstructed from the beginning by practical considerations regarding its application to routine clinical work.

That the schizophrenic psychosis is not a direct outcome of the genetic predisposition to schizophrenia is also underpinned by the following clinical and research data:

7.2.8.4.1. The Existence of Genetically Related Spectrum Disorders

"Both . . . Kraepelin and Eugene Bleuler, noted that some close relatives of patients with schizophrenia, although never psychotic, had odd or eccentric personalities that were clinically reminiscent of schizophrenia" (Riley & Kendler, 2005, p. 1360). Subsequently, controlled family and adoption studies confirmed this observation. Increased rates of personality disorders—schizoid, schizotypal, and paranoid—were found in the majority of these studies (idem, p. 1360). In these circumstances, the parsimony principle of building a scientific theory (Occam's razor) calls for postulating a common underlying genetic predisposition for both the typical schizophrenic psychotic disorder and other genetically related psychotic spectrum disorders (atypical or schizophreniform psychosis) on one hand, and, on the other, the personality disorders (which do not possess psychotic symptoms) belonging to the schizophrenic spectrum disorder category.

A prospective study from Philadelphia compared children who later developed schizophrenia with their blood relatives who had "remained healthy," as well as with a group of genetically unrelated children, for mental, behavioral, and neurological deficiencies frequently found in the prodromal phase of schizophrenia. It was found that would-be schizophrenics *and their blood relatives* "performed significantly worse than the nonpsychiatric controls (but did not differ from each other) on verbal and nonverbal cognitive tests. . . . Early social maladjustment, motor coordination deficits, and behavioral and language dysfunction (like echolalia, inappropriate laughter, or unintelligible speech) were significantly associated with both schizophrenia and sibling status" (Murray & Bramon, 2005, p. 1384).

7.2.8.4.2. Simple Schizophrenia

Schizophrenia has a contentious subtype, *simple schizophrenia*, which is characterized by negative symptoms, insidious course, and decreased social functioning, but "[d]elusions and hallucinations are not evident, and the disorder is less obviously psychotic than the hebephrenic, paranoid, and catatonic subtypes of schizophrenia. The characteristic negative features of residual schizophrenia [the end-stage of chronic schizophrenic deterioration] (e.g., blunting of affect and loss of volition) develop *without being preceded by any overt psychotic symptoms*" (ICD 10 diagnostic criteria for simple schizophrenia, emphasis added). This subcategory of schizophrenia was excluded from more recent DSM classifications but retained in the ICD classification (World Health Organization's classification of Mental and Behavioral Disorders) (Kirkpatrick & Tek, 2005, p. 1424). What makes this type of schizophrenia even more relevant for us here relates to Timothy Crow's subcategorization of schizophrenia into type I and type II. Type I is characterized by more positive, that is, florid, psychotic symptoms, a better response to anti-psychotic drug treatment, and a better prognosis. Type II schizophrenia in turn is characterized by "passive symptoms such as withdrawal and lack of emotion," and has a poorer prognosis as well as a less impressive response to drug therapy. Most importantly for our present concerns: "Type II schizophrenia appears to be more heritable than type I" (Plomin et al., 2008, p. 204). The acceptance of simple schizophrenia as a legitimate subtype of schizophrenia absolutely contradicts the idea that the innate predisposition for schizophrenia is identical with a predisposition to develop a *psychotic* mental disorder.

7.2.8.4.3. Subtypes of Schizophrenia

A further peculiarity of the genetic predisposition to schizophrenia is that, while schizophrenia as a comprehensive clinical entity has a familiar aggregation, the subtypes of schizophrenia (catatonic, paranoid, disorganized, or hebephrenic), which are much more circumscribed and clinically better-defined entities, do not (Plomin et al., 2008, p. 203). Thus, it has to be concluded that *the specific subtypes are not inherited as such; instead, it seems that only a general inclination to develop schizophrenia of any kind, including its subtypes, as well as its spectrum disorders, is inherited*, which again seems to suggest the possibility that some other, unrelated, genetic predispositions or ontogenetic influences contribute to the development of the final clinical picture. For example, an additional, unrelated genetic predisposition to paranoid disorders in general may be needed for the appearance of paranoid schizophrenia.[14]

7.2.8.5. The Transition from Covert Prepsychotic Dissociation to Overt Psychotic Behavior

Schizophrenic psychosis has basic similarities with psychoses of other etiologies, including those that can be induced in normative persons by sensory deprivation, hallucinogenic drugs, etc. This similarity is related to sharing the basic mechanism underlying psychotic states in general (discussed in section 7.2.6.4). The processing of input from the actual social/natural environment and its impact on the final mental output and overt behavior decrease considerably, and/or the impact of intrapsychic contents not directly relevant to the actual environmental circumstances increases considerably. Underlying these later intrapsychic contents are, as a rule, instinctive motives seeking outlet. In consequence, the possibility of sharing consensually surrounding realities with normative persons becomes jeopardized.

On the other hand, the schizophrenic psychosis has some *unique features* not found in other forms of psychosis. The most important reason for this is that in the schizophrenia-predisposed individual, as already noted, long before the outbreak of the overt psychosis, a considerable dissociation (instead of the coordination and integration characteristic of normative persons and nonpsychotic mental disorders) can be found between mental contents and behavior induced from without (mostly by the current social stimuli) and intrapsychic contents chiefly reflecting diffuse internal instinctive strivings of the autonomous organism.

In other words, in pre-schizophrenics a comprehensive longstanding dissociation exists precisely between those two categories of mental contents whose considerable imbalance—leading to a preferential expression of internal at the expense of externally induced contents—underlies the basic mechanism of psychosis of any kind. In the pre-schizophrenic individual (or in schizophrenics during good remission), this dissociation remains unobservable or inconspicuous so long as the idiosyncratic, identified-with mental contents remain concealed during social interactions. This in turn depends on the power balance of the underlying energetic forces that fuel these two dissociated aspects of mental life. As long as the pressure to express the internal, identified-with contents does not become excessively strong, the individual manages to express them only during lonely (mental or physical) activities, while during interpersonal encounters she conforms, as much as she is able, to social expectations. In addition, as we have seen, the pronounced diffuseness of the instinctive strivings prevents selective perception and clear, goal-oriented behavior, and may further hinder overt behavioral expression, whether or not it is socially approved.

The dissociation may manifest itself in different forms, as we have seen. It seems that the best way to approach the process of transition from pre-psychotic dissociation to overt schizophrenic psychosis is to take model child behavior as our starting point (section 7.2.8.2.1).

In the transition from seemingly normative to clearly psychotic behavior in incipient schizophrenia, Laing describes three stages or behavioral sequences, which he calls the "good," "bad," and "mad" phases. He illustrates them beautifully with clinical examples, one of which I will use here to make the reasoning more perceptible. The interested reader is advised to read Laing's original text (Laing, 1960, pp. 197–213).

The *"good" phase* is by and large the equivalent of Laing's concept of "false self" expressed in the model child behavior. Its characteristics in an actual clinical case were related by the respective patient's mother: As a baby, "she never cried really for her feeds. . . . *She was weaned without any trouble. . . . She always did what she was told. . . . She was never a trouble. . . .* Even when encouraged . . . she would not express her own wishes . . . she had to have her mother to buy her clothes, and she showed no initiative in making friends. She would never take a decision of any kind" (Laing, 1960, pp. 201–206, emphasis in the original).

Here, Laing, in spite of his existentialistic orientation, makes a remark that, in my opinion, is close to this text's proposal that the genetic predisposition to schizophrenia consists of drives that are too diffuse to give a fitness-enhancing direction to the behavior: "It seems to me quite possible that, owing to some *genetic factor*, this baby was born with its organism so formed that *instinctual need and need-gratification did not come easily to it*." Complementary to the above deficiency in bringing about the clinical disease, in Laing's opinion, was the family's inclination to see this deficiency as "goodness," as a desirable behavioral attribute instead of an ominous sign of a mental disorder: "The combination of almost *total failure of the baby to achieve self-instinctual gratification*, along with the mother's failure to realize this, can be noted as one of the recurrent themes in the early beginnings of the relation of mother to schizophrenic child" (idem, p. 202, emphasis added).

However, as detailed in the section on the pre-psychotic period (7.2.8.3), the fragile mental equilibrium—excessively conforming in actual social interactions and expressing active instinctive motives outside the interpersonal sphere—becomes more and more overstrained during puberty, adolescence, and early adulthood as a result of biological and social maturational processes, including awakening sexuality, increasing social demands, and decreasing parental protection and guidance, etc. As a consequence, the individual is compelled to express instinctive motives (active as well as self-defensive reactive ones) *during* interpersonal transactions in a way not integrated enough with accepted norms and rules of behavior in the respective social milieu: This change ushers in Laing's second, "bad," sequence of prepsychotic behavior.

In the clinical case just mentioned, this period was characterized mainly by endless accusations toward the mother. The progressively increasing intensity of active and self-protective reactive drives awakened or strengthened in the daughter the realization that her behavior is under external guidance that suppresses her own, identified-with active motives (the "real self" in Laing's terminology). "Julie's diatribes against her mother were endless and were always on the same theme: She would accuse her mother of not having wanted her, of not letting her be a person, of never having let her breathe, of having smothered her" (idem, p. 207).

These accusations seemed to the mother clearly inappropriate, given that they were seemingly triggered by her efforts to convince Julie to be more independent and autonomous: At the age of fifteen, "she [the mother] began to urge her to get out more, have friends, go to pictures and even to dances, and have boyfriends. All these things the patient 'obstinately' refused to do" (idem, p. 207).

The accusations of the patient toward her mother, in our terminology, have to be attributed to self-defensive reactive instinctive motivations in their aggressive form. They aimed, in fact, to remove two opposing kinds of disturbing influences, both coming from the mother. The first was her previous intrusive, overprotective guidance, which led the daughter to adopt a kind of behavior (Laing's "false self") that she felt was alien from her real, internal motives; the second originated from the mother's recent promptings for her to be more independent, to behave as a normative teenager, something the patient felt totally unable to do. (The mother's unrealistic promptings toward her daughter to behave like a normative teenager originated most probably from a badly misplaced empathizing mechanism. She encouraged her daughter to behave in the same way she herself did at the same age, while disregarding in fact all three stages of the empathizing process.)[15] Perhaps we can get a glimpse here of the "dead-end" or no-win situation of a pre-schizophrenic lifestyle. Neither the mother's overintrusiveness nor her promptings to her daughter to be self-dependent in a normative way were acceptable options. On the other hand, the patient's own diffuse, active instinctive tensions could not guide her toward some kind of third alternative, a kind of idiosyncratic self-realization, better suited to her atypical instinctive configuration. I think it can be assumed that the patient's strengthening diffuse *active* drives in their aggressive form of frustration (boosted by the physiological and behavioral changes characteristic to adolescence) also contributed to her inappropriately strong hostility toward the mother.

The genetic roots of the disorder (as well as the nonexistent or negligible role of the shared family environment) become evident when the daughter's behavior is compared to that of her older sister, brought up by the same "dominating" mother. "Julie's sister . . . was a rather forthright, assertive married woman, not, however, without femininity and charm. According to her mother, she had been 'difficult' from birth: demanding and always 'a trouble.' In short, she seems to have been a relatively 'normal' child of whom the mother never very much approved. But they appeared to get on well enough together. The sister regarded her mother as a rather dominating person if one did not stand up to her" (Laing, 1960, pp. 208–209).

While the "bad" phase of the patient's behavior was highly distressing for the family, it was not interpreted as psychotic behavior, but as an expression of ingratitude in view of the mother's efforts to bring her up. The whole family—mother, father, and sister—responded with indignation to the patient's "ridiculous" accusations. The formerly positive, protective relationship with the family changed into a disapproving, inimical one.

However, for the patient, retreating in order to recapture her former "model child" relationship with the mother was not an option anymore. Rather, in her upset mental state, she went one step further in order to make the message about her actual subjective experience more perceptible, more penetrating. At the same time the increasing tension of the active and self-defensive reactive drives, augmented by increasing isolation from the offended family members (remember the psychosis-inducing effect of sensory/social deprivation), further suppressed that important mental function that ascertains, through feedback loops, that what we want to communicate is understood correctly by the other party. The patient's verbalizations progressively lost their communicative value (in the sense of being successful as exchanges of information), while their subjective, experience-expressing effect intensified. She made recourse to shocking metaphors in order to get through her intended message. However, with this (not too big) step, she transgressed the boundary between behavior that can still be interpreted on socially shared moral grounds as strongly improper and behavior that is no longer understandable by consensual validation. Julie began to accuse her mother of trying to kill her after she "was told by a voice that a child wearing her clothes had been beaten to [a] pulp by the mother, and she proposed to go to the police to report this crime" (idem, p. 213).

It is worth noting that the outbreak of the overt psychotic episode coincided with the disappearance of the patient's favorite doll with whom, from infancy on, she had frequently played. Maybe she lost it, or the mother disposed of it, since she believed that the daughter had outgrown the age of playing with dolls. In fact, "no one knew in what way" she played with that doll. "It was a secret enclave in her life" (idem, pp. 212–213). It seems reasonable to assume that, with that doll's disappearance, an important route of acting out active and self-protective reactive instinctive motivations, outside the sphere of interpersonal transactions, had been lost.[16]

I have described this case in detail because it seems to illustrate well the secondary nature of the psychotic state in schizophrenia, which emerges through several steps during interactions between an inherited predisposition (unrelated directly to psychosis) and the social milieu, augmented by (sexual) maturation. Furthermore, it seems equally clear that *those steps (good, bad, mad) are not transitional steps from normative to dysfunctional behavior to disconnection from reality. Instead, all three steps are strongly dysfunctional regarding their fitness-enhancing ability in the face of normative social pressures.* What makes it seem to outsiders like a transition from relative normality to grave mental disease is the strongly conforming nature of the "good" phase; the condemnable but still comprehensible nature of the "bad" phase from a moral standpoint; and the impossibility of interpreting the statements and behavior of the "mad" phase within a consensually validatable framework. The realization that these three phases are different aspects of the same mental disorder, and not transitional steps from sanity to insanity, will assist us in our efforts to make sense of the deteriorating course so typical of this disease.

7.2.8.6. Characteristic Features of the Schizophrenic Psychosis

7.2.8.6.1. The Indistinct and Undifferentiated Nature of the Delusions

The delusions in schizophrenia are indistinct and undifferentiated in the sense of lacking detail and not being restricted to particular interpersonal relationships, at least at the disorder's more advanced phases. This quality can be best illustrated by comparing the delusions character-istic of schizophrenia with those that are characteristic of the *delusional disorder*. Two out of the five distinct forms of delusional disorder, the jealous and the erotomanic subtypes, evidently cannot appear outside a particular interpersonal relationship. The persecutory type, in my ex-perience, frequently has the same quality. One of the recurring themes of my delusional outpatients, particularly the older, solitary ones, was of being harassed by a particular neighbor (for example, by being poisoned by toxic gases introduced through the water pipes of their apartment). In another case, one such solitary male individual was hospitalized after physically attacking his neighbor. He explained that he did it because he believed that the neighbor had sent a young female social worker to visit him in order to force him to marry her.

Now contrast these kinds of delusions with those characteristic of schizophrenia, described in the following excerpt: "The conviction of being controlled by some unseen mysterious power that exercises its in-fluence from a distance is almost pathognomonic for schizophrenia. . . . The patient who is convinced that he is being persecuted by powerful agencies often harbors delusions of grandeur; he must be a very im-portant person if so much effort is spent on his persecution. . . . Also typical for many schizophrenics are delusional fantasies about the de-struction of the world" (Lehmann, 1980a, pp. 1156–1157). To me, it seems quite obvious that *underlying these impersonal, indistinct, obscure, or all-encompassing delusions are excessively diffuse instinctive drives, es-pecially of the active type in their different forms of frustration (vacuum behavior, aggression, dysphoria).*

Grandiose self-evaluation; the intention to influence other people; the feeling of being influenced by them (according to good or bad inten-tions); and the dysphoric or anxious expectation of harmful events are in fact psychic themes common to schizophrenics, other psychotic or nonpsychotic mental patients, as well as normative persons. However, in schizophrenics, these common intrapsychic motives do not arise in the context of concrete interpersonal or other kinds of social interactions. As a result, they lack detail and, since the cognition is not restricted to particular real instances, they tend to be expressed by such obscure products of the imagination as "some unseen mysterious power that ex-ercises its influence from a distance" (idem).

7.2.8.6.2. Partial Response to Antipsychotic Drugs

Another peculiarity of the schizophrenic psychosis (shared, however, with the delusional disorders) is its *partial* response to antipsychotic therapy. As argued in the chapter on psychotropic drugs, these drugs' antipsychotic action consists exclusively of the reduction of the *intensity* of the overall instinctive activity (predominantly the active one) underlying an acute psychotic episode. Consequently, *only the symptoms induced directly by the high intensity of these drives will be relieved by the drug*. Thus, in a good prognosis manic or hypomanic episode, the antipsychotic drug effect reverts the clinical picture (including delusions when present) into the former, more-or-less normative, state, or, in the case of overtreatment, may induce a state resembling lethargic depression.

In schizophrenia, however, the situation is different from that in manic psychosis. The prepsychotic behavior, as discussed previously, is not a normative one, but is characterized by a comprehensive dissociation between the aspects of the behavior that are fueled by active instincts, on one hand, and the behavior induced by compliance (or other kinds of dissociated reactive tactics) with social influences, on the other. When the active-instinct-induced behavior *strengthens* and overpowers the dissociated responses to social stimuli, an acute schizophrenic psychotic episode ensues. When these strong active instinctive intensities are *suppressed* by the antipsychotic medication, the power relationship changes in favor of the social-influences-induced behavior, leading frequently either to decreased defiance and improved conformity with social norms or to increased avoidance or withdrawal. This change induces the false impression of "remission" of the disease. (Gelfand, 1980, p. 3353.) However, as is well recognized in clinical practice, the active-instincts-induced mental contents dissociated from the actual social realities in schizophrenics *do not as a rule disappear*. Only their emotional charge, the intensity with which they are expressed or enacted in observable behavior, is reduced. This change is labeled in clinical psychiatry as the "deemotioning" effect. When the power-relationship changes again from whatever cause in favor of the active instinctive expression, the acute psychotic episode returns, and its symptomatology (its perceptual, cognitive, and behavioral content) remains quite unchanged when compared with former acute episodes.

It has to be mentioned in this context that the delusions of a delusional disorder or those of a grave agitated depression (contrary to the delusions of a pure manic attack) may respond to antipsychotic drug treatment like schizophrenic delusions. They do not disappear completely; only their energetic and emotional charge diminishes. The possible explanation, in my opinion, is the following: Unlike a pure bipolar disorder, in these conditions, besides the intensification of the energetic charge, other dysfunctional aspects of the instinctive configuration relevant to eliciting delusions remain unaffected by the antipsychotic effect. For example, in the persecutory and jealous subtypes of the delusional disorder, certain active instinctive motives are expressed overwhelmingly and for dysfunctionally protracted periods in their aggressive form of frustration, as we have seen (section 7.2.6.3). The same is true in the case of grave agitated depression, in which the characteristic delusions are induced by an overwhelming propensity of the active drives to be expressed in their dysphoric form of frustration. Since these pathological mechanisms—the differential and longstanding overaccentuation of the aggressive or dysphoric form of active instinctive frustration—remain unaffected by the antipsychotic effect, only the respective pathology's intensity will be reduced.

7.2.8.6.3. Schizophrenic Language Disorder

A further unique characteristic of schizophrenic psychosis is the peculiar schizophrenic *language*. A detailed discussion of the disturbances in schizophrenic speech and thinking is beyond the scope of the present text. However, it must be mentioned briefly, because it is, in my opinion, an additional, secondary manifestation of the peculiar dissociation between socially induced and instinctively fueled mental contents in schizophrenia.

Bizarre disturbances in the speech of a schizophrenic that may interfere gravely with communicative function of speech are among the most characteristic, albeit not pathognomonic, signs of schizophrenia. They are especially prevalent and striking in the disorganized or hebephrenic subtype, while less characteristic in other subtypes, least of all the paranoid one. (Paranoid schizophrenics express their thoughts, both the delusional as well as those needed for routine communication concerning everyday matters, clearly and coherently. Our concern here is with formal speech disturbances, not their content.)

The most desirable property of language, its coherence, is well described by Silvano Arieti: "When a [normative] person thinks logically, he organizes his thoughts according to a pattern or structure that leads toward an end or conclusion" (1974, p. 258).

The following short discussion on schizophrenic language disorder uses as its main source psychiatrist Norman Cameron's contribution to a collection of papers on this subject entitled *Language and Thought in Schizophrenia* (Kasanin, 1944/1964). In the speech of many schizophrenics, especially those with the disorganized (hebephrenic) subtype, the logical thread is often lost. Instead, it consists of a "half-organized collection of fragments instead of a functional unit." The schizophrenic loses the ability to "restrict, eliminate, and focus on the task at hand" (Cameron, 1944/1964, p. 53). Frequently, instead of a precise term, an approximation is used, one that is unusual in normative persons' conversations. When a schizophrenic is asked a specific question, his reply may contain some elements related to the external stimulus interspersed with fragments of speech that are completely unrelated. Thus, their speech can be overinclusive, containing elements that are irrelevant or peripheral to the central theme, generalizations that are too all-encompassing to be useful, and statements on shared realities interspersed with unrelated products of the fantasy.

Cameron's interpretation of the disorders of schizophrenic language seems to me instructive and relevant to the concerns of this text, so it will be quoted at some length: "Every child is born into a predetermined social organization with a language system already in operation." And "perhaps his most important achievement is that of *learning to respond to his own behavior with the same attitudes and reactions which other persons show toward him. . . . By it he learns how to be self-critical.* When necessary, he can see himself as others see him, simply by taking their attitudes. He can then, if it seems good, modify his own conduct so as to fit himself better into the social pattern" (idem, pp. 60–61, emphasis added). This maturational path seems to be entirely dependent on the ability to read others' (frequently covert) opinions, emotions, and intentions through the mechanism of empathy or "theory of mind." This does not necessarily mean an overcompliance with social norms. The person can choose between a more submissive, a more defiant, or a more idiosyncratic approach that disregards the relevant social expectations.

However, he acquires the necessary awareness of his behavior's impact on his social milieu, and this impact's possible repercussions.

However, schizophrenic patients, and especially the disorganized (hebephrenic) ones, seem "well satisfied . . . with their very inadequate communication, showing little or no evidence of concern over its unintelligibility [that is, no intrapsychic conflict is evident, only dissociation]. . . . He is no longer able to take your role in this situation, to put himself in your place and then speak more from that vantage point; and his own asocial patterns have reached a point where you are not able to take his role, either. This disorganization, in addition to being a result of growing isolation from social participation, is itself helping to accelerate the process and to perpetuate it by cutting off all effective communication by word of mouth" (idem, p. 55).

These excerpts describing schizophrenic language may be interpreted in the more evolutionarily oriented spirit of the present text in the following way: In the schizophrenic process, mental contents and their verbal expression learned through a *conforming* reactive behavioral mechanism during explicit social interactions, such as formal education, become dissociated from mental contents and their verbal or behavioral expression that are induced by diffuse active and self-protective reactive instincts.

Moreover, it seems reasonable to assume that persons with pathologic levels of diffuseness of their instinctive predispositions will have great difficulty properly comprehending the precise meaning of words expressing subjective emotional or intentional experience induced by normatively differentiated drives. Consider that what is communicated verbally is the sound, the phonetic aspect, of words with an arbitrary relationship to their meaning, which has to be reconstructed anew by each individual mind. This may be done by "joint attention" (section 2.3); by the empathizing mechanism (in the case of words expressing emotions); by explanation of the meaning of unknown words with the aid of already known ones; by heeding the context in which a word is used, etc. It follows that if the empathizing mechanism is deficient, the schizophrenia-predisposed person cannot properly understand what an emotionally charged word means for a normative person.

So long as the individual is able to keep the dissociated aspects of her mental life apart and expresses in social circumstances only the aspect acquired from social sources, her behavior is accepted by others and is not considered as considerably disturbed. During the acute psychotic process, however, this situation becomes reversed—that is, the active and self-defensive reactive motivations are uninhibitedly expressed in overt behavior, including in social encounters, suppressing the former conforming behavior. The adolescent or young adult refuses, for example, to continue her studies or work, breaks off her former social relationships, and publicly expresses psychotic contents—that is, "the schizophrenic language . . . [becomes] primarily a means of self-expression rather than a means of communication" (Lehmann, 1980a, 1159).

During schizophrenic deterioration, however, the two dissociated mental domains (especially in disorganized [hebephrenic] patients) may get mixed up and become expressed in speech in this mixed-up form (of course, without any attempt at their integration). This mixing up of dissociated mental contents leads to the intrusion of idiosyncratic products of fantasy into the flow of verbalizations that may otherwise be to the point, confusing and frustrating the listener's intentions to make sense of the hebephrenic speech.

Some of the simpler ways the dissociation between language's self-expressive and communicative functions manifests itself in schizophrenic speech may be exemplified by the results of the following clinical experiment: To study their formal thinking disturbances, institutionalized schizophrenics were asked to interpret well-known proverbs, such as: "When the cat's away, the mice will play" (Benjamin, 1944/1964, pp. 73–75).

■ *Answer number one*: "If there isn't any cat around, the mice will monkey around, and maybe get into things." A literal, concrete (not figurative or metaphoric) interpretation of this kind is expected from persons with mild mental retardation. However, the individual who gave this answer had an IQ slightly above normal. We may get closer to an explanation if we consider that a proverb is a piece of folk-wisdom, which may express in a metaphorical way common situations encountered in social coexistence. Every schoolboy knows from his own experience the difference between the frame of mind and behavior of a class of pupils when an authoritative teacher is present as opposed to that teacher's absence. Similarly, most employee or staff members have experienced the difference between the presence of a stern superior and his absence. Because this is a common social situation, most people will easily make the connection between the real situation and its metaphoric representation by the proverb. However, during such circumstances, a schizophrenic or pre-schizophrenic, with a dissociated "false self" (in Laing's term), will have an entirely different mental experience. As a schoolboy, for example, he will not participate in the relieved, noisy pandemonium when a stern teacher does not show up in the class at the expected time. Even more distant from his experience is the mental state of a person whose job is to keep order and discipline in a group of reluctant people. Because it has no personal significance for him, the schizophrenic may easily disregard the proverb even if he has heard it repeatedly in the past. The connection between the concrete situation and its metaphoric expression simply is irrelevant for him. Consequently, when asked to give an interpretation of it, he may give a literal one.

▪ *Answer number two:* "As applied to what? Just give the mice more liberty." This schizophrenic patient had an IQ of 138! We cannot know for sure whether he knew the right answer or not, but it seems quite clear to me that his present confinement in a mental hospital and his longing for more freedom intruded on his mental effort, thus supplanting a more apt response. The experimental stimulus played the role only of triggering the expression of the intrapsychic content, which was irrelevant to the experimental task.

▪ *Answer number three:* "The last supper of Jesus, all those that kissed the novicia, the covicia. The political world is too much, we can't fight it, we can't see murder." In this case the actual stimulus (the proverb to be interpreted) is completely disregarded. Instead, phrases with strong emotional connotations (religious, sexual, aggressive) are uttered in a highly disorganized way to express the patient's own confused emotional state (the psychotic state of the disorganized type, in clinical parlance).

Silvano Arieti formulated the diffuse, unfocused nature of schizophrenic thinking, especially when it tries to express internally originating contents, in the following way: "The ideas of the schizophrenic have *no definite limits*, are *diffused*, and overlap. The world of objects does not consist of separate or distinct things, but of *diffused* and disorganized complexes" (Arieti, 1974, p. 295, emphasis added). In my opinion, this quotation conveys quite accurately the impact of pathologically diffuse instincts on the ways schizophrenics perceive surrounding realities and express them verbally.

7.2.8.7. Schizophrenic Deterioration

The mixing up (without integration) of the two dissociated aspects of mental functioning, which makes the speech of schizophrenics incomprehensible, accelerates the process of disengagement from social interactions and may lead to a quick deterioration in functioning in a social milieu (as was argued in the quotation from Norman Cameron in section 7.2.8.6.3). However, while this kind of deterioration can still be observed in modern psychiatric establishments (employing, besides drug therapy, various methods of social and occupational rehabilitation), in my experience another, less malign, form of deterioration seems to be the predominant one. In this latter kind of deterioration, the two dissociated aspects of the mental contents and behavior remain kept apart (that is, are not mixed up) but each becomes progressively more *stereotypic and impoverished.*

As a head of chronic psychiatric wards for many years, I had plenty of opportunities to observe this later form of deterioration in schizophrenic patients, as well as its stabilized end product. I interpreted this process in the following way: In normative persons, engaged adequately in social life by direct interactions with various people, groups, or institutions, the mental contents are continuously enriched with new material by keeping up to date on what is happening in the more restricted, as well as the larger, human community one belongs to. This may happen indirectly, too, through listening to the media and reading professional and other kinds of literature. The variability and complexity of social life supplies continuously new challenges to the decisionmaking, coordinative, or integrative functions of the mental apparatus, new frustrations to tackle, and new opportunities for emotional (active-instinctive) gratification—for example, adjusting to changing conditions at the workplace; becoming acquainted with the new pleasures of married life and childrearing, along with its new worries and tensions; adapting to changes in one's more extended community, especially when they affect one directly; and so on. Even old age brings with it new challenges, new worries, and new sources of satisfaction (like grandchildren).

It is true that people differ in how inclined they are to seek new kinds of stimuli. One of the four basic temperamental dimensions in contemporary personality theory is "novelty seeking," which may vary from low to high (Svrakic & Cloninger, 2005, p. 2065). However, even persons whose lives are based mainly on rigid, unchangeable routine have to cope on a daily basis with situations that cause tensions between external social expectations and internal mental strivings. The mental mechanisms that have to deal with these tensions are in constant use, like the muscles of an active person, keeping them in good shape and preventing their atrophy. Consider that even in rats the quantity of neural synapses varies considerably with the level of complexity of their (laboratory) environment (Plotkin, 2007, pp. 195–196).

Neurotic and personality disorders may restrict the intensity and scope of participation in social life. Think of such disorders as social phobia, agoraphobia, and schizoid, paranoid, and dependent personality disorders. However, these restrictions are not of a kind that altogether preclude intrapsychic conflicts—which need to be handled—and limited participation in normative social life.

The schizophrenic's situation in this respect, however, is entirely different. As stressed repeatedly in this section, his mental and behavioral responses to social stimuli, based mostly on submissively learning the expected behavior, have become from an early age considerably dissociated from his active and self-defensive (but not submissive) reactive-instincts-induced behavior and mental states. When the schizophrenic eventually attempts to express his true inner motivations (in a way that is not well enough integrated and coordinated with social expectations), these attempts inevitably are met with incomprehension, guardedness, derision, or even hostility. The schizophrenic thus is forced sooner or later to live an inner life considerably *disconnected* from the normative, socially shared one in an environment in which the external stimuli are considerably simplified, routinized, and reduced in their intensity, complexity, and scope: a chronic psychiatric ward, a hostel or other sheltered environment, or perhaps a home with an old mother who cares for the schizophrenic individual's basic needs.

In such conditions, the internally originating mental contents (Ronald Laing's "real self") undergo a progressive impoverishment, a process well recognized in psychiatry. The following quotation expresses it in the impressionistic language of existential psychiatry: "Fantasy, without being either in some measure embodied in reality, or itself enriched by injections of 'reality,' becomes more and more empty and volatilized"

(Laing, 1960, p. 90). In the descriptive language of clinical psychiatry, the idiosyncratic aspects of behavior will become more and more repetitive, "stereotypic," and the delusions will be expressed in the same shortened phrases time after time, without emotional charge, and without any willingness to enlarge on them or to attempt to substantiate them.

The social environment–induced conforming behavior becomes impoverished, too. This is understandable if we consider the restricted amount of variability of the environmental stimuli in the establishments in which chronic schizophrenics are kept. Add to this the schizophrenic patient's unwillingness or inability to conform to expectations to the extent she was able to in the pre-schizophrenic period (during her school years, for example), as well as her inability to adjust quickly to changing social situations. Chronic schizophrenics prefer unchanged routine over frequent novelties: "They avoid more and more any unpredictable situation or any spontaneous response" (Arieti, 1974, p. 399). In these conditions, the fluctuating course of the disorder, that is, *the alternation of exacerbations and remissions, frequently subsides, and the two impoverished, stereotypic mental and behavioral aspects may coexist side by side for long periods without having any interactions between them.* This condition has been called "the peculiar double book-keeping" of the chronic schizophrenic (Lehmann, 1980a, p. 1175), or (by Eugen Bleuler) "double orientation" (Slater & Roth, 1969, p. 276): "They manage to live in two worlds at the same time, to fulfill the demand of reality and to follow their daily occupation . . . while at the same time secretly believing in the most fantastic delusions" (idem, p. 276).

One of my middle-aged female patients, hospitalized for about thirty years in a chronic psychiatric ward, believed that she was a great physician and a renowned pianist, that her age was "twenty-two springs" (instead of twenty-two years, her approximate age at her transfer to the chronic ward), that the lice in her hair were tiny diamonds, and that the ward's kitchen was equipped with concealed dishwashing machines. In her daily behavior, however, she was quite compliant with the nurses' wishes that she aid them in simple duties and perform daily hygienic activities. She washed the dishes after meals in the ward's kitchen sink, submitted herself to daily showers, periodically watered the ward's potted plants, etc., in exchange for some candies or a pack of cigarettes. During my more than twenty years of acquaintance with her, these behavioral and cognitive patterns remained very stable (under small amounts of

Haloperidol), without acute exacerbations and remissions, and without improvement in her delusional beliefs either.

However, I suspect that some of her delusions were not entirely dissociated from her daily activities in the ward, induced mostly by the nurses' promptings. Her insistence that the lice in her hair were in fact tiny diamonds probably contained a covert reproach—namely, that it is a big waste to wash them out with daily showers; and her delusions of the existence of concealed dishwashers in the kitchen could express her covert disapproval of the nurses, who continued (in spite of these imaginary machines) to demand that she perform kitchen duties. In the context of the present text, it seems that the dissociated active-drives-induced fantasies (her delusions) were influenced to a certain extent by reactive avoidant motives originating in real-life situations.

7.2.8.8. Catatonia

The concept of catatonia was first proposed by Kahlbaum in 1863 as a distinct psychiatric disorder, which was later included as a subcategory within the comprehensive concept of schizophrenia. More recently, however, catatonia has been recategorized as a heterogeneous syndrome, probably in both its etiology and pathogenesis. Besides schizophrenia, it is encountered in bipolar disorder, depression, and neurological disorders of various etiologies: toxic encephalopathy, encephalitis, cerebral tumors, and as a neuroleptic side effect (Fletcher, 2001, p. 1076). A short discussion of the catatonic syndrome is included here, since it seems to provide powerful observable evidence in support of the central assumptions on schizophrenia proposed in this text.

Catatonia refers to disturbances of the *motor aspects* of behavior and thus is directly observable. They include disturbances of an all-encompassing nature and seemingly of opposite kinds: catatonic excitement and stupor; as well as symptoms affecting more circumscribed aspects of the behavior: negativism, echolalia, automatic obedience, posturing, waxy flexibility, etc.

The mechanisms proposed to underlie catatonic states relevant to the causation of schizophrenia in the context of the present work are the following:

1. Fluctuations in the *intensity* of the active and egocentric reactive instinctive activity

2. More than normative *diffuseness* of these instinctive drives
3. As a secondary consequence of point 2, the *dissociation of behavior into two separate aspects*, one that expresses active and self-defensive reactive instincts in a manner dissociated from the surrounding realities, as well as one that is induced exclusively by external influences (without being integrated with intrapsychic contents)

In the following, I discuss discrete catatonic symptoms or syndromes, in relation to these instinctive mechanisms.

7.2.8.8.1. Excited Catatonia

Excited catatonia or "general hyperkinesis" manifests itself as a state of grave agitation, mostly of an aggressive nature: "The patient cries, hits, bites, breaks, and destroys everything he can lay a hand on, runs up and down, fights everybody and keeps moving day and night. It's impossible to establish any rapport with him, he continues to rage when left alone, independently of any stimulation. The impression is that of an uncontrolled, instinctive motor discharge" (Slater & Roth, 1969, p. 285). In this state, to my mind, all three instinctive mechanisms just mentioned are discernible. The active instinctive activity, mainly in its aggressive form of frustration, is extremely *intense*; it is also highly *diffuse*, lacking any organization, goal-directedness, or purpose, and expresses instinctive incentives (probably exclusively active ones) in a way that is completely *disrupted* from stimuli coming from the surrounding social world. By comparison, in *manic states,* where, according to our scheme, only the *intensity* of the instinctive activity (which lacks excessive diffuseness) is increased, and where considerable disruption from the social reality is not present, the behavior is more "versatile and adaptable . . . well prepared or cunningly planned" (idem, p. 286).

It has to be mentioned that catatonic excitement is rarely seen nowadays, probably as a result of the widespread use of antipsychotic medication, which, as argued in section 6.2.1, is highly effective in reducing the overall *intensity* of the instinctively underpinned behavior.

7.2.8.8.2. Catatonic Stupor

Catatonic stupor, at least in its motor aspects, seems to be the complete opposite of catatonic excitement. These patients "lie motionless for weeks or months, their eyelids flickering, saliva drooling from their mouths, unresponsive to any stimulus" (Lehmann, 1980a, p. 1163). These individuals (as well as those suffering from catatonic excitement and other catatonic syndromes) were considered unquestionably schizophrenic. However, recent epidemiologic studies found that only up to 35 percent of these patients are schizophrenic and the majority suffer from depressive or bipolar disorder (DSM-5, p. 120). These drastic changes in the distribution of catatonia among various mental disorders (instead of its being an exclusively schizophrenic pathology) are probably a result of several developments in psychiatry.[17]

While I have no extensive experience with stuporous patients, I have good reasons to assume that the underlying brain mechanism that induces the psychomotor slowing in lethargic depression is the *opposite* of that encountered in the catatonic stupor of schizophrenics, and the resemblance between these two mental conditions exists only on a descriptive level. In lethargic depression, the psychomotor slowing (as discussed in section 5.1) is the expression of *low intensities* of instinctive activity as an evolutionarily adaptive response to low levels of life-supporting environmental resources in the winter months. In schizophrenic catatonic stupor, on the other hand, I suspect that the extreme slowing down of motor and most mental activities is the result of *high intensities* of considerably *diffuse* instinctive activity. As we have seen in the section on antipsychotic drug effects (sections 6.2.1 and 6.2.2), high intensities of instinctive activity may either activate or inhibit (including paralyze) overt behavior in both normative and mentally affected persons.

In schizophrenia, catatonic immobility (but not stupor) can instantaneously switch into motor excitement and vice versa, hinting at the possibility that in this condition inhibition of overt behavior versus its disorganized overactivation are closely related mental mechanisms (that is, only a small alteration in some circumscribed brain function is needed to switch between these two seemingly opposite states). In other words, both inhibition and overactivation of the behavior are caused by *high* instinctive intensities. (It is difficult to imagine such a quick transition between low and high levels of instinctive activity. The transition between mania and lethargic depression in bipolar disorder may take weeks to be clearly apparent in overt behavior.)

That catatonic stupor may be the result of excitation rather than inhibition of mental activity has already been proposed: "The total inhibition of the patient in catatonic stupor may be the result of *excessive cerebral excitatory processes* ... that excess prevents the person from performing and behaving in a normal manner until his excessive cerebral functions have been *reduced* by a chemical agent that has a *depressing* effect on brain metabolism and nervous impulses" (Lehmann, 1980a, p. 1166, emphasis added). The chemical agent referred to is a short-acting barbiturate administered intravenously.[18]

7.2.8.8.3. More Circumscribed Catatonic Symptoms

The behavior patterns of the rest of the catatonic symptoms are more narrowly circumscribed than they are in stupor or catatonic excitement. The following represent, in my opinion, at a simple, motor level, the two dissociated aspects of behavior in schizophrenia: those symptoms induced by active or self-protective reactive instincts, on one hand, and those induced by submission to external, social stimuli without any attunement to or integration with internal mental contents, on the other. These motor symptoms are clearly observable evidence of the dissociation. Symptoms of the latter kind are: *automatic obedience*, in which the patient follows in a robotlike fashion the instructions of the examiner without any attempt at either understanding them, effecting some individual change in carrying them out, or expressing a kind of personal opinion on or an emotional reaction to them; and *echopraxia* and *echolalia*, in which the individual imitates the examiner's gestures or speech, again without any apparent reason or personal contribution. Ronald Laing saw in these symptoms the extremes of the "false-self's" automatic compliance with social requirements (Laing, 1960, p. 109).

Negativism is the complete opposite of automatic obedience. The individual refuses to cooperate with the instructions of the examiner or may even do the opposite of what has been required of him, again without any apparent reason. He does not seem fatigued, angry, or suspicious (Lehmann, 1980a, p. 1163). In our terminology, self-protective, reactive instinctive motives of the *opposing or defying type* (mild counteraggressive reactive tactics, the opposite of submission or conformity, in my opinion, which may appear only in the context of social life) are discharged at the simplest possible level of sophistication, without any traces of being integrated with other mental contents or adapted to the overall interpersonal situation. *The contraposition of automatic obedience and negativism illustrates, to my mind, in the most palpable way, the far-reaching dissociation in schizophrenic pathology between external influences and behavior induced by internal mental contents.*

Other catatonic symptoms are caused, in my opinion, by brain output directed toward the voluntary musculature that is so diffuse that it makes the flexible alternations between contraction of some muscle groups and simultaneous relaxation of the contra-lateral ones impossible, thus leading to motionlessness. Subjectively, these states are most probably experienced as an extreme diffuseness of intentionality or volition, and not as an inability to move due to a somatic disturbance. (Eugen Bleuler's notion of schizophrenic "ambivalency," expressed in behavior as "ambitendency," was probably referring to a similar but less extreme pathologic mechanism.)

Physical immobility, rigidity, bizarre posturing—these symptoms consist of sustaining uncomfortable and dysfunctional body postures, such as standing on one leg for prolonged periods or lying on the bed with the head kept a short distance above the pillow, without any discernible reason. The individual may be stiff and tense, resisting attempts by the examiner to move his body. In other individuals, the stiffness is not so pronounced, allowing the moving of the limbs against some resistance (known as "waxy flexibility," or "lead-pipe rigidity" in clinical parlance). The new positions induced by the examiner are maintained for long periods even if they are uncomfortable (Lehmann, 1980a, p. 1163).

It merits mention here that grave catatonic symptoms and syndromes were widespread in hospitalized patients till the second half of the twentieth century, when their prevalence diminished dramatically, with severe catatonia becoming an uncommon condition in schizophrenia. The most accepted explanation for this change concerns the widespread use of antipsychotic medication in schizophrenia (Sadock & Sadock, 2005, p. 1131).[19]

7.2.8.9. The Role of Unrelated Genetic Predispositions, as Well as of Ontogenetic Influences in Shaping the Final Clinical Picture

7.2.8.9.0. Introduction

The more than desirable diffuseness of the instinctive intentions in their appetitive phase, which may cause dysfunctional behavior, is only a predisposition. It cannot be called schizophrenia. In order to become what, in common clinical usage, is called schizophrenia or one of its spectrum disorders, the innate predisposition, as we have seen, has to interact with:

- Other innate behavioral predispositions
- Environmental influences of a physical or biological nature (intrauterine and perinatal disturbances, for example)
- Interpersonal interactions and cultural influences of the social environment

While a more detailed treatment of the effects of these factors on the final clinical outcome is beyond the scope of this text, a short account on them seems desirable in order to provide a more complete picture of the subject of schizophrenia.

Before beginning this topic, however, I want to stress again that instinctive diffuseness is a dimensional category along a scale of continuous gradations, and the extent of this excessive diffuseness itself is probably one of the most important factors, if not *the* most important, in determining the form of the clinical condition, its gravity, and its course. However, schizophrenia and its genetically related spectrum disorders are so widely heterogeneous that the quantitative variations of the gravity of the basic genetic predisposition by themselves, in my opinion, cannot satisfactorily account for them. Remember, for example, that "although schizophrenia runs in families, the particular subtype [hebephrenic, paranoid, etc.] does not" (Plomin et al., 2008, p. 203). This means that the subtypes are not inherited exclusively through the same genetic mechanism that underlies the comprehensive clinical category of schizophrenia and its spectrum disorders. In addition, ontogenetic influences, both somatic and social in nature, may play an important role in shaping the clinical picture, especially with regard to its onset, gravity, and course.

7.2.8.9.1. The Co-occurrence of Unrelated Genetic Predispositions

It seems quite possible, according to Mendel's second law (of independent assortment), that two or more innate predispositions concerning behavior, each at the dysfunctional extreme of its diversified, quantitative continuum, may co-occur in the same schizophrenic patient, while the predominant one, which determines the diagnosis, is a predisposition to excessive diffuseness of the instinctive drives. On the other hand, it is possible, too, that behavioral predispositions (unrelated to the decisive one in schizophrenia) lying in the normative range of their diversified continuum may exert a beneficial or ameliorating influence on the respective clinical disorder. The ultimate clinical picture will be the outcome of the complex interactions between these different genetic predispositions and ontogenetic factors.

Perhaps the most conspicuous condition reflecting the co-occurrence of two different dysfunctional behavioral predispositions is the *schizoaffective disorder*. As mentioned in section 5.1.4.1, this disorder has two subtypes, one in which the schizophrenic symptoms predominate and another in which the affective symptoms predominate. Genetic studies have found different patterns of inheritance regarding these subtypes. In the families of the schizophrenic subtype, an increased prevalence of schizophrenia was found, while in the families of the affective subtype, increased prevalence of mood disorders prevailed, without an increased prevalence of schizophrenia (Fennig, Fochtmann, & Carlson, 2005, p. 1533).

Another independently varying genetic predisposition influencing the clinical picture of schizophrenia is related to the three forms of frustration of the active instinctive strivings: vacuum behavior or displacement accompanied by euphoria; aggressiveness or possible paranoid ideation; and dysphoria predisposing to self-harm. Because of the great difficulty—or sheer impossibility—of expressing excessively diffuse drives in their original unfrustrated form in a fitness-enhancing interaction with the social environment, it seems quite clear that these excessively diffuse drives are expressed most of the time in one of their forms of frustration. *Whether the transitions between these three forms of frustration are elicited easily or, alternatively, whether a predisposition exists for a stable overrepresentation of one form of frustration may considerably influence the clinical picture.* I think that typical stable paranoid schizophrenia belongs to the latter group. In this subtype, in my opinion, the active instinctive frustration is stabilized, as a rule, somewhere in between the megalomanic and the paranoid attitude. In consequence, the delusions are stable and express an intermingling of the respective emotional motives. The patient believes she is in possession of some extraordinary qualities, and as a result, obscure agencies are persecuting her in bizarre ways. The specific *content* of the megalomanic delusions are determined by those cultural influences that impressed the patient most (religious, scientific, artistic, etc.). This subtype of schizophrenia, as a rule, begins at a later age than the other subtypes, and the premorbid social adaptation is better. In addition, it possesses no disorganized (hebephrenic) or catatonic symptoms. It frequently stabilizes at a better level of social functioning without grave deterioration (Lehmann, 1980a, p. 1168). (Some authors consider this subtype as a different nosological entity than other forms of schizophrenia.)

It can be assumed that frequently occurring (mostly spontaneous— that is, not externally triggered), complete *transitions between the three forms of frustration of active drives*, as well as fluctuations in their intensity, must have a grave disorganizing effect on schizophrenic cognition and behavior, a behavior already deleteriously influenced by the more than normative diffuseness of these drives. The frequent switches between the different forms of frustration of the active drives, dissociated from external, observable incentives and expressed in bizarre ways, cannot be comprehended by others through the mechanism of empathy (whereas persecutory fears and inflated self-esteem are accessible to the normative empathizing process). Perhaps this is the mechanism frequently underlying such schizophrenic symptoms and signs as incongruent affect, bizarre behavior or verbalizations, unexplainable outbursts of laughing or aggression, unexpected impulsive acts of self-mutilation, or suicidal attempts. While these "untamed" fluctuations are the primary or only pathological mechanism in borderline personality disorder (section 7.2.4.3), in schizophrenia, they play a secondary role. This kind of pathology is most characteristic in the disorganized or hebephrenic subtype of schizophrenia. In this condition, coherent, stable delusions, as in paranoid schizophrenia, cannot develop.

The additional innate predisposition leading to the *catatonic subtype* of schizophrenia is less clear to me. However, it can be presumed that it consists of an accentuated inclination to express diffuse instinctive tensions directly through the voluntary musculature rather than through cognitive elaboration or behavior induced by diffuse emotions.

The predominant *reactive tactic*, as well as the characteristic interrelationship between the active and reactive behavior—that is, their measure of coordination or dissociation—may also influence the overt clinical picture. Naturally, neither flexible, adaptive transitions between the reactive tactics (freezing, avoidance, aggression, submission) nor useful, fitness-enhancing interactions between active and reactive behavior appropriate to the current environmental situation exist in schizophrenia and its spectrum disorders. No instance is known to me in which a schizophrenic or a would-be schizophrenic was successful as a politician, businessperson, or social leader, or performed well in any other occupation that requires frequent flexible transactions with various people (sometimes simultaneously, with several persons), no matter how much abstract intelligence she may possess. However, the preponderance of a specific reactive tactic over other forms may be frequently observed, and this aspect of the instinctive constitution may have a powerful impact on the clinical picture and interpersonal relations.

If the reactive behavior accentuates the *submissive* tactic, either in the periods of remission or in chronic states with "double orientation" (section 7.2.8.7), the individual, if institutionalized, will be more manageable by the staff, and his interpersonal relations with both them and other patients will be better. These individuals are preferred for rehabilitation programs, which in turn may further improve the course of the disorder.

In the case of those whose reactive mental apparatus accentuates self-defensive tactics, avoidance or aggression at the expense of submission, worse interpersonal relationships than those in the previous category should be expected. Those individuals with predominantly *avoidant* reactions, being unable to live an independent solitary life, will remain mostly detached and withdrawn inside an inpatient or a tolerant rehabilitation unit. If the patient is predisposed to readily react with considerable *defiance or aggressiveness* to the therapist's attempts to bring him to conform with the disciplinary requirements of a chronic ward, this will create the least desirable interpersonal situation, a constant (more or less covert) inimical interrelationship. These kinds of patients are frequently kept in closed wards.

The active/self-centered-reactive/submissive behavior interaction, as mentioned, is always compromised in schizophrenia. However, the dysfunctional form of the interaction, or the dissociation between these aspects of the instinctive predispositions, varies, and that has considerable impact on the clinical picture. As we have seen, a frequent and relatively less damaging form of dissociation is the "model child" kind, which, during the deterioration of the disease, may take the form of "double orientation." Another, relatively benign kind of dissociation occurs when short periods of intense expression of dissociated intrapsychic contents (acute psychotic episodes) alternate with long periods of low-level active and self-defensive reactive instinctive activity, that is, remissions (with or without accentuated submissiveness). The least desirable kind of interrelationship between active/reactive/submissive behavior occurs when strong active instinctive activity considerably overpowers avoidance or conforming for protracted periods, leading to uninterrupted boisterous psychotic behavior, which precludes any form of psychotherapeutic or rehabilitative effort.

7.2.8.9.2. The Role of Ontogenetic Factors in Schizophrenia

7.2.8.9.2.0. Introduction

The most important indication that environmental factors do play an important role in schizophrenia comes from monozygotic twin studies: "It has been known for many years that the rate of discordance for schizophrenia among MZ twins is approximately 50 percent. Since it can be presumed that these twin pairs share all of their genes, it is likely that environmental factors account for the discordance" (Brown, Bresnahan, & Susser, 2005, p. 1372; see also Plomin et al., 2008, p. 199).

However, the present theory offers a somewhat different rationale for the discordance between MZ twins: It tries to be more specific in distinguishing between the various roles played by inheritance, psychosis, and environmental influences in generating this disorder category. As we have already stressed repeatedly, the relevant genetic predisposition does not inevitably lead to the *psychotic* schizophrenic disorder (considered exclusively as "schizophrenia" in genetic studies). Other possibilities for pathologic developments include the nonpsychotic schizophrenia spectrum disorders, subclinical behavioral disturbances, and simple schizophrenia (which lacks psychotic symptoms). Taking these into account may considerably increase the concordance rates in MZ twins. While the prevalence of psychotic schizophrenia is less than 1 percent worldwide, including schizophrenia spectrum disorders increases that number to 5 percent (Buchanan & Carpenter, 2005, p. 1330). This consideration may also suggest an explanation for the puzzling epidemiologic finding that the offspring of the *unaffected* MZ twin (that is, unaffected by the narrowly specified psychotic schizophrenia) have rates of psychotic schizophrenia similar to those of the schizophrenic twin (Brown, Bresnahan, & Susser, 2005, p. 1378).

It must be kept in mind that, since the clinical picture represents the final outcome of a long-lasting interplay between the direct behavioral manifestations of the hypothesized genetic predisposition and the ontogenetic influences (physical and social), no clinical disorder on the schizophrenia spectrum can be imagined without considerable input from the social/physical environment. In the following, the environmental factors influencing the clinical picture are discussed in two separate sections: The first deals with *somatic causes* that deleteriously influence brain maturation; the second with the *relevance* of *psychosocial factors* in shaping schizophrenic symptomatology, its gravity, and its course.

7.2.8.9.2.1. Somatic Causes Leading to Brain Dysfunction

It is well known that brain abnormalities of different kinds, both functional and structural, are overrepresented in schizophrenics. A small sample of these include enlarged brain ventricles suggesting a decrease in the brain tissue volume (Buchanan & Carpenter, 2005, p. 1334); "abnormal patterns of glucose metabolism or blood flow" in functional imaging studies (idem, p. 1334); and signs of "*disorganized timing and integration of neurological maturation*" suggesting an "*inherited congenital neuro-integrative defect*" (Murray & Bramon, 2005, p. 1384, emphasis added). The efforts to clarify how these brain abnormalities are caused and how they affect brain functioning has led to hypothesizing two possible kinds of brain pathology, *neuro-developmental* and *neuro-degenerative*. In other words: "Is the cause of schizophrenia to be found in the failure of the normal development of the brain, or is it to be found in a disease process that alters a normally developed brain" (Buchanan & Carpenter, 2005, p. 1331). I do not possess the knowledge to enter into a detailed discussion of this topic. My intention here is to relate some of these brain abnormalities to the inherited predisposition to the schizophrenia spectrum of disorders.

7.2.8.9.2.1.1. The Neuro-Developmental Hypothesis

It has to be mentioned in advance that the *neuro-developmental* hypothesis seems to be much better substantiated by facts than the neuro-degenerative one. Researchers found an overrepresentation of various harmful pathological influences potentially affecting early brain maturation in would-be schizophrenics when compared to normal controls or patients with nonschizophrenic mental disorders. These pathogenic influences occurring during the intrauterine, perinatal, and early postnatal periods of brain maturation include pregnancy complications (bleeding, diabetes, preeclampsia, Rh incompatibility); low birth weight; complications at delivery possibly leading to hypoxic brain damage; birth during late winter and early spring, when some viral infections are more frequent; malnutrition, stress, or obesity in the pregnant mother; and so on (Brown, Bresnahan, & Susser, 2005, pp. 1374–1377).

How can such diverse deleterious influences during the perinatal period contribute to a mental disorder that appears fifteen, twenty, or more years later? In my opinion, two theoretical possibilities exist:

1. The simpler presupposition has to be, in the context of our theoretical framework, that *they act as a nonspecific additional burden on the maturation of a mental apparatus already loaded with a specific genetic predisposition to the schizophrenia spectrum of disorders.* The brain insult may possibly lead to slightly reduced IQ and to learning disabilities—for example, to some discrete neurological or neuro-psychological dysfunction, etc. It was found that "attention, verbal memory, and gross motor skill deficits were substantially more prevalent among children destined to develop schizophrenia" (idem, p. 1386). (To what extent these findings reflect the effect of excessively diffuse drives or alternatively an unspecific brain insult, I cannot judge.) Since the genetic predisposition to schizophrenia is hypothesized here to be of a quantitative, gradated nature, it can be presumed that a genetic predisposition of subclinical proportions may become intensified by additional brain pathologies to a degree that warrants a clinical diagnosis. Or, alternatively, a genetic predisposition that by itself would induce a milder schizophrenia spectrum disorder (schizoid personality, for example) will also manifest itself in psychotic symptoms as a result of additional burden on brain functioning, thus justifying the diagnosis of schizophrenia in research studies.

2. While such reasoning seems to me the most straightforward, I would like to propose another, more strongly hypothetical, scenario, one that may be more in tune with our theoretical construct. In this

presupposition, only some of these perinatal insults, acting at a specific phase of fetal brain maturation, when some of the brain circuits are more sensitive to harmful influences than other brain areas, contribute to the development of schizophrenia. This hypothesis presupposes a harmful environmental effect that may have repercussions *resembling those of the genetic predisposition* hypothesized to underlie the schizophrenia spectrum of disorders. The advancement of this possibility needs some preliminary discussion.

Taking as self-evident that the level of diffuseness of human instinctive predispositions is much more pronounced than those of even the most evolved social animals, it may be intuitively concluded that increased diffuseness of drives in humans represents an evolutionary progress. However, strictly from the point of view of the instinctive phenomenon alone, this is not the case. The primary prerequisite of the life phenomenon has to be the striving for reproduction—the basic incentive of the life processes, which depends upon survival till sexual maturity—but *in a crude, diffuse form, without any kind of adaptation to a concrete environment* (remember the example of the RNA virus, chapter 4, section 4.5). This basic quality of the life phenomenon, however, seems inaccessible to scientific investigation. Nevertheless, *only when these diffuse, undifferentiated "aspirations" of the living matter are present can natural selection build on these foundations the evolutionary adaptations (both somatic and behavioral) to the particular environmental niche the respective animal lives in.* Alternatively, natural selection needs these diffuse "intentions" of the living matter when environmental conditions change and the previous adaptations become useless or even harmful. They then have to be "dismantled" and new adaptations constructed, again according to the guiding influence of these diffuse life-sustaining and life-perpetuating "strivings." From this it follows that, *strictly from the point of view of the instinctive behavior, human phylogenetic evolution, from more differentiated instinctive predispositions toward more diffuse ones, represents a regression to an earlier, more immature state of brain maturation, instead of progression.* That is, the respective neural circuits *progressively lose* much of their innate potential to guide by themselves (that is, unassisted by learning) mature behavior toward fitness-enhancing avenues.

This does not mean, either, that the more diffuse stages of organismic strivings have disappeared completely in lower animals. They are only complemented with more differentiated behavioral strategies, fitness-enhancing in the environment to which they are adapted. To illustrate this reasoning, I refer to the quotation in section 2.2 from Niko Tinbergen on the peregrine falcon's hunting behavior.[20] To me, it seems clear that this example shows a quite diffuse, directionless, instinctive striving switching to a more differentiated, goal-oriented innate behavioral pattern when an appropriate releaser is perceived.

While humans do possess the first stage of the instinctive activity in this example—that is, when hungry they become aroused in a quite undifferentiated manner to seek food (undifferentiated from the instinctive, not cognitive, point of view)—they have lost the second stage in the present example: *more specialized hunting-gathering behavioral tactics.* As we have seen in the case of antisocial personality disorder, for example, in complex social circumstances drives that are too differentiated become a drawback, hampering social adjustment. Therefore, it can be presumed that, during human evolution, instinctive predispositions present in our ancestors that were too differentiated gradually atrophied to their present state of reduced differentiation (or increased diffuseness). Instead of specialized, inherited behavioral tactics, a vast amount of acquired knowledge was needed on how to secure a steady supply of food in a certain social/natural environment, how to prepare it, how to consume it (especially if it is done in the company of others), and so on. To my mind, for this adaptation to work properly, the critically important factor is the successful *synchronization* or *integration* of the vague general direction toward which the organismic need pushes ("obtain and consume food") with the relevant knowledge acquired during ontogeny of how to successfully materialize the instinctive striving's appetitive phase in a concrete social milieu. It seems clear that this is the evolutionary rationale of the considerably extended period of maturation in humans (as well as, to a lesser extent, other highly evolved social animals).

As this text has hypothesized, the ability of instinctive arousal to effectively guide the appetitive phase of active behavior in social circumstances is deficient in the disorders of the schizophrenia spectrum. In other words, the regressive evolutionary process from well-differentiated toward adequately diffuse instinctive guiding went too far, leading in this instance to a reduced ability or inability to integrate this instinctive guiding (more precisely, some of its aspects, such as the intentional, emotional, and energetic ones) with acquired knowledge. Most likely, individual reactive instinctive behavior also underwent a similar process in most individuals predisposed to the schizophrenia spectrum of disorders.

Scientific (monistic) thinking presumes we must look for the somatic events underlying this (pathological) mental and behavioral development in intrauterine and postnatal brain maturation. More specifically, by somatic events, I refer to neural cell migration; some aspects of selective cell death during brain maturation; the growth of and meetings between axons and dendrites; multiplication of synapses, as well as their differential destruction (pruning); the myelination of axons; and so on. It can be further presumed that these events are much more complex in human than they are in animal brains, whose behavioral strategies and releasers are largely genetically preprogrammed. This complexity of human brain maturation, however, comes at a price—namely, increased vulnerability to noxious environmental influences: "The adaptively decelerated brain maturation, particularly the late myelination of the prefrontal, temporal and parietal cortices, and the evolution of greater openness of brain circuits that allow for increased behavioral flexibility may come at the cost of greater vulnerability to harmful events, regardless whether emotional, infectious or toxic in nature" (Brüne, 2008, p. 98). And again: "Biological theorists . . . describe *critical periods*, which refer to the presence of increased activity in certain parts of the system within a short, particular period of time. Such periods suggest windows of opportunity for maximal growth in that area, as well as times of particular vulnerability" (Gordon, 2005, pp. 3021–3022; emphasis in the original).

Therefore, certain brain functions may become especially vulnerable to noxious influences at particular maturational stages. Relevant to the present discussion, for example, may be the finding that certain intrauterine brain insults in the second trimester of the pregnancy are more common in schizophrenics (Buchanan & Carpenter, 2005, p. 1333).

In addition to these physical and biological factors, it was found that stimuli of a psychosocial nature may also influence organic brain maturation: "In view of the possibility that experience and learning may influence the selective survival of certain synapses, it is conceivable that the social environment may also help shape brain maturation" (idem, p. 1390). This biological mechanism was already observed, as we have seen, in rather simple laboratory animals, who showed dramatic increases in synapses per neuron in the visual cortex and other areas of the brain after exposure to complex environmental stimuli. As a result of this observation, the existence of two different mechanisms guiding synapse formation was proposed—one predetermined by phylogenesis, the other dependent on experience during ontogenesis (Plotkin, 2007, pp. 195–196).

With this preliminary discussion in mind, we will return to our second hypothesis on how perinatal brain damage may possibly contribute to an increased rate of schizophrenia: a brain injury, acting at a specific phase of fetal brain maturation when some of the brain circuits are more sensitive to harmful influences than other brain areas—an injury, that may *simulate* this hypothetical genetic predisposition. In this respect, we have two theoretical possibilities, of which the second seems much more plausible.

- The first possibility is built on the presupposition that certain noxious stimuli (of a generalized, not localized, nature) at a certain stage of brain maturation halt or hinder the development of those brain structures underlying the maturation of instinctive predispositions from their primary state of absolute diffuseness (that is, lack of any adaptive differentiation whatsoever) toward the normative amount (that is, a level of differentiation more or less optimal in a complex social environment). This assumed scenario reproduces, in fact, exactly the same mental dysfunction hypothesized in this text to underlie the genetic predisposition to the schizophrenia spectrum of disorders. However, this noxious agent has to lead to this disturbance without considerably affecting the maturation of other central mental functions, such as abstract (but not social) intelligence, learning ability, and language acquisition.

It is known that certain genetic disorders do possess such selectivity. For example, in *Asperger syndrome* a profound inability to express emotions and form emotional ties with others (a domain pertaining to the instinctive activity) coexists with normative language acquisition and learning abilities, although the way intellectual curiosity manifests itself is highly peculiar. These individuals can learn "extraordinary amounts of factual information . . . about very circumscribed topics" but "without a genuine understanding of the broader phenomena involved" (Volkmar, Klin, & Schultz, 2005, p. 3179). In *Williams Syndrome*, a rare genetic disorder, mental retardation (IQ around 50) coexists with relatively well-developed language abilities and a friendly, social disposition (Sadock & Sadock, 2005, p. 3094; Campbell, 2004, p. 694).

It seems to me improbable that environmental insults damaging the maturing brain in an overall manner may lead to a similar kind of selective damage regarding central brain functions. The only theoretical possibility I can imagine relates to the *timing* of this damage. That is, the brain function in question has to be at a stage of rapid maturation relative to other brain functions, and, in consequence, becomes differentially vulnerable to nonspecific damage to its underlying somatic processes. A finding (mentioned above) that may hint at such a possibility is that exposure to influenza during the second trimester of pregnancy (a period of active neural cell migration) may lead to an increased risk of schizophrenia in the offspring (Kalverboer & Gramsbergen, 2001, p. 73; Buchanan & Carpenter, 2005, p. 1333).

Our second presupposition regarding the possibility that a nonspecific brain insult may imitate the inherited predisposition to the schizophrenia spectrum of disorders seems more plausible than the first. It does not reproduce the predisposition itself (like in the first alternative) but imitates its *main direct consequence*—namely, a difficulty or inability with regard to synchronizing or integrating innate instinctive predispositions with ontogenetically acquired knowledge of the environment (and possibly of oneself). This mental ability, which is indispensable for normative mental and behavioral maturation, is not an exclusively human characteristic, of course, but is encountered in other animals, too, the extent of which being by and large proportional with the place of the respective species on the evolutionary ladder. The more evolved an animal is, the more it relies on this mechanism. In humans, however, this evolutionary trend attained extreme proportions. As argued repeatedly, in humans beyond the early stages of infancy, no adaptive behavioral manifestation is entirely innate. Instead, more or less diffuse innate "intentions" have to be integrated by the mental apparatus with vast amounts of acquired knowledge on almost every aspect of the social and natural environment.

From a neuro-developmental point of view, it follows from this reasoning that the human brain must be subjected to social influences at a very immature state and must develop innumerable neural connections between brain regions underlying instinctive strivings (the limbic system and structures at a lower level) and those dealing with cognitive functions (that is, the extended neocortical areas). At the same time, it must destroy (prune) dysfunctional or unfunctional connections. It can be hypothesized further that, because this is a very intense neuro-developmental process, beginning shortly after birth and lasting (at decreasing intensity) for the whole life span, different kinds of nonspecific brain insults can affect it deleteriously (as stated in the quotation from M. Brüne earlier in this section). I refer here to such pathologic influences as the postnatal repercussions of intrauterine and perinatal brain insults (found excessively in the anamnesis of schizophrenics), or early postnatal brain damage of a generalized kind. While these pathologic influences hinder the integration of instinctive and acquired aspects of behavior, the two participants in this process—the normatively diffuse (or normatively differentiated) drives and the learning abilities—are assumed to remain relatively intact, thus simulating the characteristic dissociation of mental contents of the schizophrenia spectrum (as understood in this work). Consequently, the damage induced by the ensuing deficiencies in behavior and mental life may remain relatively hidden (that is, they do not lead to serious psychopathology) for many years, due to the preservation of the formal learning abilities, as well as the protected conditions in which human children are as a rule kept.

7.2.8.9.2.1.2. The Neuro-Degenerative Hypothesis concerning Organic Brain Pathology in Schizophrenia

The second hypothesis on the nature of possible brain pathology in schizophrenia, the neuro-degenerative one, has much less appeal than the neuro-developmental hypothesis. "There is no evidence of the classic signs of neuro-degeneration in the brains of people with schizophrenia, and the general absence of gliosis in schizophrenia strongly argues against an adult-onset degenerative process" (Murray & Bramon, 2005, p. 1389).

However, a different pathology, that is, *brain atrophy* as a direct consequence of disuse caused by the disease process itself, which reduces strongly the amount of social stimulation, especially that arising from complex interpersonal interactions, seems to me much more plausible.

Therefore, it seems justified to hypothesize that the increase of ventricular size and decrease in gray matter volume found frequently in schizophrenics, especially in those with "poorer premorbid functioning, more severe symptoms, and worse outcome" (Gur & Gur, 2005, p. 1398), may be the secondary consequence of the paucity of interactions with the social environment.

7.2.8.9.2.2. Psychosocial Factors Influencing Schizophrenia's Prevalence, Gravity, or Course

It is a well-known epidemiologic finding that certain psychosocial factors may lead to an increased risk of developing schizophrenia, while others may influence its gravity or course. Here is a list of the commonly mentioned ones:

1. *Urban environments.* Persons brought up and living in a city have a greater risk of developing schizophrenia than those living in rural areas. It has to be mentioned that the cause-effect relationship in this finding (that is whether living in a city is a consequence or cause of schizophrenia) is a contentious one (Brown, Bresnahan, & Susser, 2005, p. 1373).

2. A phenomenon called *neighborhood effect* (related to the urban/rural difference in the incidence of schizophrenia). This refers to the finding that living in central areas of big cities, in conditions of social disorganization, low socioeconomic status, or social isolation, may lead to elevated rates of schizophrenia (idem, p. 1373). Again, the cause-effect relationship is contentious (idem, p. 1377).

3. *Immigration.* It is known for a long time that immigrant populations have higher rates of schizophrenia. Sometimes the difference in this respect between the immigrant and indigenous populations may be considerable, with an incidence of up to seven times higher in immigrants (Murray & Bramon, 2005, p. 1392). The risk was found to be higher in the *second generation* of immigrants already born or brought up in the host country.

4. *Poor relationship or reduced amounts of contact between mother and child.* A British epidemiologic study found a sixfold increase in the risk of developing schizophrenia in children rated at the age of four as experiencing a poor mother-child relationship. It is unknown to what extent the mother's poor parenting abilities or the pre-schizophrenic

child's inability to form emotional bonds, or both, was the decisive factor in this finding (idem, p. 1391).

5. Children with a known genetic risk for schizophrenia, when reared in a traditional Israeli kibbutz environment (in which the amount of contact between children and parents is reduced considerably in comparison with the rest of the population), were more likely to develop schizophrenia than the control group (idem, p. 1391). Since we are dealing here with children with a known genetic risk, this finding has to be interpreted as an aggravating factor in the presence of an innate predisposition.

6. *The effect of "expressed emotion" in the families of schizophrenic patients.* It was found (and repeatedly replicated in subsequent research) that intense emotional expression in schizophrenics' families, particularly of a negative kind, increases the relapse rate of remitted schizophrenics discharged from the hospital into their family environment. In particular, unrestrained critical comments concerning the patient's behavior, hostility, and emotional overinvolvement were found to negatively influence the quality and length of the remissions. However, it was found that the deleterious effect of "expressed emotion" is not restricted to schizophrenia. The same effect was demonstrated in many other conditions, such as depression, bipolar disorder, obesity, anorexia, mental handicap, senile dementia, Parkinson's disease, etc. These findings indicate that excessive amounts of "expressed emotion" by family members or other caretakers constitute a *nonspecific* environmental aggravating factor in the case of individuals suffering from various mentally or physically debilitating conditions (Bebbington & Kuipers, 1993, pp. 85–93).

7. Last but not least, we have to mention here a particular kind of parent-child relationship that led to *the concept of the schizophrenogenic mother or schizophrenogenic family,* which in turn gave birth in the late 1940s to an influential psychodynamic trend according etiological status to the behavior of the mother in the causation of schizophrenia. This concept was discarded in the 1970s, and along with it some useful observational data about the early social environment's role in shaping the clinical picture, and the course and gravity of the disorder. Maybe not entirely discarded, however. In an attenuated form (that is, as an aggravating, not etiological factor) it survived in the works of such psychodynamically oriented psychiatrists as

Silvano Arieti and R. D. Laing; and it reappears in the concept of "expressed emotion" in schizophrenics' families, this time well substantiated by clinical research. We look more closely at this topic next.

It seems clear that none of the features of the social milieu already mentioned play an *etiological role* in generating schizophrenia. These stressful circumstances are quite widespread, and the overwhelming majority of people succeed in adapting to them without developing schizophrenia (or frequently any other mental disorder of clinical proportions). However, if we take into account the basic genetic predisposition to the schizophrenia spectrum of disorders as proposed in this theory—namely, diffuse instinctive predispositions unable to guide the instinctively based strivings effectively in a social setting—we may find that predisposed individuals are especially sensitive to these forms of environmental stress. In addition, one has to take in account that the overwhelming majority of the research on deleterious psychosocial circumstances measured the incidence of the classic (psychotic) schizophrenic disorder and not the whole comprehensive schizophrenia spectrum. It follows that *both those environmental factors that magnify the effect of the proposed innate predisposition for schizophrenia and its spectrum disorders, as well as those environmental factors that facilitate the appearance of a psychosis of any kind (social isolation, for example), may increase the incidence of schizophrenic psychosis in predisposed individuals.*

In the following, I discuss these psychosocial aggravating factors according to what seems to be the primary way in which they are responsible for the transformation of a subclinical genetic predisposition, a milder spectrum disorder (schizoid personality, for example), or a prodromal state into overt schizophrenic psychosis; or the transformation of a milder form of schizophrenic psychosis into a graver one with a more deteriorating course. Some psychosocial stressors may involve more than one of these effects.

7.2.8.9.2.2.1. Social Isolation

Social isolation in schizophrenia may be a more complex issue than it seems at first glance. It can be hypothesized that in schizophrenics or pre-schizophrenics too much exposure to social stimuli may be as deleterious as considerable sensory/social deprivation. According to our theoretical framework, this is related to these individuals' reduced ability to clearly sense their own internal motivations; to sense the relevant internal motivations of people around them; to process selectively incoming stimuli; and to build, accordingly, fitness-enhancing behavioral strategies. R. D. Laing, for example, describes a defense mechanism of one of his schizoid patients (schizoid personality disorder in clinical parlance) that may be relevant in this context. This individual's strategy was to remain anonymous and continue a precarious existence without entirely breaking off social contacts: "He could, he felt, be himself with others if they knew nothing about him. . . . He would go from place to place, never staying long enough to be known" (Laing, 1960, p. 139). It seems that by his own initiative this person reduced the complexity of the social stimuli and the closeness of interpersonal relations to a level he felt he could cope with. For individuals who aren't prepared to physically move from place to place, this kind of defense against more intense or intrusive interpersonal interactions can work much better in a big city than in small rural or suburban communities.

On the other hand, we can presuppose that extreme intensities of sensory/social deprivation in a schizoid or pre-schizophrenic person, or for that matter in anyone else, can lead to psychotic symptoms or may aggravate existing ones. Whatever the amount of social exposure, the *nature* of interpersonal relatedness and transactions have paramount importance in the presence of an inherited predisposition to the schizophrenia spectrum of disorders—a topic further discussed later in this section.

7.2.8.9.2.2.2. Radical and Brusque Changes in the Social and Cultural Environment: Immigration; Military Service; Hospitalization and Discharge When Remitted; a Journey Abroad (Alone or with an Organized Group of Previously Unknown People), etc.

It seems that individuals predisposed to the schizophrenia spectrum of disorders have more difficulty adapting to a fundamentally changed social environment than those with a normative range of instinctive configuration. The requirement to get acquainted with a new environment, many of whose characteristics are too subtle to be learned by clearly formulated verbal instructions and explanations; the (temporary) disruption of formerly well-established relationships and the close contact with hitherto unknown persons; changes in the rules of conduct and the nature of the relationships; the requirement to adopt new adaptive strategies, tasks that may also represent considerable stress to normative persons, may pose insurmountable difficulties for a schizophrenia-predisposed individual's reduced coping and adaptive abilities.

In the case of immigrants, the older generation, as a rule, retains its family framework, native language, and ethnic customs and may establish more-or-less demarcated ethnic communities, which may buffer to a considerable extent the effects of a radically changed socio-cultural milieu. It can be presumed, on the other hand, that the *younger generation* is more exposed to the culture of the host country because it must better master its language, receive an education in its institutions, and come in close contact with the host country's youth. However, the more dependent members of the second generation may become torn between adapting to the new conditions, on one hand, and retaining the native country's culture, adhered to by the parents, on the other. This internal conflict may be especially problematic in the presence of a predisposition to the schizophrenia spectrum of disorders. Moreover, the younger generation seems to be more exposed to (and involved in) possible conflicts between the immigrants and the host nation, conflicts of racial or religious nature, competition for employment, and so on. These considerations, in my opinion, may explain, at least partly, the increased risk of schizophrenic psychosis in the second generation of immigrants.

Enrollment in military service may represent an even greater and probably brusquer environmental change than immigration. Schizoid or pre-schizophrenic youth who succeed in establishing a precarious mental equilibrium in the sheltered conditions of the parental home—that is, an equilibrium between conforming outwardly to social requirements while discharging active motivations in solitary activities—have no chance to do something similar within the framework of military life. The possibility of solitary retreat from time to time from the intense interactions in a highly knit and organized unit in order to express dissociated active mental contents is simply nonexistent. During my professional career, I encountered from time to time cases in which the first psychotic episode of a schizophrenic or schizophreniform illness appeared during military service, especially in the first months after enrollment.

The phenomenon of "expressed emotion" in the families of schizophrenics after the latter's discharge from hospitalization also creates a brusque change in the nature of their immediate environment.

7.2.8.9.2.2.3. Socioeconomic Status

It can be hypothesized that a higher socioeconomic status may have a protecting or shielding effect for a schizophrenia-predisposed individual. Living in a spacious private home set well apart from the neighbors (in contrast to the crowded conditions of a slum, for example); having the privacy of a separate room; being free from economic pressures to begin working at an early age; having the privilege to choose from a wider assortment of social niches (during education, when choosing a profession or living area, and so on) are all advantages that may ensure a better fit between an individual's problematic mental constitution and his living conditions. In other words, they can protect a schizophrenia-predisposed individual from many stressful, undesirable aspects of social coexistence.

Immigrants frequently have a low socioeconomic status, adding to the environmental stresses described in section 7.2.8.9.2.2.2.

7.2.8.9.2.2.4. Specific Constellations of the Parent-Child Relationship in Schizophrenia

Between the late 1940s and early 1970s, probably influenced by a specific cultural climate in the USA, it was hypothesized that the mother's way of relating to the child, and later that of the whole family, has an *etiological* role in generating schizophrenia. Thus, the concept of "schizophrenogenic mother" or "schizophrenogenic family" emerged (Neill, 1990, pp. 499–505). The concept became strongly charged both ideologically and emotionally before it was discarded. However, as mentioned, less extreme interpretations of similar observational phenomena (according it aggravating, not etiological, importance) survived in psychodynamic literature and in the concept of "expressed emotion."

My intention here is to examine these observations from the present theory's perspective. Frequently the behavior of one of the parents toward the pre-schizophrenic child, usually the mother, is described as overprotective, domineering, intrusive. We have seen an illustration of this kind of relationship in connection with the "model child" phenomenon (section 7.2.8.2.1). The mother's intrusive behavior, besides her own temperamental constitution, was encouraged by the child's inability to express her own instinctively underpinned intentions. It can be presumed that *the resulting overprotectiveness, and especially the intrusive, domineering guiding of the child's behavior, will deepen the dissociation*

between the environmentally induced conforming behavior, on one hand,
and the active instinctive expression outside the social sphere, on the other.

We should make a short digression here to discuss problems with use of the empathizing mechanism in a normative person's relatedness to a schizophrenia-predisposed one. Even when the mother is endowed with good empathizing capabilities in relating to the behavior of normative persons, this ability will be of little help with an individual whose instinctive drives are too diffuse, in this case her own child. Excessive diffuseness of instinctive drives, in contrast with other comprehensive instinctive mechanisms discussed in this text (fluctuations in overall intensity, etc.), cannot be sensed directly through empathizing. Excessively diffuse drives can only be *inferred* from their secondary, tertiary, etc., repercussions: dissociation between active-instinct-fueled and socially induced behavior; withdrawal, overinclusiveness, confusion; obscure metaphors provided by the patient (remember R. Laing's patient who described strong diffuse instinctive tensions as the pressure of hot air in a container, section 7.2.8.2.1); overgeneralized statements or delusions, etc. Therefore, while empathy can reconstruct the observed person's emotional states and intentions in the other comprehensive instinctive mechanisms and, in consequence, enables one to grasp pathological mental states in terms of quantitative differences between aspects of normative and disturbed mental functioning, empathizing alone in the case of a schizophrenia-predisposed individual is not sufficient for detecting considerable quantitative differences in the extent of diffuseness (or differentiation) of drives. In these conditions, it seems to me very difficult for a mother to adopt an optimal relatedness to the child, a relatedness that can minimize the deleterious effects on maturation of these excessively diffuse drives.

While one parent, usually the mother, is overinvolved emotionally with the schizophrenia-predisposed child, the other parent, usually the father, may be found to be remote, unavailable, emotionally distant. Both Arieti and Laing provide impressive illustrations of this family constellation (Arieti, 1974, pp. 623–624; Laing, 1960, pp. 208–210).

If overinvolvement, that is, intrusive guiding, domineering, and criticizing, is undesirable in relating to a schizophrenia-predisposed child, then its opposite, lack of involvement (which ensures more freedom of choice), would seem to be more desirable. However, this is not the case. Lack of involvement is deleterious, too, for the same reasons overinvolvement and overprotection is deleterious. The more than optimally diffuse drives by themselves are incapable of guiding the behavior toward fitness-enhancing avenues. In fact, it can be presumed that, in these circumstances, the individual needs *more* external guidance, not less, than a person with normatively differentiated drives. In this respect, it seems that the predisposition to schizophrenia is a no-win situation. Silvano Arieti formulated this dilemma in the following way: The schizophrenic's "ability to will, to make a choice will always remain impaired. They always experience indecision and ambivalent attitudes" (perhaps except in some kind of acute psychotic periods) (Arieti, 1974, p. 159). And in consequence: "schizophrenics feel 'pushed around' throughout their lives. The push will more probably be accepted and not resented if it is soft, if it is accompanied by tenderness and by a hopeful glimmer of success. The patient must be encouraged to take his own initiative whenever possible" (idem, p. 602).

It is likely that no easily standardizable, satisfactory solutions exist with regard to a therapeutically desirable manner of relating to a person with a disorder on the schizophrenia spectrum. This topic will be discussed in the final chapter.

Summary and Implications

8.1. Summary of the Theory and Suggestions on Its Preliminary Testing

The present work is built on the well-recognized and documented postulate that mental disorders, like any other aspect of human (as well as most animal) behavior, represent the intertwining of inherited predispositions with ontogenetic influences of the natural/social milieu. However, I have tried to build a more detailed picture of the nature of the innate mechanisms underlying human behavior, and especially those involved in common mental disorders; how their dysfunction-inducing potential evolved during human phylogeny, and (to the extent I have been able to do so) how they influence psychosocial maturation. This picture is built on several basic (albeit contentious) ideas, well recognized in behavioral sciences:

1. Throughout the evolutionary trajectory from simpler to more complex animals, the rigidly predetermined innate behavioral programs progressively became more malleable inclinations, predispositions modifiable to an ever-increasing extent by experience and learning during ontogeny. This trend attained extreme proportions in humans. However, it is very important to stress that, in spite of their extreme malleability, the instinctive predispositions retained their capacity to channel normative behavior (in an approximate way)

into biologically and socially adaptive, as well as fitness-enhancing, avenues.

2. With the increasing complexity and effectiveness of human social coexistence, a shift occurred in the predominance of different kinds of selective pressures. While *natural* selective pressures, as a result of increasing ecological dominance, progressively lost their strength to shape human behavior, the selective pressures that ensured better adjustment to a complex *social* way of life progressively intensified.

3. With further social and technological advances, and with the appearance of more tolerant, humanitarian political systems, the intensity of the selective pressures for improved social functioning also began to decrease, leading to an accentuated diversification of the inherited aspects of human behavior related to adaptation to both the natural and social environment. This diversification in most cases is contained and even exploited by the emergence of increasingly diversified "social niches," as well as increased social mobility, which allowed for a better match between inherited mental and physical traits and the optimal available social niche. Close cooperation between persons and groups of persons with different mental (as well as physical) assets and drawbacks secured cultural, economic, technological, etc., achievements that would be unimaginable without this mechanism.

4. However, this evolutionary trajectory also has a dark side. The accentuated diversification of the instinctive predispositions at the extreme ends of their diversified quantitative continuums have lost their capacity to guide behavior into biologically or socially adaptive avenues, even amid optimal social conditions, leading to progressively more dysfunctional patterns of behavior. These dysfunctional behavioral patterns in turn have been categorized in psychiatric nomenclature as discrete mental disorders.

5. Besides stressing relaxation of selection pressures and consequent diversification of the inherited predispositions to behave, the present theory differs from the prevailing evolutionary hypotheses in interpreting mental disorders in another important respect. Instead of stressing possible dysfunctional phylogenetic developments with regard to discrete, circumscribed instinctive entities, or the mismatch between discrete phylogenetic adaptations to the conditions of our prehistoric environment and the requirements of modern social coexistence, it proposes as the target of investigation comprehensive

instinctive mechanisms that modulate or alter a wide range of discrete instinctive predispositions. In particular, the present study proposes four such comprehensive instinctive mechanisms predisposing to maladaptive behavior at the extremes of their diversified continuum and attempts to reconstruct the relationship between them and the descriptive clinical disease categories to which they predispose. In some of them, this relationship is quite straightforward (as between seasonal variations in instinctive intensities and seasonal affective disorders), while in others (especially those involving the extreme diffuseness of instinctive predispositions and schizophrenic psychosis) a complex chain of events leading from the innate predisposition to the clinically detectable mental disorder is proposed.

6. As a more general proposition, it seems quite clear that the diversification of these comprehensive instinctive mechanisms at the extreme ends of their diversified continuums affect behavior more severely and deleteriously than the diversification of one *discrete* instinctive predisposition (feeding, sexual, parenting drive, etc.).

While this scenario may seem logically persuasive, without being subjected to rigorous, impartial testing, as stressed by the philosopher of science Karl Popper, a scientific theory has no practical value whatever. Logically tenable interrelations between different phenomena, or between certain phenomena and relevant theoretical formulations, may frequently turn out to be mistaken. Alternatively, it may be found that the same phenomenon can be explained equally well by different hypotheses (Dear, 2006, p. 185). Therefore, to materialize its inherent explanatory potential, or its ability to guide scientific research toward new, promising avenues, a theory has to succeed in arousing the interest of professionals who are able and willing to properly evaluate its merits and shortcomings.

The *preliminary steps* concerning the evaluation of a theory are: the assessment of its logical coherence or "internal consistency"; whether it is consistent with the widely accepted facts (data) in the respective scientific field; whether it explains these facts better than already existing theories on the same topic; and whether it is compatible with other widely accepted relevant theories. These considerations were briefly discussed in section 1.1. Another desirable quality of a theory is its ability to offer explanations for established data that contradict expectations according to the current state of understanding of the respective phenomenon—in our case, for example, the strange pattern of inheritance in schizophrenia (sections 7.2.8.9.0 and 7.2.8.9.1). It must also be established whether a theory is a truly empirical one, not intended for a formal science (mathematics or logic), and not a metaphysical theory. If it turns out that it is indeed intended to be an empirical theory, it must contain implications that can be tested by the methods of empirical science. Other desirable qualities of a theory are its ability to *bridge* between two separate scientific fields (in our case the evolution of the animal world and psychiatry), as well as its ability to predict the existence of hitherto unknown facts or unknown relationships between known facts.

It is important to stress again that theories, since they contain abstract concepts, cannot be tested directly; they can be tested only through whichever of their implications lend themselves to scientific experimentation. Hoping that the present theory will successfully pass these preliminary tests, I would like to propose those implications I am aware of that satisfy this criterion. I will discuss separately the four hypothesized comprehensive instinctive mechanisms in order of their complexity and according to the ease of their approachability by the scientific method. I would like to mention in advance that, since these instinctive mechanisms evolved in the animal kingdom to fulfill specific adaptive tasks, it is reasonable to assume that their somatic substrate in brain structure and functioning must be relatively discrete and more easily identifiable as a cohesive system, a unitary whole, than the expression of *whole descriptive mental disease categories*, entities that are the outcome of complex, longstanding interactions between innate predispositions and variable natural/social influences during ontogeny.

8.2. Possible Research and Treatment Implications According to the Four Comprehensive Instinctive Mechanisms

8.2.1. Fluctuations in the Overall Intensity of the Instinctive Activity

It seems to me that among the four comprehensive instinctive mechanisms discussed in this text the one for which the best chance exists of elucidating its underlying brain structure and neurophysiology is that which deals with the fluctuations of the overall instinctive intensities that evolved under the selection pressures of seasonal changes in the abundance of environmental resources. As mentioned in section 5.1.2.2, the somatic mechanisms underlying diurnal rhythms in behavior (the "circadian clock"), which seem to be related to seasonal variations triggered mainly by the changes in the daily hours of light, have been clarified in many of its details by scientific research (Foster & Kreitzman, 2009, pp. 35–37).

Whether relationships can be proved between the long-lasting episodes of high and low instinctive intensities in animals living at higher latitudes of the globe, on one hand, and on the other some common psychiatric disorders (winter depression, bipolar disorder, cyclothymic personality) and the effects of certain psychotropic drugs (antipsychotics, antidepressants) can be explored by searching for resemblances and correlations with regard to their physiological, physiopathological, and neuropathological foundations—more concretely, in this respect, for similarities or correlations between:

- Hibernation and its milder forms in animals.
- Winter depression and related nonseasonal depressive disorders.
- The main neurochemical effect of antipsychotic drugs (hypothesized to primarily reduce the overall instinctive intensity), as well as of the noradrenergic antidepressants (hypothesized to primarily increase the overall instinctive intensity).
- The poikilothermic effect of some antipsychotic drugs (that is, inducing hypothermia in winter and increased body temperature in summer through inhibiting the brain mechanism of thermoregulation).

Additional subjects of research may include the following:

- The possible similarity between the relevant changes in brain physiology during hibernation and antipsychotics-induced catalepsy in laboratory animals.
- The possible parallels between changes in human brain functioning in bipolar disorders (in particular, during the transition from depression to mania) and changes detectable in

hibernating animals when winter inactivity turns into increased activity with the arrival of spring. Whether antidepressant therapy, when effective in lethargic depression, induces similar changes in brain chemistry to that observable in animals awakening from winter inactivity; whether antidepressant administration (especially the noradrenergic type) for a more prolonged period can interrupt hibernation in animals, or render it lighter.

- Whether slight changes in the metabolic rates in seasonal bipolar disorder exist and, if so, whether they parallel well-known changes in this respect in hibernating animals during seasonal fluctuations; how seasonal affective disorders compare with nonseasonal ones in this respect.

The outcome of these proposed research studies may substantiate or contradict the hypothesis concerning a causal relationship between the evolutionary mechanisms underlying seasonal behavioral and physiological changes in hibernating animals on one hand, and the clinical picture of bipolar and related disorders on the other, as well as the proposal to separate sharply the lethargic form of depression from that of the agitated dysphoric one, since they are induced by different underlying evolutionary mechanisms.

8.2.2. Forms of Frustration of Active Instincts, Displacement and Vacuum Behavior, Aggression, and Dysphoria

To effectively treat the frustrated forms of active instinctive behavior (and related subjective experience) when they are excessively intense or long-lasting, particularly the aggressive and dysphoric ones, has long been a therapeutic aspiration in psychiatry. However, the presently available biological treatments (mostly drugs) are nonspecific agents that reduce or increase the energetic charge of behavior conferred by the instinctive activity in an overall, unselective manner. I refer here to the antipsychotic and antidepressant drugs. In spite of advertisements to the contrary by drug companies, these drugs cannot directly and differentially address the frustrated forms of active drives and cannot transform one form of frustration into another (for example, a dysphoric depressive mood into euphoria), or one of the frustrated forms into the unfrustrated one (for example, anger or dysphoria into euthymia).

However, it seems that an increased awareness exists of the need for biological treatments that *differentially* target the pathologically intense, frustrated instinctive activity, while the rest of the mental functioning (its inherited aspects) are left unaffected. We must remind ourselves that the respective form of active instinctive frustration is apparent as a rule within one single active instinctive striving at a time (sexual, parental, etc.). If, however, the frustrated drive is excessively strong, it may inhibit both unrelated instinctively based strivings and cognitive functions (think of a person in intense grief).

With regard to the aggressive form of frustration, researchers have attempted to develop drugs that target the aggressive response differentially. Jaak Panksepp describes such an example (2005, pp. 189–190; pp. 201–202). Panksepp defines the aggression as a result of active instinctive frustration in a way that is in keeping with the one proposed in this text: "The aim of anger is to increase the probability of success in the pursuit of one's ongoing desires and competition for resources" (idem, pp. 189–190). In the following, he remarks that, while antipsychotic drugs in high doses are effective antiaggressive agents, their effect consists of nonspecific "sedation" (idem, pp. 201–202). Subsequently, he mentions an experimental drug (Eltoprazine), a serotonin receptor agonist that is related to the group of drugs called "serenics." This drug "[n]ot only does . . . decrease aggression without any sedation, but it actually increases other friendly and social exploratory behaviors [in rats]" (idem, p. 202). This effect may be interpreted in our theoretical framework as the restitution of unfrustrated active behavior after the differential suppression of the aggressive transformation.

This example may suggest that, since the aggressive form of active instinctive frustration is a deep-rooted evolutionary mechanism also present in animals with less evolved central nervous systems, its brain substrate is relatively simple and circumscribed, possibly located in the ancient "reptilian brain," according to Paul MacLean's "Triune Brain" model (idem, pp. 70–72). Accordingly, it seems likely that biological research will clarify its neurophysiological details, and the development of effective, specific biological treatment modalities seems within reach.[1]

8.2.2.1. Afterthought

I would like to mention here a more far-reaching intervention concerning the frustration of active drives, despite the fact that it transcends
the scope of conventional psychiatric practice. This intervention involves the possibility of artificially modulating the ease with which an
active instinctive striving transforms into one of its forms of frustration
(that is, by decreasing or, more importantly, increasing the threshold at
which this striving transforms into one of its forms of frustration). It
is evident that a great diversity, a wide scale of continuous gradations,
exists in this respect in human populations. At one end of this spectrum
are individuals who are capable of persevering in their instinctively,
emotionally founded aims or projects in spite of grave adversities.
Disturbing circumstances, recurrent failures, and dangers of different
kinds cannot divert them from their chosen route (for example, great
political, military, or religious leaders, renowned scientists, etc.). At
the other end are those persons who may be diverted easily toward a
less difficult displaced or "vacuum" route in such circumstances, or get
easily entangled in petty disputes or quarrels that are irrelevant to their
basic, emotionally fueled strivings or projects, thus needlessly endan-
gering their successful accomplishment. Or, alternatively, they may
respond to relatively minor difficulties with discouragement, disappointment, or despair, thus renouncing prematurely some cherished,
emotionally fueled goal. While one extreme of this diversified scale
may represent or be part of diagnosable mental disorders (dysphoric
personality traits, for example, or impulse-control disorder), most of
this diversified scale represents variations inside the normative range.
Theoretically, artificial intervention that, for example, could raise the
threshold at which an active instinctive striving switches from its original intent into one of its forms of frustration (through drugs or some
other kind of biological therapy) could increase considerably the efficiency and perseverance of the behavior (with highly desirable or undesirable consequences, depending on how the respective individual's
instinctively underpinned projects relate to socially cherished values).
However, at this point, we transgress psychiatry's legitimate boundaries, and enter into an ideologically charged, problematic domain,
which is not my intention in this text.

8.2.3. The Differentiated-Diffuse Spectrum of Human Instinctive Predispositions

As detailed in section 7.2.8.9.2.1.1, a diffuse, directionless "striving" to reproduce (which presupposes survival till reproductive maturity) has to be the primary "aspiration" of living matter. Only when this shapeless aspiration is present can natural selection build upon it differentiated behavioral mechanisms that ensure its materialization in a given environment at the level of the behavioral sophistication of the respective organism.

The biological sciences are much less successful at explaining the true nature of these diffuse strivings than they are at deciphering the actual mechanisms through which they materialize in a certain environment. In more general terms, it has been argued that science is much more successful at answering "how" and "why" questions (that is, how it works, how it manifests itself, how and why different aspects of a phenomenon are interrelated) than "what" questions (that is, what it is): "In our description of nature the purpose is not to disclose the *real essence* of the phenomena but only to track down, so far as it is possible, *relations* between the manifold aspects of our experience" (Niels Bohr, quoted in van den Hooff & van den Hooff, 1995, pp. 82–83, emphasis added). Himmelfarb makes the same point with regard to evolution: "Provoked by the criticism of friends and enemies, Darwin fell back upon historical precedent: If Newton was not obliged to show *what* gravity is, apart from *how* it manifests itself, so he was not called upon to go *behind the mechanism* of natural selection" (Himmelfarb, 1962, p. 322, emphasis added). In the same way, I suggest that the somatic base of this "primal striving" of the life phenomenon, as well as its interconnections with the mechanisms that materialize it in a certain organism living in particular environmental conditions, are outside the reach of scientific inquiry, at least for the present.

It has been argued throughout this work that the instinctive mechanisms that materialize these primeval strivings at a behavioral level in lower animals consist mostly of well-differentiated, rigid behavioral programs directly observable in overt behavior. *As we ascend on the evolutionary ladder toward more evolved animals, and especially highly evolved social ones, these rigid instinctive behavioral programs undergo retrograde evolution—that is, the expression of these strivings at a purely instinctive level moves progressively closer to their diffuse, primordial state, which makes their scientific exploration currently very problematic.* These vague, normatively diffuse instinctive strivings (normative for the respective species) need to be integrated with a vast amount of acquired knowledge and personal experience during ontogeny in order to become adaptive and fitness-enhancing behavior amid complex, quickly changing social circumstances. In humans, in addition, a wide scale of *diversification* appeared in this instinctive trait as a result of relaxation from natural and intragroup selection pressures, leading to the differentiated-diffuse spectrum. While these innate predispositions at the middle section of their gradated scale retain the ability to guide behavior in an approximate manner toward adaptive and fitness-enhancing avenues, the excessively differentiated end of this scale interferes with social adaptation, while the excessively diffuse one becomes unable to channel learned behavior into the adaptive and fitness-enhancing avenues.

Because we are unable to imagine how this differentiated-diffuse scale may be represented in known brain structures or functions, we have to be content, at least for the present, with more modest therapeutic and research aspirations in this domain. I have in mind the conceiving of certain "therapeutic adjustments" at an early stage of maturation, which may counteract to a certain extent the deleterious consequences of considerably diffuse or differentiated drives. To make this project possible, however, we have to find first a method that can reliably identify this aspect of the instinctive configuration (first of all, in individuals genetically related to known patients) as early as possible. The only practicable way I can imagine to achieve this aim is to try to identify the *most direct early behavioral manifestations* induced by these too diffuse or differentiated drives, and at the same time to separate these manifestations as much as possible from secondary or tertiary consequences (which may become progressively more unreliable for the purpose), and to build rating scales or psychological tests accordingly. The rationale for identifying these potentially detrimental predispositions as early as possible is to try to prevent their integration with ontogenetically acquired elements, leading to ingrained maladaptive behavior patterns.

The nature of the specifically constructed therapeutic social environment I propose, in case the too diffuse or differentiated instinctive predispositions can be identified early, can be imagined only in very tentative and general terms. Nor can I judge to what extent they are practicable and effective. However, the theoretical framework offered here may add an additional, hopefully useful, perspective to the treatment and research efforts that have been used up to this point (efforts, in my opinion, that are based largely on a quite random search for empirical regularities).

The therapeutic environment proposed here would have to contain specific stimuli that address the basic problems the respective instinctive configuration poses to social coexistence. For example, in the case of drives that are too differentiated, one central aim has to be to interfere with, to frustrate, their direct gratification when they bypass or disregard social prohibitions (in a more intense form than in normative children of the same age). I would add that another aim should be to provide explanations, to the extent the cognitive maturational state of the child permits it, of the rationale for these prohibitions, as well as of the more distant gains the restriction or recanalization of instinctive pursuits might ensure. I believe these (educational) measures are already practiced in fact, in a way less substantiated theoretically, by some parents and institutions that have to deal with children with sociopathic inclinations.

Similarly, the therapeutic milieu in children with excessively diffuse drives has to compensate as much as possible for the harmful effects of their problematic hereditary endowment. The respective individual's ability to compensate to a certain extent by cognitive means for what she lacks in her genetic endowment may be explored. In other words, aspects of interpersonal relations that in normative persons' encounters remain as a rule *implicit* since the participants are similar enough in their instinctive constitutions to understand each other well through the mechanism of empathy have to be *explicitly* explained to them.

It may be worthwhile to explain to the more psychologically minded patients the essential difference between their own and the normative persons' instinctive constitution, and the deleterious influence of excessively diffuse drives on social coexistence. These explanations may help the milder schizophrenia spectrum patients become more sensitive to and more successful in decoding the covert emotional and intentional messages of normative persons.

Since the excessive diffuseness of instinctive predispositions is a dimensional (not categorical) entity, the possibility cannot be excluded—taking into account the immature brain's plasticity—that explanations may help some individuals to perfect their emotional, intentional expressions through imitation, making them more similar to those of normative persons. Even if these emotional expressions remain contrived and awkward to a certain extent, they may still improve interpersonal communication.

Better acquaintance with their own unusual instinctive constitution, as well as gentle, insightful therapeutic guidance, may at later stages of maturation orient these individuals toward available social niches or subcultures that may better contain their problematic behavior and reward possible inherited assets (artistic, intellectual, athletic, etc.), as well as guarantee their increased need for privacy.

8.2.4. Interactions between Active, Reactive, and Conforming Instinctive Predispositions

It seems evident that this aspect of the current theory deals with more complex, innately underpinned behavioral categories than the former three comprehensive mechanisms. The individual active and reactive instincts refer to basic instinctive behavioral patterns or predispositions characteristic of all living organisms that possess what may be called behavior, while their modification by social coexistence is present only in animals that live in cooperative groups.

8.2.4.1. Interactions between Individualistic Active and Reactive Instinctive Predispositions

Individual active and self-defensive reactive instincts, as well as their basic way of interacting, have to be very old evolutionary developments. Even relatively primitive organisms possessing behavior express *active* innate behavioral programs in order to fulfill life's basic motives: survival, maturation, and reproduction, as well as basic reactive ones against harmful or dangerous environmental influences: freezing, avoidant, or aggressive tactics. Their interrelatedness—that is, the initial priority of reactive over active behavior when they interfere with each other, and the gradual reduction in reactive behavior's priority with progressive intensification of the suppressed active drives—also has to be an old evolutionary development. It therefore seems reasonable to assume that they (more precisely, their innate aspects) are represented in older structures of the human brain—for example, according to the triune brain model

(Panksepp, 2005, pp. 42–43), in the "reptilian" section of the brain (basal ganglia) in discrete, well-circumscribed localizations, brain circuits, or neurotransmitter systems. I don't think that the basic relationship between active and reactive behavior has been changed dramatically by diversification resulting from relaxation of natural selection pressures in humans. However, quantitative diversification in the proportions, that is, in the differential weight of the participant motives, of this relationship is clearly present. The power relationship between them, that is, the extent and length of the active drives' suppression by reactive instinctive arousal, or alternatively the ease with which active instinctive motives are recaptured, and the time period needed for their reappearance after a disturbing or dangerous event or confrontation is over, vary widely in human populations. At one extreme, this variability predisposes to certain anxiety disorders, while at the other extreme of the continuum, it may produce such mental traits as a preference for dangerous sports or occupations, or accident proneness.

This diversification with regard to the power relationship between active and reactive behavior may determine whether the clinical picture will gravitate toward an anxiety disorder or toward its corresponding personality disorder counterpart (as detailed in chapter 7—that is, for example, toward a social phobia or an avoidant personality disorder, or toward separation anxiety or dependent personality disorder).

8.2.4.2. Interactions between Active-Reactive Instinctive Predispositions and Instinctive Predispositions to Conform to the Requirements of Social Coexistence

The second kind of evolutionary development that powerfully shaped human behavior and led to far-reaching modifications of the self-centered active/reactive instinctive predispositions and their bidirectional interaction—that is, the need to conform to the imperatives of social coexistence, which, in turn, being exceptionally successful, led to a far-reaching relaxation of natural selection pressures—occurred at a later stage of human evolution.

Simple coordinated motor patters can be observed in groups of conspecifics whose social life in other respects is very simple or nonexistent, for example in birds and ungulates during migration, and in fish shoals or insect swarms, reflecting probably the predation-reducing effect of moving in dense groups (Williams, 1974, pp. 212–217). Simple submissive gestures can also be seen in solitary animals—for example, in immature offspring reacting to maternal education.

However, with the evolution of behaviorally advanced animals with a highly developed central nervous system (elephants, dolphins, lower primates, apes), and in those environmental circumstances in which living in groups confers fitness advantages, as in the case of most apes (but not the orangutan or gibbon), complex patterns of social interaction evolved. As we have seen, a prerequisite for this development was the accelerated "atrophying" or disappearance of rigidly preprogrammed behavior patterns, a development that enabled the incorporation of increasing amounts of learned modifications into the progressively more blurred, more malleable innate behavioral propensities. This evolutionary path in humans, especially after the agricultural revolution, attained high levels of sophistication. While the increase in learning abilities and intelligence have important roles in acquiring the knowledge of how one is expected to behave in complex social situations, other innate factors (mostly not well understood) that powerfully influence the ability to socialize must also exist. Good learning abilities do not necessarily predict good socializing abilities. Many highly intelligent persons with excellent learning abilities are not especially social beings, and vice versa—people with modest or even subnormal intellectual and learning abilities may be highly social creatures (remember Williams syndrome, section 7.2.8.9.2.1.1).

Because it is a complex phenomenon, this interaction between individualistic drives, on one hand, and innate predispositions to adapt to, enjoy, and exploit the opportunities of social coexistence, on the other, cannot have as its somatic base a relatively simple, circumscribed structural and functional unit in the brain (contrary to the first two comprehensive mechanisms discussed in this text). All three levels of the "triune brain"—the "reptilian" (basal ganglia), the "old mammalian" (limbic system), and the "neo-mammalian" (neocortex) (Panksepp, 2005, pp. 42–43)—have to participate actively in inducing and regulating this behavioral repertoire.

8.2.4.3. Therapeutic Possibilities

What are the contemporary possibilities of influencing the active-reactive-conforming behavior interrelationship with psychoactive drugs? It seems to me that, by their ability to influence directly one circumscribed aspect or constituent of this complex mechanism, they may indirectly affect other aspects. For example, as we have seen in section 6.5.1, sedatives or alcohol at anxiolytic doses primarily decrease the intensity of arousal, fear, or anxiety in the presence of potentially harmful or dangerous environmental stimuli, thus enabling, secondarily, the expression of formerly suppressed active drives (including in their various forms of frustration). In addition, the decreased sensitivity to environmental danger encourages the use of counteraggressive reactive tactics (instead of avoidance). This effect is most evident in alcohol intoxication or in the paradoxical effect of sedatives.

Another example is the antipsychotic effect. These drugs, by reducing the active and individual reactive instinctive intensities, may indirectly increase the extent to which a person tends to conform to social pressures or expectations.

In my opinion, awareness of the distinction between the direct and indirect effects of psychotropic drugs (instead of clinging to the belief that they possess specific effects for ameliorating whole clinical disease entities) may lead to the development of drugs with more circumscribed effects on behavior (thus avoiding comprehensive suppression or disinhibition of an extended gamut of innate predispositions), thus rendering their clinical employment more discriminating in improving discrete aspects of dysfunctional behavior.

It is possible, I think, to design clinical research studies that compare the efficacy of these drugs on whole clinical disease entities with their efficacy on circumscribed behavioral parameters, as proposed. The results of such experiments could support or refute some of the foregoing theoretical assumptions.

In addition to drugs, a therapeutic milieu can also be used to alter or equilibrate undesirable configurations of active/reactive/ conforming behavior interactions. Since the pressure to conform comes from the social environment (as well as, more rarely, the encouragement not to conform, to be "genuine"), one can imagine a therapeutic milieu in which these or similar social influences can be purposefully manipulated to suit a specific therapeutic goal.

8.3. How the Present Theory May Contribute to the Genetic Research in Psychiatry

Genetic studies in psychiatry today mainly seek correlations between genetic closeness (that is, the extent to which genes are shared by biological relatives), the prevalence of clinical mental disorder entities, and, when possible, try to identify the involved genes. The strength of descriptive disease categories consists of the ease with which they can be identified in clinical and research work and the resultant good inter-rater correlations—that is, the extent to which raters agree on the diagnosis in spite of differences in cultural background, personality, biasing interests concerning the research outcome, etc. However, as mentioned repeatedly in this text, this research method also has a very problematic side: The ultimate clinical pictures standardized by the commonly used diagnostic classifications (DSM, ICD) do not directly reflect the innate (genetic) predisposition but are the end products of a long-lasting behavioral maturation. During this period, vaguely defined innate predispositions to behave interact with ontogenetic influences of a physical and social nature, resulting in a pattern of integrated behavior in which the innate and acquired elements are inseparably intertwined or welded together. More precisely, as was discussed previously, a final clinical picture may contain three different elements:

1. The relevant main genetic predisposition at the extreme of its diversified continuum (see also Brüne, 2008, pp. 37–38).
2. Additional genetic predispositions to behave that are irrelevant to the respective mental disorder's main genetic foundation.
3. The effects of ontogenetic events of a somatic or psychosocial kind.

Accordingly, the pronounced instinctive diffuseness, for example, may manifest itself clinically as schizophrenic psychosis (including its various subtypes), schizoaffective disorder, schizoid or schizotypal personality disorder, etc., depending on the assortment of and interactions between these three kinds of causative factors. As a rule, the most conspicuous or behaviorally most harmful genetic influence determines the diagnosed mental disorder category, while an unrelated genetic predisposition may determine the subtype of the disorder (such as paranoid schizophrenia) or an additional syndrome (obsessive/compulsive disorder, for example) in a schizophrenic patient, as discussed in the introduction to chapter 7.

Consequently, it seems highly probable that unrelated genetic as well as ontogenetic factors will confuse or considerably bias the results of genetic studies relying on standardized full-blown clinical pictures.

The present work represents a different, almost opposing, approach to the clinical one, reflecting the contrast between the inductive and hypothetico-deductive methods of theory building, discussed in chapter 1. Instead of trying to discover the genetic predisposition by taking as a starting point the full-blown (observable) clinical picture, it begins its argument from the hypothesized genetic predisposition and its evolutionary causes and mechanisms, that is, from its original fitness-enhancing roles and why these functions have gone astray during human evolution in a portion of the population. It then attempts to build from this starting point a chain of events that leads ultimately to the full-blown clinical picture—that is, it attempts to show how the hypothesized innate predisposition in interaction with unrelated genetic predispositions and relevant social or physical influences led to the observable clinical pathology. This strategy, however, also has a big drawback, which seems to be the opposite of that encountered in current genetic studies. As we have seen repeatedly in this text, *in humans, the genetic predispositions to behave in a certain way are almost never expressed directly in overt behavior in a form unaltered by environmental influences.* That situation occurs only in the case of animals low on the evolutionary ladder. As we ascend this ladder, however, the innate predispositions to behave are altered, modified, and enriched more and more extensively by acquired knowledge and personal experience. Therefore, that covert, diffuse, but vitally important guiding influence that innate predispositions confer on behavior in humans is never directly observable.

How can we overcome this serious obstacle to employing scientific methodology in behavioral genetic research, a methodology that demands an exclusive reliance on observable material (that is, material that is intersensorily corroborable, in the sense that Karl Popper advocated)? I can think of such a possibility, but this alternative has also serious shortcomings.

The method I have in mind, which may best approximate the hypothesized innate predisposition to behave (as mentioned in the discussion of the diffuse/differentiated scale of human instinctual strivings at the end of section 5.3.1), consists of attempts to identify the *direct, primary* effects of these predispositions on overt behavior, while separating them as much as possible from secondary, tertiary, etc., repercussions during the respective innate predisposition's interaction with ontogenetic influences or with other, unrelated genetic predispositions to behave. The more remote the secondary consequence that a behavioral expression represents in this respect, the more variable and uncertain its relevance to the original innate predisposition will be (as well as to the possible genes involved). For example, when trying to identify the basic innate predisposition for schizophrenia, it seems desirable to single out those behavioral traits common to the whole, genetically related, schizophrenia spectrum of disorders, to the prepsychotic behavior of would-be schizophrenics, as well as to commonly encountered behavioral peculiarities in genetically related family members, which are insufficient for a clinical diagnosis. At the same time, we have to *exclude* those symptoms or peculiarities that are the exclusive characteristic of only one, or a restricted number, of subcategories—for example, the psychotic symptoms, the prominent affective fluctuations, and a stable, high level of suspiciousness and paranoid ideation.

It seems to me that these direct consequences of excessively diffuse drives (as discussed repeatedly throughout this text) may contain the following symptoms:

■ A restricted ability to spontaneously express, in a flexible, nuanced way, emotions and intentions representing underlying biological needs or strivings in overt behavior, body language, intonation (prosody) of speech, or the use of emotionally charged words. Instead, when the need arises for this kind of behavior, the examiner can witness indecisiveness, clumsiness, inhibition, shyness, and confusion. When the patient attempts to express emotions (probably as a result of an amorphous internal tension), he does so in a strained, unnatural, exaggerated, or inflexible way. The other party, in consequence, will have difficulty figuring out with the aid of the empathizing mechanism the individual's emotional, intentional attitudes.

■ The patient's behavior expresses naiveté in social matters (especially in their practical and "self-evident" aspects); she is unable to participate in spontaneous social interactions, and to grasp and to react properly to body language and other aspects of emotionally charged messages.

I realize the great difficulties that exist in attempting to build rating scales for genetic research from these elements of behavior; rating scales that contain only observable phenomena can be easily applied in routine work, and ensure good inter-rater correlation. My lack of experience in building rating scales for psychiatric research prevents me from giving more specific suggestions in this respect. However, I am hopeful that professionals with more knowledge and experience in this domain may be able to overcome, at least partially, the difficulties involved. At any rate, no easy choices in this research domain exist, and I think my proposal is worth attempting.

The outcome of such research (which targets the direct behavioral expressions of the hypothesized genetic predisposition but which are difficult to clearly discriminate, standardize, and systematize), when compared to the results of currently used methodology in genetic studies (that is, research which relies on more easily observable behavioral manifestations, but which may be *remote* consequences of the original genetic predisposition), may be interesting, even instructive. It may suggest a better compromise for the following dilemma: Must the subject of scientific experimentation in psychiatry be tailored to the strict requirements of the method of the exact sciences, or must the method of inquiry be modified and adapted to better suit the nature of the phenomena under study?

An additional approach to circumventing this dilemma in psychiatric genetic research (more precisely, an approach that may provide a complementary perspective to it) concerns the physiological, neurophysiological, and neurochemical events accompanying the direct behavioral manifestations of the inherited predisposition—in our present example, that of the diffuse instinctive discharge. If the identification of the direct behavioral effects of excessively diffuse drives proves reasonably successful, as a next step, the physiological (mostly cerebral) events that correlate with these behavioral manifestations could be investigated. The most exciting research I can think of in this respect is to try to identify in functional brain-imaging studies direct neurophysiological signs of this diffuse, adaptively ineffective arousal. For example, can advanced imaging methods document a kind of "damming up" of excitation in brain areas (maybe the brain stem), underlying incipient diffuse instinctive arousal concurrently with *insufficient* arousal of those brain areas responsible for the *elaboration* of this diffuse arousal, that is, its differentiation into more discrete intentional, emotional behavioral dispositions (most probably the limbic system)? Or maybe this diffuse arousal has a tendency to activate directly brain regions underlying cognitive functions (that is, the neocortex) while bypassing the limbic system?

Functional imaging studies have identified a great variety of functional cerebral abnormalities in schizophrenics (Gur & Gur, 2005, pp. 1400–1403) whose relevance to our theoretical framework I cannot interpret. However, one finding known to me does merit a mention in this context. Researchers compared the responses of schizophrenic and normative persons who observed facial displays of different emotions, looking for changes in the functional state of brain structures during these stimuli with regard to "mood induction effects and . . . emotional discrimination," capacities that resemble empathizing in our theoretical scheme. The interesting finding was that while "patients and healthy controls did not differ in performance . . . in schizophrenics, *a reduced limbic activity* was found." The conclusion of the experimenters was that "failure to activate limbic regions during emotional valence discrimination may explain emotion-processing deficits in patients with schizophrenia. Although the lack of limbic recruitment did not significantly impair simple valence discrimination performance in this clinically stable group, it may impact performance of more demanding tasks" (idem, p. 1403, emphasis added). Whether the patients' adequate performance in this study was achieved by mobilizing *acquired* knowledge dependent mostly on neocortical functions while bypassing the limbic system, which "primarily elaborates . . . uniquely mammalian emotional tendencies" (Panksepp, 2005, p. 70), remains an open question.

In summary, a promising alternative for psychiatric genetic research may be to seek correlations between the material substrate of inheritance with regard to behavior (genes), on one hand, and discrete innate predispositions to behave (and their neurobiological accompaniments), on the other, instead of correlations between genes and full-blown clinical pictures, even when these discrete inherited predispositions manifest in behavior in a less clear-cut way. If this kind of research proves feasible, the rates of heritability of all four comprehensive instinctive mechanisms discussed in this text, more exactly their respective extremes on a continuous scale, can be compared with the rates of heritability of corresponding whole clinical-disease entities. If such a research program can be successfully carried out, the results may serve as indicators of the tenability or untenability of the central proposals of the present theory.

8.4. Dimensional versus Categorical Approach to Dysfunctional Behavior

The *categorical* approach presumes that clinical mental disorders are discrete entities with well-defined boundaries, are qualitatively distinct from normal functioning, remain stable over time, and have discrete pathologies and etiological, treatment, prevention, and other implications. However, it has been recognized for quite a long time that psychiatric disorders do not satisfactorily fulfill these requirements. The clinical picture, course, and response to treatment is variable (the extreme example is probably schizophrenia); comorbidity—the co-occurrence of nosologically different disorders in the same individual—is very frequent; the boundaries between clinical disorder entities are by no means clear-cut; mixed states exist in which symptoms of different disorders co-occur to an extent that makes the clinical diagnosis of the most prominent disorder problematic (for example, schizoaffective disorder); and the therapeutic means, both biological as well as psychotherapeutic, are not specific for discrete clinical entities (Widiger & Samuel, 2005, pp. 494–504). The growing dissatisfaction with the present categorization of psychiatric disorders is clearly reflected in the titles of such publications as "The beginning of the end for the Kraepelinian dichotomy" (Craddock & Owen, 2005) or "The incredible insecurity of psychiatric nosology" (Kendler & Zachar, 2008).

The alternative to that rigid separation of clinical entities is the *dimensional* approach, which presupposes that what underlies mental disorders are continuous gradated variations of behavioral traits that span the range between normality and disease. An example of such an approach in the field of personality disorders is C. R. Cloninger's four "Temperamental Dimensions": harm avoidance, novelty seeking, etc. Each in turn consists of four qualities, and each quality has a high and low degree of intensity; for example, "harm avoidance" has the following pairs at its extremes: pessimistic/optimistic; fearful/daring, etc. (Svrakic & Cloninger, 2005, p. 2065).

A lot of research speaks to the usefulness of this approach. However, in spite of attempts to relate these parameters to animal behavior and evolution, many of these behavioral categories are clearly problematic outside human psychology. It is very difficult to extrapolate such attributes as optimistic/pessimistic; extravagant/thrifty; affectionate/independent; determined/spoiled; enthusiastic/underachiever; and perfectionist/pragmatic to animal behavior, or to conceive their evolutionary pathway into human behavioral traits. However, other parameters like fearful/daring; exploratory/reserved; impulsive/deliberate may have more convincing counterparts in behaviorally highly evolved animals. This issue is important when we see these traits as inherited predispositions affecting human behavior and try to interpret them according to their fitness-enhancing value in an evolutionary context, and when, as the targets of biological psychiatric research and somatic treatment modalities, we try to localize their somatic substrate in circumscribed brain structures and functions.

As a matter of fact, the controversy between the categorical and the dimensional approach is neither new nor limited to the domain of psychiatry. The historian of science Peter Dear, discussing eighteenth- and nineteenth-century medicine, describes this controversy as follows: "Nosology, which typified and classified diseases, was of little help to the practicing physician, but held significant philosophical content regarding *theories* of disease. Those who regarded nosology as worthwhile regarded diseases themselves as specific, distinct entities that the classification codified. By contrast, those medical theorists who regarded disease as a variable affliction that expressed some kind of imbalance in the body saw no legitimacy to nosology, because they reckoned that it arbitrarily broke up what was in reality a continuous spectrum of symptoms" (Dear, 2006, p. 53, emphasis in the original).

We should also mention in this respect the conflicting views of two great naturalists of the Enlightenment era, Buffon and Linnaeus: "Buffon recognized a pattern of variation in the human species quite different from that promoted by Linnaeus [the founder of classification in botany]: namely, continuity, the absence of discrete boundaries between any distinct formal categories of humans" (Marks, 2002, p. 60).

How does the present text on the evolutionary roots of mental disorders relate to the controversy over discrete categories versus dimensional continuums? My view is that the categorical and dimensional approaches are not antithetical but complementary. Without distinct categories (which in common usage are abstract classes to which similar concrete entities belong), no effective communication between persons (in this case mental health professionals) can exist. For example, the above dimensional approach to personality disorders first established large categories such as harm avoidance, novelty seeking, reward dependence, etc.; these are distinct areas, not simply different loci on a dimensional scale. However, each category has its own quantitative continuum, which bridges normative and dysfunctional behavior in its respective domain.

The same is the case with the present theory. The categorical entities in this text, instead of whole clinical disease categories, are the four comprehensive instinctive mechanisms, as well as their subcategories (the forms of frustration of active instinctive strivings, the different reactive behavioral tactics, etc.).[2] As argued repeatedly, these categories, being adaptive behavioral mechanisms that evolved during phylogeny of the animal kingdom, have to possess discrete genetic, somatic, and physiological substrates that in principle can be elucidated by scientific research. In addition, every one of these categorical entities also possesses a dimensional or quantitative aspect, a continuous gradated scale of variations that extends from normality to highly maladaptive behavioral inclinations, the latter constituting the predisposition to circumscribed psychiatric symptoms, syndromes, or disorders—for example, the differentiated/diffuse continuum of instinctive strivings, a continuum whose middle portion underlies normative behavior, and whose extremes predispose to different (opposing) kinds of psychopathology.

However, we have to remember that we are talking about discrete innate predispositions, not clinical disorders. The final clinical picture reflects the outcome of the interactions between a person's whole inherited instinctive configuration, on one hand, and ontogenetic influences of a physical and social nature, on the other. Since the innate predispositions to behave possess the above dimensional quality, and since the social environment (being complex and variable) may have aggravating, ameliorating, or neutral effects on the expression of the dysfunctional inherited tendency in overt behavior, the final clinical picture will always possess the features of a dimensional condition as well. (In fact, the diagnostic criteria of many mental disease categories include, besides the characteristic symptomatology, the requirement of deficient social, occupational, etc., functioning, which constitutes a dimensional aspect of the disorder.)

I would like to end with some emotionally charged expectations (perhaps wishful thinking), something I have tried to avoid throughout the writing of this work.

My furthest-reaching hope regarding the influence of the present theory on future psychiatric practice is the following: If the brain mechanisms underlying the comprehensive instinctive categories discussed in this text can be successfully identified, and the positions of the innate behavioral predispositions (or their neuro-physiologic equivalents) on their respective dimensional scales become quantitatively measurable at an early maturational phase, it may become feasible to construct a "map" or profile (a quantitative distribution of the respective innate elements) of the instinctive configuration in children at increased risk for mental disorder. As a next step, this map would guide attempts to alter (by biological or psychotherapeutic means) the dysfunctional instinctive predispositions in a manner that brings them closer to their normative range on their respective diversified continuum. At the same time, the social environment (especially emotionally charged interpersonal interactions and the provision of relevant knowledge) should be manipulated in ways that reduce the deleterious effect of the dysfunctional instinctive constellation on maturation. In democratic political systems concerned with individual needs, some social niches may be altered or deliberately created in order to better contain dysfunctional innate behavioral predispositions and more effectively exploit concomitant innate special abilities. (An example of progress in this direction is the employment of persons with autism spectrum disorders in jobs requiring sustained attention to fine details, like detecting small flows in computer software or decoding computerized satellite images, a task in which some of these individuals may be very good.)

The last remark I wish to make here transcends the concerns of psychiatry. If it turns out that the idea of a far-reaching relaxation of selection pressures with regard to the innate determinants of behavior in humans, which results in a wide scale of diversification, is a valid one, this realization should no longer be overlooked by the scientific community. In spite of its negative ideological connotations (I mean here the aftereffect of the attempts to replace weak or nonexistent natural selection with enforced artificial selective practices inspired by the eugenics ideology; see section 3.2.2), in view of the considerable desirable as well as deleterious consequences of this evolutionary process, it must be understood as thoroughly as possible. Since understanding in science frequently entails the ability to manipulate the understood phenomenon, it may lead to successful attempts to maximize its desirable consequences and minimize its deleterious ones.

Acknowledgments

Since evolutionary psychiatry is an immature, relatively new field in psychiatry dealt with by a restricted number of professionals, and since I do not belong to any academic body related to this domain, I sought no expert opinion or advice while preparing this book.

However, during the editing process, I got substantial aid from Radius Book Group's staff. I would like to thank especially Mark Fretz, the editorial director, for guiding me (a novice in book publishing) thoughtfully through the intricacies of the editing process. Besides, I want to thank Ed Levy, the editor of the book, for reshaping my less-than-adequate English, as well as the structure of the text, into a much higher standard.

And, last but not least, I would like to thank my daughter, Michal, for helping me to use the more advanced software programs of my PC during the editing process, programs with which I had no previous experience.

References

Adriaens, P. R., & De Block, A. (2011). Why philosophers of psychiatry should care about evolutionary theory. In P. R. Adriaens & A. De Block (eds.), *Maladapting minds: Philosophy, psychiatry, and evolutionary theory* (pp. 1–32). Oxford: Oxford University Press.

Akiskal, H. S. (2005a). Mood disorders: Historical introduction and conceptual overview. In B. J. Sadock & V. A. Sadock (eds.), *Kaplan & Sadock's comprehensive textbook of psychiatry*, 8th ed., vol. 1 (pp. 1559–1575). Philadelphia: Lippincott Williams & Wilkins.

Akiskal, H. S. (2005b). Mood disorders: Clinical features. In B. J. Sadock & V. A. Sadock (eds.), *Kaplan & Sadock's comprehensive textbook of psychiatry*, 8th ed., vol. 1 (pp. 1611–1652). Philadelphia: Lippincott Williams & Wilkins.

Alcock, J. (1979). *Animal behavior: An evolutionary approach*. Sunderland, MA: Sinauer Associates.

Alexander, R. D. (2013). Evolution of the human psyche. In K. Summers & B. Crespi (eds.), *Human social evolution: The foundational works of Richard D. Alexander* (pp. 244–304). Oxford: Oxford University Press.

American Psychiatric Association. (1980). *Diagnostic and statistical manual of mental disorders*, 3rd ed. Washington, DC: American Psychiatric Association. (DSM-III)

American Psychiatric Association. (2000). *Quick reference to the diagnostic criteria from DSM-IV-TR*[R]. Arlington, VA: American Psychiatric Association. (DSM-IV-TR)

American Psychiatric Association. (2013). *Diagnostic and statistical manual of mental disorders*, 5th ed. Washington, DC: American Psychiatric Publishing. (DSM-5)

Andreasen, N. C. (1980). Posttraumatic stress disorder. In H. I. Kaplan, A. M. Freedman, & B. J. Sadock (eds.), *Comprehensive textbook of psychiatry/III*, 3rd ed., vol. 2 (pp. 1517–1525). Baltimore: Williams & Wilkins.

Arana, G. W., & Rosenbaum, J. F. (2000). *Handbook of psychiatric drug therapy*, 4th ed. Philadelphia: Lippincott Williams & Wilkins.

Arieti, S. (1974). *Interpretation of schizophrenia*, 2nd ed. New York: Basic Books.

Audi, R. (ed.) (2005). *The Cambridge dictionary of philosophy*, 2nd ed. Cambridge: Cambridge University Press.

Barkow, J. H., Cosmides, L., & Tooby, J. (1995). *The adapted mind: Evolutionary psychology and the generation of culture.* New York: Oxford University Press.

Barlow, N. (1969). *The autobiography of Charles Darwin 1809–1882.* New York: W. W. Norton.

Bebbington, P., & Kuipers, L. (1993). Social causation of schizophrenia. In D. Bhugra & J. Leff (eds.), *Principles of social psychiatry* (pp. 82–98). Oxford: Blackwell Scientific.

Becker, E. (1975). *The denial of death.* New York: Free Press.

Becker, E. (1976). *The structure of evil: An essay on the unification of the science of man.* New York: Free Press.

Beer, M. D. (trans.) (1992). The manifestations of insanity: Introduction. *History of Psychiatry,* 3, no. 12,4, 504–508. http://dx.doi.org.hartzler.emu.edu/10.1177/0957154X9200301207. Originally published in 1920 by E. Kraepelin (1920), Die Erscheinungsformen des Irreseins. *Zeitschrift für die gesamte neurologie und psychiatrie,* vol. 62, pp. 1–29.

Behe, M. (1998). *Darwin's black box: The biochemical challenge to evolution.* New York: Simon & Schuster.

Bell, G. (2008). *Selection: The mechanism of evolution,* 2nd ed. Oxford: Oxford University Press.

Belmaker, R. H. (2015). Editorial: The Future of Randomized Clinical Trials. *Israel journal of psychiatry,* 52(1), 4.

Benedict, R. (1968). *Patterns of culture.* London: Routledge & Kegan Paul.

Benjamin, J. (1964 [1944]). A method for distinguishing and evaluating formal thinking disorders in schizophrenia. In J. S. Kasanin & American Psychiatric Association (eds.), *Language and thought in schizophrenia: Collected papers presented at the meeting of the American Psychiatric Association, May 12, 1939* (pp. 65–90). Berkeley: University of California Press; New York: W. W. Norton.

Beveridge, W. I. B. (1961). *The art of scientific investigation.* London: Mercury Books.

Birney, R. C., & Teevan, R. C. (1961). *Instinct: An enduring problem in psychology.* Princeton, NJ: D. van Nostrand Company.

Boddy, J. (1978). *Brain systems and psychological concepts.* Chichester, England: John Wiley & Sons.

Bolton, D., & Hill, J. (2007). *Mind, meaning, and mental disorder: The nature of causal explanation in psychology and psychiatry,* 2nd ed. Oxford: Oxford University Press.

Brown, A. S., Bresnahan, M., & Susser, E. S. (2005). Schizophrenia: Environmental epidemiology. In B. J. Sadock & V. A. Sadock (eds.), *Kaplan & Sadock's comprehensive textbook of psychiatry,* 8th ed., vol. 1 (pp. 1371–1381). Philadelphia: Lippincott Williams & Wilkins.

Brown, D., & Associates. (2002). *Career choice and development,* 4th ed. San Francisco: Jossey-Bass.

Brüne, M. (2008). *Textbook of evolutionary psychiatry: The origins of psychopathology*. Oxford: Oxford University Press.

Buchanan, R. W., & Carpenter, W. T., Jr. (2005). Concept of schizophrenia. In B. J. Sadock & V. A. Sadock (eds.), *Kaplan & Sadock's comprehensive textbook of psychiatry*, 8th ed., vol. 1 (pp. 1329–1345). Philadelphia: Lippincott Williams & Wilkins.

Burns, J. (2011). From "evolved interpersonal relatedness" to "costly social alienation": An evolutionary neurophilosophy of schizophrenia. In P. R. Adriaens & A. De Block (eds.), *Maladapting minds: Philosophy, psychiatry, and evolutionary theory* (pp. 289–307). Oxford: Oxford University Press.

Buss, D. M. (2008). *Evolutionary psychology: The new science of mind*. Boston: Pearson, Allyn and Bacon.

Butcher, H. J. (1968). *Human intelligence: Its nature and assessment*. London: Methuen & Co.

Calvin, W. H., & Ojemann, G. A. (1994). *Conversations with Neil's brain: The neural nature of thought and language*. Reading, MA: Addison-Wesley.

Cameron, N. (1944/1964). Experimental analysis of schizophrenic thinking. In J. S. Kasanin & American Psychiatric Association (eds.), *Language and thought in schizophrenia: Collected papers presented at the meeting of the American Psychiatric Association, May 12, 1939* (pp. 50–64). Berkeley: University of California Press; New York: W. W. Norton.

Campbell, R. J. (2004). *Campbell's psychiatric dictionary*, 8th ed. Oxford: Oxford University Press.

Carlson, G. A., & Goodwin, F. K. (1973). The stages of mania. *Archives of General Psychiatry, 28*, pp. 221–228.

Chance, M. R. A., & Silverman, A. P. (1964). The structure of social behavior and drug action. In H. Steinberg, A. V. S. Reuck, & J. Knight (eds.), *Animal behavior and drug action* (pp. 65–79). London: J. & A. Churchill.

Cochran, G., & Harpending, H. (2010). *The 10,000 year explosion: How civilization accelerated human evolution*. New York: Basic Books.

Colp, R., Jr. (1980). Psychiatry and the creative process. In H. I. Kaplan, A. M. Freedman, & B. J. Sadock (eds.), *Comprehensive textbook of psychiatry/III*, 3rd ed., vol. 3 (pp. 3112–3121). Baltimore: Williams & Wilkins.

Craddock, N., & Owen, M. J. (2005). The beginning of the end for the Kraepelinian dichotomy. *British Journal of Psychiatry, 185*, pp. 364–366.

Craig, W. (1918). Appetites and aversions as constituents of instincts. *Biological Bulletin, 34*(2), pp. 91–107.

Crow, T. J. (1985). The two syndrome concept: Origins and current states. *Schizophrenia Bulletin, 11*, pp. 471–486.

Daly, M., & Wilson, M. (1988). *Homicide*. New York: Aldine De Gruyter.

Daly, M., & Wilson, M. (1995). The Man Who Mistook His Wife for a Chattel. In J. H. Barkow, L. Cosmides, & J. Tooby, *The adapted mind: Evolutionary psychology and the generation of culture* (pp. 289–322). New York: Oxford University Press.

Darwin, Ch. (1859/1976). *The origin of species by means of natural selection or the preservation of favoured races in the struggle for life*. London: John Murray; New York: Collier Books.

Darwin, Ch. (1871/1998). *The descent of man*. London: William Clowes and Sons; New York: Prometheus Books.

Darwin, Ch. (1872/1965). *The expression of the emotions in man and animals*. London: John Murray; Chicago: University of Chicago Press.

Davis, J. M. (1980). Minor tranquilizers, sedatives, and hypnotics. In H. I. Kaplan, A. M. Freedman, & B. J. Sadock (eds.), *Comprehensive textbook of psychiatry/III*, 3rd ed., vol. 3 (pp. 2316–2333). Baltimore: Williams & Wilkins.

Dawkins, R. (1987). *The blind watchmaker: Why the evidence of evolution reveals a universe without design*. New York: W. W. Norton.

Dawkins, R. (1989). *The selfish gene*. Oxford: Oxford University Press.

Dear, P. (2006). *The intelligibility of nature: How science makes sense of the world*. Chicago: University of Chicago Press.

Denton, M. (1986). *Evolution: A theory in crisis*. Bethesda, MD: Adler & Adler.

Derryberry, D., & Rothbart, M. K. (2001). Early temperament and emotional development. In A. F. Kalverboer & A. Gramsbergen (eds.), *Handbook of brain and behavior in human development* (pp. 967–987). Dordrecht: Kluwer Academic Publishers.

Dobzhansky, T. (1962). *Mankind evolving: The evolution of the human species*. New Haven, CT: Yale University Press.

Dobzhansky, T. (1970). *Genetics of the evolutionary process*. New York: Columbia University Press.

Dobzhansky, T. (1973). *Genetic diversity and human equality*. New York: Basic Books.

Dobzhansky, T., Ayala, F. J., & Stebbing, G. L. (1977). *Evolution*. San Francisco: W. H. Freeman.

Dubovsky, S. (2005). Benzodiazepine receptor agonists and antagonists. In B. J. Sadock & V. A. Sadock (eds.), *Kaplan & Sadock's comprehensive textbook of psychiatry*, 8th ed., vol. 2 (pp. 2781–2791). Philadelphia: Lippincott Williams & Wilkins.

Dunbar, R. (1997). *Grooming, gossip and the evolution of language*. London: Faber and Faber.

Ehrlich, P. (2000). *Human Natures; Genes, Cultures and the Human Prospect*. Washington, DC: Island Press; Covelo, CA: Shearwater Books.

Eibl-Eibesfeldt, I. (2010). *Human Ethology*. New Brunswick, NJ: Aldine Transaction.

Ekman, P., & Davidson, R. J. (eds.). (1994). *The Nature of Emotion: Fundamental Questions*. New York: Oxford University Press.

Esquire, B. M. (1857). *The works of Francis Bacon, Lord Chancellor of England*, vol. 1. Philadelphia: Parry & McMillan.

Fawcett, J. (2005). Sympathomimetics and dopamine receptor agonists. In B. J. Sadock & V. A. Sadock (eds.), *Kaplan & Sadock's comprehensive textbook of psychiatry*, 8th ed., vol. 2 (pp. 2938–2945). Philadelphia: Lippincott Williams & Wilkins.

Fennig, S., Fochtmann, L. J., & Bromet, E. J. (2005). Delusional disorder and shared psychotic disorder. In B. J. Sadock & V. A. Sadock (eds.), *Kaplan & Sadock's comprehensive textbook of psychiatry*, 8th ed., vol. 1 (pp. 1525–1533). Philadelphia: Lippincott Williams & Wilkins.

Fennig, S., Fochtmann, L. J., & Carlson, G. A. (2005). Schizoaffective disorder. In B. J. Sadock & V. A. Sadock (eds.), *Kaplan & Sadock's comprehensive textbook of psychiatry*, 8th ed., vol. 1 (pp. 1533–1536). Philadelphia: Lippincott Williams & Wilkins.

Feynman, R. P. (1998). *The meaning of it all: Thoughts of a citizen scientist.* Reading, MA: Perseus Books.

Fletcher, N. (2001). Movement disorders. In M. Donaghy (Ed.), *Brain's Diseases of the Nervous System* (pp. 1015–1095). Oxford: Oxford University Press.

Fletcher, R. (1966). *Instinct in man in the light of recent work in comparative psychology.* New York: Schocken Books.

Flinn, M., & Alexander, R. (2007). Runaway social selection in human evolution. In S. W. Gangestad & G. A. Simpson, *The evolution of mind: Fundamental questions and controversies* (pp. 249–255). New York: Guilford Press.

Focquaert, F., & Braeckman, J. (2011). Mirroring the mind: On empathy and autism. In P. R. Adriaens & A. De Block (eds.), *Maladapting minds: Philosophy, psychiatry, and evolutionary theory* (pp. 241–263). Oxford: Oxford University Press.

Foster, R. G., & Kreitzman, L. (2009). *Seasons of life: The biological rhythms that enable living things to thrive and survive.* New Haven, CT: Yale University Press.

Fowler, W. S. (1962). *The development of scientific method.* Oxford: Pergamon Press.

Freeman, D., & Freeman, J. (2008). *Paranoia: The twenty-first century fear.* Oxford, Oxford University Press.

Freud, S. (1966). Source, aim, and object of instinctive energy. In D. Bindra & J. Stewart (eds.), *Motivation: Selected Readings* (pp. 20–22). Baltimore: Penguin Books.

Frijda, N. H. (1994). Emotions are functional, most of the time. In P. Ekman & R. J. Davidson (eds.), *The nature of emotion: Fundamental questions* (pp. 112–122). New York: Oxford University Press.

Geary, D. C. (2007). The motivation to control and the evolution of general intelligence. In S. W. Gangestad & G. A. Simpson, *The evolution of mind: Fundamental questions and controversies* (pp. 305–312). New York: Guilford Press.

Gelfand, R. (1980). Glossary. In H. I. Kaplan, A. M. Freedman, & B. J. Sadock (eds.), *Comprehensive textbook of psychiatry*, 3rd ed., vol. 3 (pp. 3306–3365). Baltimore: Williams & Wilkins.

Glynn, I. (2000). *An anatomy of thought: The origin and machinery of the mind.* London: Phoenix.

Goleman, D. (1995). *Emotional intelligence.* New York: Bantam Books.

Gomez, J. C. (2004). *Apes, monkeys, children, and the growth of mind.* Cambridge, MA: Harvard University Press.

Goodall, J. (1988). *In the shadow of man,* revised edition. London: Weidenfeld and Nicolson.

Goodall, J. (2000). *Through a window: My thirty years with the chimpanzees of Gombe.* Boston: Houghton Mifflin Harcourt.

Goodwin, B. (1996). *How the leopard changed its spots, The evolution of complexity.* New York: Simon & Schuster.

Gordon, M. F. (2005). Normal child development. In B. J. Sadock & V. A. Sadock (eds.), *Kaplan & Sadock's comprehensive textbook of psychiatry,* 8th ed., vol. 2 (pp. 3018–3035). Philadelphia: Lippincott Williams & Wilkins.

Gottesman, J. I., & Gould, T. D. (2003). The endophenotype concept in psychiatry: Etymology and strategic intentions. *American Journal of Psychiatry, 160,* pp. 636–645.

Gould, Stephen Jay. (2007). *The richness of life: The essential Stephen Jay Gould.* P. McGarr & S. Rose (eds.). New York: W. W. Norton.

Grandin, T., & Deesing, G. (1998). Behavioral genetics and animal science. In T. Grandin (ed.), *Genetics and the behavior of domestic animals* (pp. 1–30). San Diego: Academic Press, Elsevier.

Gregory, R. L. (1987). *The Oxford companion to the mind.* Oxford: Oxford University Press.

Gur, R. E., & Gur, R. C. (2005). Neuroimaging in schizophrenia: Linking neuropsychiatric manifestations to neurobiology. In B. J. Sadock & V. A. Sadock (eds.), *Kaplan & Sadock's comprehensive textbook of psychiatry,* 8th ed., vol. 1 (pp. 1396–1408). Philadelphia: Lippincott Williams & Wilkins.

Harris, M. (1974). *Cows, pigs, wars & witches: The riddles of culture.* New York: Random House.

Harris, M. (1989). *Our kind: Who we are, where we came from, where we are going.* New York: Harper & Row.

Hawking, S., & Mlodinov, L. (2010). *The grand design.* London: Bantam Press.

Hayek, F. A. (1955). *The counter-revolution of science: Studies on the abuse of reason.* Glencoe, IL: The Free Press.

Healy, D. (1999). *The antidepressant era.* Cambridge, MA: Harvard University Press.

Healy, D. (2002). *The creation of psychopharmacology.* Cambridge, MA: Harvard University Press.

Healy, D. (2009). *Psychiatric drugs explained.* Edinburgh: Churchill Livingstone, Elsevier.

Heird, J. C., & Deesing, M. J. (1998). Genetic effects on horse behavior. In T. Grandin (Ed.), *Genetics and the behavior of domestic animals* (pp. 203–234). San Diego: Academic Press, Elsevier.

Himmelfarb, G. (1962). *Darwin and the Darwinian revolution*. Garden City, NY: Anchor Books.

Hinde, R. A. (1966). Critique of energy models of motivation. In D. Bindra & J. Stewart (eds.), *Motivation: Selected readings* (pp. 34–45). Baltimore: Penguin Books.

Hinde, R. A. (1970). *Animal behaviour, A Synthesis of ethology and comparative psychology*, 2nd ed. New York: McGraw-Hill.

Howard, J. (1982). *Darwin*, Oxford: Oxford University Press.

Hrdy, S. B. (1977). *The Langurs of Abu: Female and male strategies of reproduction*. Cambridge, MA: Harvard University Press.

Hull, D. (1974). *Philosophy of biological science*. Englewood Cliffs, NJ: Prentice-Hall.

Insel, T. R., & Fenton, W. S. (2005). Future of psychiatry. In B. J. Sadock & V. A. Sadock (eds.), *Kapan & Sadock's comprehensive textbook of psychiatry*, 8th ed., vol. 2 (pp. 4060–4064). Philadelphia: Lippincott Williams & Wilkins.

Jablonka, E., & Lamb, M. J. (2005). *Evolution in four dimensions: Genetic, epigenetic, behavioral, and symbolic variation in the history of life*. Cambridge, MA: MIT Press.

Kalverboer, A. F., & Gramsbergen, A. (eds.). (2001). *Handbook of brain and behaviour in human development*. Dordrecht: Kluwer Academic Publishers.

Kaplan, H. I., Freedman, A. M., & Sadock, B. J. (eds.). (1980). *Comprehensive textbook of psychiatry*, 3rd ed. Baltimore: Williams & Wilkins.

Kendler, K. S., & Parnas, J. (2008). *Philosophical issues in psychiatry: Explanation, phenomenology, and nosology*. Baltimore: Johns Hopkins University Press.

Kim, J. (2011). *Philosophy of mind*, 3rd ed. Boulder, CO: Westview Press.

Kirkpatrick, B., & Tek, C. (2005). Schizophrenia: Clinical features and psychopathology concepts. In B. J. Sadock & V. A. Sadock (eds.), *Kaplan & Sadock's comprehensive textbook of psychiatry*, 8th ed., vol. 1 (pp. 1416–1436). Philadelphia: Lippincott Williams & Wilkins.

Klee, R. (1977). *Introduction to the philosophy of science: Cutting nature at its seams*. New York: Oxford University Press.

Klein, J., & Takahata, N. (2002). *Where do we come from? The molecular evidence of human descent*. Berlin: Springer.

Klerman, G. L. (1980). Other specific affective disorders. In H. I. Kaplan, A. M. Freedman, & B. J. Sadock (eds.), *Comprehensive textbook of psychiatry*, 3rd ed., vol. 2 (pp. 1332–1338). Baltimore: Williams & Wilkins.

Kramer, P. D. (1993). *Listening to prozac*. New York: Viking, Penguin Group.

Kübler-Ross, E. (1969). *On Death and Dying*. New York: Collier Books.

Kuhn, T. S. (1970). *The Structure of Scientific Revolutions*, 2nd ed., enlarged. Chicago: University of Chicago Press.

Laing, R. D. (1960). *The divided self: A study of sanity and madness*. London: Travistock Publications.

Laland, K. N., & Brown, G. R. (2011). *Sense and nonsense: Evolutionary perspectives on human behaviour*, 2nd ed. Oxford: Oxford University Press.

Lampert, A. (1997). *The evolution of love*. Westport, CT: Praeger.

Lavoisier, A. (1965 [1790]). *Elements of chemistry*. Trans. Robert Kerr. Edinburgh: William Creech; New York: Dover Publications.

Lehmann, H. E. (1980a). Schizophrenia: Clinical features. In H. I. Kaplan, A. M. Freedman, & B. J. Sadock (eds.), *Comprehensive textbook of psychiatry*, 3rd ed., vol. 2 (pp. 1153–1192). Baltimore: Williams & Wilkins.

Lehmann, H. E. (1980b). Schizophrenia: History. In H. I. Kaplan, A. M. Freedman, & B. J. Sadock (eds.), *Comprehensive textbook of psychiatry*, 3rd ed., vol. 2 (pp. 1104–1113). Baltimore: Williams & Wilkins.

Lieberman, P. (2006). *Toward and evolutionary biology of language*. Cambridge, MA: Belknap Press of Harvard University Press.

Linn, L. (1980). Clinical manifestations of psychiatric disorders. In H. I. Kaplan, A. M. Freedman, & B. J. Sadock (eds.), *Comprehensive textbook of psychiatry*, 3rd ed., vol. 1 (pp. 990–1034). Baltimore: Williams & Wilkins.

Lipowski, Z. J. (1980). Organic mental disorders: Introduction and review of syndromes. In H. I. Kaplan, A. M. Freedman, & B. J. Sadock (eds.), *Comprehensive textbook of psychiatry*, 3rd ed., vol. 2 (pp. 1359–1392). Baltimore: Williams & Wilkins.

Lorenz, K. (1961). *King Solomon's ring: New light on animal ways*. New York: Thomas Y. Crowell Company.

Lorenz, K. (1964). *Man meets dog*. Harmondsworth, Middlesex, England: Penguin.

Lorenz, K. (1965). *Evolution and modification of behavior*. Chicago: University of Chicago Press.

Lorenz, K. (1966). An energy model of instinctive actions. In D. Bindra & J. Stewart (eds.), *Motivation: Selected readings* (pp.23–27). Baltimore: Penguin Books.

Lorenz, K. (1974). *On aggression*. London: Methuen & Co.

Lorenz, K. (1975). The conception of instinctive behavior. In C. H. Schiller (Ed.), *Instinctive behavior: The development of a modern concept* (pp. 129–175). New York: International Universities Press.

Lorenz, K. Z. (1982). *The foundations of ethology: The principal ideas and discoveries in animal behavior*. New York: Simon & Schuster.

Lumsden, C. G., & Wilson, E. O. (1981). *Genes, mind and culture: The coevolutionary process*. Cambridge, MA: Harvard University Press.

Lynn, R. (2001). *Eugenics: A reassessment*. Westport, CT: Praeger.

Maestripieri, D. (2012). *Games primates play: An undercover investigation of the evolution and economics of human relationships*. New York: Basic Books.

Mandler, G. (1984). *Mind and body: Psychology of emotion and stress*. New York: W. W. Norton.

Marks, J. (1995). *Human biodiversity: Genes, race and history*. New York: Aldine de Gruyter.

Marks, J. (2002). *What it means to be 98% chimpanzee: Apes, people, and their genes*. Berkeley: University of California Press.

Maynard Smith, J., & Szathmáry, E. (1995). *The major transitions in evolution*. Oxford: W. H. Freeman.

Mayr, E. (1974). Behavior programs and evolutionary strategies. *American Scientist, 62*(6), pp. 650–659.

Mayr, E. (1982). *The growth of biological thought: Diversity, evolution and inheritance*. Cambridge, MA: The Belknap Press of Harvard University Press.

Mayr, E. (1991). *One long argument: Charles Darwin and the genesis of modern evolutionary thought*. Cambridge, MA: Harvard University Press.

Mayr, E. (2001). *What evolution is*. New York: Basic Books.

McDougall, W. (1966). On the nature of instinct. In D. Bindra, & J. Stewart (eds.), *Motivation: Selected reading* (pp. 17–19). Baltimore: Penguin.

McGuire, M., & Troisi, A. (1998). *Darwinian psychiatry*. New York: Oxford University Press.

Medawar, P. B. (1969). *Induction and intuition in scientific thought*. London: Methuen & Co.

Medawar, P. B., & Medawar, J. S. (1977). *The life science: Current ideas in biology*. London: Wildwood House.

Miller, J. G. (1980). General living systems theory. In H. I. Kaplan, A. M. Freedman, & B. J. Sadock (eds.), *Comprehensive textbook of psychiatry*, 3rd ed., vol. 1 (pp. 98–114). Baltimore: Williams & Wilkins.

Mirsky, A. F., Bielianskas, L. A., French, L. M., van Kammen, D. P., Jönsson, E., & Sevdall, G. (2000). A 39-year follow up of the Genain quadruplets. *Schizophrenia Bulletin, 26*(3), pp. 699–708.

Mithen, S. (1996). *The prehistory of the mind: A search for the origins of art, religion and science*. London: Thames and Hudson.

Mithen, S. (2007). Key changes in the evolution of human psychology. In S. W. Gangestad & J. A. Simpson, (eds.), *The evolution of mind: Fundamental questions and controversies* (pp. 256–266). New York: Guilford Press.

Mora, G. (1980). Historical and theoretical trends in psychiatry (4–134). In Kaplan, H. I., Freedman, A. M., & Sadock, B. J. (eds.), *Comprehensive Textbook of Psychiatry*, 3rd ed., vol. 1 (pp. 4–134). Baltimore: Williams & Wilkins.

Morange, M. (2008). *Life explained*. New Haven, CT: Yale University Press; Paris: Editions Odile Jacob.

Mrazek, D., & Mrazek, P. J. (2005). Prevention of psychiatric disorders in children and adolescents. In B. J. Sadock & V. A. Sadock (eds.), *Kaplan & Sadock's comprehensive textbook of psychiatry*, 8th ed., vol. 2 (pp. 3513–3519). Philadelphia: Lippincott Williams & Wilkins.

Murakami, H. (2002). *Sputnik sweetheart*. London: Vintage Books.

Murray, R. M. & Bramon, E. (2005). Developmental model of schizophrenia. In B. J. Sadock & V. A. Sadock (eds.), *Kaplan & Sadock's comprehensive textbook of psychiatry*, 8th ed., vol. 2 (pp. 1381–1396). Philadelphia: Lippincott Williams & Wilkins.

Nagel, E. (1961). *Structure of science: Problems in the logic of scientific exploration*. New York: Harcourt, Brace & World.

Nagel, T. (2002). What is it like to be a bat? In D. J. Chalmers (ed.), *Philosophy of mind: Classical and contemporary readings* (pp. 219–226). New York: Oxford University Press.

Neill, J. (1990). Whatever became of the schizophrenogenic mother? *American Journal of Psychotherapy, 44*(4), pp. 499–505.

Nelson, C. J. (2005). Tricyclics and tetracyclics. In B. J. Sadock & V. A. Sadock (eds.), *Kaplan & Sadock's comprehensive textbook of psychiatry*, 8th ed., vol. 2 (pp. 2956–2967). Philadelphia: Lippincott Williams & Wilkins.

Nemeroff, C. B. & Putnam, J. S. (2005). Barbiturates and similarly acting substances. In B. J. Sadock & V. A. Sadock (eds.), *Kaplan & Sadock's comprehensive textbook of psychiatry*, 8th ed., vol. 2 (pp. 2775–2781). Philadelphia: Lippincott Williams & Wilkins.

Nemiah, J. C. (1980). Obsessive-compulsive disorder (Obsessive-compulsive neurosis). In H. I. Kaplan, A. M. Freedman, & B. J. Sadock (eds.), *Comprehensive textbook of psychiatry*, 3rd ed., vol. 2 (pp. 1504–1517). Baltimore: Williams & Wilkins.

N.A. (1981). *New Larousse Encyclopedia of Animal Life*. New York: Bonanza Books.

Nissen, H. W. (1966). Instinct as seen by a psychologist. In D. Bindra & J. Stewart (eds.), *Motivation: Selected readings* (pp. 46–51). Baltimore: Penguin Books.

Nowak, R. M. (1999). *Walker's mammals of the world*. Baltimore: Johns Hopkins University Press.

Odling-Smee, F. J., Laland, K. N., & Feldman, M. W. (2003). *Niche construction: The neglected process in evolution*. Princeton, NJ: Princeton University Press.

Panksepp, J. (1994). The basics of basic emotion. In P. Ekman & R. J. Davidson (eds.), *The Nature of Emotion: Fundamental Questions* (pp. 20–24). New York: Oxford University Press.

Panksepp, J. (2005). *Affective neuroscience: The foundations of human and animal emotions*. Oxford: Oxford University Press.

Pearce, Eiluned, Stringer, Chris, & Dunbar, R. I. M. (2013). New insights into differences in brain organization between Neanderthals and anatomically modern humans. *Proceedings of the Royal Society B, 280* (March 13): 20130168. DOI: 10.1098/rspb.2013.0168. Accessed July 23, 2018 at http://rspb.royalsocietypublishing.org/content/280/1758/20130168.

Perry, S. (2011). Manipulative monkeys: The capuchins of Lomas Barbudal. Cambridge, MA: Harvard University Press.

Piaget, J. (1953). *The origins of the intelligence in the child*. London: Routledge and Kegan Paul.

Pine, D. S., & McClure, E. B. (2005). Anxiety disorders: Clinical features. In B. J. Sadock & V. A. Sadock (eds.), *Kaplan & Sadock's comprehensive textbook of psychiatry*, 8th ed., vol. 2 (pp. 1768–1780). Philadelphia: Lippincott Williams & Wilkins.

Pinker, S. (1997). *How the mind works*. London: Penguin Books.

Pinker, S. (2007a). *The language instinct: How the mind creates language.* New York: Harper Perennial.

Pinker, S. (2007b). *The stuff of thought: Language as a window into human nature*. New York: Penguin Books.

Plomin, R., De Fries, J. C., McClearn, G. E., & McGuffin, P. (2008). *Behavioral genetics*, 5th ed. New York: Worth Publishers.

Plotkin, H. (2007). *Necessary knowledge*. Oxford: Oxford University Press.

Popper, K. (1968). *The logic of scientific discovery*. New York: Harper & Row.

Popper, K. (1973). *Objective knowledge: An evolutionary approach*. Oxford: The Clarendon Press.

Popper, K. (2002). *Conjectures and refutations: The growth of scientific knowledge*. London: Routledge.

Popper, K. R., & Eccles, J. (1986). *The self and its brain*. London: Routledge; New York: Kegan Paul.

Price, E. O. (2002). *Animal domestication and behavior*. New York: CABI Publishing.

Prinz, J. (2004). Embodied emotions. In R. C. Solomon (Ed.), *Thinking About Feeling, Contemporary Philosophers on Emotions* (pp. 44–58). Oxford: Oxford University Press.

Provencio, I. (2005). Chronobiology. In B. J. Sadock & V. A. Sadock (eds.), *Kaplan & Sadock's comprehensive textbook of psychiatry*, 8th ed., vol. 2 (pp. 161–171). Philadelphia: Lippincott Williams & Wilkins.

Quinton, A. (1973). *The nature of things*. London: Routledge.

Quinton, A. (1980). *Francis Bacon*. Oxford: Oxford University Press.

Redlick, F. C., & Freedman, D. X. (1966). *The theory and practice of psychiatry*. New York: Basic Books.

Richerson, P. J., & Boyd, R. (2005). *Not by genes alone: How culture transformed human evolution*. Chicago: University of Chicago Press.

Riley, B. P., & Kendler, K. S. (2005). Schizophrenia: Genetics. In B. J. Sadock & V. A. Sadock (eds.), *Kaplan & Sadock's comprehensive textbook of psychiatry*, 8th ed., vol. 1 (pp. 1354–1371). Philadelphia: Lippincott Williams & Wilkins.

Rogers, C. R. (1980). Client-centered psychotherapy. In H. I. Kaplan, A. M. Freedman, & B. J. Sadock (eds.), *Comprehensive textbook of psychiatry*/III, 3rd ed., vol. 2 (pp. 2153–2167). Baltimore: Williams & Wilkins.

Rossor, M. (2001). Neuropsychological disorders, dementia, and behavioral neurology (733–74). In Donaghy, M. (Ed.), *Brain's Diseases of the Nervous System*. Oxford: Oxford University Press.

Russell, B. (2002). Analogy. In D. J. Chalmers (Ed.), *Philosophy of mind: Classical and contemporary readings* (pp. 667–669). New York: Oxford University Press.

Ryle, G. (1963). *The concept of mind*. Harmondsworth, Middlesex, England: Penguin Books.

Sadock, B. J., & Sadock, V. A. (eds.). (2005). *Kaplan & Sadock's comprehensive textbook of psychiatry* (8th ed.), 2 vols. Philadelphia: Lippincott Williams & Wilkins.

Sadock, B. J., Sadock, V. A., & Sussman, N. (2006). *Kaplan & Sadock's pocket handbook of psychiatric drug treatment*, 4th ed. Philadelphia: Lippincott Williams & Wilkins.

Sass, L. A. (2001). Self and world in schizophrenia: Three classic approaches. *Philosophy, Psychiatry and Psychology, 8*(4), pp. 251–270.

Selye, H. (1964). *From dream to discovery: On being a scientist*. New York: McGraw-Hill Company.

Simpson, G. G. (1964). *This view of life: The world of an evolutionist*. New York: Harcourt, Brace & World.

Skinner, B. F. (1976). *About Behaviorism*. New York: Vintage Books.

Slater, E., & Roth, M. (1969). *Mayer-Gross, Slater and Roth, clinical psychiatry*, 3rd ed. London: Bailliere, Tindall & Cassell.

Soanes, C., & Stevenson, A. (eds.). (2009). *Concise Oxford English dictionary*, 11th ed. Oxford: Oxford University Press.

Solomon, P., & Kleeman, S. T. (1980). Sensory deprivation. In H. I. Kaplan, A. M. Freedman, & B. J. Sadock (eds.), *Comprehensive textbook of psychiatry*, 3rd ed., vol. 1 (pp. 600–608). Baltimore: Williams & Wilkins.

Solomon, R. C. (Ed.). (2004). *Thinking about feeling: Contemporary philosophers on emotions*. Oxford: Oxford University Press.

Stevens, A., & Price, J. (2000). *Evolutionary psychiatry: A new beginning*, 2nd ed. London: Routledge.

Stournaras, S. N. (1990). *Meteora: The sacred rocks and their history*. Athens: Stephanos N. Stournaras, Lenorman.

Summers, K., & Crespi, B. (eds.). (2013). *Human social evolution: The foundational works of Richard D. Alexander*. Oxford: Oxford University Press.

Svrakic, D. M. & Cloninger, R. C. (2005). Personality disorders. In B. J. Sadock & V. A. Sadock (eds.), *Kaplan & Sadock's comprehensive textbook of psychiatry*, 8th ed., vol. 2 (pp. 2063–2104). Philadelphia: Lippincott Williams & Wilkins.

Tasman, A., Kay, J., & Lieberman, J. A. (1997). *Psychiatry*. Philadelphia: W. B. Saunders Company.

The New Larousse Encyclopedia of Animal Life. (1981). New York: Bonanza Books.

Tinbergen, N. (1966). Hierarchical organization of instinctive actions. In D. Bindra & J. Stewart (eds.), *Motivation: Selected readings* (pp. 28–33). Baltimore: Penguin Books.

Tinbergen, N. (2003). *The study of instinct.* Oxford: The Clarendon Press.

Tsuang, M. T., & Tohen, M. (2002). *Textbook in psychiatric epidemiology,* 2nd ed. New York: Wiley-Liss.

Van den Hooff, A., & Van den Hooff, P. (1995). *A propos of science : Quotations and aphorisms.* Amsterdam: AmstelScience.

Van Praag, H. M. (1975). *On the origin of schizophrenic psychosis.* Amsterdam: de Erven Bohn.

Victor, M., & Ropper, A. H. (2001). *Adams and Victor's principles of neurology,* 7th ed. New York: McGraw-Hill.

Volkmar, F. R., Klin, A., & Schultz, R. T. (2005). Pervasive developmental disorders. In B. J. Sadock & V. A. Sadock (eds.), *Kaplan & Sadock's comprehensive textbook of psychiatry,* 8th ed., vol. 2 (pp. 3164–3182). Philadelphia: Lippincott Williams & Wilkins.

Waal, F. de (1982/2007). *Chimpanzee politics: Power and sex among apes.* New York: Harper & Row; Baltimore: Johns Hopkins University Press.

Waal, F. de (1996). *Good natured: The origins of right and wrong in humans and other animals.* Cambridge, MA: Harvard University Press.

Waal, F. de (2002). Apes from Venus: Bonobos and human social evolution. In F. B. M. de Waal (Ed.), *Tree of origin: What primate behavior can tell us about human social evolution* (pp. 40–69). Cambridge, MA: Harvard University Press.

Waal, F. de (2005). *Our inner ape: A leading primatologist explains why we are who we are.* New York: Riverhead Books.

Waal, F. de, & Lanting, F. (1997). *Bonobo: The forgotten ape.* Berkeley: University of California Press.

Watches, The Collector's Corner. (1999). London: Grange Books.

Watson, J. B. (1957). *Behaviorism.* Chicago: University of Chicago Press.

Wehr, T. A., & Rosenthal, N. E. (1989). Seasonality and affective illness. *American Journal of Psychology, 146* (4), p. 836.

Weiner, H. (1980). Schizophrenia: Etiology. In H. I. Kaplan, A. M. Freedman, & B. J. Sadock (eds.), *Comprehensive textbook of psychiatry,* 3rd ed., vol. 2 (pp. 1121–1152). Baltimore: Williams & Wilkins.

Weisman, A. D. (1980). Thanatology. In H. I. Kaplan, A. M. Freedman, & B. J. Sadock (eds.), *Comprehensive textbook of psychiatry/*III, 3rd ed., vol. 2 (pp. 2042–2056). Baltimore: Williams & Wilkins.

Whyte, L. L. (1965). *Internal factors in evolution.* London: Travistock Publications .

Widiger, T. A., & Samuel, D. B. (2005). Diagnostic categories or dimensions? A question for the diagnostic and statistical manual of mental disorders, fifth edition. *Journal of Abnormal Psychology, 114* (4), pp. 494–504.

Williams, G. C. (1974). *Adaptation and natural selection: A critique of some evolutionary thought.* Princeton, NJ: Princeton University Press.

Wilson, E. O. (1978). *On human nature.* Cambridge, MA: Harvard University Press.

Wilson, E. O. (1994). *Naturalist*. Washington, DC: Island Press.

Wilson, E. O. (1998). *Consilience: The unity of knowledge*. New York: Alfred A. Knopf.

Yager, J., & Gitlin, M. J. (2005). Clinical manifestations of psychiatric disorders. In B. J. Sadock & V. A. Sadock (eds.), *Kaplan & Sadock's comprehensive textbook of psychiatry*, 8th ed., vol. 1 (pp. 964–1002). Philadelphia: Lippincott Williams & Wilkins.

Youngson, R. M. (1998). *Scientific blunders: A brief history of how wrong scientists can sometimes be*. New York: Carroll & Graf Publishers.

Zahavi, A., and A. Zahavi. (1997). *The handicap principle: A missing piece of Darwin's puzzle*. Oxford: Oxford University Press.

Endnotes

Introduction

1. In chapter 2, I will develop some aspects of the topic of instinct-learning inter-action, arguing that instinct and learning, far from being mutually exclusive, strongly depend on each other. Learning evolved slowly on the foundations of instincts, enabled by this evolutionary development regarding instinctive behavior.

Chapter 1

1. A more detailed enumeration of them, together with comments on the ways this situation leads to confusion according to Kuhn's views as delineated above, can be found in M. McGuire and A. Troisi., *Darwinian Psychiatry* (1998), in the section on "conceptual pluralism" (6–7).
2. "Emergence" here means that more complex levels of organization of living matter create new properties that cannot be deduced from the properties of their constituent parts; for example, mental functioning cannot be deduced from the summing-up of the functioning of the individual neurons.

Chapter 2

1. The molecular clock uses mutation rates of biomolecules, usually DNA or amino acid sequences, to determine when two or more life forms diverged. According to this method, it took hundreds of millions of years for the evolution of different classes of vertebrates and several million years for the evolution of cetaceans from terrestrial mammals (Klein &, Takahata, 2002, p. 247).
2. Pinker explains his choice of the word *instinct* in the following way: "Language is a complex, specialized skill, which develops in the child spontaneously, without conscious effort or formal instruction, is deployed without awareness of its underlying logic, is qualitatively the same in every individual.... [Therefore for its designation] I prefer the admittedly quaint term 'instinct.' It conveys the idea that people know how to talk in more or less the sense that spiders know how to spin webs.... Although there are differences between webs and words, I will encourage you to see language in this way, for it helps to make sense of the phenomena we will explore" (2007A, p. 4–5).
3. "Insight" here means problem-solving in the context of spatial orientation (Lorenz, 1982, pp. 237–241).
4. Specificity here can be interpreted in the light of ethology as a more or less clearly determined, partly inherited, course of action toward a genetically more-or-less narrowly defined environmental agent.

5. The mechanism of empathizing as a means of gaining knowledge of others' mental states is recognized by fiction writers, too: "Sumire might very well have been pondering the sexual desire she felt. The same way I thought about my own desire when I was with her. It wasn't hard for me to understand how she felt" (Murakami, 2002, p. 140).

6. Jane Goodall described the same phenomenon in the case of male fraternal chimpanzee twins several weeks old (2000, pp. 184 and 187).

7. It has to be mentioned that this mechanism in certain instances may have adaptive value. Empathizing inhibits agonistic or inimical attitudes and therefore has to be suppressed when aggressive behavior (group or individual) is needed. Indeed, when watching an action movie intensely and identifying with the good characters, who cares about the injured or killed bad guys and the suffering of their families? Furthermore, men's greater involvement in aggressive activities (like war and hunting) during our phylogeny may have accentuated gender differences in readiness for empathizing.

8. In the field of primatology, however, this attitude is not entirely unequivocal. It is a matter of contention, for example, whether apes under observation should be given human forenames [which may encourage anthropomorphizing tendencies in interpreting their behavior] or should be identified by numbers, a practice that is seen as more neutral or "objective" in this respect.

Chapter 3

1. It has to be noted, however, that during group excitement in chimpanzees (as a result of reunion with other chimps or finding a rich food source) outbursts of sexual activity or fighting were observed (Goodall, 2000, p. 105). The "sparking over" of energetic and intentional aspects of the instinctive activity from one instinctive domain to another, as advanced by Nico Tinbergen, and its implications concerning human behavior, are further elaborated in section 5.2.1.

2. For that matter, in the case of a proportion of kindergarten teachers, for example, it could include a classroom full of children; in the case of nurses and other therapeutic personnel, it might include disabled children who are institutionalized; even pets, who are in effect "adopted," become the recipients of human parental feelings.

Chapter 4

1. This mechanism differentiates humans from lower animals only quantitatively. It is known, for example, that the number of brain synapses in laboratory rats is greatly influenced by the level of environmental complexity (Plotkin, 2007, p. 195).

2. Other forms of selection exist, such as *frequency-dependent selection*, which means "the tendency of predators and pathogens to prefer the most abundant kind of cryptic prey" (Bell, 2008, p. 349), or *runaway selection*, in the framework of sexual selection, which means the exaggeration of some male features preferred by female conspecifics, even if they pose a considerable cost to fitness. These categories of selection have no relevance to the issues addressed by this text.

3. Compare, for example, the cost to a bacteria of simply dividing into two similar cells (replication), or the pollination of a flower that possesses both pollen and

stigma (uniparental reproduction), with the costs of finding a mate, competing with other conspecifics, pregnancy, bringing up the offspring, and so on.

4. Blind evolutionary forces evidently cannot discern between categories of good and bad, of desirable and undesirable, in accordance with human interests or moral principles (see Richard Dawkins's *The Blind Watchmaker*).

5. In one of her classic accounts of chimpanzee life in the wild, Jane Goodall describes movingly how intolerant these highly evolved social animals can be toward conspecifics with physical disabilities (in this instance, paralysis from poliomyelitis), even in the case of high-ranking members of the group. The healthy group members' first reaction to the sight of paralyzed limbs and its dysfunctional consequences was fear, aggression, and even physical violence. Later, they kept their distance from these unfortunate cripples and excluded them completely from everyday social encounters, not to mention any provision of help (Goodall, 1988, pp. 221–222).

6. Holland hypothesized six personality types: Realistic, Investigative, Artistic, Social, Enterprising, and Conventional, as well as six corresponding environments with the same designations: Realistic, Investigative, Artistic, etc.

7. It has to be mentioned, however, that because of the complexities of the concerned organ's functions, new evolutionary events are not always preceded by considerable relaxation of former selective pressures. For example, during the evolution of the human vocal apparatus, the selection pressures for the respiratory and digestive functions of the throat structures were maintained for obvious reasons, leading to a kind of optimal compromise between its old and new functions. This compromise led to an increase in the risk of aspiration and asphyxia (choking) while vastly improving articulating abilities.

8. (We have to mention, however, that there is recognized a kind of evolution, *mosaic evolution*, which argues in favor of different rates of evolution for different organs or systems of the organism. For a short discussion of mosaic evolution and its relatedness to the principle of cohesion of the genotype, see Mayr, 1991, pp. 158-161.)

Chapter 5

1. A branch of the evolutionary behavioral sciences named *behavioral ecology* studies optimization strategies in the trade-offs between life-sustaining aims and environmental conditions (but not regarding the overall energy level of behavior, to my knowledge) in circumscribed behavioral spheres. (See, for example, the paragraph on "optimal foraging theory" in Laland & Brown, 2011, pp. 79-80.)

2. Another theoretically possible route of learning-instinct interaction in the animal kingdom, diametrically opposite to that just described, may be of interest. It is known as the "Baldwin effect" or Waddington's "genetic assimilation." This hypothesis proposes that when animals are faced with a new environmental challenge they first adapt to it by learning. Those animals able to learn a new and effective behavior pattern more easily and quickly (either from other animals or by discovering it through experimentation) will have a fitness advantage that, through natural or sexual selection, gradually changes the genotype in order to incorporate the preprogramming of the acquired behavior. In consequence, that activity will eventually be discharged as a result

of minimal learning, or without learning at all, thus becoming instinctive (Jablonka & Lamb, 2005, pp. 286–290). Since this mechanism can work only in simple, constant, or very predictably changing environmental conditions, it does not concern us in the context of the present work. As a matter of fact, Jablonka and Lamb also consider the possibility of the reverse order of events, too. When the environment is "not stable but changes and fluctuates . . . we would expect an increased reliance on individual and social learning to evolve from more instinctive responses" (idem, p. 290). This scenario is obviously in keeping with the evolutionary mechanism proposed in this section.

3. As shown in the earlier example of ground squirrels' "circannual clock" preserving overeating and hibernation in spite of constant laboratory conditions (section 5.1), regularly recurring external influences can induce in animals behavior with the characteristics of active instinctive behavior. Even generations born in the laboratory who have never experienced a natural photoperiod displayed that circannual clock.

4. Reactive behavior's evolution from its phylogenetic roots into its expression in modern human populations is a very complex topic, and, beyond some general guidelines, was barely touched on in this discussion. However, a more detailed account of this topic (or of the same evolutionary development in the domain of active behavior) is beyond the scope of the present text.

Chapter 6

1. As detailed in the following, by antidepressant action I mean chiefly the *noradrenergic* effect of this drug category. The *serotonergic* effect seems more complex, making its assimilation into the conceptual framework of this text more problematic.

2. Henri Laborit (1914–1995) was a French military surgeon, often responsible for patients' anesthesia, and one of the founders of neuropsychopharmacology.

3. It is likely that other professionals in the field have arrived at the same simple hypothesis; in the section on catatonia (7.2.8.8), I quote a passage that assumes a similar mechanism in the case of catatonic stupor.

4. We have to mention here an additional complication. Most tricyclic antidepressants possess both noradrenergic and serotonergic effects. The tertiary amines in tricyclics, which block the reuptake of *serotonin* at the nerve synapses, are demethylated in the brain to secondary amines, which block the reuptake of *noradrenaline* (Nelson, 2005, p. 2956). In the case of antidepressants affecting more selectively the serotonergic system, the picture is equally problematic. Most SSRIs are *agonists*—that is, they increase the serotonin's effect, while some other serotonergic drugs, also marketed as antidepressants (mianserin, mirtazapine, trazodone), are *antagonists* at another kind of serotonin receptor (S2 instead of S1) (Healy, 2009, p. 64).

5. Suicidality in these cases is explained in the psychiatric literature with the presupposition that depressive thought content is still present when the energizing effect of antidepressants begins to take place. However, thoughts alone, without an underlying strong emotionally charged intention, are powerless, especially in such a fateful decision. (For an illustration of the confusion existing in this domain, see Kramer, 1993, p. 304.)

6. In clinical psychiatry, a drug effect that is contrary to the expected therapeutic effect is named "paradoxical," as when a sedative drug that is expected to induce calm leads to agitation or violence. In my opinion, this denomination is a mistake. In the explanatory scheme proposed here, both responses (the therapeutic and the paradoxical) are caused by the same effect of the drug on behavior.

Chapter 7

1. These are (1) long-lasting fluctuations in the intensity of the energetic aspect of behavior; (2) the three forms of frustration of active instinctive strivings; (3) the diffuse-differentiated scale of instinctive predispositions; and (4) the dichotomy and interrelationship of active and reactive behavior, as well as its modification by social influences.

2. We have previously mentioned some of these dangers, which include small poisonous animals emerging unexpectedly from a hiding place; falling from high places; closed places from which escape is problematic or impossible once the need arises; injury with bleeding when the continuation of risky or exhausting bodily activity could lead to further physiologic damage; and so on.

3. A person is considered legally accountable, at least in Western cultures I am familiar with, only if he is *aware* of the wrongfulness of his act and/or able to distinguish between right and wrong. An acute psychotic patient or one with considerable mental retardation is not legally accountable, but one with a personality disorder, such as antisocial personality disorder, is, in spite of having a psychiatric diagnosis (see Sadock & Sadock, 2005, pp. 3983–3985).

4. These are, again, the relatively uninhibited expression of a dysfunctional instinctive constitution in spite of social dissent, opposition, or the considerable loss of resources that successful social coexistence may secure.

5. Schizophrenia appears as a rule at quite an early age (adolescence, young adult), that is, at an age when a personality disorder "crystalizes," and the diagnosis of schizophrenia precludes a concomitant diagnosis of schizoid personality (which is not the case with obsessive-compulsive personality versus obsessive-compulsive disorder).

6. These include depression, bipolar disorder, schizophrenia, attention-deficit/hyperactivity disorder, eating disorders, Tourette and other tic disorders (DSM-5, p. 242), post-encephalitic Parkinsonism, and other neurologic disorders (Rossor, 2001, p. 763).

7. To review, the "bird type" mating strategy refers to a long-lasting monogamous mating relationship that includes, besides the sexual act, a more comprehensive sharing of reproductive and other life-perpetuating activities, like feeding, protecting, and teaching the offspring.

8. This third group of factors that may cause psychotic symptoms is more problematic. At first I tended to include here all the *known organic causes* of psychosis—that is, those that remain after the exclusion of sensory deprivation and "functional" (nonorganic) psychotic disorders (grave psychotic mania or depression, delusional disorders, or schizophrenia)—but I realized that some *organic causes* belong to the first group, such as organic factors that reduce the acuity of the sensory organs or interfere either with the respective information's transfer along neural projections in the brain or, like cortical blindness, interfere

with the primary processing of the sensory information at the appropriate cortical centers, thus reducing also the external input's effect on the overall brain functioning. Other organic factors, like the hallucinogenic drugs (LSD, mescaline, etc.) or temporal-lobe epilepsy, belong to the second group, which differentially excites certain brain regions, thus inducing psychotic symptoms.

9. It should be mentioned that the need for external stimulation, presumably in order to improve integrative brain functioning, was also demonstrated in animals (monkeys) in the case of light deprivation (Miller, 1980, p. 110). In humans, darkness—meaning in this context the drastic reduction in visual sensory stimuli—is well recognized for its ability to intensify fear or anxiety and to induce frightening illusions.

10. With regard to the organic causation of hallucinations (the nosological category of Organic Hallucinations was included in the DSM-III and IIIR but not in its newer versions), two neuropathological mechanisms were hypothesized, both compatible with the mechanisms of causation suggested in this text. The first is: "stimulation of specific cerebral sites or sensory pathways . . . by irritative cortical or subcortical lesions," as is the case with "hallucinogens, cocaine and migraine [which cause] excitation of the central nervous system, with resulting release of stored perceptions." The second psychopathologic mechanism proposed to lead to hallucinations is "disinhibition of brain structures subserving the storage of information," probably as a result of "lesions, either peripheral or central, that had removed the customary sensory input and thus allowed perceptual release . . . diseases of the eyes and ears seem to represent release phenomena." Furthermore some "drug induced hallucinations result from disinhibition of brain areas mediating visual sensation" (Lipowski, 1980, p. 1384.) Stimulation of the appropriate brain centers and pathways belongs to the present group of causative factors, while disinhibition as a result of the "removal of customary sensory input" was discussed in the first group of factors leading to psychosis.

11. I have previously noted perhaps the most famous of these formulations, Ernst Mayr's computer analogy. To reiterate, it states that the "closed genetic programs" of lower animals gradually evolved into more and more "open genetic programs" that in turn became capable of incorporating knowledge and experience to an ever-increasing degree (section 2.2).

12. It has to be mentioned, however, that this reasoning on behalf of a conflict-free dissociation between the internally originating and externally induced mental contents in schizophrenia has not been unanimously accepted in the professional literature. For example, Silvano Arieti, a recognized authority in the field of schizophrenia with a psychodynamic orientation, formulated the same theme in the following way: "These [pre-schizophrenic] children are unable to accept their environmental conditions, and at the same time they are unable to fight them. The situation that produces the least anxiety in them is one of *ostensible acceptance*, that is, compliance in spite of themselves" (Arieti, 1974, p. 159, emphasis in the original). It seems to me that Arieti could not detach himself from the concept of ubiquitous conflict between "superego" and "id," proposed by Freud originally to account for the pathology of neurotic disorders.

13. Such drives may include both the "mammal type" mating strategy directed toward sexual access and the "bird type," which aspires toward an exclusive and

more comprehensive relationship with a potential long-term partner and which in its appetitive phase presumes intense emotionally charged interactions with the other party, as well as confrontations with potential competitors.

14. The Genain quadruplets were monozygotic females, all four of whom suffered from schizophrenia. These sisters were intensely studied and followed for almost forty years. In spite of being identical in their genetic inheritance, their clinical symptoms varied considerably, as did the course and severity of the mental disorder and their level of intelligence and social functioning (Plomin et al., 2008, pp. 198–199, 203; Mirsky et al., 2000, pp. 699–708).

15. The reader will recall that the three stages are 1) observing the behavior and its eliciting circumstances; 2) employing empathy; and 3) using the corrective or compensatory mechanisms whose function is to assess possible dissimilarities between one's subjective experience and that of the observed person's and to compensate or correct accordingly.

16. A case resembling this one, that of a young woman who was hospitalized after behaving violently toward her parents, is described in Lehmann, 1980a, p. 1166. Before the outbreak of the psychotic illness, she had been a good student at high school but always shy and avoided dating with men. Her acute psychotic disorder broke out when at her first workplace "[a]nother girl . . . told her about boys and petting and began to exert a great deal of influence over the patient."

17. The motor aspects of severe, retarded depressive states (in our terminology lethargic depression) were included in the category of catatonic slowing or stupor, while the motor expressions of grave, agitated, dysphoric depressive, as well as manic, states were placed in the category of catatonic excitement. On the other hand, the prevalence of serious catatonic syndromes in schizophrenic patients was reduced drastically after the introduction of neuroleptics. For example: "Psychomotor slowing in young persons is sometimes so extreme that patients may slide into *a stupor*, unable to participate even in basic biological functions, such as feeding themselves. Such an episode is often the precursor of bipolar disorder, which later declares itself in a manic episode" (Akiskal, 2005b, p. 1616).

18. A reflection of the *opposing* instinctive intensities—*low* levels underlying lethargic depression versus *high* levels of instinctive activity in catatonic stupor—may be the different states of tonicity of the voluntary muscles in each case. In grave lethargic depression, the musculature has to be completely *relaxed* in accordance with its original adaptive role of energy preservation. In catatonic stupor, while I could not find descriptions of overall muscle tonicity, "rigidity" of the body is mentioned repeatedly. However, other, more circumscribed, catatonic states consist of immobility in which increased tonicity of the musculature is more evident: bizarre posturing, rigidity combined with posturing against gravity, "lead pipe rigidity," "waxy flexibility," and resistance to attempts to make changes in body posture. Some of these symptoms are detailed in the following. The hypothesis on the various states of tonicity of the voluntary musculature in lethargic depression versus catatonic stupor could be easily tested by clinical research studies employing electromyographic recording.

19. This presupposition is consistent with the antipsychotic effect proposed in this text, which is the reduction of instinctive intensities (active and self-defending reactive ones), which secondarily enhance the openness and reactivity to and better compliance with social influences. On the other hand, and contrary to

this interpretation of the antipsychotic effect, catatonic stupor does not respond to antipsychotic treatment, but may dramatically improve after intravenous administration of sedatives (diazepam, short-acting barbiturates). This effect is instantaneous and short-lived, and it seems to me that it concerns a more all-encompassing suppression of central nervous system activity than the anxiolytic effect of sedatives (taken orally in a smaller dose). At any rate, the discrepancy between the impression that antipsychotic drugs are ineffective in catatonic stupor, on the one hand, and the claim that the near disappearance of grave catatonic syndromes in schizophrenia is the outcome of widespread antipsychotic use, on the other, still awaits clarification.

20. "The hunting of the peregrine falcon usually begins with *relatively random roaming* around its hunting territory" (emphasis added), that is, a relatively directionless activity aroused by hunger or by the need to feed offspring. Only when a specific stimulus situation is found—a flock of teal, a sick gull, or a running mouse—does the falcon appeal to a more specialized instinctive maneuver adapted to the specific environmental situation. For example, if a flock of teal is found, the falcon "releases a series of sham attacks serving to isolate one or a few individuals from the main body of the flock" (Tinbergen, 2003, pp. 106–107).

Chapter 8

1. It may be mentioned in this context that the behavioral mechanisms underlying the three forms of frustration are not uniform. When the original active motive is discharged in a displaced or vacuum form, no need exists to switch to a different motor pattern; only the nature or presence of the releaser changes. Its underlying mechanism can be imagined as a kind of "overflow phenomenon." When the adequate releaser is unavailable, the growing urgency to discharge the respective behavior pattern overpowers some inhibitory mechanism whose role is to postpone or suppress the discharge until the adequate releaser is contacted. The aggressive and dysphoric transformations, on the other hand, are clearly different from the original unfrustrated active behavior pattern. The aggressive form is similar to reactive behavior's counteraggressive tactic, while the dysphoric form makes use of the immature offspring's innate behavior pattern in frustrating conditions. Therefore, it can be assumed that they are more complex brain mechanisms than displacement or vacuum behavior.

2. To review, these are 1) seasonal variation in instinctive intensity as the result of changing amounts of environmental resources for the sustainment of life processes, 2) basic changes in active instinctive behavior in frustrating conditions, 3) the regressive evolution of instinctive activity from well-differentiated behavior patterns toward more diffuse predispositions, and 4) the active/reactive behavior dichotomy and active/reactive/conforming behavior interrelationship.

Index

Contents

2

3

4

Relaxation of Natural Selection Pressures   153

5

Four Comprehensive Instinctive Mechanisms and the Relevance of Their Excessive Diversification to Human Dysfunctional Behavior

6

The Effects of Psychotropic Drugs in the Context of the Present Theory

7

Discrete Clinical Disorders in Evolutionary Perspective   373

8

Summary and Implications   531